D1451988

MICROSCALE ENERGY TRANSPORT

Series in Chemical and Mechanical Engineering
G. F. Hewitt and C. L. Tien, *Editors*

Carey, Liquid-Vapor Phase-Change Phenomena: An Introduction to the Thermophysics of Vaporization and Condensation Processes in Heat Transfer Equipment
Collier and Hewitt, Introduction to Nuclear Power, Second Edition
Dincer, Heat Transfer in Food Cooling Applications
Diwekar, Batch Distillation: Simulation, Optimal Design and Control
Raal and Mühlbauer, Phase Equilibria: Measurement and Computation
Tien, Majumdar and Gerner, Microscale Energy Transport
Tong and Tang, Boiling Heat Transfer and Two-Phase Flow, Second Edition
Tzou, Macro- to Microscale Heat Transfer: The Lagging Behavior

MICROSCALE ENERGY TRANSPORT

Edited by

Chang-Lin Tien
University of California
Berkeley, California

Arunava Majumdar
University of California
Berkeley, California

Frank M. Gerner
University of Cincinnati
Cincinnati, Ohio

USA	Publishing Office:	Taylor & Francis 1101 Vermont Avenue, NW, Suite 200 Washington, DC 20005-3521 Tel: (202) 289-2174 Fax: (202) 289-3665
	Distribution Center:	Taylor & Francis 1900 Frost Road, Suite 101 Bristol, PA 19007-1598 Tel: (215) 785-5800 Fax: (215) 785-5515
UK		Taylor & Francis Ltd. 1 Gunpowder Square London EC4A 3DE Tel: 0171 583 0490 Fax: 0171 583 0581

MICROSCALE ENERGY TRANSPORT

1 2 3 4 5 6 7 8 9 0 EBEB 9 8 7

This book was set in Times Roman. The acquisitions editor was Lisa Ehmer. Composition and editorial services by TechBooks. Cover design by Curtis Tow, Curtis Tow Graphics.

A CIP catalog record for this book is available from the British Library.
⊗ The paper in this publication meets the requirements of the ANSI Standard Z39.48-1984 (Permanence of Paper)

Library of Congress Cataloging-in-Publication Data

Microscale energy transport / edited by Chang-Lin Tien, Arunava
Majumdar, Frank M. Gerner.
p. cm. — (Series in chemical and mechanical engineering)
Includes bibliographical references and index.
ISBN 1-56032-459-7 (alk. paper)

1. Heat–Transmission. 2. Mass transfer. 3. Micromechanics.
4. Microelectronics I. Tien, Chang L., 1935– . II. Majumdar, Arunava. III. Gerner, F. M. (Frank, M.) IV. Series.
TJ260.M476 1997
621.402′2—dc21 97-23695
CIP

ISBN 1-56032-459-7 (case)

CONTENTS

Preface ix
List of Contributors xiii

Part I FUNDAMENTALS 1

1 MICROSCALE ENERGY TRANSPORT IN SOLIDS
Arunava Majumdar 3

1-1 Introduction 3
1-2 Microstructures of Solids 5
1-3 Crystal Vibrations and Phonons 14
1-4 Electronic Structure of Solids 28
1-5 Interactions of Photons with Electrons and Phonons 53
1-6 Particle Transport Theories 67
1-7 Nonequilibrium Energy Transfer 83
1-8 Summary 92
References 93

2 HEAT TRANSPORT IN DIELECTRIC THIN FILMS AND AT SOLID–SOLID INTERFACES
David G. Cahill 95

2-1 Introduction 95
2-2 Measurement Techniques 96
2-3 Heat Transport in Strongly Disordered Materials 103
2-4 Solid–Solid Interfaces 109
Nomenclature 115
References 116

3 MICROSCALE RADIATION PHENOMENA
Jon P. Longtin and Chang-Lin Tien 119

3-1 Introduction 119
3-2 Preliminaries 120
3-3 Radiation Phenomena on the Spatial Microscale 122
3-4 Radiation Phenomena on the Temporal Microscale 127
3-5 Radiation Phenomena on the Structural Microscale 140
3-6 Summary 143
Nomenclature 143
References 145

4 MELTING AND FREEZING PHENOMENA
R. Stephen Berry 149

4-1 Introduction 149
4-2 Solid and Liquid Clusters and Their Equilibria 151
4-3 "Surface-Melted" Clusters and Coexistence of Multiple Phases 159
4-4 Some Unsolved and Open Questions 161
4-5 Summary 162
Acknowledgments 162
References 163

5 MOLECULAR CLUSTERS
Susumu Kotake 167

5-1 Introduction 167
5-2 Clusters and Clustering 168
5-3 Thermophysical Properties of Clusters 177
5-4 Control of Clustering and Condensation 183
References 185

6 INTERFACIAL FORCES AND PHASE CHANGE IN THIN LIQUID FILMS
Peter C. Wayner, Jr. 187

6-1 Introduction 187
6-2 Thermodynamics of Thin Films: Interfacial Properties 189
6-3 Thermodynamics of Thin Films: Meniscus Properties 204
6-4 Quasi-Thermodynamics of Thin Films: Interfacial Mass Flux 209
6-5 An Evaporating Extended Meniscus 210
6-6 Applications 216
6-7 Summary 222
Acknowledgment 222
Nomenclature 223
References 224

Part II APPLICATIONS

7 THERMAL PHENOMENA IN SEMICONDUCTOR DEVICES AND INTERCONNECTS
Kenneth E. Goodson, Yongho Sungtaek Ju, and Mehdi Asheghi 229

7-1 Introduction 229
7-2 Basics of Charge Transport and Heat Generation 236
7-3 Thermal Transport Properties 253
7-4 Simulation 264
7-5 Thermometry 271
7-6 Summary and Recommendations 282
Acknowledgments 283
Nomenclature 284
References 288

8 MICRO HEAT PIPES
G. P. Peterson, L. W. Swanson, and Frank M. Gerner 295

8-1 Fundamental Operating Principles 295
8-2 Modeling Micro Heat Pipe Performance 297
8-3 Experimental Investigations 306
8-4 Construction Techniques and Issues 323
8-5 Summary 334
Nomenclature 334
References 335

9 MICROSCALE HEAT TRANSFER IN BIOLOGICAL SYSTEMS AT LOW TEMPERATURES
Boris Rubinsky 339

9-1 Introduction 339
9-2 Life at Low Temperatures Above the Freezing Temperature of Water 340
9-3 Life at Low Temperatures Below the Freezing Temperature of Water 344
9-4 Applications 359
References 364

10 SILICON MICROMACHINED THERMAL SENSORS AND ACTUATORS
Norman C. Tien 369

10-1 Introduction 369
10-2 MEMS Technology 370
10-3 Microscale Thermal Sensors 377
10-4 Microscale Thermal Actuators 380
10-5 Conclusion 384
References 384

Index 387

PREFACE

Science and engineering have always striven to explore, control, and exploit the extremes of length and time scales. The last few decades have witnessed a trend in miniaturization, emphasizing phenomena and devices that are small and/or fast. Microelectronics and molecular biology, for example, are two disciplines that illustrate this emphasis. A common feature in these disciplines is the *transport and interactions of matter in confined geometrical structures*. Whereas microelectronics utilizes controlled charge transport for digital computations, molecular biology studies reactions and transport of biological molecules. It must be remembered, however, that transport and interactions of matter must involve flow and/or exchange of energy due to fundamental reasons. The second law of thermodynamics suggests that any irreversible transport phenomenon must dissipate some energy in the form of heat. In addition, any molecular reconfiguration in the form of chemical reaction or phase transformation must involve some energy exchange with the environment. Despite the ubiquitous nature of energy transport and exchange, it has not received as much attention in the context of small and fast phenomena and devices. With the birth of micromechanical systems that integrate electronics, acoustics, optics, fluid mechanics, heat transfer, and biological processing, there is a need for a fundamental background in microscale energy transport. This need is further endorsed by the design requirements for high-power-density microelectronic and optoelectronic devices and systems, design and processing of novel materials, and the control of biological reactions and processes.

Due to the interdisciplinary nature of the research, most topics cannot be found in one textbook. Since new knowledge is generated and developed through research, it is difficult to expose students to this knowledge without a textbook or a course. In view of the strong research interest and the lack of a good book on this topic, we decided to collect contributions from experts in the field to develop a textbook on microscale energy transport. By bringing in experts from a wide range of science and engineering disciplines, the book attempts to cultivate cross-fertilization of ideas and background that are necessary for innovations and progress. We feel that dynamism in the field will come if we tie fundamental scientific understanding with engineering applications to develop new systems or improve old ones. Therefore, we have divided the book into two sections: fundamentals and applications. Since the field is relatively new, several topics are developing faster than others. The book may not cover all the topics that are of current interest, but it attempts to provide a foundation for people who want to learn a large variety of topics in microscale energy transport.

The book is intended for graduate students and researchers who are active or intend to become active in this field. It covers a wide range of topics and hence will be of interest to (i) mechanical engineers, who are involved in heat transfer, fluid mechanics, micromechanical systems (MEMS), electronic packaging, or cryogenics; (ii) electrical engineers, who work with electronic devices, electronic packaging, and MEMS; (iii) bioengineers, who are involved in cryobiology, cryosurgery, and metabolism, as well as phenomena related to energy transport and/or phase transformations at the level of single cells and molecules; (iv) materials scientists and engineers, who are interested in designing new materials and material structures; and (v) physicists and chemists, who are interested in energy transport in condensed matter, properties of nanostructures, and phase transformation in micro- and nanostructures. Each chapter presents the state of the art and assumes that the reader is a novice in the field. The topics are built from simple macroscopic concepts which gradually lead into microscopic concepts.

In Chapter 1, "Microscale Energy Transport in Solids," the reader will first be introduced to different types of natural and fabricated microstructures encountered in solids. Next, a detailed description of the physics of the three energy carriers in a solid—phonons, electrons, and photons—is provided. The interactions of these carriers between themselves and each other are discussed, and this discussion leads to the thermal, electrical, and optical properties of metals, semiconductors, and ceramics. The chapter then discusses different particle transport theories and introduces a fundamental framework for developing governing equations involving mass, momentum, and energy conservation, as well as the equation of radiative transfer for phonons and photons. This forms the basis for more advanced topics on nonequilibrium energy transfer, which are described in the context of submicrometer electronic devices and interactions of ultra-short pulsed lasers with materials.

Chapter 1 provides the fundamental background for two more chapters: Chapter 2, "Heat Transport in Dielectric Thin Films at Solid–Solid Interfaces," and Chapter 7, "Thermal Phenomena in Semiconductor Devices and Interconnects." Both of these chapters contain descriptions of microscale thermal measurement techniques. Chapter 2 discusses the mechanisms and models of energy transport in thin films and interfaces and provides an elaborate study of thermal boundary resistance. In addition, it also provides a state-of-the-art study of energy transport in strongly disordered materials. Chapter 7 gives an elaborate introduction to thermal phenomena, such as heat generation and transport, in different types of modern electronic devices and structures. It also provides detailed discussion on methods of modeling and measuring thermal characteristics and properties of microstructures with high spatial and temporal resolution.

Microelectromechanical systems (MEMS) is a rapidly growing field that has shown tremendous technological promise in the recent past. The key lies in the ability to integrate several diverse physical effects into miniaturized engineering systems that either enable existing technology or exhibit unique functionality. Chapter 10, "Silicon Micromachined Thermal Sensors and Actuators," shows how thermal effects can be effectively used for microsensors and actuators by utilizing membranes, cantilevers, and bridges that are surface micromachined on silicon substrates.

Energy transport by electromagnetic radiation as well as photon–matter interactions are fundamental both in science and several modern engineering systems. Chapter 3, "Microscale Radiation Phenomena," establishes a fundamental background on the microscale interactions of radiation. In particular, micro-length scale studies concentrate on the interaction of radiation with microstructures, whereas discussions on micro-time scales focus on the behavior of matter in the presence of ultra-short laser pulses. This chapter shows that traditional macroscale approaches may break down at these length and time scales, requiring the use of more general and fundamental microscopic theories. These are provided in detail in this chapter.

Melting, freezing, evaporation, and condensation have been studied traditionally at macroscopic scales. In the recent past, the study of phase transformations in confined structures such as atomic and molecular clusters and thin liquid films have provided new insights into phases and phase transformation of bulk matter. Because of the interplay between intermolecular and surface forces, these clusters and thin films can exist in phases that are not encountered in bulk matter. The book contains two chapters on molecular clusters—Chapter 5, "Molecular Clusters," and Chapter 4, "Melting and Freezing Phenomena"—and one chapter on thin liquid films, Chapter 6, "Interfacial Forces and Phase Change in Thin Liquid Films." All three chapters provide both a strong foundation and a new outlook for studying phase transformations in the context of intermolecular forces, which may have impact on analyzing phase change phenomena in a wide variety of engineering problems. Two such applications are considered in this book. The first is Chapter 9, "Microscale Heat Transfer in Biological Systems at Low Temperatures," which is about microscale heat transfer and phase change in biological systems. This chapter provides an in-depth introduction to freezing in single cells and groups of cells, as well as information regarding the effect of proteins and enzymes on phase change. Several interesting examples are used to illustrate how molecular-level phenomena involving proteins can drastically alter phase change phenomena. This chapter also provides an introduction to the energetics of metabolism. The second application of microscale phase change is discussed in Chapter 8, "Micro Heat Pipes." These devices contain highly confined channels that are partially filled with a liquid. Energy is efficiently transported by such devices due to phase change and mass transport. The performance limits of micro heat pipes are discussed in this chapter.

LIST OF CONTRIBUTORS

MEHDI ASHEGHI
Department of Mechanical Engineering
Stanford University
Stanford, CA 94305

R. STEPHEN BERRY
Department of Chemistry
University of Chicago
Chicago, IL 60637

DAVID G. CAHILL
Department of Materials
Science and Engineering
University of Illinois
at Urbana-Champaign
Urbana, IL 61801

FRANK M. GERNER
Department of Mechanical Engineering
University of Cincinnati
Cincinnati, OH 45221

KENNETH E. GOODSON
Department of Mechanical Engineering
Stanford University
Stanford, CA 94305

SUSUMU KOTAKE
Department of Mechanical Engineering
Toyo University
Kawagoe, Saitama 350, Japan

JON P. LONGTIN
Department of Mechanical Engineering
State University of New York at Stony Brook
Stony Brook, NY 11794

ARUNAVA MAJUMDAR
Department of Mechanical Engineering
University of California
Berkeley, CA 94720

G. P. PETERSON
Department of Mechanical Engineering
Texas A&M University
College Station, TX 77843

BORIS RUBINSKY
Department of Mechanical Engineering
University of California
Berkeley, CA 94720

YONGHO SUNGTAEK JU
Department of Mechanical Engineering
Stanford University
Stanford, CA 94305

L. W. SWANSON
Department of Mechanical Engineering
Texas A&M University
College Station, TX 77843

CHANG-LIN TIEN
Department of Mechanical
Engineering
University of California
Berkeley, CA 94720

NORMAN C. TIEN
School of Electrical Engineering
Cornell University
Ithaca, NY 14853

PETER C. WAYNER, JR.
Isermann Department of Chemical
Engineering
Rensselaer Polytechnic Institute
Troy, NY 12180

PART ONE

FUNDAMENTALS

CHAPTER

ONE

MICROSCALE ENERGY TRANSPORT IN SOLIDS

Arunava Majumdar

Department of Mechanical Engineering,
University of California,
Berkeley, CA 94720

1-1 INTRODUCTION

Recent trends in science and technology have indicated an increasing emphasis on miniaturization of engineering systems and study of microscale phenomena. Examples of such systems include modern microelectronic devices, the development of micromechanical devices and systems, and microsensors. Many such devices and systems utilize controlled transport of mass, energy, and/or charge. Therefore, the study of microscale transport phenomena has become an integral part of not only understanding the performance and operation of miniaturized systems, but also designing new devices. Microscale charge transport has been studied extensively for many years, espcially in the context of modern microelectronic devices. Mass transport and chemical reactions also have received attention for understanding processes such as chemical vapor desposition and biological reactions. In contrast, microscale energy transport has received limited attention in the context of miniaturized engineering systems. The second law of thermodynamics requires that any irreversible transport of mass or charge must involve entropy generation and thereby some exchange of energy with the surroundings. Energy transport controls the performance of several miniaturized systems and is an important ingredient in the processing of materials. It is clear that microscale energy transport is both fundamental from a scientific viewpoint and important for engineering applications. In this chapter, we will investigate the different mechanisms of energy transport in solids.

There exists two different ways to study energy transport in materials: the macroscopic approach and the microscopic approach. The macroscopic approach uses phenomenological models that require no knowledge of the mechanism of energy transport or the microstructure of solids. It focuses on the overall large-scale effects and requires only a constitutive relation or a simple transport law. Such a relation usually contains

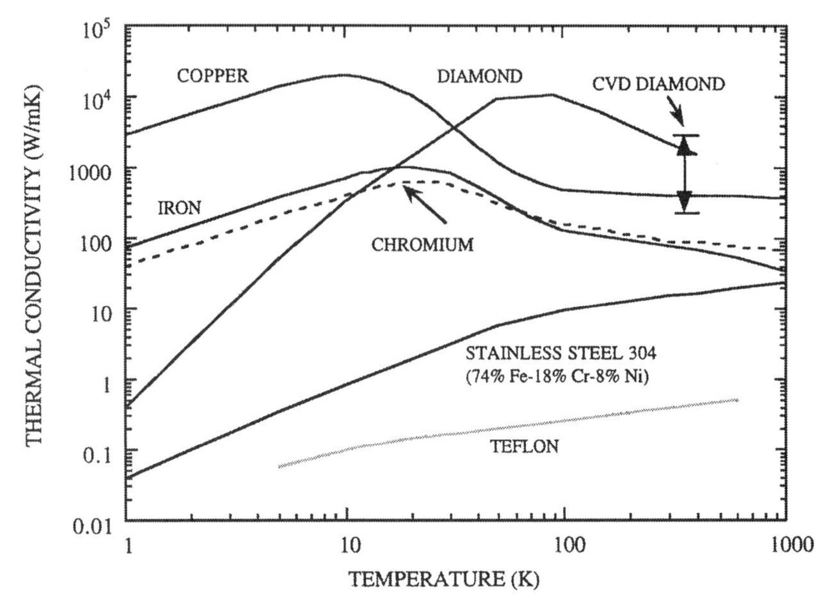

Figure 1-1 Thermal conductivity of different materials as a function of temperature. (Courtesy of T.Q. Qiu.)

material constants. For example, the Fourier law of heat conduction includes thermal conductivity as a material property. Data on thermal conductivity are provided as a function of temperature, as shown in Fig. 1-1. It is evident that the thermal conductivity can vary over five orders of magnitude and is not a constant. Moreover, it often depends on the microstructure of the solids, which such a plot cannot show. For example, as shown in Fig. 1-1, the thermal conductivity of diamond can span an order of magnitude, depending on the type of microstructure obtained by chemical vapor deposition. Finally, a plot such as Fig. 1-1 does not reveal the mechanism of energy transport but instead gives a number that can be used in calculations.

Although this is a simple approach that has been used to solve many engineering problems, it is often not sufficient to understand and solve problems that involve miniaturized engineering systems. Such systems usually demand the knowledge of the mechanisms of energy transport and interactions of energy carriers between themselves and with solid microstructures. This approach provides a fundamental understanding of energy transport in materials, and it becomes increasingly indispensable for developing novel materials and high-performance devices. Although physicists, in the past, have explored the science of energy transport in solids, applications to miniaturized engineering systems require a new outlook. It is the goal of this chapter and others in this book to tie fundamental knowledge of microscale energy transport with the modern technology of miniaturized systems.

The first step towards understanding microscale energy transport in solids is to study the different microstructures that one may encounter in engineering systems. The second step is to understand the physics of individual energy carriers: electrons, lattice vibrations (called *phonons*), and photons. The third step is to study energy transport by each of these individual carriers, as well as to investigate the mechanisms of energy transfer between these carriers. A general framework will be developed from fundamental principles to analyze energy, charge, and mass transport. The final step is to apply this basic knowledge to contemporary engineering systems. This chapter follows this sequence and provides a systematic introduction to microscale energy transport in solids.

1-2 MICROSTRUCTURES OF SOLIDS

1-2-1 Natural Microstructures

Crystals and chemical bonds. Solids are made of atoms that are bound together by chemical bonds. Sometimes these atoms are arranged in an ordered and periodic array called a crystal. A crystal is formed by repeating a unit cell in all directions. Figure 1-2 shows the shapes and sizes of different types of unit cells that are found in solids. The arrangement of atoms in a unit cell and the chemical bonds between the atoms are of vital importance, since they control the electrical, thermal, optical, and mechanical properties of the solid. Several books can be found with detailed discussions of the crystal structure of solids and the nomenclature used to describe the directions and surfaces of crystals (Ashcroft and Mermin, 1976; Kittel, 1986). The purpose of this chapter is to provide a working knowledge of crystal structures and its influence on energy transport mechanisms.

Perhaps the most important and relevant crystal structures are derivatives of the simple cubic and the hexagonal closed packed structures. Derivatives of the simple cubic structure are shown in Fig. 1-3 and include the body-centered cubic (bcc), the face-centered cubic (fcc), and the diamond structure. Table 1-1 lists some elements that are found in the bcc crystal structure. The size of a unit cell in a crystal is called the lattice constant, a, which is also provided for the bcc elements. When the atom at the center of the bcc structure is different from the atoms at the corners of the cube, it is called a *cesium chloride* structure, named after the configuration of crystals formed of these elements. Table 1-2 lists some compounds and alloys which form the cesium chloride structure.

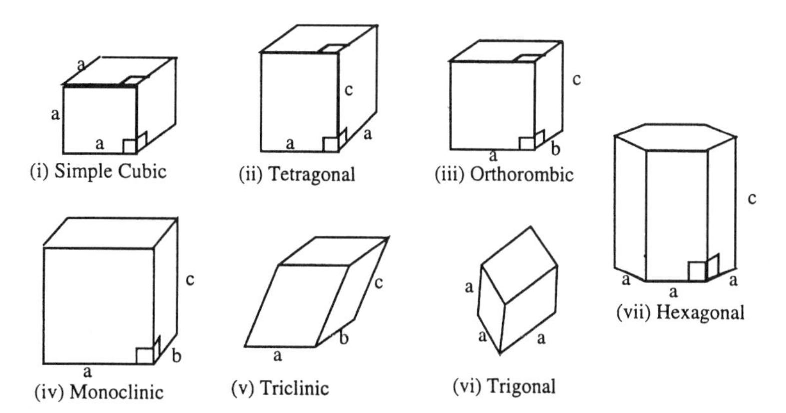

Figure 1-2 Shapes and sizes of the unit cells of seven crystal systems.

Table 1-3 lists some elements that form the fcc structure. When the fcc structure of two different types of atoms are enmeshed with each other as shown in Fig. 1-3, the crystal is called the sodium chloride structure. Examples of the sodium chloride structure are listed in Table 1-4. The diamond structure consists of two interpenetrating fcc lattices that are displaced along the body diagonal of the cubic cell by one-quarter the length of the diagonal. When all the atoms of the lattice are the same, this structure is called the diamond structure. Table 1-5 lists the elements that form this structure. These

Table 1-1 Some elements with body-centered cubic crystal structure

Element	a (Å)	Element	a (Å)	Element	a (Å)
Ba	5.02	K	5.23	Nb	3.30
Cr	2.88	Li	3.49	Ta	3.31
Cs	6.05	Mo	3.15	V	3.02
Fe	2.87	Na	4.23	W	3.16

Table 1-2 Some compounds with cesium chloride crystal structure

Crystal	a (Å)	Crystal	a (Å)	Crystal	a (Å)
BeCu	2.70	CsI	4.57	AgMg	3.28
AlNi	2.88	CuZn(β-brass)	2.94	TlCl	3.83
CsCl	4.12	AgMg	3.28	TlBr	3.97
CsBr	4.29	NH_4Cl	3.87	TlI	4.20

Table 1-3 Some elements with face-centered cubic crystal structure

Element	a (Å)	Element	a (Å)	Element	a (Å)
Ag	4.09	Ir	3.84	Rh	3.80
Al	4.05	La	5.30	Sc	4.54
Au	4.08	Ni	3.52	Sr	6.08
Ca	5.58	Pb	4.95	Th	5.08
β-Co	3.55	Pd	3.89	Yb	5.49
Cu	3.61	Pt	3.92		

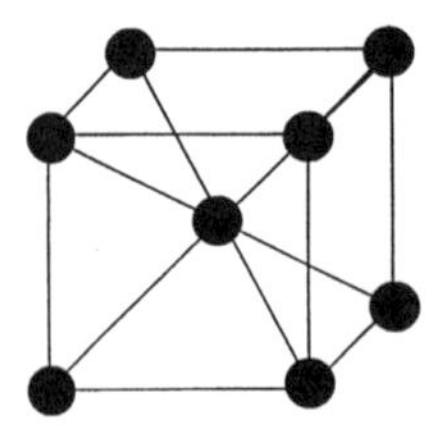

(i) Body-centered cubic

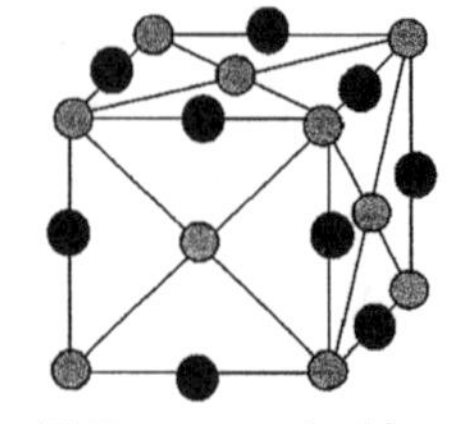

(ii) Face-centered cubic

(iii) Diamond Structure

Figure 1-3 Derivatives of the simple cubic structures.

Table 1-4 Some compounds with sodium chloride crystal structure

Crystal	a (Å)	Crystal	a (Å)	Crystal	a (Å)
LiH	4.08	AgBr	5.77	CaO	4.81
LiBr	5.50	MgO	4.21	CaS	5.69
NaF	4.62	MgS	5.20	CaSe	5.91
NaCl	5.64	PbS	5.92	SrO	5.16
NaBr	5.97	KF	5.35	SrS	6.02
AgF	4.92	KCl	6.29	BaO	5.52
AgCl	5.55	KBr	6.60	BaS	6.39

Table 1-5 Elements with diamond crystal structure

Element	a (Å)	Element	a (Å)
C (diamond)	3.57	Ge	5.66
Si	5.43	a-Sn	6.49

Table 1-6 Some compounds with the zincblende crystal structure

Crystal	a (Å)	Crystal	a (Å)	Crystal	a (Å)
CuCl	5.41	CdS	5.82	GaP	5.45
AgI	6.47	CdTe	6.48	GaAs	5.65
BeS	4.85	HgS	5.85	GaSb	6.12
BeTe	5.54	HgSe	6.08	InP	5.87
MnS	5.60	HgTe	6.43	InAs	6.04
ZnS	5.41	AlP	5.45	InSb	6.48
ZnSe	5.67	AlAs	5.62	SiC	4.35
ZnTe	6.09	AlSb	6.13		

happen to be the first four elements of Group IV of the periodic table. These elements can form sp^3 tetrahedral bonds[1] which result in the diamond structure. When two different types of atoms are distributed on the diamond lattice, it is called the *zincblende* or *zinc sulfide structure*. Table 1-6 lists several compounds that are found to have the zincblende structure. Many of these are formed by bonding between elements of groups III and V of the periodic table, as well as those of transition metals and group VI.

The chemical bond between the atoms in a crystal plays an extremely important role. In conjunction with the crystal structure, it controls the electrical, optical, thermal, magnetic, and mechanical properties of a solid. Chemical bonds can be classified into several categories: (i) van der Waals, (ii) hydrogen, (iii) covalent, (iv) ionic, and (v) metallic.

[1]The single s and the three p electronic orbitals are hybridized to form four sp^3 orbitals that form a symmetric tetrahedral structure.

A *van der Waals bond* is formed due to fluctuations in electric dipoles of two atoms (Israelachvili, 1992). For example, if two atoms of helium are brought close together, despite their inert nature they will attract each other due to van der Waals interactions. Here, charge in one atom will induce a dipole in the other atoms, which will make the two atoms attract each other. It is a weak bond with an energy[2] of about 0.01 eV and is found in solids formed by organic compounds, inert gases, and other gases such as oxygen and nitrogen. The bond energy is comparable to and often lower than the thermal vibrational energy, which is on the order of $k_B T$, which at room temperature is 0.026 eV. Here, k_B is the Boltzmann constant equal to 1.38×10^{-23} J/K and T is the absolute temperature. Hence, van der Waals solids easily melt and evaporate at room temperature.

When a hydrogen atom is bonded to a highly electronegative atom such as oxygen, the electron belonging to the hydrogen is strongly pulled towards the oxygen. Thus the hydrogen atom is left with a net positive charge of the proton, and the oxygen atom contains a net negative charge. This hydrogen atom can bond with the oxygen atom of another molecule that has a net negative charge. Such a bond is called a *hydrogen bond.* An example is the intermolecular bond between water molecules in liquid water and ice. Biological molecules often contain hydrogen bonds such as the bond between the two strands of a DNA molecule. The hydrogen bond is a relatively weak bond with the bond energy on the order of $k_B T$.

Covalent bonds are formed by atoms that share their valence electrons. When the two atoms come close together, the electronic orbitals of the individual atoms overlap to form two molecular orbitals. One is called the *bonding orbital*, and other is the *antibonding orbital.* The valence electrons of the two individual atoms initially occupy the bonding orbital of lower energy. Once this orbital is filled, it starts to occupy the antibonding orbital. Electrons in the bonding orbital have a higher probability to reside in the region between the two atoms, and thus the electrons are said to be shared. Therefore, the bonds are localized and highly directional, and thus they determine the geometrical structures of atoms in a crystal. Core electrons usually are not involved in the bonding, although they do affect chemical bonds by their influence on the valence electrons. The bond energy of covalent bonds falls typically 1–10 eV. Covalent solids are hard and brittle since the bonds are strong, directional, and short range. Examples include diamond, silicon, and germanium.

Ionic bonds, such as those in sodium chloride, are formed when a net charge is exchanged between two or more atoms. Here, one atom (e.g., sodium) can easily give up its valence electron, whereas the other atom (e.g., chlorine) has a high electron affinity such that it can attract this electron. The charge is highly localized around the electronegative atom and can be thought to be an extreme case of a covalent bond. Coulombic or electrostatic attraction binds the oppositely charged atoms. The bonds are not directional since Coulomb forces have spherical symmetry. The bond energy is 1–10 eV, and hence ionic solids are very stable at room temperature. Many covalent bonds display ionicity, in that the shared electrons are localized more towards one of the atoms. This ionicity of covalent bonds plays an important role in its optical and electronic properties.

[2] 1 electron volt (eV) is equal to 1.6×10^{-19} Joules.

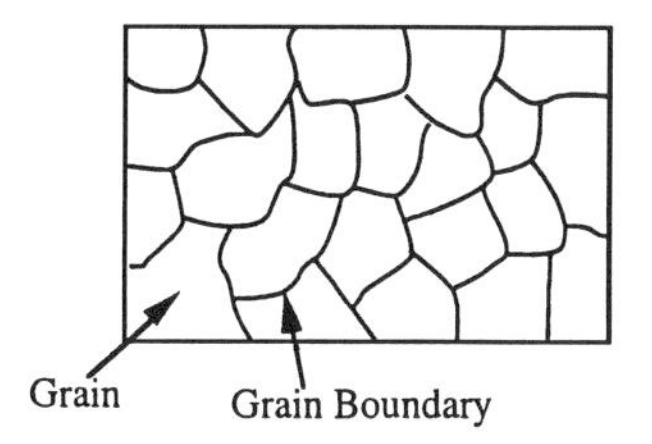

Figure 1-4 A polycrystalline solid containing single crystal grains separated by grain boundaries.

Metallic bonds can be thought of as a limiting case of covalent bonds in which electrons are shared by all the ions in the crystal. This occurs when the atoms have weakly bound electrons in the outermost shell. The freed electrons move in the combined potential of all the positive ions. Thus the term "electron gas" can be used since electrons are free to wander in the solid. The bond energy is typically on the order of 1–10 eV.

Single crystals and polycrystals. When the unit cell of a crystal repeats itself in three-dimensional space and forms a macroscopic object without any break in periodicity or presence of interfaces, it is called a *single crystal*. Such single crystals can be found in nature (such as in gem stones) or grown under carefully controlled conditions. The growth of single crystals of high purity is one of the major accomplishments of modern technology. It is difficult to imagine the development of electronics and optoelectronics without high-purity single crystals.

Despite the importance of single crystals, they are sometimes difficult to grow and are not always necessary. Solids often appear in the form of polycrystals, as schematically shown in Fig. 1-4. Here, the solid is made up of units of single crystals called *grains*, which are separated by interfaces called *grain boundaries*, much like an array of soap bubbles. The orientation of the crystals in neighboring grains can be different. The grain boundaries are regions where the two crystal orientations are accommodated within a very short distance, typically on the order of a few nanometers. This region can contain many crystal imperfections and defects, created to relieve the stress that results from trying to match unit cells with different orientations. The grain boundary plays a very important role in mechanical, chemical, electrical, and thermal properties of a solid. It forms a channel for mass diffusion, imposes resistance to charge and energy transport by electrons and crystal vibrations, and influences the plasticity of materials.

When a polycrystalline solid is heated or annealed, the grains grow in size and coalesce in an attempt to form a single crystal with the lowest free energy. Therefore, the question remains as to why polycrystals are formed if the free energy is higher than that of single crystals. The formation of grains is a kinetically controlled process. For example, if a molten metal is suddenly solidified, small crystals nucleate in different regions of the melt. These crystallites need not have the same orientation. During solidification, the crystallites grow in size and maintain their crystal orientation. After a while, however, the boundaries of these crystallites meet. Since the crystal orientations of these individual crystallites are different, the interface remains and forms the grain boundary. Similar processes occur during the deposition of a polycrystalline film. Here, the heterogeneous nucleation of the deposited atoms forms islands that grow in size during deposition, as shown in Fig. 1-5. However, the crystal orientation of the islands can be different, which eventually results in grains and grain boundaries.

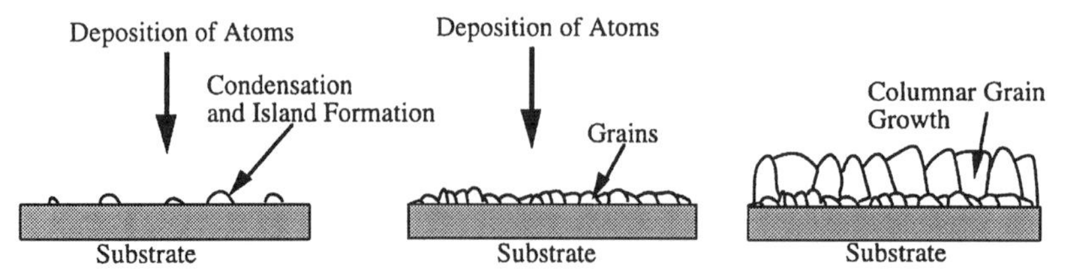

Figure 1-5 Evolution of grain structure during deposition of a thin film on a substrate. First is the adsorption and surface diffusion of atoms on the substrate surface. Next is the heterogeneous nucleation of small stable islands on the surface. These islands join to form the first layers of grains. Subsequent deposition of atoms leads to further grain formation. The grains can grow in many shapes, columnar being one of them. The grain size and shape can vary across the film thickness.

The size and shape of grains depend strongly on the deposition conditions as well as the chemistry between the deposited atoms and the substrate surface. The grain geometry influences several properties of the solids, and their control and prediction are important aspects of materials research (Ohring, 1992). Deposition of single crystal thin films can be achieved, but only under precise control of the deposition conditions (Ohring, 1992).

Point defects and dislocations. Other than grain boundaries, crystal imperfections such as point defects and dislocations have significant influence on the properties of solids. *Point defects* can be in the form of vacancies or missing atoms, of extra atom not in a regular lattice position, or of an atom with a different chemical composition. The simplest point defect is a *lattice vacancy* or a *Schottky defect*, where the atom is removed from the bulk and placed on the surface. Lattice vacancies are always present in a perfect crystal in thermal equilibrium, since the entropy is increased due to disorder in the crystal. The number of vacancies, n, follows the Boltzmann distribution

$$\frac{n}{N-n} = \exp(-E_\nu/k_B T) \tag{1-1}$$

where N is the total number of atoms, E_ν is the energy needed to take a atom from the bulk lattice to a surface (typically about 1 eV), k_B is the Boltzmann constant, and T is the temperature. A *Frenkel defect* is one where the atom is removed from its lattice position and placed within the lattice but at an interstitial position which is not a regular lattice position. The number of defects can be calculated from Eq. (1-1) with E_ν replaced by E_I and $N - n$ replaced by $\sqrt{NN'}$, where N' is the number of interstitial sites.

Point defects can diffuse in the solid and follow Fick's law of diffusion. They can also move under thermodynamic potential gradients. For example, they are responsible for electrical conductivity in ionic crystals. Point defects also move under stress gradients and give rise to some degree of plasticity in a material. When the defect atom donates an extra electron to the crystal or traps an electron from the lattice, it influences the electronic conductivity of the material. Point defects containing chemical impurities also change the optical properties of solids and produce colors in crystals that are otherwise transparent. Point defects can strongly influence charge and energy transport in solids since they can act as scattering sites for electrons and crystal vibrations. This will be studied later in more detail.

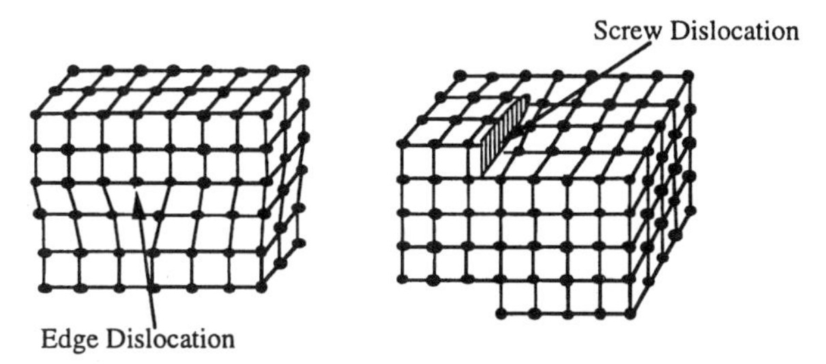

Figure 1-6 Illustrations of edge and screw dislocations.

A *dislocation* is a line imperfection in a crystal. Two types of dislocations, edge and screw dislocations, are found in crystals and are shown in Fig. 1-6. An *edge dislocation* is a line that marks the end of a row of atoms in a crystal. Under applied stress, dislocations can move within the crystal as shown in Fig. 1-7. The direction of motion is called the *slip direction*. Edge dislocations move perpendicular to the slip direction, whereas *screw dislocations* move parallel to it. The latter type creates an atomic step on the surface and therefore has influence on surface phenomena such adsorption, nucleation, and thin film growth. The nucleation of dislocations is still not completely understood, although they are observed to form under high stresses. Such stresses can be externally induced or can be produced at interfaces of solids with large mismatch in lattice sizes. Once a dislocation is nucleated, more dislocations can be generated from the parent one under applied stress. This is called *work hardening*. The nucleation and the multiplication of dislocations under stress is thermodynamically driven since it reduces the internal strain energy of a solid. However, in a stress-free solid it is thermodynamically favorable for the crystal to have no dislocations. Therefore, annealing reduces the number of dislocations. The *dislocation density* is the number of dislocation lines that intersect a unit area in the crystal. They can vary from 10^2 dislocations/cm^2 in the best silicon crystals to 10^{12} dislocations/cm^2 in heavily deformed metal crystals. Detailed discussions on dislocations can be found in Hull and Bacon (1984).

Dislocations have strong influence on mechanical properties of solids. As shown in Fig. 1-7, they move under applied stress. Such dislocation motion is largely responsible for plasticity in solids. For example, dislocations have high mobility at room temperature in metals, which leads to their plasticity. The mobility reduces with decreasing temperature, making metals brittle at cryogenic temperatures. In covalently bonded solids, such

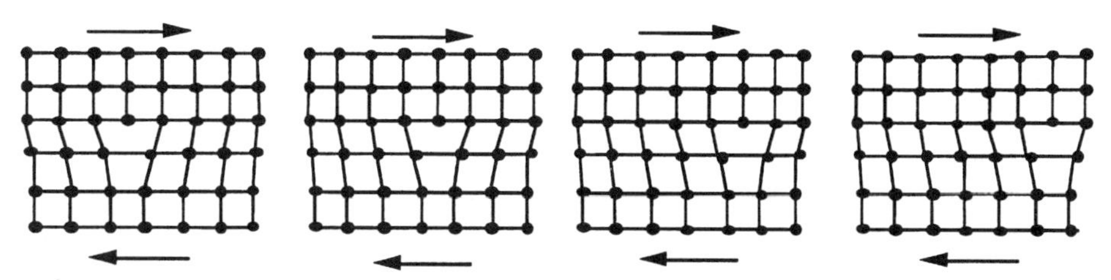

Figure 1-7 Motion of an edge dislocation under shear stress.

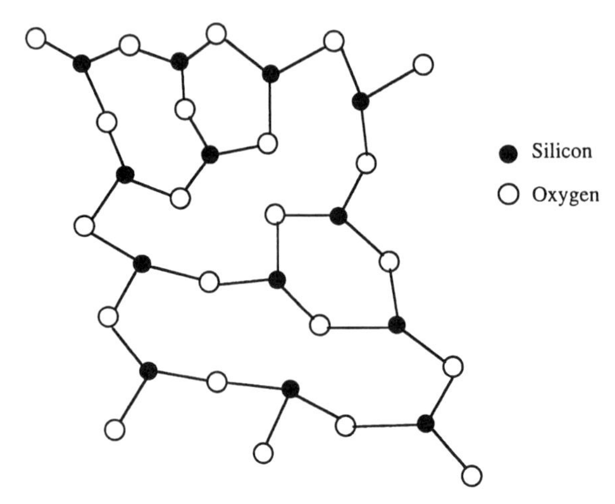

Figure 1-8 Two-dimensional projection of amorphous silica.

as semiconductors and ceramics, the bonds are highly directional and localized. This reduces dislocation mobility and thus makes such solids brittle.

Since dislocations are crystal imperfections, the chemical bonding is different at dislocations than the rest of the crystal. The presence of an extra row of atoms or steps can produce dangling bonds. Such aberrations in bonding alter the charge distribution and can thereby influence the electronic and optical properties of a solid. Dislocations also produce a stress field around it, which can act as a scattering site for crystal vibrations. Therefore, in addition to plasticity, dislocations have a strong influence on electrical, optical, and thermal properties of solids.

Amorphous solids. The discussions so far have focused on crystalline solids, where the atoms are arranged in a periodic array over a length scale much larger than the lattice spacing. Amorphous or glassy solids, on the other hand, exhibit no long-range order, although they do show some order in the short range, that is, within few lattice spacings. This is schematically shown in Fig. 1-8. A convenient way to look at their structure is to consider it as an instantaneous snapshot of a liquid. Therefore, they are are sometimes called *supercooled liquids*. However, the mobility of atoms is highly restricted compared to that of liquids.

Amorphous solids are formed by cooling a liquid sufficiently fast such that crystal nuclei are formed but do not have time to grow. This is because the viscosity of the liquids rises sharply as the temperature is decreased. Amorphous thin films are also grown by depositing the film at an extremely high rate on a substrate that is maintained at low temperature. This also reduces surface mobility of the adsorbed atoms and essentially freezes the atoms wherever they are adsorbed. The conditions needed to form amorphous solids depends on the material. Although the crystalline state has a lower free energy, the amorphous state is not unstable since the activation barrier for structural change can be large. However, when amorphous solids are annealed, they can recrystallize to reach a lower free energy.

Since long-range order does not exist in amorphous solids, they exhibit interesting properties. Some crystal vibrations are strongly scattered by the long-range disorder, and hence this affects heat conduction. Since the angle and spacing between atoms are different in amorphous solids than in crystalline solids, the charge distribution within the bonds is altered. Consequently, the electronic properties are modified. Finally, the optical properties are also changed since the electronic structure of the material is different from that of the crystalline state.

1-2-2 Fabricated Microstructures

The discussion in Section 2.1 was focused on naturally occurring microstructures. These microstructures cannot be controlled as precisely in size, orientation, shape, and property as one would desire. Modern technology has, however, provided several methods of fabricating solid microstructures that do just that. In fact, fabricated microstructures can have properties that may not be observed in naturally occurring ones. Such developments have led to many advances in microdevices. It is therefore important to have a basic knowledge of the sizes of microstructures that one can fabricate with current technology.

Microstructures, by definition, must have confinement in size in at least one of the three dimensions. The following discussion will describe the confinement in one, two, and even three dimensions.

Thin films. A *thin film* is a fabricated microstructure in which the thickness of the film is much smaller than the lateral dimensions. Thin films of metals, ceramics, and semiconductors can be deposited on a substrate by a variety of techniques, such as: (i) physical vapor deposition, (ii) chemical vapor deposition, (iii) sputtering, (iv) laser ablation, (v) liquid-phase epitaxy, and (vi) molecular beam epitaxy. These are described in detail elsewhere (Ohring, 1992). In the context of fabricated microstructures, the thickness of the film is the most important quantity. Depending on the application, the thickness of films is required to be in the range of 1 nm to 100 μm. For example, in semiconductor quantum devices, films that are about 1 nm thick are used to produce unique electronic and optical properties. On the other hand, films of 100 μm are required for thermal barrier coatings. The above-referenced deposition techniques often have the capability of controlling the film thickness to within a few atomic monolayers. Using reflected electron or photon beams or crystal oscillators, modern instruments have the capability of monitoring the film thickness to within a single atomic layer which is approximately 1 Å thick.

Although the film thickness can be directly controlled, often what is desired is to control the grain structure within the film. This is not a trivial problem because the grain structure can be very sensitive to the chemistry and stress at the film-substrate interface, as well as the kinetics of surface diffusion, adsorption, desorption, and so on. Epitaxial or single-crystal films are a special case of such films that are necessary for applications in optoelectronics and electronics. These films can be grown only under very restrictive conditions, which are described in detail elsewhere (Ohring, 1992).

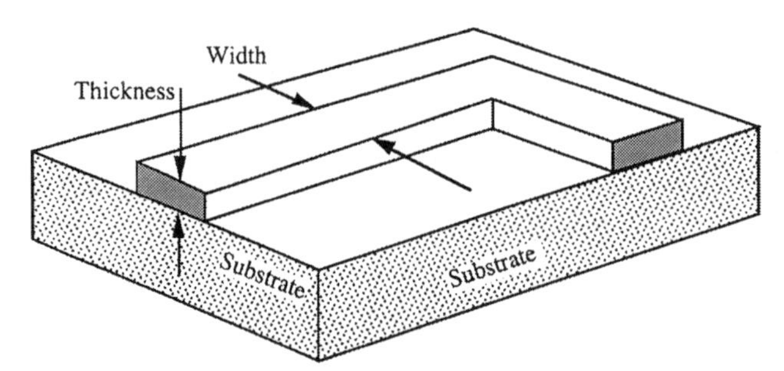

Figure 1-9 A thin film of thickness t is pattered to form a narrow line of width w. Modern deposition techniques have allowed deposition of thin films with single-atomic-layer thickness control. Lithographic techniques can be used to pattern thin films to form lines that can be 10–500 nm.

Narrow lines. After a thin film is deposited, it can be patterned to form lines or wires, as shown in Fig. 1-9. State-of-the-art photolithographic techniques are capable of making lines that have a width of about 0.25 μm. The width is limited by diffraction of the light used for exposing a polymeric resist. Electron beams also have been used to expose resists for lithography. Since the electrons used have high energy (typically 1–50 KeV), the wavelength is short, making it capable of high-resolution lithography. Electron-beam lithography has been used to fabricate structures on the order of 10 nm (Broers, 1995). Scanning tunneling (Stroscio and Eigler, 1991) and atomic force microscopes (Majumdar et al., 1992; Minne et al., 1995) have also been used to fabricate lines ranging in size from single-atom width to about 100 nm (Marrian, 1993).

Dots. When semiconductors are confined to zero-dimensional dots that are on the order of a few nanometers in size, quantum effects dominate the electronic and optical properties of these dots. In particular, the electronic energy states become size dependent, allowing one to tune the optical absorption spectrum as well as to control the electron transport across these quantum dots. Self-assembly has recently been used to fabricate semiconductor quantum dots that display luminescence in the visible spectrum (Leon et al., 1995). In addition, when silicon is electrochemically etched, it becomes porous with pores on the order of 1–10 nm. This makes silicon luminesce in the visible range (Feng and Tsu, 1994). Fabrication and properties of such zero-dimensional structures are not well understood and are currently topics of intense research.

1-3 CRYSTAL VIBRATIONS AND PHONONS

The chemical bonds between atoms in a solid are not rigid but act like springs. A crystal can be thought of as an array of atoms attached by springs. Motion of one or a group of atoms can be transmitted as waves through this spring-mass network. Such vibrations are responsible for transport of energy in many solids. Therefore, it is important to understand the nature of these vibrations and how they contribute to energy transport in solids.

A typical energy–distance diagram of a chemical bond is shown in Fig. 1-10. The equilibrium distance between atoms, r_o, is decided by the equilibrium between attractive

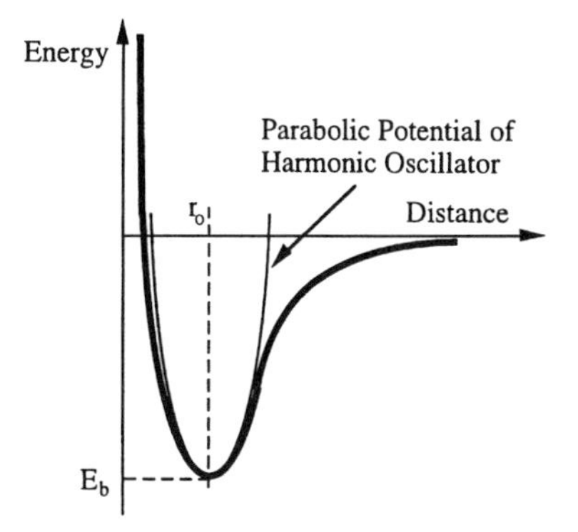

Figure 1-10 Energy–distance diagram of a typical chemical bond. The equilibrium bond length is r_o. Also shown is the parabolic potential of a harmonic oscillator. Note that for small vibrations around r_o, the potential can be approximated to be that of a harmonic oscillator, whereas for larger vibrations the actual potential deviates from the parabolic one.

and repulsive forces. The bond energy E_b is the depth of the potential well. If one studies the shape of the energy–distance curve in the neighborhood of r_o, it is clear that for small distances the curve can be represented as a parabola. This is characteristic of a harmonic oscillator where the potential energy $u = 1/2gx^2$ is related to the displacement, x, and the spring constant, g. This gives a simple linear force-distance behavior as $F = gx$. It should be noted, however, that as x increases, nonlinear or anharmonic effects can become significant, and the harmonic oscillator model breaks down. Nevertheless, the simple harmonic oscillator is a good place to start investigating crystal vibrations.

1-3-1 Waves Propagation in Crystals

Acoustic mode vibrations. A crystal can be imagined to be a three-dimensional array of masses and springs where vibrations can travel as waves. To simplify the analysis to its basic element, consider a one-dimensional array of masses and springs, as shown in Fig. 1-11, where the masses of all the atoms and the springs between them are the same, and neighboring masses are separated by an equilibrium distance, a. If the nth atom is displaced by a distance x_n, the dynamical equation for the nth atom can be written as

$$m \frac{d^2 x_n}{dt^2} = g(x_{n+1} + x_{n-1} - 2x_n) \tag{1-2}$$

where it is assumed that the force on the nth atom is only by its neighboring atoms. The solution of this equation should have a wave nature and is assumed to have the form $x_n = x_o \exp(-i\omega t)\exp(inKa)$. Here, ω is the angular frequency and $K = 2\pi/\lambda$ is

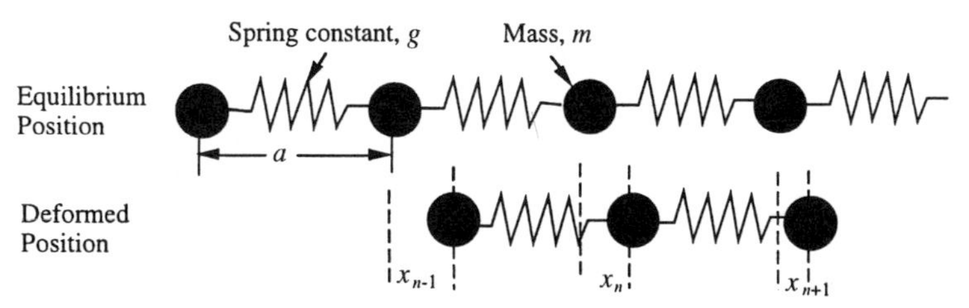

Figure 1-11 A one-dimensional array of a spring-mass system.

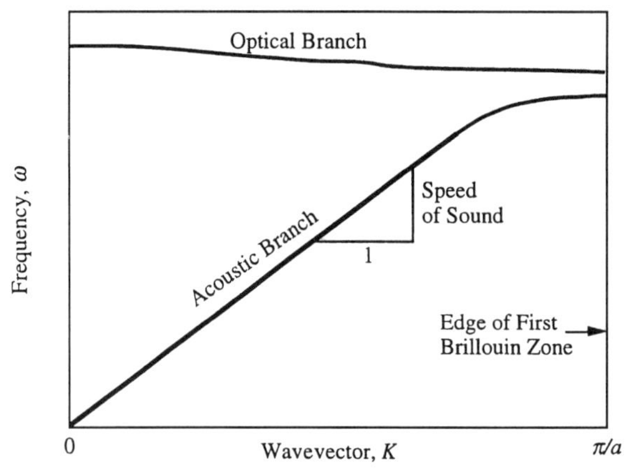

Figure 1-12 Dispersion relation for crystal vibrations.

the wavenumber, where λ is the wavelength. If this is used in Eq. (1-2), the following dispersion relation can be derived:

$$\omega^2 m = g[2 - \exp(-iKa) - \exp(iKa)] = 2g(1 - \cos Ka)$$
$$\omega = \sqrt{\frac{2g}{m}}(1 - \cos Ka)^{1/2}. \tag{1-3}$$

The ω-K plot of Eq. (1-3) is shown in Fig. 1-12 and is called the *acoustic branch.* The discreteness of the crystal lattice limits the minimum value of the wavelength to be $2a$. Hence, the range of wavenumber K is $-\pi/a < K < \pi/a$.

The dispersion relation of Eq. (1-3) is an important quantity. The group velocity, v_g, of a wavepacket can be found as $v_g = d\omega/dK$, or the slope of the ω-K curve. In the long-wavelength or *continuum* limit when $Ka \to 0$, the dispersion relation becomes

$$\omega = \sqrt{\frac{g}{m}}aK. \tag{1-4}$$

The linear ω-K behavior suggests that the speed of wave propagation is a constant and equal to $v_g = a\sqrt{g/m}$. This is the velocity of sound in a crystal. Therefore, the dispersion relation in Fig. 1-12 is called the *acoustic mode of vibrations.* Note that the resonant frequency of an individual simple harmonic oscillator is $\omega_o = \sqrt{g/m}$. Hence, $v_g = a\omega_o$. It is clear from Fig. 1-12 that v_g decreases as K increases from 0 to $\pm\pi/a$. It is interesting to study the ratio of the displacements between two neighboring atoms. From the assumed solution it is clear that $x_{n+1}/x_n = \exp(iKa)$. When $Ka = \pm\pi$, the neighboring atoms vibrate 180 degrees out of phase. Hence, when K is equal to $\pm\pi/a$, the atoms vibrate against each other and therefore do not propagate any information or energy.

The region of K, $-\pi/a < K < \pi/a$, is called the *first Brillouin zone*, which is the K-space of the crystal lattice. This implies that the Brillouin zone can be obtained by a spatial Fourier transform of the one-dimensional crystal of Fig. 1-11. Since the

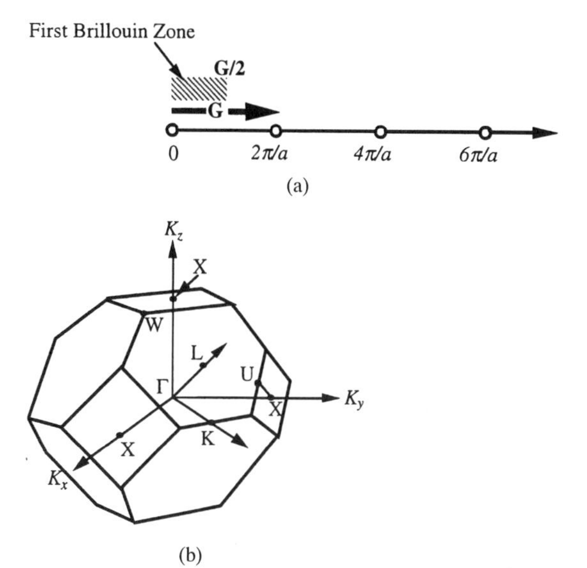

Figure 1-13 (a) The reciprocal lattice of a one-dimensional crystal which can be obtained by a spatial Fourier transform of a periodic lattice. Shown in this figure are the reciprocal lattice vector, **G**, and the first Brillouin zone, which spans between 0 and **G**/2. (b) The first Brillouin zone of a three-dimensional lattice of an fcc crystal. The zone center is called the Γ point. The point where the three axes meet the edge of the zone is called the X point. These are the crystal directions (100), (010), and (001). The body diagonal of the crystal is the (111) direction. The point where the body diagonal meets the zone edge is called the L point. The K, W, and U points are also shown.

crystal lattice is a periodic array, the atoms have a periodicity of a. Hence, any distribution ϕ in the crystal that is periodic in nature can be expressed as a Fourier series as $\phi(x) = \sum_{n=1}^{\infty} A_n \exp(i2\pi nx/a)$. The spatial Fourier transform of this Fourier series produces the wavevectors $2n\pi/a$ that are required to describe this Fourier series. Since n is an integer, these wavevectors form discrete points in K-space as shown in Fig. 1-13a. The one-dimensional lattice formed by these discrete points is called the *reciprocal lattice*. The line joining 0 to $2\pi/a$ is called the *reciprocal lattice vector*, **G**. The first Brillouin zone is formed by the region between 0 and π/a (i.e., half of the reciprocal lattice vector or **G**/2).

Of course, the one-dimensional picture can be extrapolated to three dimensions. Note that in one dimension the Brillouin zone is the K-space of the crystal lattice. Since a crystal is a periodic three-dimensional array in real space, one needs to perform a three-dimensional spatial Fourier transform of a crystal to obtain the three-dimensional reciprocal lattice. The first Brillouin zone is the unit cell of this three-dimensional lattice. Figure 1-13b shows the first Brillouin zone for an fcc crystal. Similar zones can be obtained for other crystal structures. The directions K_x, K_y, and K_z in Fig. 1-13b are the three orthogonal axes of the crystal, which are called the (100), (010), and the (001) directions, respectively. The center of the Brillouin zone is called the Γ *point*, and the intersection of the three axes with the zone edge is called the *X point*. The *body diagonal*, or the (111) direction, meets the zone edge at the *L point*. Other points of the

Brillouin zone are shown in Fig. 1-13. It must be noted that the ω-K dispersion relation can be obtained for each crystal direction within the Brillouin zone. It is also worth noting that the speed of sound in these directions can be different.

Note that the reciprocal lattice consists of such Brillouin zones placed adjacent to each other to form a lattice. The reciprocal lattice vector, **G**, is the vector joining the Γ points of the adjacent unit cells.

Optical mode of vibrations. Having studied the simple system with only one type of masses and springs, consider now two atoms per unit cell, such as in NaCl. If x is the displacement of atom 1 and y is the displacement of atom 2, then the equations of motion are

$$\begin{aligned} m_1 \frac{d^2 x_n}{dt^2} &= g(y_n + y_{n-1} - 2x_n) \\ m_2 \frac{d^2 y_n}{dt^2} &= g(x_{n+1} + x_n - 2y_n) \end{aligned} \tag{1-5}$$

where the subscript n is for a unit cell containing a pair of atoms 1 and 2. Traveling wave solutions for x and y can be assumed as $x_n = x_o \exp[i(nKa - \omega t)]$ and $y_n = y_o \exp[i(nKa - \omega t)]$. Similar to the last case, a ω-K dispersion relation can be obtained. Due to the presence of two atoms per unit cell, there are two solutions for $\omega = f(K)$, as opposed to only one solution in Eq. (1-3). These two, called the *acoustic branch* and the *optical branch*, are graphically shown in Fig. 1-12. Consider now the long-wavelength limit, $Ka \to 0$. The dispersion relation for the two branches are

$$\begin{aligned} \omega &= \sqrt{\frac{2g(m_1 + m_2)}{m_1 m_2}} && \textit{optical branch} \\ \omega &= \sqrt{\frac{2g}{(m_1 + m_2)}} Ka && \textit{acoustic branch.} \end{aligned} \tag{1-6}$$

In the short-wavelength limit (i.e., at the edge of the Brillouin zone, $Ka = \pm\pi$), the relation for optical and acoustic branches are $\omega_{op} = \sqrt{2g/m_2}$ and $\omega_{ac} = \sqrt{2g/m_1}$, respectively, for $m_1 > m_2$. It is important to note that this splitting into two branches can occur if the masses are the same but the spring constants are different. Consider if $m_1 = m_2 = m$ and the two spring constants are g and j. The long-wavelength limits for the dispersion relation are (Ashcroft and Mermin, 1976)

$$\begin{aligned} \omega &= \sqrt{\frac{2(g + j)}{m}} && \textit{optical branch} \\ \omega &= \sqrt{\frac{gj}{2m(g + j)}} Ka && \textit{acoustic branch} \end{aligned} \tag{1-7}$$

whereas the short wavelength limits are $\omega_{op} = \sqrt{2g/m}$ and $\omega_{ac} = \sqrt{2j/m}$ when $g > j$.

It is clear from Fig. 1-12 that the group velocity of the optical branch is close to zero or negligible compared to that of the acoustic branch. For the optical branch, it

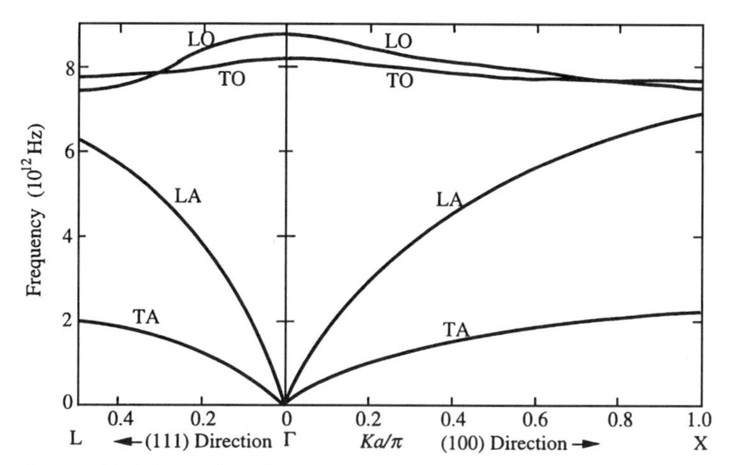

Figure 1-14 Phonon dispersion relation for a GaAs crystal along the (100) direction from the Γ point to the X point, as well as the (111) direction from the Γ point to the L point.

can be shown that the two atoms within a unit cell vibrate against each other, whereas for the acoustic branch at low values of wavevector, the two atoms remain in phase. Therefore, if there is an electrical dipole created by an uneven charge distribution in a chemical bond (due to some ionicity), then the optical vibrational mode corresponds to an oscillating dipole. From radiation theory, it can be shown that an oscillating dipole scatters radiation. Therefore, the optical branch is so named because it influences some of the optical properties of a crystal.

It is worth noting that waves in a crystal can have three polarizations: longitudinal and two orientations of transverse. In the longitudinal mode, the atoms vibrate in the direction of wave propagation; in the transverse mode, they vibrate perpendicular to the propagation direction. There are two transverse (TA) and one longitudinal (LA) for acoustic mode vibrations, and two transverse (TO) and one longitudinal (LO) for optical mode vibrations. The dispersion relations for GaAs are shown in Fig. 1-14 (Waugh and Dolling, 1963). Table 1-7 lists the vibrational frequencies of several semiconductors at the symmetry points of the Brillouin zone (Ferry, 1991). Note that the optical mode vibrations correspond to frequencies on the order of 10^{13} Hz. This relates to radiation wavelength of 30 μm, which is in the infrared regime of the radiation spectrum.

1-3-2 Phonons

Energy quantization, total energy, and heat capacity. Having discussed the wave motion of crystal vibrations, consider now the amplitude of vibrations. For a *macroscopic* simple harmonic oscillator with an energy–distance relation $u = 1/2gx^2$, the oscillator is allowed to occupy all values of energy. Hence, u and x are continuous variables. However, for a quantum mechanical oscillator, the energy u is discrete such that the oscillator is allowed to occupy only certain discrete energy states. This is shown in Fig. 1-15. The discrete value or quantum of vibrational energy is called a *phonon*. The energy quantum depends on the frequency of vibration, ω, and is equal to $\hbar\omega$, where $\hbar$

Table 1-7 Vibrational frequencies at different symmetry points of the Brillouin zone (in 10^{12} Hz) [13]

	Γ(100)		X(100)				L(111)			
	LO	TO	LO	TO	LA	TA	LO	TO	LA	TA
C	39.9	39.9	35.8	32.3	35.8	24.1	37.3	36.3	31.1	16.6
Si	15.6	15.6	12.3	13.9	12.3	4.5	12.6	14.7	11.3	3.42
Ge	9.03	9.03	6.9	8.25	6.9	2.46	7.41	8.4	6.45	1.95
AlP	15.03	13.2				3.36				4.53
AlAs	12.1	10.8				3.1				2.43
AlSb	5.91	5.55			8.9	2.4	7.0	9.2	6.4	1.86
GaP	12.1	11.0	10.6	10.9	7.44	3.24	11.1	10.6	6.24	2.5
GaAs	8.76	8.07	7.22	7.56	6.8	2.36	7.15	7.84	6.26	1.86
GaSb	7.3	6.9								
InP	10.35	9.1	7.1	8.3	6.9	3.2	7.5	8.9	7.1	2.6
InAs	7.3	6.57	4.9	6.4	4.4	3.3	5.8	6.5	4.4	2.2
InSb	6.0	5.37	3.9	5.3	3.6	1.1	4.8	5.1	3.0	1.0
CdTe	5.13	4.2	4.5	4.2	2.9	1.0				
HgTe	4.17	3.5								

is Planck's constant divided by 2π. The total energy of an oscillator can be written as the sum of all the energy quanta as

$$u = \left(n + \frac{1}{2}\right)\hbar\omega \tag{1-8}$$

where n is an integer $(0, 1, 2, \ldots)$ and $\hbar\omega/2$ is called the zero-point energy. As represented in Eq. (1-8) the oscillator of energy u contains n phonons.

Since the amplitude and energy of crystal waves are quantized, the propagation of a wave of vibrational energy u can be thought to be the propagation of a packet of energy containing n phonons. This brings some elements of a particle-like behavior since particles in motion can also be thought of as a packet of kinetic energy. The analogy goes beyond that, since collisions between particles is similar to scattering of waves. The particle and wave duality breaks down during wave interference, where the particle approach has no analog. However, this is something that will not be considered

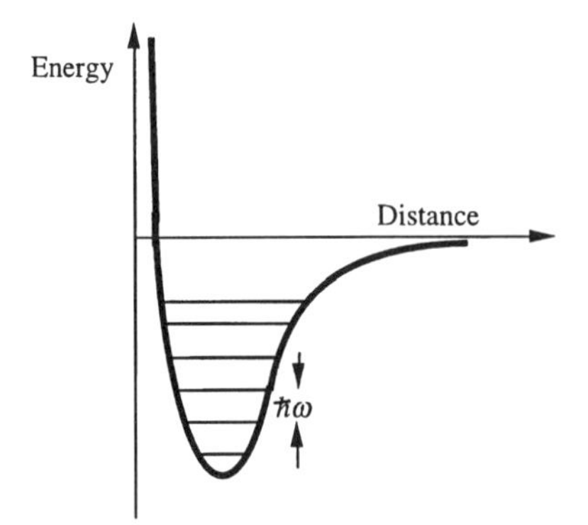

Figure 1-15 Quantization of energy of crystal vibrations.

here, although interference optics with phonons has been studied. Since a particle has momentum, the question is how to assign a momentum to a wave. If one considers a classical particle of kinetic energy $E = mv^2/2$ and momentum $p = mv$, then the particle velocity can be found as $v = dE/dp$. When applied to a quantum-mechanical particle whose energy is $\hbar\omega$, the ω-K dispersion relation suggests that $\hbar K$ must be phonon momentum. Although phonons do not actually have any momentum, they interact with each other and with electrons and photons in a way that phonon energy $\hbar\omega$ and momentum $\hbar K$ are usually conserved.

At a given temperature T, what is the equilibrium number of phonons in a solid? This question was answered by Bose, who used quantum statistics to derive the distribution of a class of particles called Bosons, examples of which are photons and phonons. The equilibrium distribution of phonons, $\langle n \rangle$, follows the Bose-Einstein distribution and is given as (Eisberg and Resnick, 1985)

$$\langle n \rangle = \frac{1}{\exp\left(\frac{\hbar\omega}{k_B T}\right) - 1} \tag{1-9}$$

Therefore, the total energy in a lattice, E_l, can be written as

$$\mathrm{E}_l = \sum_{\mathbf{K},p} \left(\langle n \rangle_{\mathbf{K},p} + \frac{1}{2} \right) \hbar\omega_{\mathbf{K},p} \tag{1-10}$$

where the subscript K stands for the wavevector and the subscript p stands for the particular branch of the phonon mode. The summation is over the whole wavevector space in a three-dimensional Brillouin zone and over all phonon branches. In the limit of a large crystal, the wavevector space becomes so dense that the summation in Eq. (1-10) can be replaced by an integral as

$$\mathrm{E}_l = \sum_p \int \left(\langle n \rangle_{\mathbf{K},p} + \frac{1}{2} \right) \hbar\omega_{\mathbf{K},p} \frac{d\mathbf{K}}{(2\pi)^3}. \tag{1-11}$$

The integration of the wavevector space is difficult because it involves the direction and magnitude of the wavevector. It is possible to transform this into an integral over frequency as

$$\mathrm{E}_l = \sum_p \int \left(\langle n \rangle_p + \frac{1}{2} \right) \hbar\omega_p D_p(\omega) d\omega. \tag{1-12}$$

The transformation introduces $D_p(w)$, which is the density of phonon states. This means that the number of phonon states $\omega_{\mathbf{K},p}$ between frequency ω and $\omega + d\omega$ for phonon branch p is $D_p(\omega)d\omega$. Using this expression for the total energy of the crystal, the heat capacity can be found to be

$$C_l = \frac{\partial \mathrm{E}_l}{\partial T} = \sum_p \int \frac{\partial \langle n \rangle_p}{\partial T} \hbar\omega_p D_p(\omega) d\omega. \tag{1-13}$$

Einstein model. In Eq. (1-13), all the terms inside the integral are known expect for two things: the density of states and the limits of the integral. This is where various assumptions appear. If there are N atoms or oscillators in a solid in a volume V, then each oscillator can have three modes of vibration, resulting in $3N$ total modes. The Einstein model assumes that all of the atoms oscillate at the same frequency, ω_o, such that the density of states is $D(\omega) = 3N\delta(\omega - \omega_o)/V$. Using this, the heat capacity per unit volume becomes

$$C_l = 3\eta k_B \left(\frac{\theta_E}{T}\right)^2 \frac{\exp(\theta_E/T)}{(\exp(\theta_E/T) - 1)^2} \qquad (1\text{-}14)$$

where η is the number density of oscillators or atoms and θ_E is the *Einstein temperature*, defined as $\theta_E = \hbar\omega_o/k_B$.

The Einstein model is the simplest model and is acceptable when there is only one oscillator frequency. This could be justified for the optical branch, where the ω-K curve is flat. However, it cannot be applied for the acoustic branch, where ω-K is not flat for most of the region in the Brillouin zone. In the low-temperature limit, measurements show that the phonon heat capacity of most solids vary as T^3, whereas for the Einstein model it varies as $\exp(-\theta_E/T)$. This discrepancy is due to the single-frequency assumption. However, it is interesting to note that at temperatures $T \gg \theta_E$, the heat capacity is $C_v \approx 3\eta k_B = 3R$, which is the law of Dulong and Petit for gas molecules where R is the gas constant. Therefore, the Einstein temperature can be interpreted to be a demarcation between the quantum and the classical regimes.

Debye model. A more realistic approach is adopted in the Debye model, which assumes that the dispersion relation is linear such that $\omega = v_g K$ where v_g is the group velocity or the speed of sound. This is the elastic or continuum limit since, as evident in Fig. 1-12, it is valid for low values of K in the Brillouin zone but not for high values. The model assumes that the crystal waves are coherent for spacings much larger than the lattice spacing. This is the basic difference between the Debye and the Einstein models, since the latter assumes the motion of neighboring atoms to be uncorrelated.

If one looks at the K-space of crystal vibrations, it can be shown that the number of vibrational modes in a region between 0 and K in the K-space is (Kittel, 1986)

$$N = \frac{V}{6\pi^2} K^3 \qquad (1\text{-}15)$$

where V is the volume of the solid. Hence, the density of states in general can be derived to be

$$D(\omega) = \frac{1}{V}\frac{dN}{d\omega} = \frac{K^2}{2\pi^2}\frac{dK}{d\omega} \qquad (1\text{-}16)$$

Under the Debye assumption of $\omega = v_g K$, the density of states can be derived to be

$$D(\omega) = \frac{\omega^2}{2\pi^2 v_g^3} \qquad (1\text{-}17)$$

Table 1-8 Debye temperatures of selected elements (Ashcroft and Mermin, 1976)

Element	θ_D (K)	Element	θ_D (K)
Li	400	Ar	85
Na	150	Ne	63
K	100	Cu	315
Be	1000	Ag	215
Mg	318	Au	170
Ca	230	Zn	234
B	1250	Cd	120
Al	394	Hg	100
Ga	240	Cr	460
In	129	Mo	380
Tl	96	W	310
C (diamond)	1860	Mn	400
Si	625	Fe	420
Ge	360	Co	385
Sn (grey)	260	Ni	375
Sn (white)	170	Pd	275
Pb	88	Pt	230
As	285	La	132
Sb	200	Gd	152
Bi	120	Pr	74

If there are a total of N oscillators in a volume V, then using Eq. (1-15) there must be a maximum cutoff wavenumber, K_D, which is

$$K_D = (6\pi^2\eta)^{1/3} \tag{1-18}$$

where the subscript D denotes the Debye cutoff. Note that $\eta^{-1/3}$ is on the order of the lattice spacing, a. Therefore, K_D is representative of the edge of the Brillouin zone, which comes from the facts that the lattice is discrete and that there is a minimum wavelength that can be sustained in the lattice. Associated with the cutoff wavenumber is the upper limit or *Debye cutoff frequency*, ω_D, which under the Debye assumption is

$$\omega_D = v_g K_D = v_g(6\pi^2\eta)^{1/3} \tag{1-19}$$

Associated with the Debye cutoff frequency is the *Debye temperature*, defined as

$$\theta_D = \hbar\omega_D/k_B = \frac{h v_g(6\pi^2\eta)^{1/3}}{k_B} \tag{1-20}$$

In Section 1-3-1, it was shown qualitatively that group velocity varies as $v_g = a\sqrt{g/m}$, where a is the lattice spacing, g is the spring constant, and m is the atomic mass. It is clear that θ_D increases with increasing bond stiffness and decreasing atomic mass. This is clear in Tables 1-8 and 1-9, which list θ_D for elements and compounds.

Using the Debye density of states and the upper cutoff frequency, the total energy per unit volume and the heat capacity per unit volume, E_l, can be derived from Eqs. (1-12)

Table 1-9 Debye temperatures for alhali halide crystals (Ashcroft and Mermin, 1976)

	F	Cl	Br	I
Li	730	422	—	—
Na	492	321	224	164
K	336	231	173	131
Rb	—	165	131	103

and (1-13), respectively, to be

$$\mathrm{E}_l = \left[\frac{9\eta k_B}{\theta_D^3}\int_0^{\frac{\theta_D}{T}}\frac{x^3dx}{e^x-1}\right]T^4, \tag{1-21}$$

$$C_l = 9\eta k_B\left(\frac{T}{\theta_D}\right)\int_0^{\frac{3\theta_D}{T}}\frac{x^4dx}{(e^x-1)^2}. \tag{1-22}$$

It should be noted that in the low-temperature limit when $\theta_D/T \gg 1$, the integral in Eq. (1-21) tends towards $\pi^4/15$. It is clear that in this limit, $\mathrm{E}_l \approx T^4$. The heat flux can be found as $Q = v_g\mathrm{E}_l$, which can be written as $Q = \sigma T^4$. This is exactly the same as the Stefan-Boltzmann law for photon radiation, where now in the phonon case, the constant σ is $\sigma = 9\pi^4 v_g \eta k_B/15\theta_D^3$. Due to this similarity, heat conduction by phonons is often called *phonon radiation*. The integral in Eq. (1-21) is essentially over the Planck distribution for photons. The similarity arises from the fact that both follow the Bose-Einstein distribution of Eq. (1-9). In addition, the Debye assumption of $\omega = v_g K$ for the dispersion relation is exactly that for photons, but with v_g replaced by the speed of light. The differences between photons and phonons lie in the fact that in addition to the two transverse polarizations for electromagnetic wave propagation, acoustic waves can also have longitudinal polarization. In addition, whereas photon frequencies can vary from zero to infinity, those of phonons are limited to be within the first Brillouin zone in K-space.

Since the Planck distribution is used in Eq. (1-21), it can be differentiated to get the maxima. This leads to the Wien's displacement law for phonons, $\hbar\omega_{max} = k_B T$, which can be used to calculated the frequency ω_{max}, which is dominant in phonon energy. In the low-temperature limit, $\theta_D/T \gg 1$, the integral in Eq. (1-22) also tends to a constant such that the heat capacity becomes $C_l \approx 234\eta k_B(T/\theta_D)^3$. The $C_l \approx T^3$ behavior has been observed for several solids at low temperatures. Deviations from the T^3 behavior starts at around $T = \theta_D/50$. Figure 1-16 shows a comparison between experimental results (Touloukian and Buyco, 1970) and Debye model predictions for diamond. The good agreement is a validation of the Debye model. The reason it works so well is because the assumption of $\omega = v_g K$ is valid for most of the Brillouin zone except near the edges. At low temperature, since ω_{max} by Wien's law is low, it falls in a region where the assumption of $\omega = v_g K$ is valid.

When the temperature T is on the order of or higher than θ_D, the integral of Eqs. (1-21) and (1-22) cannot be approximated as before. In addition, the density of states is

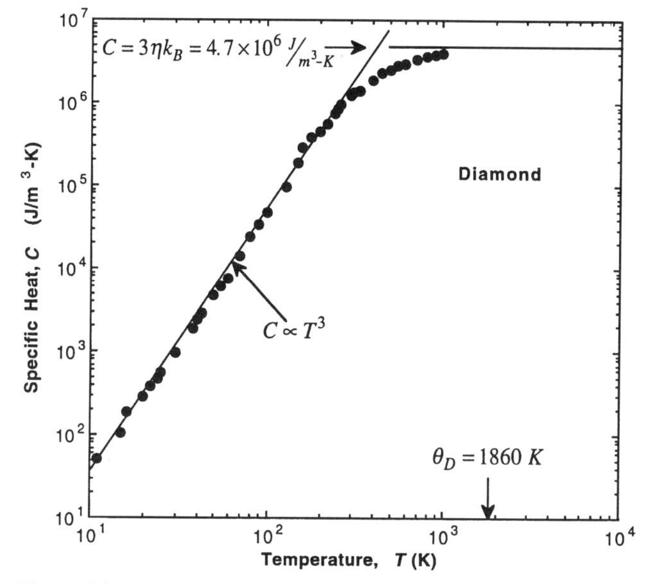

Figure 1-16 Comparison of experimental data and theoretical predictions of specific heat of diamond (Touloukian and Buyco, 1970).

not a smooth parabolic relation but is more complicated (Ashcroft and Mermin, 1976). In the high-temperature limit, $T \gg \theta_D$, the heat capacity again tends to a constant $C_l \approx 3\eta k_B = 3R$ which is the Dulong-Petit law for gases. This is clear in Fig. 1-16 where $R = 2$ cal/mole-K.

1-3-3 Phonon Scattering

Phonon scattering can be divided into two categories: (i) elastic scattering by lattice imperfections, such as defects, dislocations, and boundaries, where the energy or the frequency of the incident phonon is unchanged; and (ii) inelastic scattering between three or more phonons where the frequency is modified. In the following discussion, elastic scattering by dislocations will not be considered in this study and can be found in more detail in Berman (1976), Klemens (1958), and Ziman (1960).

Scattering by crystal imperfections. Experiments have shown that even a slight rise in impurity concentration can have substantial effect on heat conduction by phonons (Walton, 1967; Worlock, 1966). A defect can be thought of as a change in mass or spring within a crystal that can alter the local acoustic impedance or the vibrational characteristics of a crystal. Hence, when an incident crystal wave encounters a change in acoustic properties, it tends to scatter. This is similar to the scattering of an electromagnetic wave in the presence of a change of optical refractive index.

One can consider a solid to be made of a phonon gas, where the phonons can be considered traveling in all directions and colliding between themselves and with fixed impurities. Using kinetic theory of gases, the average time τ_i between collisions with imperfections can be obtained as (Vincenti and Kruger, 1977)

$$\tau_i = \frac{1}{\alpha \sigma \rho v_g} \tag{1-23}$$

where σ is the scattering cross section, ρ is the number of scattering sites per unit volume or defect density, and α is a constant of the order of unity.

The scattering cross section depends on the wavelength of the incident phonon. Let us define a size parameter $\chi = 2\pi R/\lambda = \omega R/v_g$ where R is the radius of the lattice imperfection and λ is the phonon wavelength. The limit where $\chi \rightarrow 0$ is called the *Rayleigh limit*. The scattering cross section in this limit follows the behavior $\sigma \propto \chi^4$, which is called the *Rayleigh law* (Lord Rayleigh, 1896). In the other limit, $\chi \gg 1$, the cross section is independent of χ and is equal to the projected area of the defect, which is πR^2. In the region in between the two limits, Walton and Lee (1967) have shown that the cross section can have oscillations with the wavelength. However, when integrated over all the wavelengths to determine the total heat flux, these oscillations would average out. For simplicity in calculations, Majumdar (1993) proposed that the scattering cross section varies as

$$\sigma = \pi R^2 \left(\frac{\chi^4}{\chi^4 + 1} \right) \tag{1-24}$$

which provides a smooth transition between the two limits.

It is interesting to note that the scattering cross section is not directly dependent on temperature. However, since the phonon frequency that is dominant in energy, ω_{max}, depends on the temperature, the cross section σ will be temperature dependent at low frequencies or low temperatures.

Inelastic scattering (N and U processes). Inelastic scattering between phonons occurs due to the anharmonic nature of interatomic potential energy, as shown in Fig. 1-10. Consider a lattice wave propagating in a crystal. As the wave passes by atoms in a lattice, it moves them from their equilibrium positions, creating a local strain. For a perfect harmonic oscillator, the spring stiffness g is independent of the strain. However, due to the anharmonicity of the interatomic potential, the spring stiffness changes with deformation. When another lattice wave is incident on this strain field, it encounters a change in spring stiffness or, in other words, a change in acoustic impedance, and hence the wave scatters. Such scattering processes are often called intrinsic because they can occur even in a pure crystal. Intrinsic scattering involves three or more phonons. Those involving four phonons or more are important only at temperatures much higher than θ_D (Joshi et al., 1970) and will not be considered in this study. A detailed description of inelastic scattering can be found in Ziman (1960).

There are two types of three-phonon inelastic scattering: the N process and the U process. Three-phonon processes can be shown schematically as in Fig. 1-17, where two phonons can merge to form a third phonon or a single phonon can break up into two phonons. Energy conservation in these processes requires

$$\begin{aligned} \omega_1 + \omega_2 &= \omega_3 \qquad \text{Case A,} \\ \omega_1 &= \omega_2 + \omega_3 \qquad \text{Case B,} \\ \omega_1 + \omega_2 &= \omega_3 \qquad \text{Case C.} \end{aligned} \tag{1-25}$$

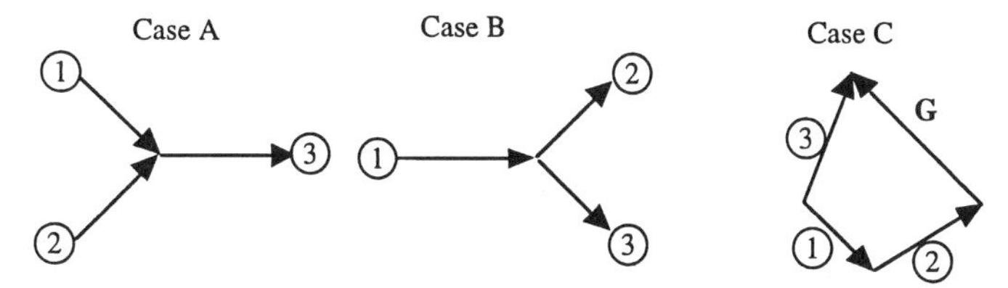

Figure 1-17 Vectorial representation of phonon–phonon scattering. Cases A and B are normal processes and case C is Umklapp process.

Momentum conservation in these interactions requires the vectorial sums

$$\begin{aligned} \mathbf{K}_1 + \mathbf{K}_2 &= \mathbf{K}_3 && \text{Case A,} \\ \mathbf{K}_1 &= \mathbf{K}_2 + \mathbf{K}_3 && \text{Case B,} \\ \mathbf{K}_1 + \mathbf{K}_2 &= \mathbf{K}_3 + \mathbf{G} && \text{Case C.} \end{aligned} \tag{1-26}$$

Cases A and B conserve phonon momentum (i.e., $\hbar\mathbf{K}$) during scattering. This type of scattering is called a *normal* or *N process*. It is possible, however, that the magnitude of the sum $|\mathbf{K}_1 + \mathbf{K}_2|$ can exceed the maximum limit set by the edge of the first Brillouin zone, which is π/a. Since the smallest wavelength of a crystal wave is $2a$, a crystal wave of wavevector larger than π/a cannot physically occur. So when $|\mathbf{K}_1 + \mathbf{K}_2| > \pi/a$, the third phonon is "flipped over" and has a lower frequency and wavevector as graphically shown as case C in Fig. 1-17. The magnitude and direction of the third phonon is decided by adding the reciprocal lattice vector **G** in Eq. (1-26). As explained earlier, the reciprocal lattice vector **G** is the vector joining the Γ points of adjacent Brillouin zones in the reciprocal lattice. A three-phonon process in which the third phonon is flipped over is called an *Umklapp process* or *U process* and is usually dominant in heat conduction at temperatures close to or higher than θ_D. Both N and U processes conserve energy.

The momentum conserving N processes do not pose any resistance to heat flow (Ziman, 1960). This is because the momentum in a particular direction of phonon flow is always conserved, no matter how many particles are involved in the scattering. Consequently, nothing impedes the flow of momentum in that direction. If only N processes were to exist, a solid would have infinite conductivity. Although posing no direct resistance by themselves, N processes do play a role in resistance to heat flow. Since the frequency of incident waves is changed in inelastic scattering, N processes are responsible for distributing the phonon energy over different frequencies. Since other scattering processes are frequency dependent, they feel the effect of phonons scattered by N processes. Therefore, N processes indirectly influence the resistance to heat flow. Although N processes have been shown to influence thermal conductivity (Callaway, 1959), neglecting N processes may not be totally inaccurate in some temperature ranges. In contrast, U processes do not conserve momentum and therefore directly resist phonon flow in a particular direction. They are responsible mainly for the thermal conductivity of semiconductors and insulators at room temperature.

Several expressions for the relaxation times for U processes are used in the literature; the correct one is still a matter of discussion (Berman, 1976). An approximation commonly used for U-process relaxation time is (Klemens, 1958)

$$\tau_U = A \frac{T}{\theta_D \omega} \exp\left(\frac{\theta_D}{\gamma T}\right) \tag{1-27}$$

where A is a nondimensional constant that depends on the atomic mass, the lattice spacing, and the Grüneisen constant (Ashcroft and Mermin, 1976); γ is a parameter representing the effect of crystal structure. Typical phonon–phonon scattering rates in solids at room temperature are on the order of 10 ps.

As indicated in Eq. (1-27), there is a very strong temperature dependence in the relaxation time for U processes. This is because the phonon scattering rate depends on the density of phonons in the phonon gas. Since the number of phonons is highly temperature dependent through the Bose-Einstein distribution, the phonon gas density is also temperature dependent. The phonon gas density is very low at temperatures much lower than the Debye temperature, such that phonon–phonon scattering is frozen out. As the temperature increases, impurity scattering becomes important, and a further increase in temperature makes U processes dominant and the N processes negligible.

Effective relaxation time. The effective relaxation time arises from a combination of all phonon scattering mechanisms. We have discussed scattering by crystal imperfections and by other phonons. Another scattering mechanism is by the boundary of a crystal. This becomes important for nanostructures that are smaller than the mean free path of phonons or even for larger structures at low temperatures, where other scattering modes are frozen out. An assumption often used is the Matthiessen rule, which adds up the scattering rates or the reciprocal of the relaxation times for different scattering processes as

$$\frac{1}{\tau} = \frac{1}{\tau_i} + \frac{1}{\tau_b} + \frac{1}{\tau_U}. \tag{1-28}$$

Here τ is the effective relaxation time, and τ_b is due to boundary scattering, $\tau_b = L/v_g$, where L is the size of the crystal. Physically, Eq. (1-28) means adding up the resistances posed by each scattering process. The assumption is valid under the condition that the scattering processes do not interact with each other, which clearly is not true for inelastic scattering. However, the error involved in applying the Matthiessen rule is usually quite small (Ziman, 1960). The effective mean free path of phonons can be found from the relaxation time as $\ell = v_g \tau$. This will be used to calculate the phonon thermal conductivity of solids.

1-4 ELECTRONIC STRUCTURE OF SOLIDS

It is important to understand the electronic structure of solids for two reasons. First, it forms the basis for studying energy transport by electrons; second, it is directly related to the optical properties of a solid in the near infrared, visible, and ultraviolet regions of the spectrum. There are several books in solid state physics that provide a strong mathematical foundation for the electronic structure of solids. This section will rely more on physical intuition and introduce the mathematical techniques only when necessary.

The emphasis will be more on trying to understand the origins of the electronic structure from chemical and electronic viewpoints as well as the influence of electronic structure on electron motion and optical properties.

1-4-1 Origin of Energy Bands: A Chemical View Point

The nucleus of the atoms is positively charged and provides an attractive potential for the electrons. For a single atom, the potential is symmetrical and has the form $V(r) \approx -1/r$, where r is the radius from the nucleus. According to quantum mechanics, the energy states that the electrons can occupy within this potential are discrete, as shown in Fig. 1-18. The quantum state of an electron is characterized by the set of four quantum numbers: principal, angular momentum, magnetic, and spin. Associated with each set of quantum numbers is an energy of the state. The electrons first occupy the lowest energy levels. But according to Pauli's exclusion principle, two electrons cannot occupy the same quantum state. Therefore, once a quantum state is occupied by an electron, the other electrons must occupy a different quantum state of higher energy until all the electrons of the atom are used up to ensure charge neutrality of the atom. The electrons in the outermost atomic orbital or the highest energy state are the valence electrons and those interior are called the core electrons.

Now consider two atoms that are initially far apart, as shown in Fig. 1-18. The potential energies of the two atoms do not overlap, and hence the electrons of the individual atoms are distinguishable. They can, therefore, have the same quantum numbers and occupy exactly the same quantum states. If the atoms are brought closer so that the outer

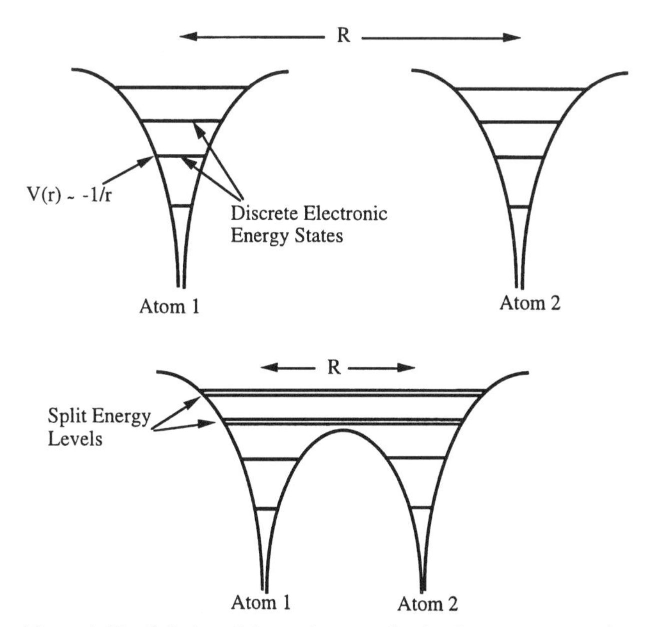

Figure 1-18 Splitting of electronic energy levels when two atoms are brought sufficiently close such that the atomic orbitals overlap.

atomic orbitals overlap, the overlapping orbitals must occupy different energy levels in order for the electrons to occupy different energy states, as shown in Fig. 1-18. Hence, a given energy level of the two-atom system is split into two distinct energy levels as the overlap commences, and the splitting increases as the separation of the atoms, R, decreases (Eisberg and Resnick, 1985). If a third atom is brought into the picture, each overlapping orbital is split into three energy states. Note that the core energy states do not split, since the orbitals of the core electrons do not overlap. Now if we consider N atoms, the same energy level of the N separate atoms must split into N distinct energy levels when the separation is reduced to the point that the atomic orbitals overlap. The spread in energy between the lowest and the highest level of a particular atomic orbital depends on the separation between the atom, R, since the separation determines the amount of overlap that causes the splitting. However, the spread does not depend so much on the number of atoms in the system if the separation is maintained. Thus, as N increases in value, each atomic orbital is split finer and finer and spread between a certain energy range. The range of energy depends on R. Therefore, a particular atomic orbital is said to have formed an *energy band.*

Figure 1-19 shows the band formation for the energy levels of sodium (Eisberg and Resnick, 1985), whose ground state atomic configuration is $1s^2 2s^2 2p^6 3s^1$. As the separation R is decreased, the outer orbitals split first, since they experience the maximum overlap. These form bands of allowable energy states. The inner orbitals start to split when R is further reduced. For a given separation R, the diagram shows bands that can be occupied and some bands that are forbidden. Note that at the equilibrium distance of 3.67 Å, the $2p$ level is not split, whereas the higher orbitals are split and form bands.

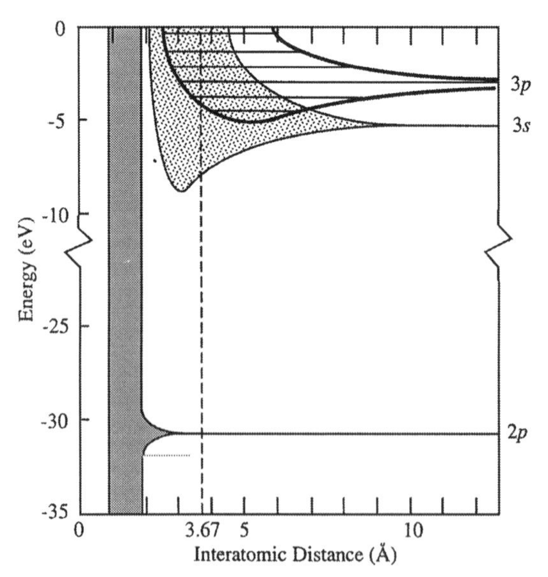

Figure 1-19 Formation of energy bands from the energy levels of isolated sodium atoms as the distance between atoms decreases. The dashed line shows the equilibrium interatomic spacing (Eisberg and Resnick, 1985).

Also worth noting is the fact that at 3.67 Å, the different bands overlap providing a continuous band of electronic energy. If we now consider the occupation of electrons in these energy states, the electrons will occupy first the inner core orbitals, which do not overlap. The $3s^1$ valence electrons of all the atoms in the solid can occupy the $3s$ band, which overlaps strongly with $3p$ and higher orbital bands. Since there are only a finite number of $3s$ electrons, the energy band is partially filled, leaving the rest of the band empty. Therefore, when an electric field is applied across a solid like sodium, the $3s$ electrons have higher energy states available for them to occupy and move. This makes sodium a good conductor of electricity. If there had been a large forbidden gap, the electrons would not be able to gain energy, and the solid would behave like an insulator. The electronic band structure of a solid is determined by the crystal structure and the valence of the atoms forming the solid.

1-4-2 Free Electron Theory of Metals

Electronic energy states. It was argued in the previous section that when the electrons are allowed to occupy any higher energy levels without any forbidden gap, the electrons can gain energy from an electric field and thereby conduct electricity. In trying to determine the motion of electrons, it must be realized that these electrons should interact with each other and with the positive ions in the crystal. However, in many cases the electron–ion interactions can be ignored, giving rise to the term electron gas. The theory of electronic structure and motion under this assumption is called the *free electron theory*. It has been observed that it applies best to metals, although some key effects cannot be explained by this theory. Nevertheless, it is simple and forms the first step in understanding more complex behavior encountered in band structure.

Under the free electron theory, all the valence electrons of a solid of volume V are assumed to be placed in a combined attractive potential created by all the positive ions[3] in that volume. What is unknown is the density of electrons in each energy level, as well as the total energy of the free electron gas. The Schrödinger equation for the free electron in this volume is given as (Eisberg and Resnick, 1985)

$$-\frac{\hbar^2}{2m}\nabla^2\psi(\mathbf{r}) = E\psi(\mathbf{r}) \tag{1-29}$$

where E is the electron energy, m is the electron mass, and $\psi(\mathbf{r})$ is the electron wavefunction. The probability of finding an electron at a position $\mathbf{r}$ is $|\psi(\mathbf{r})^2|$. The Schrödinger Eq. (1-29) has a wave-like solution of the form

$$\psi_{\mathbf{k}}(\mathbf{r}) = \frac{1}{\sqrt{V}}e^{i\mathbf{k}\cdot\mathbf{r}} \tag{1-30}$$

where $\mathbf{k}$ is the wavevector of the electron. The wavelength of the electron is given as $\lambda = 2\pi/k$. The energy of the electron of wavevector $\mathbf{k}$ can be determined from the

[3]Note that the core electrons shield the nucleus from the valence electrons giving rise to a net positive charge.

Schrodinger equation to be

$$E = \frac{\hbar^2 k^2}{2m}. \tag{1-31}$$

Noting that the energy of a particle is related to its momentum p as $E = p^2/2m$, the momentum of an electron can be related to its wavevector as

$$\mathbf{p} = \hbar \mathbf{k}. \tag{1-32}$$

Note that this is analogous to the momentum of phonons. In fact, this relation holds true for photons also.

To make the calculations simpler, consider the volume V to be a cube of side L. Since the origin of the cube and its length, L, is arbitrarily chosen, the wavefunctions must satisfy the cyclic boundary condition in one dimension as $\psi(x+L) = \psi(x)$. When applied to three dimensions, it gives rise to the relation

$$e^{ik_x L} = e^{ik_y L} = e^{ik_z L} = 1 \tag{1-33}$$

such that the wavevector, **k**, must satisfy the conditions

$$k_x = \frac{2\pi n_x}{L}; \qquad k_y = \frac{2\pi n_y}{L}; \qquad k_z = \frac{2\pi n_z}{L}; \quad n_x, n_y, n_z \text{ integers.} \tag{1-34}$$

Thus, in a three-dimensional **k**-space with axes k_x, k_y, and k_z, the allowed wavevectors form discrete points separated by a distance $2\pi/L$ along each axis direction. This is shown as a two-dimensional projection (for simplicity) in Fig. 1-20. The question remains as to how many of these states will be filled by N electrons.

According to Pauli's exclusion principle, each state in k-space can be occupied by two electrons with opposite spins. The electrons first start filling the lowest energy states and then occupy the higher energy levels. The relation between electron energy and wavevector (Eq. (1-31)) suggests that the smallest values of k get occupied first. It is evident from Fig. 1-20 that the volume associated with each state containing two

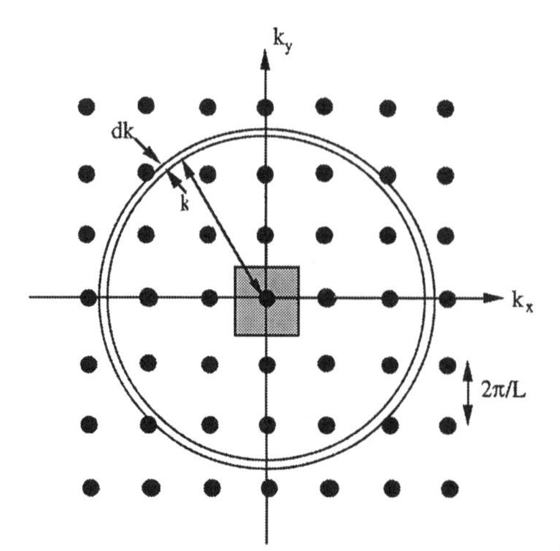

Figure 1-20 Two-dimensional projection of the k-space of electrons in the free electron theory.

electrons is equal to $(2\pi/L)^3$. From the origin (0, 0, 0), let the largest wavevector be at a radius k_F. Therefore, the total volume of the k-space occupied by N electrons is equal to $(4\pi k_F^3/3)$. Hence, we can derive the relation

$$N = 2 \cdot \frac{(4\pi k_F^3/3)}{(2\pi/L)^3} = \frac{k_F^3}{3\pi^2} V. \tag{1-35}$$

The wavevector k_F is called the *Fermi wavevector*. The Fermi wavevector forms a sphere in the k-space such that all wavevectors within this sphere are occupied with electrons, whereas those outside this sphere are unoccupied. At least, this is what happens at 0 K; we will later see how temperature affects the distribution. From Eq. (1-35), it is clear that k_F is related to the electron density, $\eta = N/V$, of a solid. Associated with the Fermi wavevector, we can derive the Fermi energy E_F and the Fermi velocity v_F of the electron as follows:

$$\begin{aligned} E_F &= \frac{\hbar^2 k_F^2}{2m} = \frac{\hbar^2}{2m}(3\pi^2\eta)^{\frac{2}{3}} \\ v_F &= \frac{\hbar k_F}{m} = \frac{\hbar}{m}(3\pi^2\eta)^{\frac{1}{3}}. \end{aligned} \tag{1-36}$$

If one were to use the analogy between gas molecules and electrons, one could determine the Fermi temperature, T_F, by the relation $T_F = E_F/k_B$, where k_B is the Boltzmann constant. Table 1-10 lists the values of electron density and the Fermi parameters for some metals. Note that the Fermi temperature is on the order of 10^4 K. This is not the real temperature of the metal, but rather a measure of the electron velocity had it been a particle that followed Maxwell-Boltzmann statistics like gas molecules. At room temperature, for example, electrons move much faster than gas molecules because of Pauli's exclusion principle. Only two electrons of opposite spin can occupy a given quantum state under this principle. Hence, to accommodate all the electrons, higher energy levels become occupied. Also worth noting is the Fermi wavelength of electrons, which is typically on the order of 1–10 Å. If one looks at the lattice constants, a, in Tables 1.1–1.6, it is clear that the Fermi wavelength of electrons is about the same size as lattice spacing. This has bearing on the electronic structure of solids, as we shall see later. The ion cores in metals are well shielded by the electrons such that electrons at the Fermi level do not quite feel the effect of periodic lattice. (However, this is not true in many cases.) The Fermi velocities are on the order of 10^6 m/s, which is about a factor 100 less than the speed of light. Hence, relativistic effects need not be included for electron transport unless spin interactions are involved. The Fermi velocity is an important quantity for transport calculations since only the Fermi level electrons have unoccupied energy levels above them and therefore can be excited for electrical or heat conduction.

The picture that emerges of a free electron solid is that of a potential well (or a bucket) created by the positive ions, as shown in Fig. 1-21. The electronic states in this potential well are discrete in momenta and energy. However, there are so many electrons in metals and there is so much of degeneracy of energy in these states that these electronic states are nearly continuous in energy. So if the electron density η has to be accommodated in these energy states, the states with lowest energy start to fill

Table 1-10 Free electron densities, Fermi energy, Fermi temperature, Fermi wavelength, Fermi velocity, and work function of selected metals (Ashcroft and Mermin, 1976)

Element	Electron Density, η [10^{28} m^{-3}]	Fermi Energy, E_F [eV]	Fermi Temperature, T_F [10^4 K]	Fermi Wavelength, λ_F [10^{-10} m or Å]	Fermi Velocity, v_F [10^6 m/s]	Work Function, Φ [eV]
Li	4.70	4.74	5.51	5.59	1.29	2.38
Na	2.65	3.24	3.77	6.85	1.07	2.35
K	1.40	2.12	2.46	8.36	0.86	2.22
Rb	1.15	1.85	2.15	8.98	0.81	2.16
Cs	0.91	1.59	1.84	9.68	0.75	1.81
Cu	8.47	7.00	8.16	4.65	1.57	4.44
Ag	5.86	5.49	6.38	5.22	1.39	4.3
Au	5.90	5.53	6.42	5.22	1.40	4.3
Be	24.7	14.3	16.6	3.27	2.25	3.92
Mg	8.61	7.08	8.23	4.65	1.58	3.64
Ca	4.61	4.69	5.44	5.66	1.28	2.80
Sr	3.55	3.93	4.57	6.15	1.18	2.35
Ba	3.15	3.64	4.23	6.41	1.13	2.49
Nb	5.56	5.32	6.18	5.34	1.37	3.99
Fe	17.0	11.1	13.0	3.67	1.98	4.31
Mn	16.5	10.9	12.7	3.70	1.96	3.83
Zn	13.2	9.47	11.0	3.98	1.83	4.24
Cd	9.27	7.47	8.68	4.49	1.62	4.10
Hg	8.65	7.13	8.29	4.59	1.58	4.52
Al	18.1	11.7	13.6	3.59	2.03	4.25
Ga	15.4	10.4	12.1	3.79	1.92	3.96
In	11.5	8.63	10.0	4.16	1.74	3.8
Tl	10.5	8.15	9.46	4.30	1.69	3.7
Sn	14.8	10.2	11.8	3.83	1.90	4.38
Pb	13.2	9.47	11.0	3.98	1.83	4.0
Bi	14.1	9.90	11.5	3.90	1.87	4.4
Sb	16.5	10.9	12.7	3.70	1.96	4.5

up first. Due to Pauli's exclusion principle, each state can be occupied by only two electrons. Hence, as more and more electrons are accommodated in these states, higher energy states are filled up. Finally, when all the electrons are accommodated, the highest energy reached the electrons is the Fermi energy, E_F. Note that we have not considered thermal excitations of electrons so far; this will be considered next. The Fermi energy, however, is below the vacuum energy level by an energy called the *work function*, Φ. The work function is the energy required to remove an electron from the metal. Table 1-10 lists values of work functions of some metals.

Fermi-Dirac statistics. The discussion so far has focused on what would happen at 0 K, where there is no thermal energy to excite electrons above the Fermi energy, E_F. As the temperature is increased, there exists a finite probability for such excitations. The probability f that a energy level E is occupied by an electron is given by the Fermi-Dirac

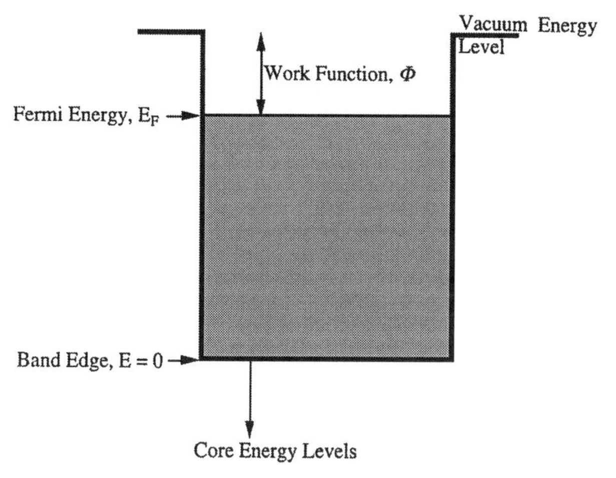

Figure 1-21 A simplified illustration of the electronic structure of a metal.

distribution:

$$f(E) = \frac{1}{1 + \exp\left(\frac{E-\mu}{k_B T}\right)} \tag{1-37}$$

where μ is the thermodynamic chemical potential, k_B is the Boltzmann constant, and T is the temperature. For free electrons, the chemical potential is related to the Fermi energy as (Ashcroft and Mermin, 1976)

$$\mu = E_F \left[1 - \frac{1}{3}\left(\frac{\pi k_B T}{2E_F}\right)^2\right]. \tag{1-38}$$

At room temperature ($T = 300$ K), $k_B T$ is 0.026 eV, whereas the Fermi energy is in the range of 1–10 eV (see Table 1-10). Therefore, the chemical potential is nearly equal to the Fermi energy at room temperature $\mu \cong E_F$. In future discussions, although the chemical potential will be replaced by the Fermi energy without explanation, one should remember the approximation.

Figure 1-22 shows the Fermi-Dirac distribution as a function of temperature. It should be noted that as the temperature is increased, the distribution spreads around the Fermi level, indicating that electrons get excited to higher energy states and so vacant states get created below the Fermi level. Note that the spread of the distribution is on the order of $k_B T$. At room temperature, this is 0.026 eV. The work function of metals is in the range of 2–5 eV (see Table 1-10) and hence is much higher than the spread of the Fermi-Dirac distribution. Therefore, we do not find electrons being emitted from metals at room temperature. However, when the temperature of a metal is raised to 2000–3000 K, the tail of the Fermi-Dirac distribution reaches the vacuum level and hence electrons are emitted. This process is known as *thermionic emission.* It is used extensively to produce electron beams that are widely used in electron microscopes and for thin film deposition.

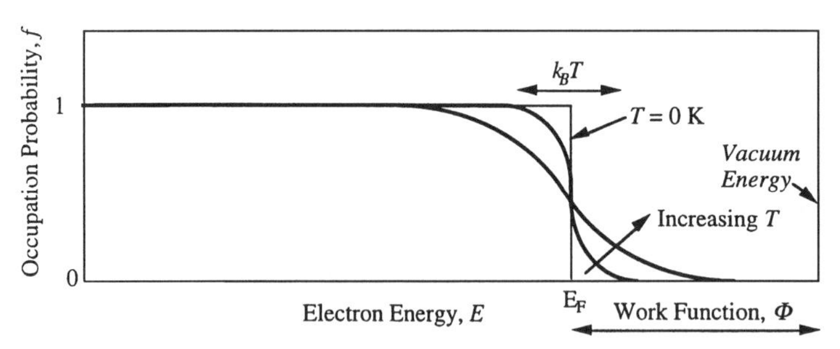

Figure 1-22 The Fermi-Dirac probability distribution.

The electron number density, η_e, and the total energy density of the electrons, E_e, can be found by integrating over all the energy states in **k**-space as follows:

$$\eta_e = \int f(E(\mathbf{k}))\frac{d\mathbf{k}}{4\pi^3},$$
$$\mathrm{E}_e = \int E(\mathbf{k})f(E(\mathbf{k}))\frac{d\mathbf{k}}{4\pi^3} \tag{1-39}$$

Here, $d\mathbf{k}$ is an element in the three-dimensional **k**-space. In spherical coordinates, this can be written as $4\pi k^2 dk$, which is the thin shell of **k**-space shown in Fig. 1-20. Now, using the relation $E = \hbar^2k^2/2m$, it is possible to transform the integration in Eq. (1-39) to an integral over energy as follows:

$$\eta_e = \int_0^\infty f(E)D(E)dE$$
$$\mathrm{E}_e = \int Ef(E)D(E)\,dE \tag{1-40}$$

where $D(E)$ is the density of states such that the number of electronic states between E and $E + dE$ is

$$D(E) = \frac{m}{\hbar^2\pi^2}\sqrt{\frac{2mE}{\hbar^2}}. \tag{1-41}$$

Note the $D(E) \propto \sqrt{E}$ dependence, which suggests that the density of states increases with electron energy. However, beyond the Fermi energy E_F, the probability of occupation goes to zero. Hence, the important quantity for determining the electron density at a certain energy level is the product $f(E)D(E)$. The electron heat capacity, C_e, can be derived from the relation

$$C_e = \frac{\partial \mathrm{E}_e}{\partial T} = \int_0^\infty E\frac{\partial f}{\partial T}D(E)dE. \tag{1-42}$$

The only temperature dependence is in the Fermi-Dirac distribution. If one studies Fig. 1-22, it becomes apparent that the only change of the distribution with temperature occurs near the Fermi level, whereas for energies much below or above it, the distribution is unchanged. Hence, a good approximation of Eq. (1-42) can be written as

$$C_e = k_B^2 T D(E_F)\int_{-E_F/k_BT}^\infty \frac{z^2e^z dz}{(e^z+1)^2} \tag{1-43}$$

where $z = E/k_BT$. For $E_F/k_BT \gg 1$ (which is true at room temperature), the heat capacity follows the relation

$$C_e = \frac{\pi^2}{3} k_B^2 T D(E_F) = \frac{\pi^2}{2}\left(\frac{k_B T}{E_F}\right)\eta_e k_B. \tag{1-44}$$

It is worth noting that the heat capacity of a monoatomic ideal gas is $3\eta k_B/2$. The heat capacity of electrons is multiplied by a factor k_BT/E_F, which at room temperature is on the order of 10^{-2}. In contrast, we saw that for phonons the heat capacity asymptotes to $3\eta k_B$ when the temperature is equal to or higher than the Debye temperature, whereas it varies as T^3 at temperatures much lower than the Debye temperature. Note that η is the number density of oscillators, whereas η_e is the number density of electrons. The total heat capacity of a metal can be written as

$$\begin{aligned} C &= AT + BT^3; \quad T \ll \theta_D \\ C &= AT + 3\eta k_B; \quad T \geq \theta_D \end{aligned} \tag{1-45}$$

where the first term represents the electron contribution and the second term represents the lattice contribution. At low temperatures, the electronic term will dominate, although at room temperature the lattice term usually dominates.

1-4-3 Periodic Potentials and Electronic Band Structure

The free electron theory assumes that there are no interactions between the electrons and individual ions, and that the electrons move around in a combined potential of all the ions. This is not a bad assumption for electrons at the Fermi level in a *metal* since the ions are well shielded by the core electrons as well as by the electrons in the conduction band below the Fermi level. However, the shielding is not always perfect, and in many cases the periodic potential of the ions is encountered by the electrons at the Fermi level. This is shown schematically in Fig. 1-23. The amplitude of the periodic potential, U, is much less than the potential of the ion core; this reduction is due to the shielding effect. In the free electron theory, however, $U = 0$.

Physical arguments. The motion of electrons in a periodic potential produces interesting effects, which are first described by simple physical explanations and then derived

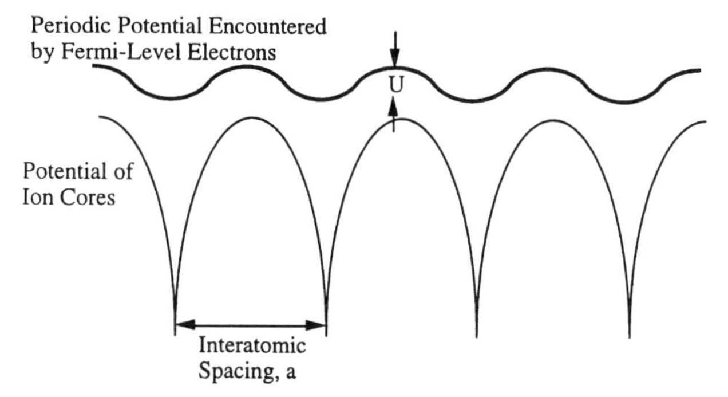

Figure 1-23 Periodic potential that is encountered by Fermi-level electrons.

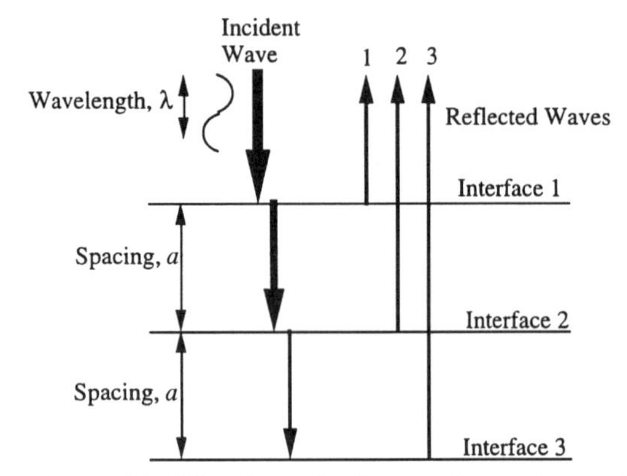

Figure 1-24 When the spacing between the interfaces, a, is equal to $n\lambda/2$, where λ is the wavelength of the incident wave and $n \geq 1$ is an integer, the reflected waves will have the same phase. This will produce constructive interference of the reflected waves, which can be thought of as an impedance for wave propagation in the incident direction.

mathematically. Consider this motion from purely a wave aspect. A periodic potential represents a periodic array of scattering sites for the electron wavefunction. Note that the Fermi wavelength of electrons (see Table 1-10) is of the same order as the lattice constant given in Tables 1-1–1-6.

Consider only two periods of a one-dimensional lattice and let us study its effect on electron motion. An analog for this can be found in optics, where one considers an electromagnetic wave of wavelength λ, normally incident on a film that is sandwiched between two bulk media, as shown in Fig. 1-24. The three interfaces represent changes in refractive index and therefore are scattering sites for the electromagnetic wave. Ignoring any change in phase of the wave during reflection, one would estimate that if film thickness a is equal to $n\lambda/2$, where $n \geq 1$ is an integer, the reflected waves from the two interfaces interfere constructively. For example, if $a = \lambda/2$, the transmitted wave from the first interface travels a distance a to the second interface. The reflected wave from this interface again travels a distance a before it interferes with the first reflected wave. Hence, the second wave travels a distance $2a$. If $2a = \lambda$, the second wave is of the same phase as the first reflected wave. This produces constructive interference for the reflected waves. For example, if the reflected waves 1, 2, and 3 in Fig. 1-24 are of the same phase, they will constructively interfere, producing a strong reflected signal and a weak transmitted one. This can be thought of as an impedance towards transport in the incident direction. If now we consider a whole array of scattering sites, we will find that the wave can never propagate in its incident direction and is always reflected back due to constructive interference. This is exactly what happens for electrons in a periodic potential. In terms of electron wavevectors, one can write that when the wavevector $k = \pi/a$, the electron cannot propagate in its incident direction. Similarly, the electron wavefunction cannot propagate in the reverse direction. Hence, one gets standing waves of electrons. Two standing waves are formed—one with peaks at the nuclei that is of lower energy and the other with troughs at the nuclei that is of higher energy. Hence, an energy bandgap is formed at $k = \pi/a$. Note that $k = \pi/a$ is also the edge of the Brillouin zone. If we look at the whole energy spectrum of the electrons, we find that there are forbidden gaps in the energy where the electrons are reflected due to constructive interference. Therefore, the origin of energy gaps can be explained from such a wave interference point of view.

Mathematical derivation. Now we will show this through a mathematical derivation. For a periodic potential shown in Fig. 1-23, we can write $U(r+a) = U(r)$, where a is the separation between two ions. In terms of the reciprocal lattice vector **G**, the potential can be written as a Fourier series:

$$U(\mathbf{r}) = \sum_{\mathbf{G}} U_{\mathbf{G}} e^{i\mathbf{G}\cdot\mathbf{r}}. \tag{1-46}$$

Using this potential, the Schrödinger equation for the electron becomes

$$-\frac{\hbar^2}{2m}\nabla^2\psi + U(\mathbf{r})\psi = E\psi. \tag{1-47}$$

When $U = 0$, Eq. (1-47) reduces to Eq. (1-29), used for free electrons. Bloch showed that the wavefunction of an electron in a periodic potential must have the form

$$\psi(\mathbf{r}) = e^{i\mathbf{k}\cdot\mathbf{r}} u_{\mathbf{k}}(\mathbf{r}) \tag{1-48}$$

where $u_{\mathbf{k}}(\mathbf{r})$ is a periodic function with the periodicity of the lattice. Such a function is called the *Bloch function*. The periodicity of the lattice can be used in a way that the electron wavefunction is written as a Fourier series as

$$\psi(\mathbf{r}) = \sum_{\mathbf{k}} C(\mathbf{k}) e^{i\mathbf{k}\cdot\mathbf{r}}. \tag{1-49}$$

When Eqs. (1-46) and (1-49) are used in Eq. (1-47), the Schrödinger equation becomes

$$\sum_{\mathbf{k}} \left[\frac{\hbar^2 k^2}{2m} C(\mathbf{k}) + \sum_{\mathbf{G}} U_{\mathbf{G}} C(\mathbf{k}) e^{i\mathbf{G}\cdot\mathbf{r}} - E C(\mathbf{k})\right] e^{i\mathbf{k}\cdot\mathbf{r}} = 0. \tag{1-50}$$

Now we make a change of variables, $\mathbf{k}' = \mathbf{k} + \mathbf{G}$, and drop the prime to get the following form:

$$\sum_{\mathbf{k}} \left[\frac{\hbar^2 k^2}{2m} C(\mathbf{k}) + \sum_{\mathbf{G}} U_{\mathbf{G}} C(\mathbf{k} - \mathbf{G}) - E C(\mathbf{k})\right] e^{i\mathbf{k}\cdot\mathbf{r}} = 0. \tag{1-51}$$

This produces an entire set of equations that provides the energy spectrum of the electrons. A representative of this set is

$$\left(\frac{\hbar^2 k^2}{2m} - E\right) C(\mathbf{k}) + \sum_{\mathbf{G}} U_{\mathbf{G}} C(\mathbf{k} - \mathbf{G}) = 0. \tag{1-52}$$

Note that when $U = 0$, we get back the energy–wavevector relation $E = \hbar^2 k^2/2m$ for free electrons.

Now consider the simplest case in which the summation in Eq. (1-52) contains only one term. This amounts to saying that the periodic potential of Eq. (1-46) has only one term in the Fourier series, that having the periodicity of the lattice spacing. From the set of equations (1-52), consider only two of them:

$$\begin{aligned} (E - E_{\mathbf{k}}) C(\mathbf{k}) &= U_{\mathbf{G}} C(\mathbf{k} - \mathbf{G}) \\ (E - E_{\mathbf{k}-\mathbf{G}}) C(\mathbf{k} - \mathbf{G}) &= U_{-\mathbf{G}} C(\mathbf{k}) = U_G^* C(\mathbf{k}) \end{aligned} \tag{1-53}$$

where $E_{\mathbf{k}} = \hbar^2(\mathbf{k} \cdot \mathbf{k})/2m$. In the second equation above, we have used a transfer of variables $\mathbf{k} = \mathbf{k} - \mathbf{G}$ for the right-hand side. The term U_G^* is the complex conjugate of $U_{\mathbf{G}}$. Equation (1-46) contains two simultaneous equations that can be solved through the determinant

$$\begin{vmatrix} E - E_{\mathbf{k}} & -U_{\mathbf{G}} \\ -U_G^* & E - E_{\mathbf{k}-\mathbf{G}} \end{vmatrix} = 0. \tag{1-54}$$

This yields the following relation for the energy E:

$$E = \frac{E_{\mathbf{k}} + E_{\mathbf{k}-\mathbf{G}}}{2} \pm \left[\left(\frac{E_{\mathbf{k}} - E_{\mathbf{k}-\mathbf{G}}}{2}\right)^2 + |U_{\mathbf{G}}|^2\right]^{\frac{1}{2}} \tag{1-55}$$

Note that at the edge of the Brillouin zone, $\mathbf{k} = \mathbf{G}/2$. At this value of $\mathbf{k}$, we find that the energy of the electron can have two levels:

$$E = E_{\mathbf{G}/2} \pm |U_{\mathbf{G}}|. \tag{1-56}$$

This is illustrated in Fig. 1-25.

It should be noted that this analysis was made only for the region $\mathbf{k} \leq \mathbf{G}$. In contrast to phonons, electron wavevectors can take values higher than $\mathbf{G}$. But for all values of $\mathbf{k} = n\mathbf{G}/2$ $(n = 1, 2, \ldots)$, the electron energy is split into two different levels, just as in Fig. 1-25. The number n is called the *band index*. The formation of electronic band structure is illustrated in Fig. 1-26. It can be seen from this diagram that it is worth considering the electronic structure only in the first Brillouin zone since the structure in the second and third zones are repetitions of the one in the first zone. Note that although the above analysis was one-dimensional, it can be extended for three dimensions. In such a case the electronic structure is studied in the first Brillouin zone of the crystal, with each direction having different structure.

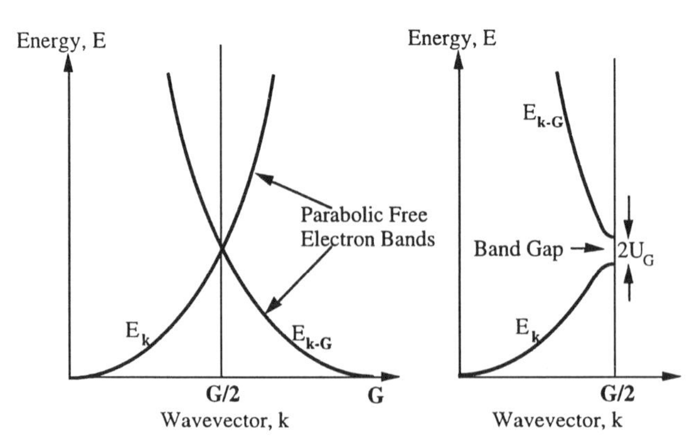

Figure 1-25 Formation of a band gap of $2U_G$ at the Brillouin zone edge. The parabolic bands $E \propto k^2$ are for free electrons. Near the zone centers at $\mathbf{k} = 0$, $\mathbf{G}$, $2\mathbf{G}$ etc., the bands remain parabolic. However, at the zone edge, the energy levels are split by $2U_G$ due to interference of the electron wavefunction in a periodic potential of amplitude U_G. Note that in the figure on the right, the band $E_{\mathbf{k}-\mathbf{G}}$ is shown in the first Brillouin zone ($\mathbf{k} < \mathbf{G}/2$), whereas the band $E_{\mathbf{k}}$ is not shown outside this zone. Electronic band structure often is shown only in the first Brillouin zone.

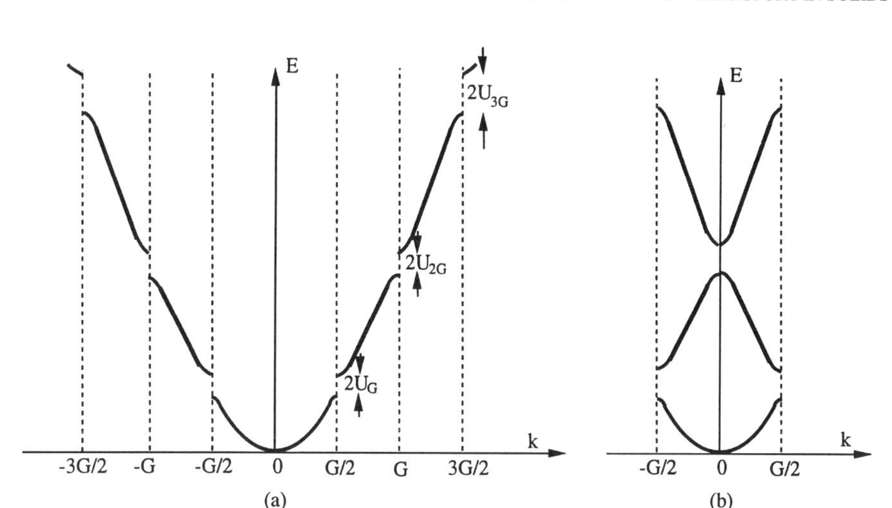

Figure 1-26 (a) Effect of periodic potential on the structure of a single free-electron parabolic band, $E_{\mathbf{k}}$. When $\mathbf{k} = n\mathbf{G}/2$ ($n = 1, 2, 3, \ldots$), the interaction of the electron wavefunction and the periodic potential creates a forbidden energy gap. (b) The band structure within the first Brillouin zone contains all the information of the total band structure, if all the other bands $E_{\mathbf{k}}$, $E_{\mathbf{k+G}}$, $E_{\mathbf{k-G}}, \ldots$ are considered.

Band structure of common semiconductors. Having explained the origin of the electronic band structure in materials, it is worth studying the band structure of two common semiconductors, Si and GaAs, which have a basic difference. The band structures are shown in Fig. 1-27. The band that is concave downwards is called the *valence band* and is formed by the bonding orbitals of the valence electrons. The top of the valence band is arbitrarily chosen to be of zero potential. The *conduction band* is concave upwards. The *energy gap*, E_g, of a semiconductor is the difference between the top of the valence band and the bottom of the conduction band. Note that these two locations need not be at the same value of k within the Brillouin zone. This occurs in the case of Si. In such cases, the material is called an *indirect gap semiconductor*. When the bottom of the conduction band and the top of the valence band occur at the same value of k, it is called a *direct gap semiconductor*. Direct and indirect gap semiconductors have vastly different optical properties, as we will see later. Table 1-11 lists the energy band gaps of some semiconductors (insulators), noting whether they are direct or indirect.

When the valence band is completely full, as in semiconductors and insulators, electrons can be excited from the valence to the conduction band through thermal or optical excitations. For direct and indirect gap semiconductors, such optical excitations are shown in Fig. 1-28 and will be explained in detailed later. An incident photon is absorbed, and its energy is used to excite an electron from the valence to the conduction band. This leaves a vacant orbital in the valence band called a *hole*. A simple way of thinking about a hole is to consider a bucket of water where the surface of the water is the top of the valence band. Consider a small amount of water to be extracted from the bucket and raised above the surface, thereby leaving a bubble inside the bulk of the water. Despite the action of gravity in the direction pointing downwards, the bubble will try to move upwards. One could think of the bubble as having negative energy. Hence, the lower

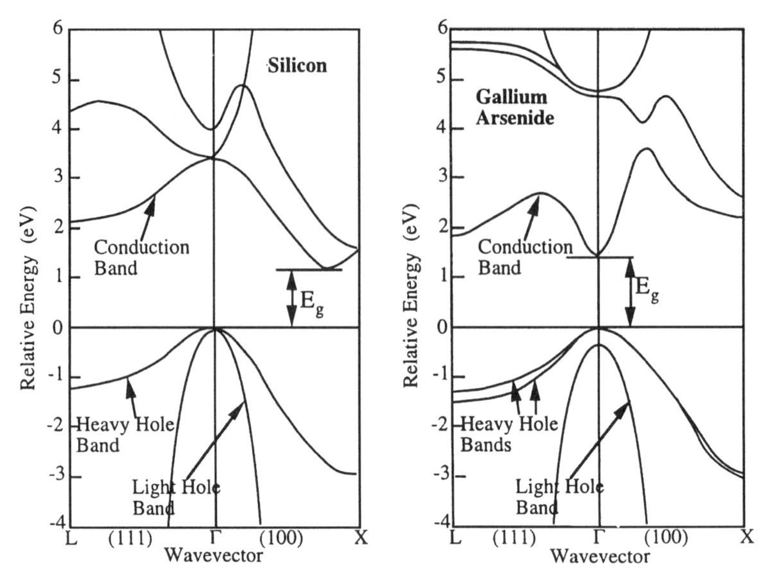

Figure 1-27 Electronic band structure of Si and GaAs in two directions (Γ-L and Γ-X) within the first Brillouin zone (Chelikowsky and Cohen, 1976). The valence and conduction band edges for GaAs occur at the same position in the Brillouin zone (Γ point or zone center in this case), whereas for Si it occurs at different positions in the Brillouin zone. Thus GaAs is called a direct gap semiconductor and Si is called an indirect gap semiconductor.

the original position of the bubble, the higher the system energy. If there is a pressure gradient in the fluid that induces mass flow, one would find that, ignoring advection effects, the fluid will flow in the direction of negative pressure gradient, whereas the bubble will flow in the opposite direction. This gives rise to the concept of a negative mass. So in comparison to an electron, a hole has opposite momentum, negative mass, same velocity, and negative energy.

The curvature of the conduction and valence bands has special significance. For a classical particle, the energy E and momentum p are related by the equation $E = p^2/2m$. Therefore, the mass of a particle can be found from the relation $1/m = d^2E/dp^2$. Using $p = \hbar k$ as the momentum for a quantum particle, we find that the effective mass, m^*, can be written as

$$\frac{1}{m^*} = \frac{1}{\hbar^2}\frac{d^2E}{dk^2}. \tag{1-57}$$

Hence, the effective mass is related to the curvature of the band structure. The valence band is concave downwards, leading a negative effective mass. The higher the curvature of the band, the lower the effective mass. Table 1-11 lists the ratio of the effective mass of electrons and holes to the electron rest mass for various semiconductors. Note that the effective mass can be less than the electron rest mass. The physical interpretation of this is as follows. If an electric field, E, is applied across a conducting material, the charges are accelerated according to Newton's law $a = e\mathrm{E}/m$. For a free electron the acceleration is based on the rest mass of the electrons (unless relativistic conditions exist).

Table 1-11 Energy band gap E_g (in eV) at 300 K, the type of band gap (i = indirect; d = direct) between the valence and conduction bands, and the effective masses of electrons in the conduction band (m_c^*/m) and of holes in the valence band (m_v^*/m) of different semiconductors (Ashcroft and Mermin, 1976; Kittel, 1986; Sze, 1981)

Crystal	Gap	E_g	m_c^*/m	m_v^*/m	Crystal	Gap	E_g	m_c^*/m	m_v^*/m
Diamond	i	5.47	0.20	0.25	α-SiC	i	3.0	0.6	1.0
Si	i	1.11	0.98[2]	0.16[4]	ZnSb		0.56		
			0.19[2]	0.49[5]					
Ge	i	0.66	1.64[2]	0.04[4]	HgTe[1]	d	<0.0		
α-Sn	d	0.00	0.08[3]	0.28[5]	PbS	d	0.41	0.25	0.25
InSb	d	0.17	0.015	0.40	PbSe	i	0.27		
InAs	d	0.36	0.023	0.04	PbTe	i	0.31	0.17	0.2
InP	d	1.35	0.077	0.64	CdS	d	2.42	0.21	0.80
GaP	i	2.26	0.82	0.60	CdSe	d	1.70	0.13	0.45
GaAs	d	1.42	0.067	0.082	CdTe	d	1.56	0.077	0.64
GaSb	d	0.72	0.042	0.40	ZnO	d	3.35	0.27	
AlSb	i	1.58	0.12	0.98	ZnS	d	3.68	0.40	
GaN		3.36	0.19	0.60	SnTe	d	0.18		
BN	i	7.5			Cu_2O		≈2.0		
BP		2.0			TiO_2		≈3.0		

[1]HgTe is a semimetal where the bands overlap.

[2]Longitudinal effective mass: The constant energy surface at the bottom of the conduction band is not spherical in three-dimensional space but ellipsoidal. Therefore, the curvature is different along the two principle axes directions. Longitudinal effective mass is along the major axis, which happens to be along the crystal symmetry direction.

[3]Transverse effective mass: This is due to the curvature of the conduction band along the minor axis of the ellipsoid, which is perpendicular to the crystal symmetry direction.

[4]Light-hole effective mass: The valence band contains two hole bands where the light hole band has the higher curvature, although the band energy falls slightly below that of the heavy hole band.

[5]Heavy-hole effective mass: This is due to the higher curvature of the heavy-hole valence band.

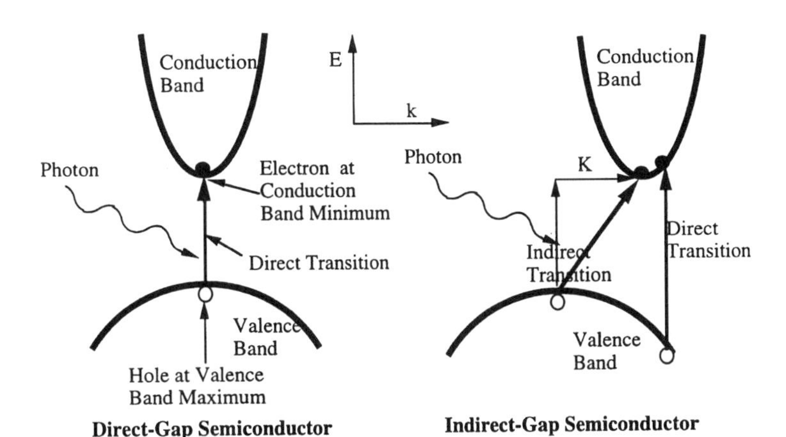

Figure 1-28 Optical excitations in direct and indirect gap semiconductors. Direct transition occurs in direct gap semiconductors, whereas in indirect gap semiconductors an additional wavevector K, involving a phonon, is required.

However, in a semiconductor the acceleration can be higher than that for a particle with rest mass of an electron. This again has to do with the effect of the periodic potential. The interactions of the periodic potential and the electron wavefunction can be such that the constructive interference is stronger in the direction of the electron motion and that the electron is accelerated more in the presence of the periodic potential than in its absence. The origin of a lower effective mass, therefore, is a quantum mechanical effect.

1-4-4 Charge Carrier Densities in Semiconductors

The difference between semiconductors and insulators is only in the size of the energy band gap. Semiconductors are those materials for which the band gap is sufficiently low that electrons can be thermally excited from the valence to the conduction band at room temperature. If the band gap is larger than about $100k_BT$, the material is usually called an insulator.

First, we will study the carrier density in the conduction and the valence bands. Consider a pure (with no impurities) or *intrinsic* semiconductor with a band gap E_g, at temperature T. The densities of electrons in the conduction band, η_c, and of holes in the valence band, η_v, can be written as

$$\begin{aligned} \eta_c(T) &= \int_{E_c}^{\infty} D_c(E) f(E) dE \\ \eta_v(T) &= \int_{-\infty}^{E_v} D_v(E)[1 - f(E)] dE \end{aligned} \tag{1-58}$$

where $f(E)$ is the Fermi-Dirac distribution of Eq. (1-37), E_c is the bottom of the conduction band, E_v is the top of the valence band, and D_c and D_v are the densities of states for the conduction and valence bands, respectively. Note that the chemical potential μ in the Fermi-Dirac distribution is still an unknown. Also note that the probability of a hole present at an energy state E is equal to $1 - f(E)$.

It must be noted that in most semiconductors, the band gap $E_g = E_c - E_v$ is much larger than k_BT, and hence the Fermi-Dirac distribution can be approximated as

$$\begin{aligned} \frac{1}{\exp\left(\frac{E-\mu}{k_BT}\right)+1} &\approx \exp\left(-\frac{E-\mu}{k_BT}\right); \qquad \text{for } E > E_c \text{ (conduction band)}, \\ \frac{1}{\exp\left(\frac{\mu-E}{k_BT}\right)+1} &\approx \exp\left(-\frac{\mu-E}{k_BT}\right); \qquad \text{for } E < E_v \text{ (valence band)}. \end{aligned} \tag{1-59}$$

As explained in the previous section, the structure of the two bands near the band edges is nearly parabolic (i.e., $E \propto k^2$), which is a free electron behavior. We saw for free electrons that this produced the $D(E) \propto \sqrt{E}$ behavior for the density of states. Thus, the density of states for the two bands near the band edges can be approximated as

$$D_{c,v}(E) = \frac{m^{*\frac{3}{2}}_{c,v}}{\hbar^3 \pi^2} \sqrt{2|E - E_{c,v}|} \tag{1-60}$$

where m^* is the effective carrier mass in the respective band. This is qualitatively shown in Fig. 1-29. Note that the density of states in the band gap is zero. Using these

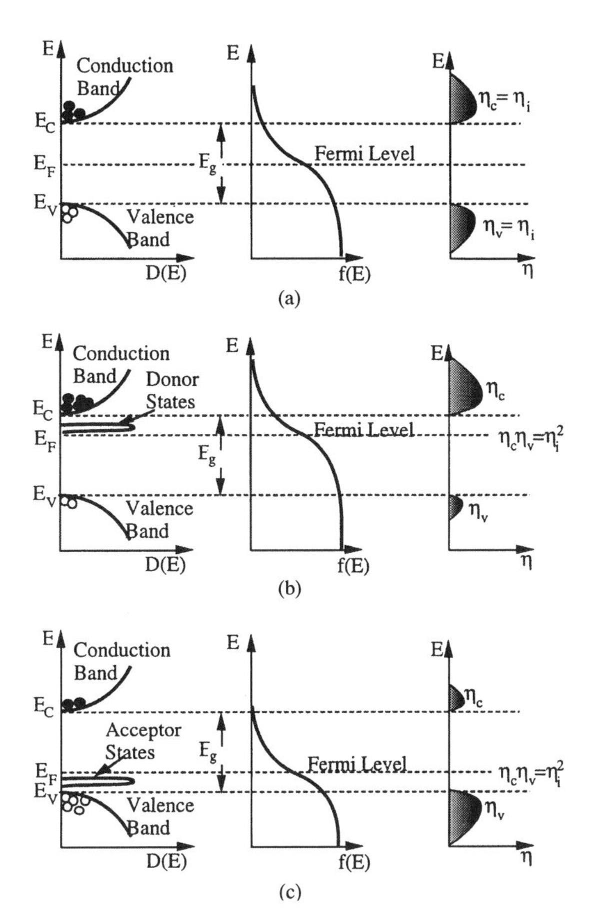

Figure 1-29 The density of states $D(E)$, Fermi-Dirac distribution $f(E)$, carrier concentrations, position of Fermi level, and donor and acceptor states for (a) intrinsic, (b) n-doped, and (c) p-doped semiconductors.

approximations, the carrier densities in the two bands can be written as

$$\begin{aligned} \eta_c(T) &= N_c(T)\exp\left(-\frac{E_c-\mu}{k_BT}\right) \\ \eta_v(T) &= N_v(T)\exp\left(-\frac{\mu-E_v}{k_BT}\right) \end{aligned} \tag{1-61}$$

where $N_c(T)$ and $N_v(T)$ are coefficients derived as

$$\begin{aligned} N_c(T) &= \int_{E_c}^{\infty} D_c(E)\exp\left(-\frac{E-E_c}{k_BT}\right)dE = \frac{1}{4}\left(\frac{2m_c^*k_BT}{\pi\hbar^2}\right)^{\frac{3}{2}} \\ N_v(T) &= \int_{-\infty}^{E_v} D_v(E)\exp\left(-\frac{E_v-E}{k_BT}\right)dE = \frac{1}{4}\left(\frac{2m_v^*k_BT}{\pi\hbar^2}\right)^{\frac{3}{2}}. \end{aligned} \tag{1-62}$$

A convenient way to write these for calculating numerical values is

$$N_c(T) = 2.5 \times 10^{19} \left(\frac{m_c^*}{m} \right)^{\frac{3}{2}} \left(\frac{T}{300} \right)^{\frac{3}{2}} \text{ cm}^{-3}$$
$$N_v(T) = 2.5 \times 10^{19} \left(\frac{m_v^*}{m} \right)^{\frac{3}{2}} \left(\frac{T}{300} \right)^{\frac{3}{2}} \text{ cm}^{-3} \qquad (1\text{-}63)$$

where T is the temperature in Kelvin. The actual concentrations, η_c and η_v, are still unknown since the chemical potential μ has not been determined. But it is worth noting that the product $\eta_c \eta_v$ is independent of the chemical potential and depends only on the band gap:

$$\eta_c \eta_v = N_c N_v \exp\left(-\frac{E_g}{k_B T} \right). \qquad (1\text{-}64)$$

In an intrinsic semiconductor, the carrier concentrations in the valence and the conduction bands must be equal: $\eta_c = \eta_v = \eta_i$. The value of η_i can be found by taking the square root of the right-hand side of Eq. (1-64). For Si and GaAs, the intrinsic carrier concentration at room temperature is on the order of 10^{10} and 10^6 cm^{-3}, respectively. Note the large effect of the band gap on the intrinsic carrier concentration. As the temperature is increased, the electron and hole concentration increases. Thus it is natural to expect the electronic behavior to be temperature dependent. Although η_i is independent of the chemical potential, we know from Eq. (1-61) that η_c and η_v are dependent. Equating these concentrations, $\eta_c = \eta_v$, we can find the chemical potential or the Fermi level for semiconductors, which is

$$\mu = E_v + \frac{1}{2} E_g + \frac{1}{2} k_B T \ln \left(\frac{N_v}{N_c} \right) = E_v + \frac{1}{2} E_g + \frac{3}{4} k_B T \ln \left(\frac{m_v^*}{m_c^*} \right). \qquad (1\text{-}65)$$

Unless the effective electron masses in the two bands are vastly different, $k_B T \ln(N_v/N_c) \ll E_g$ at room temperature. Hence, the Fermi level falls in the middle of the conduction and valence bands, as shown in Fig. 1-29.

Impurities often are deliberately included in semiconductors and have a profound effect on the carrier densities in the valence and the conduction bands. These are point defects that substitute the parent atom in a crystal lattice. The deliberate addition of impurities is called *doping*. When P, As, or Sb (Group V) are used as impurities in Si, they act as electron donors since they have an extra electron in the outer shell. The energy level of donors are only within 10–50 meV below the conduction band (see Fig. 1-29), and therefore the extra electron in these donors can be thermally excited in the conduction band. The electron concentration in the conduction band can be altered by controlling the concentration of donor impurities. Semiconductors, like Si, that contain such donor impurities are called *n-doped* or simply n-Si or n-Ge or n-GaAs. Similarly, if atoms such as B, Al, Ga, or In (Group III) are included as impurities, they become acceptors for electrons for two reasons. First they have one fewer valence electron than silicon, and second, the acceptor energy level lies only 10–70 meV above the valence band edge (see Fig. 1-29). Hence, electrons can be thermally excited from the valence band to the acceptor level, leaving a hole in the valence band. If the electron is trapped at the acceptor atom, the hole permanently remains in the valence band. Semiconductors with acceptor atoms are called *p-doped* or just p-Si.

When a semiconductor is doped, the Fermi level must adjust itself based on the concentrations of electrons and holes. For example, for an n-doped semiconductor, the Fermi level are closer to the conduction band edge, and vice versa for a p-doped semiconductor. However, note that the product of the electron and hole concentrations in Eq. (1-64) is independent of the position of the Fermi level and depends only on the band gap. Hence, Eq. (1-64) is valid for doped semiconductors, too. This is called the *law of mass action*. But one often encounters that the concentration of carriers is nearly equal to the doping concentrations. For example, in an n-doped semiconductor with doping concentration η_D, the electron concentration in the conduction band is $\eta_c \cong \eta_D$. This occurs under two conditions: first, the temperature is sufficiently high that electrons can get thermally excited from the donor levels to the conduction band; second, the donor concentration is much higher than the intrinsic electron concentration, $\eta_D \gg \eta_i(T)$. If the temperature is too low (typically below 100 K), the first condition may not be met and the donor levels are frozen out. The law of mass action states that if the electron or the donor concentration is known, the hole density in the valence band can be found as $\eta_v = \eta_i^2/\eta_D$. One may ask that since holes are created by thermal excitations why should the hole density reduce if the electron density in the conduction band is increased by donors. The reason lies in the rate of recombination of electrons and holes under dynamic equilibrium. If the electron density increases, the recombination rate between electrons and holes also increases, thereby reducing the hole concentration in the valence band. The Fermi level of a n-doped semiconductor can be found using the relation

$$E_F = E_C - k_B T \ln\left(\frac{N_c}{\eta_D}\right). \tag{1-66}$$

Hence, if the donor concentration is increased, the Fermi level moves up towards the conduction band edge. In the event that the donor concentration is sufficiently high that the Fermi level enters the conduction band, semiconductor is said to be *degenerate* and the symbol used is n^+. This occurs when $\eta_D \cong N_c$. Note that according to Eq. (1-63), N_c is on the order of 10^{19} cm^{-3}. Similarly, for a p-doped semiconductor, the Fermi level can be found as

$$E_F = E_V + k_B T \ln\left(\frac{N_v}{\eta_A}\right). \tag{1-67}$$

1-4-5 Scattering of Charge Carriers

The scattering of charge carriers is an important aspect in its transport, for it is the scattering that controls the electrical and heat conduction, as well as optical absorption. This in turn influences the operation of electronic and optoelectronic devices made of semiconductors, as well as charge and energy transport in metals. A scattering event can change the momentum and energy of a carrier. Usually, this is referred to as $\mathbf{k} \rightarrow \mathbf{k}'$, where $\mathbf{k}$ is the wavevector before scattering and $\mathbf{k}'$ is the wavevector after scattering. Since the wavevector is related to the carrier energy through the band structure, the carrier energy can change during scattering. However, since $\mathbf{k}$ and $\mathbf{k}'$ may be at the nearly same energy level, a scattering process need not involve appreciable energy transfer but does involve significant momentum change. Such processes are called *elastic* scattering, as opposed to *inelastic* scattering, where there are both appreciable energy and momentum changes.

Electron scattering comes in two types, intravalley and intervalley. Here a valley represents the valley in the electronic band structure. For example, if we look at band structure of GaAs, there are three valleys: the Γ, the L, and the X valleys. *Intravalley scattering* implies that both $\mathbf{k}$ and $\mathbf{k}'$ remain in the same valley, whereas *intervalley scattering* means that $\mathbf{k}$ and $\mathbf{k}'$ are in different valleys. *Interband scattering* for holes implies scattering between the conduction and the valence bands, or between the light and heavy hole bands.

What is important for transport processes is to determine the scattering rate of carriers. The scattering rate, however, depends on the carrier energy and, sometimes, the temperature. The calculations for scattering rates are difficult since the requirements are to obtain the scattering rate of a carrier of wavevector, $\mathbf{k}$. This implies that one has to add the probabilities of scattering over all possible $\mathbf{k}'$, which is often difficult. Nevertheless, we will try to give explanations for the basic scattering mechanisms and present expressions for scattering rates.

There are three basic categories of scattering processes: (i) carrier–carrier, (ii) carrier–phonon, and (iii) carrier–defect. *Carrier-carrier scattering* plays a minor role in solids and will not be considered here. Interested readers are referred to more advanced books (Ashcroft and Mermin, 1976) and monographs (Jacoboni and Reggiani, 1983). We will start from the simplest mechanism, impurity scattering, and then go to the more complicated mechanisms of carrier–phonon scattering.

Defect scattering. Defects can be of two types, neutral and ionized. Both scatter carriers, although through slightly different mechanisms. As we saw earlier, an electron in a periodic lattice does not really collide with the ions but instead uses the periodic potential created by the ions to propagate itself as a wave. However, when a defect is present in the lattice, the following situations may occur: (i) the periodicity of the potential is broken; and/or (ii) the amplitude of the periodic potential is altered. The first situation occurs when there is a missing or an interstitial atom. The second occurs either when a different neutral atom substitutes a parent atom or if the impurity atom is charged thereby changing the potential. Since the electron wavelength at the Fermi level is nearly equal to the lattice spacing, scattering by impurities can be strong.

From kinetic theory, we know that if the defect density is ρ, the mean time between collisions, $\tau_d(\mathbf{k})$ is

$$\frac{1}{\tau_d(\mathbf{k})} = \rho\sigma(\mathbf{k})v(\mathbf{k}) \tag{1-68}$$

where $\sigma(\mathbf{k})$ is the scattering cross section and v is the carrier velocity. For metals, v is the Fermi velocity, whereas for semiconductors it is the velocity of electrons in the conduction band and holes in the valence band. The velocity, in general, can be found from the band structure as

$$\mathbf{v}(\mathbf{k}) = \frac{1}{\hbar}\nabla_k E(\mathbf{k}). \tag{1-69}$$

For metals, the velocity is equal to the Fermi velocity, which is on the order of 10^6 m/s. For semiconductors, however, the Fermi level lies in the band gap. Note that for electrons at the bottom of a valley in the conduction band of a semiconductor, the velocity is zero. But at temperature T, thermal excitations can make electrons occupy levels slightly higher than the band edge. Since the Fermi-Dirac distribution in the conduction band contains its very tail, it can be assumed to be that of a Maxwell-Boltzmann distribution, which is used for gas molecules. Hence, the average speed of electrons in the conduction band at temperature T can be approximated as

$$v = \sqrt{\frac{3k_B T}{m^*}}. \tag{1-70}$$

This is also true for holes in the valence band. At room temperature, the carrier velocity in a semiconductor is on the order of 10^5 m/s.

The problem in finding the scattering rate, $1/\tau_d(\mathbf{k})$, boils down to determining the scattering cross section. An elegant treatment of this can be found in Ferry (1991). We will not derive all the relations here but will instead give the essential physics and final form of the scattering rate. If the defect is an ionized impurity, then associated with it is a potential field given by

$$V(r) = \frac{Ze}{4\pi\varepsilon_\infty r}\exp(-q_d r) \tag{1-71}$$

which is the screened Coulomb potential, where $1/q_d$ is the Debye screening length, Z is the level of ionization, ε_∞ is the permittivity, and r is the distance from the impurity. The scattering of a carrier from such a potential is an elastic process such that only the direction changes but not the energy. The scattering cross section can be written as (Ferry, 1991)

$$\sigma(k) = \frac{Z^2 e^4 m^{*2}}{8\pi\hbar k^4 \varepsilon_\infty^2}\left[\ln\left(\frac{1+\beta^2}{\beta^2}\right) - \frac{1}{1+\beta^2}\right] \tag{1-72}$$

where $\beta = q_d/2k$. If one puts the numbers in Eq. (1-72) and assuming the electron wavelength to be 1 Å, the coefficient in front of the square brackets is on the order of 10^{-20} m^2, which is approximately the cross-sectional area of an atom. The Debye wavevector, q_d, can be written as

$$q_d = \sqrt{\frac{(\eta_c + \eta_v)e^2}{\varepsilon_\infty k_B T}}. \tag{1-73}$$

For metals at room temperature, this falls between 10^{10} and 10^{11} m^{-1} such that the screening length is on the order of a lattice spacing. For semiconductors, the carrier density is lower, and hence the screening length is larger.

Carrier–acoustic phonon scattering. The discussion on the effect of periodic ionic potential on the electronic band structure assumed that the ions were static. However, we know from Section 1-2 that lattice ions vibrate and that the quantum of vibrational energy is a phonon. Phonons can be imagined to be deviations from the perfect periodic lattice that can alter the propagation of the carrier wavefunction, thereby scattering the carrier. There are two types of phonons, acoustic and optical. Whereas acoustic phonon energies range from zero to an upper limit at Brillouin zone edge, optical phonons occur only at higher energies. The maximum phonon energy that can occur in a solid falls typically in the range of 0.01–0.1 eV. In contrast, electron energy at the Fermi level of a metal is typically in the range of 1–10 eV. In semiconductors too, electrons can be energized by electric fields to be in the range of 1 eV. Therefore, in certain types of carrier–phonon scattering, the phonon energy may be ignored. This is called *elastic scattering*. Here the carrier energy is not significantly changed, although the direction (and the momentum) is changed. Typically these are interactions between electrons and acoustic phonons, since acoustic phonons are the low-energy ones.

Momentum conservation in carrier–phonon scattering can be written in terms of wavevectors as

$$\mathbf{k}' + \mathbf{G} = \mathbf{k} \pm \mathbf{K} \tag{1-74}$$

where $\mathbf{k}$ and $\mathbf{k}'$ are the initial and final carrier states, respectively; $\mathbf{K}$ is the phonon wavevector; and $\mathbf{G}$ is the reciprocal lattice vector. The upper sign corresponds to phonon absorption, whereas the lower sign corresponds to phonon emission. The generic form of the scattering rate $1/\tau(\mathbf{k})$ can be written as (Ferry, 1991)

$$\frac{1}{\tau(\mathbf{k})} = \frac{2\pi}{\hbar}|M(\mathbf{k},\mathbf{K})|^2 N(E_{\mathbf{k}} \pm \hbar\omega_{\mathbf{k}}) \tag{1-75}$$

where $|M(\mathbf{k},\mathbf{K})|^2$ is the coupling of the energy between the initial and final carrier states, and $N(E_{\mathbf{k}} \pm \hbar\omega_{\mathbf{k}})$ is the total number of final states available for a carrier at energy $E_{\mathbf{k}}$ to couple with a phonon of energy $\hbar\omega_{\mathbf{k}}$ and wavevector $\mathbf{K}$. For each type of scattering, the problem reduces to finding M and N.

Consider the first type of elastic scattering only due to *deformation* of the lattice. For example, when an acoustic wave passes through the lattice, it creates a local strain field that perturbs the local energy bands due to lattice distortion. The change in band energy, ΔU, can be written as $\Delta U = \zeta\varphi$, where ζ is the *deformation potential* and φ is the strain. The coupling term can be written as

$$|M(\mathbf{k},\mathbf{K})|^2 = \frac{\hbar\zeta^2 K^2}{2\rho V \omega_{\mathbf{K}}}(n_K + 1), \tag{1-76}$$

where ρ is the mass density of the solid, V is the volume under consideration (it will go away), ω_K is the frequency of the phonon of wavevector K, and n_K is the Bose-Einstein distribution for phonons. Since we are considering low phonon energies such that $\hbar\omega_K \ll k_B T$, the distribution can be approximated as

$$n_K = \frac{1}{\exp\left(\frac{\hbar\omega_K}{k_B T}\right) - 1} \cong \frac{k_B T}{\hbar\omega_K} \gg 1. \tag{1-77}$$

Using these and the density of states for a parabolic electronic band structure, the scattering rate can be written as (Ferry, 1991)

$$\frac{1}{\tau(\mathbf{k})} = \frac{\zeta^2 k_B T (2m^*)^{\frac{3}{2}}}{2\pi\hbar^4 \rho v_s^2}\sqrt{E} \tag{1-78}$$

where v_s is the speed of sound. For most semiconductors, ζ is on the order of 7 to 10 eV.

When there are two different types of atoms per unit cell such as in GaAs, there is always a *piezoelectric* effect. This means that when an acoustic wave passes by a crystal site, it creates a local strain field that induces a local electric field. Carriers encounter a change in the potential due to the local piezoelectric field and hence can be scattered. Since this is similar, in principle, to the effect of an ionized impurity (which also induces a local field), the scattering rate can be written as

$$\frac{1}{\tau(\mathbf{k})} = \frac{m^* e^2 \wp^2 k_B T (2m^*)}{\pi \varepsilon_\infty \hbar^3 k} \left[\ln \left(\frac{1+\beta^2}{\beta^2} \right) - \frac{1}{1+\beta^2} \right] \tag{1-79}$$

where $\beta = q_d/2k$, with q_d being the screening length; $\wp$ is the coupling coefficient

$$\wp^2 = \frac{d^2}{\varepsilon_\infty} \left(\frac{12}{35 c_L} + \frac{16}{35 c_T} \right) \tag{1-80}$$

where c_L and c_T are the longitudinal and transverse speeds of sound, respectively; and d is the piezoelectric constant of the solid that couples the strain to the local polarization. Piezoelectric scattering occurs mostly at low temperatures and is not significant at room temperatures.

The total scattering rate by acoustic phonons in a material such as Si at 300 K ranges between 10^{12} Hz for 10 meV electrons to 10^{13} Hz for 1 eV electrons (Jacoboni and Reggiani, 1983).

Carrier–optical phonon scattering. Optical phonon energies in most materials fall in the range of 10–100 meV and are nearly monoenergetic across the whole Brillouin zone. Exchange of energy during a scattering process can significantly change the carrier energy. Thus, these processes are known as *inelastic scattering* processes. Note that acoustic phonons can also undergo inelastic scattering, although the probabilities are low. Whereas elastic scattering results in an intravalley (electron) or an intraband (hole) transition, optical phonons can induce intervalley/interband transitions in addition to the previous ones. In a material such as GaAs, for example, the transition of an electron from the Γ valley to the X valley requires a large change in wavevector as well as sufficiently high energy. Optical phonons can provide this coupling and therefore facilitate intervalley scattering. Since optical phonons also contain the same energy for phonons at zone center ($K = 0$), they can lead to interband transitions for holes. In such processes, holes can undergo transition from the light hole to the heavy hole band.

Optical phonon scattering can be of two types, nonpolar and polar. First, consider the nonpolar process. An optical phonon is created when there are two atoms per unit cell and when these two atoms move against each other out of phase. When two neighboring atoms move apart, or closer, the charge distribution in the bonds is changed. A net positive charge is created when the atoms move apart, and the negative charge density is increased in the bond when the atoms move closer. This creates a net *deformation field*, Θ (in eV/m), such that the net change in the potential is $\Delta U = \Theta x$, where x is the net

displacement. The scattering rate for such a process can be written as (Ferry, 1991)

$$\frac{1}{\tau(\mathbf{k})} = \frac{m^{*\frac{3}{2}}\Theta^2}{\sqrt{2}\pi\rho\hbar^3\omega_o}[n_K\sqrt{E_k + \hbar\omega_o} + (n_K + 1)\sqrt{E_k - \hbar\omega_o}\nu(E_k - \hbar\omega_o)] \quad (1\text{-}81)$$

where ω_o is the optical phonon energy (assumed to be a constant for a material), n_K is the Bose-Einstein distribution of phonons, ρ is the mass density, and $\nu(x)$ is the Heaviside step function such that $\nu(x) = 0$ for $x < 0$ and $\nu(x) = 1$ for $x \geq 0$. The first term in the square brackets is the phonon absorption term, whereas the second term is the phonon emission term. The Heaviside function ensures that $E_k \geq \hbar\omega_o$ for phonon emission. Since optical phonon energies are typically on the order of or higher than $k_B T$, the approximation of Eq. (1-77) cannot be invoked, and the full Bose-Einstein distribution must be used. This makes the scattering process more sensitive to temperature. Whereas an elastic scattering process changes only the carrier momentum, such an inelastic process relaxes both energy and momentum. This makes it a highly efficient channel for energy transfer between carrier and lattice and plays a major role in Joule heating.

In materials which contain two dissimilar atoms per unit cell (such as III-V or II-VI semiconductors), the charge distribution in the chemical bond between them is not symmetric. This ionicity results in a dipole between them. An optical phonon leads to an oscillating dipole, which can strongly scatter charged particles. This type of scattering is called *polar optical phonon scattering*. The LO-phonon mode is the most effective scatterer for electrons and holes. The scattering rate can be written as

$$\frac{1}{\tau(\mathbf{k})} = \frac{m^* e^2 \omega_o}{4\pi\hbar^2 k}\left[\frac{1}{\varepsilon_\infty} - \frac{1}{\varepsilon(0)}\right]\left[G(k) + \frac{q_d^2}{2}H(k)\right] \quad (1\text{-}82)$$

where the static and high-frequency electrical permittivity are used, and where

$$G(k) = (n_K + 1)\ln\left[\frac{\left(k + \sqrt{k^2 - K_o^2}\right)^2 + q_d^2}{\left(k - \sqrt{k^2 - K_o^2}\right)^2 + q_d^2}\right]^{\frac{1}{2}} + n_K \ln\left[\frac{\left(\sqrt{k^2 - k_o^2} + k\right)^2 + q_d^2}{\left(\sqrt{k^2 - K_o^2} - k\right)^2 + q_d^2}\right]^{\frac{1}{2}}$$

$$H(k) = -(n_K + 1)\frac{4k\sqrt{k^2 - K_o^2}}{\left(K_o^2 - q_d^2\right)^2 + 4k^2 q_d^2} - n_K\frac{4k\sqrt{k^2 - K_o^2}}{\left(K_o^2 + q_d^2\right)^2 + 4k^2 q_d^2}. \quad (1\text{-}83)$$

The phonon with wavevector K_o is the called the *dominant phonon*, where K_o is given by

$$K_o^2 = \frac{2m^*\omega_o}{\hbar}. \quad (1\text{-}84)$$

Evaluation of the phonon distribution n_K is based on the optical phonon energy $\hbar\omega_o$. The Debye wavevector q_d is given by Eq. (1-73). Note that although many of the scattering rate relations are provided in terms of the carrier wavevector, k, the scattering rate is directly related to the carrier energy through the band structure. For electrons in the range 0.01–1 eV, optical phonon scattering rates range from 10^{12}–10^{13} Hz (Jacoboni and Reggiani, 1983). When the electron energy is higher than the optical phonon energy, the emission rate becomes larger than the absorption rate.

1-5 INTERACTIONS OF PHOTONS WITH ELECTRONS AND PHONONS

Besides electrons and crystal vibrations, propagation of electromagnetic radiation or photons is another mode of energy transport in solids. If incident photons pass through a solid without any interaction with the solid, then there is no net energy transfer from the photons to the solid. Energy transfer occurs only during interaction of the photons with electrons and phonons. Therefore, this section will concentrate on photon–electron and photon–phonon interactions, which form the foundation for optical properties of solids. The second issue regarding photons is the generation of photons within a solid. Blackbody radiation from a solid is an example of such emission. Since such broadband radiation is covered well in several textbooks of radiative heat transfer, we will not focus on this topic. Instead, we will look at emission of monochromatic radiation in solids, which forms the basis of light-emitting diodes, diode lasers, and other optoelectronic devices.

Electromagnetic waves are solutions of Maxwell's equations. These equations can be combined to produce the following equation, which represents the wave nature of electromagnetic radiation (Bohren and Huffman, 1983)

$$\nabla^2 \mathbf{E} = \frac{\varepsilon}{c^2}\frac{\partial^2 \mathbf{E}}{\partial t^2} + \frac{4\pi\sigma}{c^2}\frac{\partial \mathbf{E}}{\partial t}. \tag{1-85}$$

Here, $\mathbf{E}$ is the electric field vector, which is perpendicular to the direction of propagation (as required by Maxwell's equations); ε is the dielectric constant of a solid; c is the speed of light in vacuum; t is time; and σ is the electrical conductivity of a solid. A similar equation can be derived for the magnetic field, but for the sake of simplicity we will ignore any magnetic effects. Note that the electrical conductivity is a transport property that we have not yet studied from the microscopic viewpoint, but we will assume for the time being that it is known. For a frequency of the electromagnetic wave, ω, the solution of Eq. (1-85) is of the form (Bohren and Huffman, 1983)

$$\mathbf{E} = \mathbf{E}_o \exp[i(\mathbf{q}\cdot\mathbf{r} - \omega t)] \tag{1-86}$$

where $\mathbf{E}_o$ is the amplitude of the electric field vector, $\mathbf{q}$ is the wavevector related to the wavelength λ ($|\mathbf{q}| = 2\pi/\lambda$), and $\mathbf{r}$ is a position vector. If Eq. (1-86) is inserted into Eq. (1-85), we get the dispersion relation for light, which has the form

$$q = \frac{\omega}{c}\left(\varepsilon + i\frac{4\pi\sigma}{\omega}\right)^{\frac{1}{2}} = \frac{\omega}{c}\sqrt{\tilde{\varepsilon}} \tag{1-87}$$

where $\tilde{\varepsilon}$ is the complex dielectric constant ($\tilde{\varepsilon} = \varepsilon + i\varepsilon'$ and $\varepsilon' = 4\pi\sigma/\omega$). Another common way of writing this is

$$q = \frac{\omega\tilde{n}}{c} \tag{1-88}$$

where $\tilde{n}$ is the complex refractive index such that $\tilde{n} = n + i\kappa$ can be divided into a real part, n, and an imaginary part, κ. Using this, we find that the electromagnetic wave has the form

$$\mathbf{E} = \mathbf{E}_o \exp\left[i\omega\left(\frac{nz}{c} - t\right)\right]\exp\left(-\frac{\kappa\omega z}{c}\right). \tag{1-89}$$

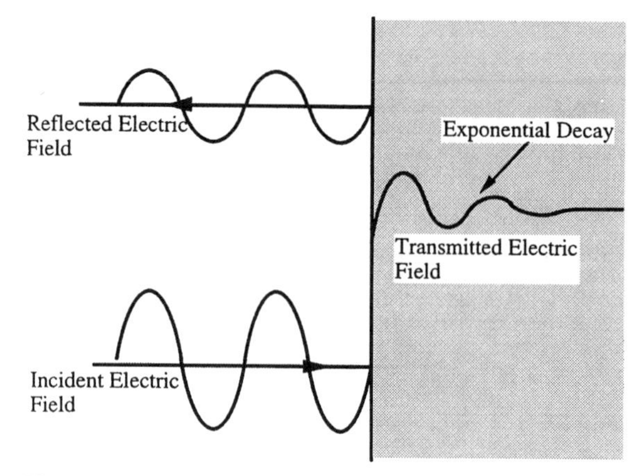

Figure 1-30 Schematic diagram showing incident, reflected, and transmitted electric fields of an electromagnetic wave. If the material absorbs radiation, the transmitted field decays exponentially in the solid.

This suggests that the electromagnetic wave travels at a speed c/n, which is directly affected by the real part of the refractive index. In addition, the field exponential decays in the material suggesting that the radiative energy, which is proportional to $|\mathbf{E}|^2$, decays with a characteristic length of $c/2\kappa\omega$, as shown in Fig. 1-30. The reduction in energy is due to absorption of radiation, which in microscopic terms involves absorption of photons by electrons and/or phonons in a solid. If we consider the interface between a solid of refractive index $\tilde{n}$ and vacuum ($\tilde{n} = 1$), the reflectivity of that interface can be shown to be (Bohren and Huffman, 1983)

$$R = \frac{(n-1)^2 + \kappa^2}{(n+1)^2 + \kappa^2}. \tag{1-90}$$

The reflectivity can be related to other surface properties such as emissivity and absorptivity through energy balance (Siegel and Howell, 1992). So determining the optical properties of a solid reduces to finding the values of n and κ, and their dependence on interactions of electromagnetic waves with electrons and phonons. Since $\tilde{n} = \sqrt{\tilde{\varepsilon}}$, we can reduce the problem to finding the real and imaginary parts of the complex dielectric constant, $\tilde{\varepsilon}$, since n and κ can be derived as

$$\begin{aligned} n &= \left[\frac{\sqrt{\varepsilon^2 + \varepsilon'^2} + \varepsilon}{2}\right]^{\frac{1}{2}}, \\ \kappa &= \left[\frac{\sqrt{\varepsilon^2 + \varepsilon'^2} - \varepsilon}{2}\right]^{\frac{1}{2}}. \end{aligned} \tag{1-91}$$

Note that if the conductivity of the solid is zero, the imaginary part of the dielectric constant becomes zero and hence $\kappa = 0$, making the solid transparent and $n = \sqrt{\varepsilon}$.

The first step in understanding optical properties is to develop a framework based on classical microscopic models of photon–oscillator interactions. The quantum mechanical approach will be developed subsequently.

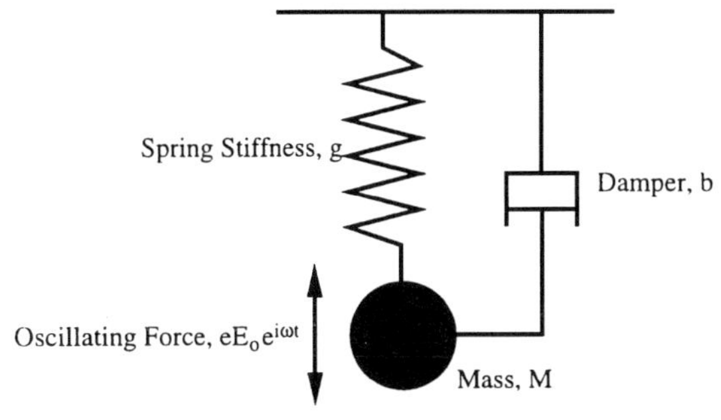

Figure 1-31 Spring-mass-damper model of a bound oscillator subjected to an oscillating force at frequency, ω.

1-5-1 Classical Microscopic Theories

Lorenz bound-oscillator model. Consider a charge oscillator of mass M and charge e connected to a stationary wall by a spring of stiffness g in the presence of an electric field, $\mathbf{E}$ (see Fig. 1-31). The dynamical equation for this oscillator is

$$M\ddot{\mathbf{x}} + b\dot{\mathbf{x}} + g\mathbf{x} = e\mathbf{E} \tag{1-92}$$

where the term $b\dot{\mathbf{x}}$ is the damping force, which is proportional to the velocity $\dot{\mathbf{x}}$, with b being the damping constant. In the presence of an electromagnetic wave, the electric field oscillates as $\mathbf{E}_0e^{i\omega t}$ at a certain location. A charged particle at that location will experience an oscillating force $e\mathbf{E}_0e^{i\omega t}$. Hence, one would expect the charged particle also to oscillate as $\mathbf{x}_0e^{i\omega t}$. Using this in Eq. (1-92), we get the relation

$$\mathbf{x}_0 = \frac{\left(\frac{e}{M}\right)\mathbf{E}_0}{\omega_o^2 - \omega^2 - i\frac{\omega}{\tau}} \tag{1-93}$$

where $\omega_o = \sqrt{g/M}$ is the resonant or natural frequency of the oscillator and $\tau = M/b$ is a time scale associated with damping. When a charge is shifted by a distance $\mathbf{x}$, the dipole moment that it generates is $e\mathbf{x}$. The net polarization $\mathbf{P}$ is the dipole moment per unit volume that is generated by this shift. If there are η oscillators per unit volume, the polarization is $\mathbf{P} = \eta e\mathbf{x}_0$, which can be written as

$$\mathbf{P} = \frac{\eta\left(\frac{e^2}{M}\right)\mathbf{E}_0}{\omega_o^2 - \omega^2 - i\frac{\omega}{\tau}}. \tag{1-94}$$

The dielectric constant is related to the polarization by the relation

$$\tilde{\varepsilon} = \frac{\mathbf{P} + \varepsilon_o\mathbf{E}}{\varepsilon_o\mathbf{E}} = 1 + \frac{\mathbf{P}}{\varepsilon_o E} \tag{1-95}$$

where ε_o is the permittivity of vacuum. Hence, the dielectric constant follows as

$$\tilde{\varepsilon} = 1 + \frac{\eta\left(\frac{e^2}{\varepsilon_o M}\right)}{\omega_o^2 - \omega^2 - i\frac{\omega}{\tau}}. \tag{1-96}$$

The numerator of the second term has units of frequency squared. It is related to the plasma frequency, ω_p, which is defined as

$$\omega_p^2 = \frac{\eta e^2}{\varepsilon_o M}. \tag{1-97}$$

The real and the imaginary parts of the dielectric constant can be extracted from Eq. (1-96) and shown to be

$$\begin{aligned} \varepsilon &= 1 + \frac{\omega_p^2(\omega_o^2 - \omega^2)}{(\omega_o^2 - \omega^2)^2 + \frac{\omega^2}{\tau^2}}, \\ \varepsilon' &= \frac{\frac{\omega_p^2\omega}{\tau}}{(\omega_o^2 - \omega^2)^2 + \frac{\omega^2}{\tau^2}}. \end{aligned} \tag{1-98}$$

Using these relations, the real and the imaginary part of the refractive index can be derived as a function of the electromagnetic wave frequency. The plots of ε, ε', n, κ, and reflectivity R, as a function of frequency ω, of the electromagnetic wave are shown in Fig. 1-32.

It is interesting to study the limiting cases of $\omega \gg \omega_o$ and $\omega \ll \omega_o$. In these cases, the dielectric constant reduces to the following relations, which can be clearly seen in Fig. 1-32:

$$\begin{aligned} \varepsilon &= 1 - \frac{\omega_p^2}{\omega^2}; \qquad \varepsilon' = \frac{\omega_p^2}{\omega^3\tau} \qquad \text{for } \omega \gg \omega_o, \\ \varepsilon &= 1 + \frac{\omega_p^2}{\omega_o^2}; \qquad \varepsilon' = \frac{\omega_p^2\omega}{\omega_o^4\tau} \qquad \text{for } \omega \ll \omega_o. \end{aligned} \tag{1-99}$$

We see that for $\omega \gg \omega_o$, $\varepsilon \to 1$ and $\varepsilon' \to 0$ as the frequency is increased. This makes the solid transparent. On the other hand, when $\omega \ll \omega_o$ in the low-frequency limit, ε is a constant and $\varepsilon' \to 0$. This produces some reflection at the solid–vacuum interface, but the solid does not absorb any radiation. So most of the absorption occurs in the neighborhood of the resonant frequency ω_o. Therefore, it is worth studying this region in more detail.

It is clear from Eq. (1-98) that when $\omega \to \omega_o$, the denominator in Eq. (1-98) goes through a minimum and ε' goes through a maximum. In addition, in this region the value of ε falls to unity at ω_o and then to zero. As a result, Eq. (1-91) suggests that the value of n decreases and that κ must go through a maximum. Therefore, this is a region of high absorption and reflection. This can sometimes be thought of as contradictory, but it is not. An increase in κ suggests that the *interface* is very reflective. However, if any radiation that penetrates the interface of the solid, it is efficiently absorbed, then this makes the solid highly absorbing too. In the region $\omega \geq \omega_o$, it is interesting to see that ε takes negative values. Let's try to determine the region where $\varepsilon < 0$. For most materials, it can be assumed that $\omega_p, \omega_o \gg 1/\tau$. Solving Eq. (1-98) for the zeros of ε, we derive the frequency range to be

$$\omega_o^2 + \frac{1}{2\tau^2}\left(1 + 4\frac{\omega_p^2}{\omega_o^2}\right) \leq \omega^2 \leq \omega_o^2 + \omega_p^2. \tag{1-100}$$

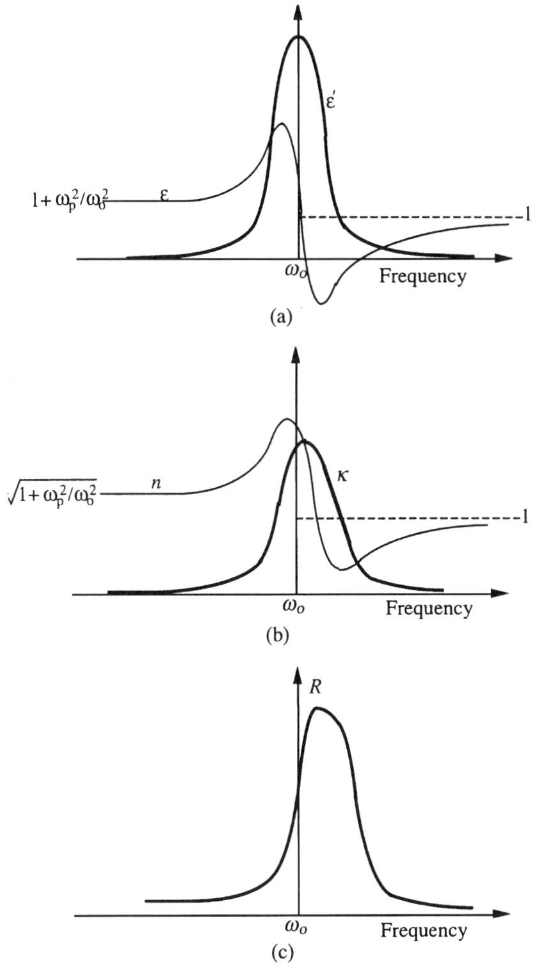

Figure 1-32 Plot of (a) real and imaginary parts of the dielectric constant; (b) real and imaginary part of the refractive index; and (c) reflectivity as a function of the frequency of the electromagnetic wave. The resonant frequency of the bound oscillator is ω_o.

Since $\omega_p, \omega_o \gg 1/\tau$, we see that the lower limit corresponds to ω_o. There is a slight correction to this relation, which takes the form (Ziman, 1972)

$$\omega_o^2 \leq \omega^2 \leq \omega_o^2 + \frac{\omega_p^2}{\varepsilon_\infty} \tag{1-101}$$

where ε_∞ is the electronic contribution to the dielectric constant at very high frequencies. Typically, electronic interactions with photons occur at much higher frequencies, and hence ε_∞ is the contribution of that mechanism at the frequencies considered here. To put this in stronger ground, the unity in Eq. (1-98) should be replaced by ε_∞. We find that in this region the value of n decreases sharply, whereas κ increases as a resonance peak. Therefore, this is a region of very high interface reflectivity and absorption in the bulk.

Drude free-electron model. The bound oscillator model cannot be applied to study the optical interactions of free charge carriers. In the free electron model, it is assumed that electrons are not bound to any ion but rather move freely through the electron gas while colliding with ions, impurities, and other electrons. Therefore, the spring constant in the single oscillator model, g (see Eq. (1-92)), must be zero. As a result, $\omega_o = 0$, which is a limiting case of the Lorenz model. Therefore, the dielectric constant can be written as

$$\varepsilon = 1 - \frac{\omega_p^2}{\omega^2}; \qquad \varepsilon' = \frac{\omega_p^2}{\omega^3 \tau}. \tag{1-102}$$

Figure 1-33 shows a plot of the dielectric constants, the refractive index, and the reflectivity as functions of frequency. It can be clearly seen that below the plasma frequency, the reflectivity is very high, whereas above the plasma frequency the reflectivity drops to zero. This sudden transition is characteristic of metals and other free charge carriers. The plasma frequency can be calculated by Eq. (1-97) where the mass, M, must be taken to be the effective carrier mass, m^*. The plasma wavelengths of some metals, $\lambda_p (= 2\pi c/\omega_p)$, are shown in Table 1-12. For most metals, the photon energy

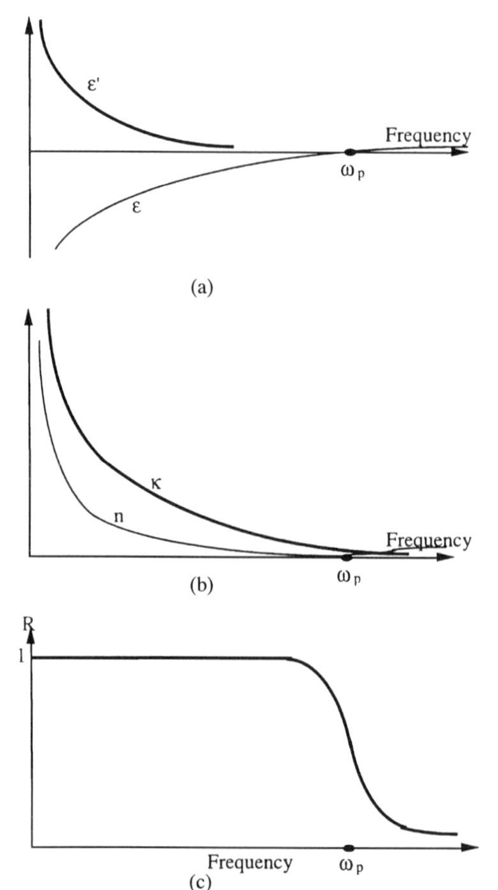

Figure 1-33 Plot of (a) real and imaginary parts of the dielectric constant; (b) real and imaginary parts of the refractive index; and (c) reflectivity as a function of the frequency of the electromagnetic wave for a free electron metal. The plasma frequency of the is ω_p.

Table 1-12 Plasma wavelength, λ_p, of some metals (Ashcroft and Mermin, 1976; Kittel, 1986)

Metal	Observed [nm]	Calculated [nm]
Li	200	150
Na	210	200
K	310	280
Rb	360	310
Cs	440	350
Mg	118	114
Al	81	79

$\hbar\omega_p$ at the plasma frequency typically falls in the 2–15 eV range, which is in the ultraviolet range. Therefore, metals appear to be highly reflective in the visible and the infrared range and are transparent in the ultraviolet regime above the plasma frequency. For semiconductors, however, the plasma frequency depends on the carrier concentration as suggested by Eq. (1-97), which in turn depends on the level of doping. Typically, this falls in the far infrared range.

The damping time scale, τ, is the mean free time between carrier scattering events. As will be discussed, damping is the only mechanism for absorption in a metal. If there were no damping, the electron would travel freely in a metal and re-radiate a photon. Damping provides a means of converting the photon energy gained by the electron to be scattered away into thermal energy. The electron is then said to be *thermalized.* For most metals and semiconductors, the scattering time is typically on the order of 10^{-13} s. Therefore, $\omega_p \gg 1/\tau$ in most materials at room temperature. However, note that the scattering rate of electrons in metals depends on temperature. This makes ε', κ, and consequently the reflectivity R below the plasma frequency temperature dependent. This forms the basis of optically measuring the temperature of a metal. At the plasma frequency, we find that $\varepsilon = 0$. As will be discussed, longitudinal oscillations in the electric field can be generated in the solid when the electromagnetic wave passes through it. The longitudinal oscillations of charged particles are called plasma oscillations or *plasmons*.

Hybrid models. So far we have discussed only free charge carriers in a solid. Even in metals, bound electrons can significantly change the optical properties within a certain wavelength regime. For example, the colors of copper and gold are quite different from those of other metals. This is the effect of bound electrons. It must be noted that the free electron theory is only a model for metals. In reality, the periodic potential of the ions do produce an electronic band structure in metals (Ashcroft and Mermin, 1976). The transition of electrons from the conduction band to a higher band (or from a lower band to the conduction band) can be optically induced. In copper, for example, a 2-eV transition takes place from the lower band to the Fermi level (Ashcroft and Mermin, 1976). This absorption lies in the orange part of the visible spectrum, giving copper a reddish color. Another transition of 4 eV occurs between the Fermi level and a higher

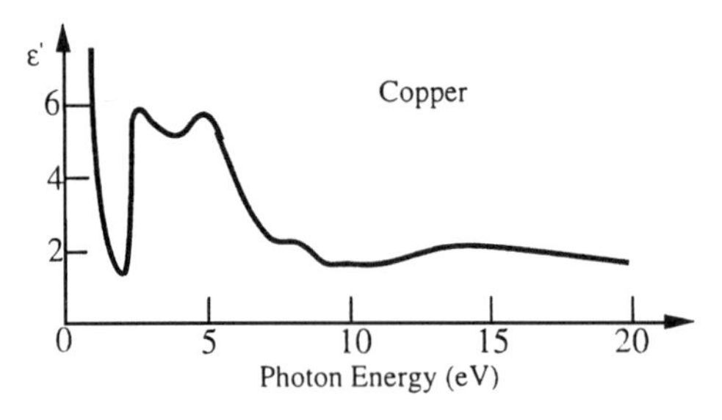

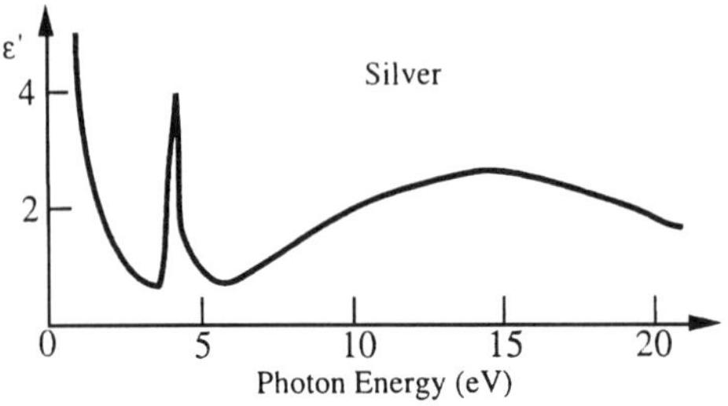

Figure 1-34 Imaginary part of the dielectric constant of copper and silver plotted against photon energy (Ehrenreich and Phillip, 1962). Note the two peaks at 2 eV and 4 eV for copper and the single peak at 2 eV for silver. Photon energy of 2 eV corresponds to a wavelength of 0.62 μm, whereas that of 4 eV corresponds to 0.31 μm.

band. The imaginary parts of the dielectric constants, ε', are shown in Fig. 1-34 for copper and silver. The 2 eV and 4 eV transitions show up as peaks in the spectrum for copper. For silver, there is a peak at about 4 eV. Below the 2 eV for copper and 4 eV for silver, note the $1/\omega^3$ behavior predicted by the free electron model.

The peak in the $\varepsilon'(\omega)$ plot is reminiscent of the bound oscillator behavior in Fig. 1-32. Therefore, such interband transitions in a metal are accounted for in a hybrid model (free electron and bound oscillator) for which the dielectric constant can be written as (Bohren and Huffman, 1983)

$$\tilde{\varepsilon} = 1 - \frac{\omega_{pf}^2}{\omega^2 + i\frac{\omega}{\tau_f}} + \sum_j \frac{\omega_{pj}^2}{\omega_{oj}^2 - \omega^2 - i\frac{\omega}{\tau_j}}. \tag{1-103}$$

where the subscript f is for free electrons, whereas j is for the jth bound oscillator which has a resonance frequency, ω_{oj}.

1-5-2 Photon Interactions: The Quantum Mechanical Picture

Absorption by phonons. The classical microscopic theory illustrated above considers a very simple model of a bound oscillator, which brings out the essence of the interactions but does not present the material in terms of photon–phonon and photon–electron interactions. But clearly, the characteristic frequencies ω_o, ω_p and time constants, τ, must be somehow related to the phonon and electron characteristic time scales. So we will explore the relationship between the single oscillator model and phonon characteristics.

A quantum of energy in an electromagnetic wave of frequency ω and wavelength λ is a *photon*. A photon has an energy of $\hbar\omega$ and a momentum $\hbar q$, where $q = 2\pi/\lambda$. This is an analog of a phonon, which is a quantum of vibrational energy. In contrast to phonons, however, the wavevector q and frequency ω are related by a linear dispersion relation $q = \omega/c$, where c is the speed of light. Note that in the linear part of the phonon dispersion relation, $q = \omega / v_s$, where v_s is the speed of sound. Typically, v_s is on the order of 10^3 m/s, whereas the speed of light is 3×10^8 m/s. Hence, for the same quantum of energy, $\hbar\omega$, the photon momentum $\hbar q$ and wavevector q are five orders smaller than those of phonons. In other words, the photon wavelength is five orders larger than the phonon wavelength for the same quantum of energy. Therefore, if a photon penetrating a solid converts to or emits a phonon, the wavevector of the phonon must be at the center of the Brillouin zone or $K = 0$. This must be satisfied in order to conserve momentum. We know from Fig. 1-12 that acoustic phonons at the zone center have zero energy. Hence, to conserve energy and momentum, photons must emit optical phonons. Such single photon–phonon interaction is perhaps the most dominant, although two-photon (Raman scattering) and two-phonon interactions are also possible. For such cases, photon absorption results in a vertical transition at any location in the Brillouin zone since the photon wavevector is negligibly small compared to that of the phonons.

There are two types of optical phonons: transverse optical (TO) and longitudinal optical (LO). The resonant frequency, ω_o, must correspond to one of these phonon frequencies. To interact with electric fields, the vibrating ions must possess some degree of ionicity or a dipole moment. Since electromagnetic waves are transverse waves, their motion through such a solid creates a transverse oscillation in the vibrating ions. Hence, the resonant frequency of the ω_o must correspond to ω_{TO}, which is the TO-phonon frequency at the zone center. As indicated in Table 1-7, the TO-phonon frequency at the zone center of some semiconductors is on the order of 10^{13} Hz. For photons, this frequency corresponds to a wavelength of about 10 μm, which is in the infrared range. Hence, phonons influence the optical properties of solids in the infrared regime.

Let us now consider longitudinal vibrations. Maxwell's equations require the electromagnetic waves to be transverse. This follows from the equation

$$\nabla \cdot (\varepsilon \mathbf{E}) = 0. \tag{1-104}$$

When $\varepsilon \neq 0$, it implies that $\nabla \cdot \mathbf{E} = 0$ which, in conjunction with the other Maxwell's equations, ensures that electromagnetic waves are transverse in nature. However, when $\varepsilon = 0$ the condition of $\nabla \cdot \mathbf{E} \neq 0$ can be achieved within a solid. Under these conditions, the electromagnetic field can travel as a longitudinal wave within the solid. If the solid contains even a weak dipole, the longitudinal electromagnetic wave will interact with these dipoles and produce a longitudinal vibration in the ions. These are of course the LO-phonons. To conserve momentum, LO-phonons are produced only at the zone center ($K = 0$). Therefore, the frequency at which $\varepsilon = 0$ must correspond to the LO-phonon frequency ω_{LO}. Comparison with Eq. (1-101) suggests that

$$\omega_{LO}^2 = \omega_{TO}^2 + \frac{\omega_p^2}{\varepsilon_e}. \tag{1-105}$$

The region $\omega_{TO}^2 \leq \omega^2 \leq \omega_{LO}^2$ in the frequency spectrum forms a forbidden gap where photons cannot propagate through the solid and are highly absorbed or reflected. Figure 1-35 shows the experimental measurement of the reflectivity of α-SiC where such a forbidden gap can be clearly seen (Bohren and Huffman, 1983).

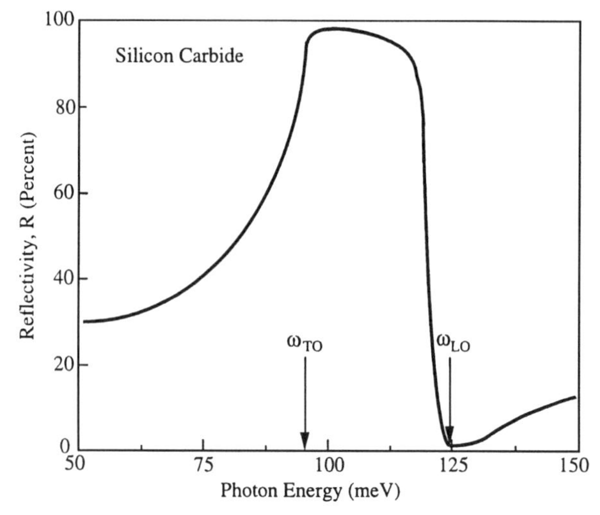

Figure 1-35 Reflectivity of α-SiC in the infrared range (Bohren and Huffman, 1983). The peak in reflectivity occurs between the energies of the transverse optical and longitudinal optical phonons. Photon energy of 100 meV corresponds to a wavelength of 12.5 μm.

If we look at the values of TO- and LO-phonon frequencies (see Table 1-7) at the Γ point (or zone center where $K = 0$), we will see an interesting trend. For the group IV elements (C, Si, and Ge), the TO- and LO-phonon frequencies coincide. This means that there is no forbidden gap for photons to travel through these materials. This is because the chemical bonds between atoms are perfectly symmetric and, therefore, produce no ionicity or dipole moment. The lack of a dipole moment prevents it from interacting with oscillating electric fields. However, if we look at the III-V compounds, we find that lower the atomic numbers, the higher the difference in the frequencies of the TO- and LO-phonons. For example, consider the AlX compounds where X can be P, As, or Sb as we go down Group V. The difference between the TO and LO frequencies decreases as we increase the atomic number. A similar trend can be seen for the GaX compounds. The nature of covalent bond is such that there seems to be a higher degree of charge transfer to P than to Sb. This is because of the well-known fact that the lower the atomic number in Groups V, VI, or VII in the periodic table, the higher the electron affinity of the atom.

An important result from this comparison of the classical model and phonon–photon interactions is the following. If we let $\omega \to 0$, the dielectric constant can be written as

$$\varepsilon(0) = \varepsilon_\infty + \frac{\omega_p^2}{\omega_o^2} = \varepsilon_\infty \frac{\omega_{LO}^2}{\omega_{TO}^2} \tag{1-106}$$

which is called the *Lyddane-Sachs-Teller relation*. This relation directly tells us how the splitting of the optical modes is related to the low- and high-frequency dielectric constants. We can express the dielectric constant of a solid as

$$\varepsilon(\omega) = \varepsilon_\infty \left(1 + \frac{\omega_{LO}^2 - \omega_{TO}^2}{\omega_{TO}^2 - \omega^2}\right) \tag{1-107}$$

assuming that the damping term is negligible.

This brings us to the damping term and the time constant, τ. Damping of a physical system inherently involves some form of dissipation. In thermodynamical terms, this produces entropy in a system. If a particle is moving in a certain direction, it dissipates its kinetic energy by colliding with other particles, which move around in a random fashion. This is the only mechanism of photon absorption in a material. If the phonon is not damped, the phonon would keep traveling and eventually re-radiate a photon. It is during the collision process that the phonon loses its energy and produces other phonons, either acoustic or optical, which may be incapable of emitting photons. In the case of phonons, the damping term must involve scattering of phonons. Therefore, τ is the relaxation time or mean free time between phonon collisions, which is discussed in Section 1-2. For phonon–phonon scattering at room temperature this is typically on the order of 10^{-11} s or 10 ps. Hence $1/\tau$ is on the order of 10^{11} Hz, which is two orders of magnitude lower than the TO-phonon frequencies, which are on the order of 10^{13} Hz.

Absorption by electrons. As discussed earlier, the wavevector of photons, $\mathbf{q}$, is negligibly small compared to that of electrons and phonons at the same energy level. Hence, any optically induced transitions for electrons must be vertical in the Brillouin zone. In mathematical terms, the absorption of a photon can be expressed as

$$\begin{aligned} E_{\mathbf{k}'} &= E_{\mathbf{k}} + \hbar\omega_{\mathbf{q}} && \text{energy conservation} \\ \mathbf{k}' &= \mathbf{k} + \mathbf{q} \quad \text{or} \quad \mathbf{k}' \cong \mathbf{k} && \text{momentum conservation.} \end{aligned} \tag{1-108}$$

For free electrons, which follow the relation $E \propto k^2$ and result in a parabolic band structure, this is clearly impossible. Unless the excited electron scatters in k-space either through elastic or inelastic scattering by impurities or phonons, as illustrated in Fig. 1-36, the photon will not be absorbed. Therefore, it is not surprising that the imaginary part of the dielectric constant in Eqs. (1-102), $\varepsilon' \propto 1/\tau$, involves the mean free time between scattering, τ. Since the imaginary part of the dielectric function, κ, is closely related to ε', we find that scattering is the only mechanism by which free electrons can absorb photons. The inability to simultaneously conserve momentum and energy in the absence of scattering prevents absorption of photons.

The only vertical transition in the electronic band structure that does not involve scattering by impurities and/or phonons is the direct interband transitions shown in Fig. 1-28. Here, only the band edge transition is shown, but direct transitions in indirect

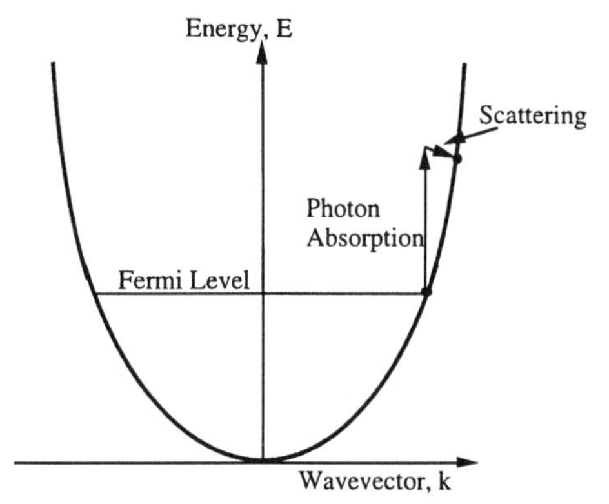

Figure 1-36 Scattering-assisted absorption of a photon in a parabolic electronic band.

gap semiconductors are also possible as shown in Fig. 1-28. The treatment of such transitions involves calculations of quantum mechanical transition probabilities, which are not discussed here but can be found elsewhere (Ferry, 1991; Ziman, 1972). The final form, however, is close to those derived by classical form and can be written as

$$\begin{aligned} \varepsilon &\approx 1 + \frac{4\pi e^2}{m\varepsilon_o} \int \frac{f(\omega) D(\omega)}{\left(\omega^2 - \omega_g^2\right)} d\omega \\ \varepsilon' &= \frac{\pi^2 e^2}{m\varepsilon_o} \frac{f(\omega) D(\omega)}{\omega} \end{aligned} \tag{1-109}$$

where $f(\omega)$ is the oscillator strength, that is, the ratio of the kinetic energy of an electron to the photon energy. This is typically on the order of unity. $D(\omega)$ is the density of electronic states available in the conduction band such that $D(\omega)d\omega$ are the number of states per unit volume in a frequency range ω to $\omega + d\omega$. Here, $\hbar\omega_g$ is the *direct* energy gap, E_g, at a certain position in the Brillouin zone. This includes the minimum energy gap at the zone center for direct gap semiconductors. Equation (1-60) suggests that the density of states follows the relation

$$D(\omega) = \left(\frac{m_r^*}{\hbar}\right)^{\frac{3}{2}} \frac{\sqrt{2(\omega - \omega_g)}}{\pi^2} \nu(\omega - \omega_g) \tag{1-110}$$

where $\nu(x)$ is the Heaviside function and m_r^* is the reduced effective mass that is related to the conduction and valence band effective masses as

$$\frac{1}{m_r^*} = \frac{1}{m_c^*} + \frac{1}{m_v^*}. \tag{1-111}$$

Note, however, that the mass m used in Eq. (1-109) is equal to the electron rest mass. Equation (1-110) suggests that states are available only when $\omega > \omega_g$. Thus, for frequencies below ω_g, the absorption of radiation is zero and the semiconductor is transparent (except for free carrier absorption and phonon absorption). Figure 1-37 shows a plot of the absorption coefficient α, defined as $\alpha = 4\pi\kappa\omega/c$, which is the reciprocal of the penetration depth of photons into a solid. Note the sharp rise in the absorption coefficient as the photon energy exceeds the direct band gaps of InSb and GaAs. Also plotted is the absorption coefficient of Si, which is an indirect gap semiconductor.

The calculations of the absorption coefficients of indirect transitions must involve phonons of energy $\hbar\omega_K$ and momentum $\hbar\mathbf{K}$. The energy and momentum conservation equations can be written as

$$\begin{aligned} E_{\mathbf{k}'} &= E_{\mathbf{k}} + \hbar\omega_{\mathbf{q}} \pm \hbar\omega_K && \text{energy conservation} \\ \mathbf{k}' &= \mathbf{k} + \mathbf{q} \pm \mathbf{K} \cong \mathbf{k} + \mathbf{K} && \text{momentum conservation} \end{aligned} \tag{1-112}$$

where the wavevector of the photon can be ignored. The plus and the minus signs correspond to phonon absorption and emission, respectively. Typically, the phonon energies are much smaller than the energy band gap of semiconductors but momentum conservation restricts the available final states. Since the number of phonons is sensitive to the temperature of a solid, the transition probabilities, and hence the optical properties, are also temperature sensitive. We will not go through the derivation of the absorption

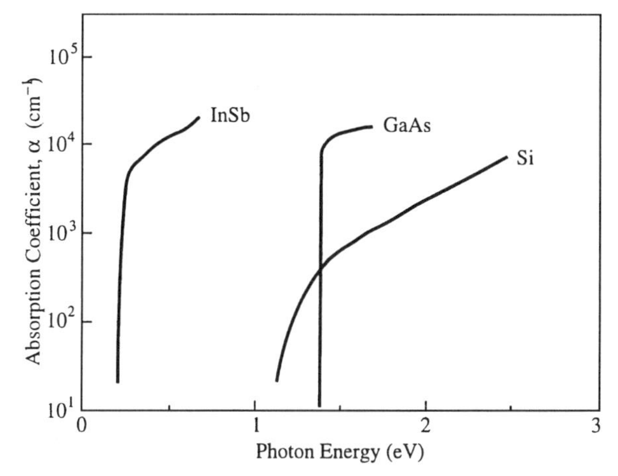

Figure 1-37 Absorption coefficient as a function of photon energy for InSb (Gobeli and Fan, 1956), GaAs (Moss and Hawkins, 1962), and Si (Dick and Newman, 1955). InSb and GaAs are direct gap semiconductors and show a sharp rise in absorption coefficient when the photon energy increases beyond the band gap. Si is an indirect gap semiconductor where photon absorption is mediated by phonons. Here the rise is more gradual and the absorption coefficient is temperature dependent.

coefficient but instead present the general form of the result:

$$
\begin{aligned}
\alpha &\propto \frac{\left(\frac{\hbar\omega_K}{k_B T}\right)}{\exp\left(\frac{\hbar\omega_K}{k_B T}\right) - 1}[\hbar\omega - E_g + \hbar\omega_K]^2 \quad \text{phonon absorption,} \\
\alpha &\propto \frac{\hbar\omega_K}{k_B T}[\hbar\omega - E_g - \hbar\omega_K]^2 \quad \text{phonon emission.}
\end{aligned}
\tag{1-113}
$$

Electron–hole radiative recombination. As discussed in the beginning of Section 1-5, the electromagnetic field decays exponentially in an absorbing solid (see Fig. 1-30). The characteristic length or the penetration depth of the field is the reciprocal of the absorption coefficient, α. This length scale is the mean free path of a photon in the solid. Since the photon travels at the speed of c/n in the solid, the mean free time or the photon lifetime in the solid, τ, is equal to $\tau = \alpha c/n$. Note that both α and n can depend on the photon energy, $\hbar\omega$. At equilibrium, the photon absorption rate is equal to the photon emission rate. Since photon absorption creates an electron pair, the recombination of an electron and a hole must also generate a photon. Hence, the electron–hole generation rate must be equal to the electron–hole recombination rate at equilibrium. Since photons follow Bose-Einstein statistics, the number of photons created in a solid depends on the number of photons already existing in the solid. In other words, the probability of photon emission increases with the number of photons existing in the solid. The distribution of photons in a solid is given by the Planck distribution $N(\omega)$. Hence, the photon emission rate can be written as

$$
G(\omega) = \frac{N(\omega)}{\tau} = \frac{2\alpha n^2 \omega^2}{\pi c^2\left[\exp\left(\frac{\hbar\omega}{k_B T}\right) - 1\right]} \tag{1-114}
$$

where $G(\omega)d\omega$ is equal to the number of photons emitted in unit time per unit volume in the frequency range ω to $\omega + d\omega$. The total photon emission rate, which is the electron–hole recombination rate, is an integral of Eq. (1-114) over all frequencies:

$$R_o = \frac{2(k_B T)^3}{\pi c^2 \hbar^3} \int_0^\infty \frac{x^2 n^2 \alpha}{e^x - 1} dx \qquad \text{where } x = \frac{\hbar\omega}{k_B T}. \tag{1-115}$$

Most of the contribution of the integral in Eq. (1-115) comes from the band edge. This is because of two factors. First, α is zero for photon energies below the band gap, and hence there is no contribution from below the band gap. Second, for energies higher than the band gap, the exponential factor in the denominator makes the integrand small. If $\alpha = \alpha_o(\hbar\omega - E_g)^p$, then Ferry (1991) has shown that the recombination rate can be written as

$$R_o \cong \frac{2\alpha_o n^2 (k_B T)^{2+p}}{\pi c^2 \hbar^3} \left(\frac{E_g}{k_B T}\right)^2 \exp\left(-\frac{E_g}{k_B T}\right) \left[\Gamma(p+1) + \frac{2k_B T}{E_g}\Gamma(p+2)\right] \tag{1-116}$$

where $\Gamma(x)$ is the Gamma function.

The recombination rates discussed above are for equilibrium conditions, that is, when thermodynamic equilibrium exists in the solid between the electrons and the holes. However, under nonequilibrium conditions, the recombination rate can be different since the carrier concentrations, which influences the rate, can be different from the equilibrium values. It was established earlier that regardless of the concentrations of carriers in the bands, the condition $\eta_{co}\eta_{vo} = \eta_i^2$ must be satisfied. Here η_{co} and η_{vo} are the equilibrium concentrations of carriers in the conduction and the valence bands, respectively. Under nonequilibrium conditions, the recombination rate must be proportional to the carrier concentrations and can be written as

$$R = R_o \frac{\eta_c \eta_v}{\eta_i^2}. \tag{1-117}$$

Highly nonequilibrium carrier distribution occurs in optoelectronic devices such as light-emitting diodes and lasers where the recombination rate is necessary to calculate the photon flux. The nonequilibrium carrier concentrations, η_c and η_v, need to be calculated from balance equations, which are derived in the next section.

Radiative recombinations can also occur at impurities. Impurities provide energy states in between the energy gap of the host semiconductor, as well as states larger than the energy gap. In insulators such as diamond and sapphire, these impurities can lend color. This occurs when photons at a higher energy are absorbed to create an electron–hole pair across an impurity state. The electron can decay down to a lower energy state and then recombine with the hole to produce a photon of a lower energy, giving the impression of a colored material. In semiconductors, impurity states can be responsible for radiative electron–hole recombination to produce photons at energies other than the band gap.

Electron–hole nonradiative recombination. Although this section focuses on photon interactions, the nonradiative recombination is still included in this section since it naturally follows radiative recombination.

In a direct gap semiconductor, electrons and holes recombine to produce photons. Phonons are usually not produced since the maximum phonon energy is typically on

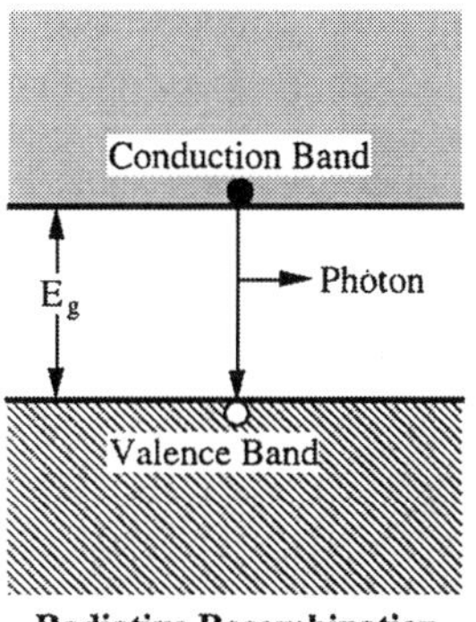

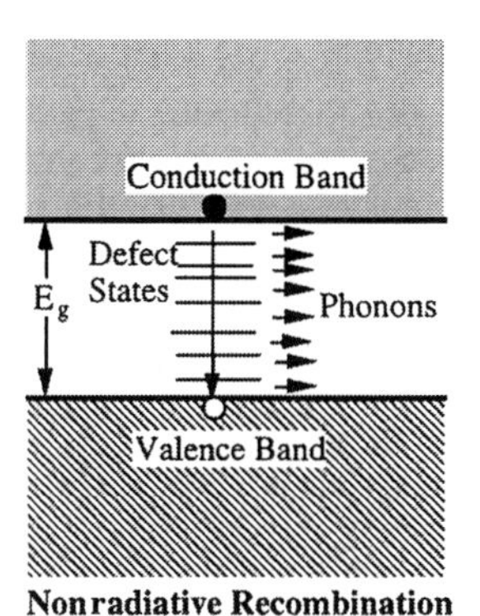

Figure 1-38 Energy diagram of radiative and nonradiative recombination of an electron–hole pair across the band gap of a semiconductor. Nonradiative recombination occurs through energy states in the band gap that are created due to crystal imperfections such as point defects, dislocations, and surfaces. Nonradiative recombination produces phonons and thereby results in heat generation as opposed to photon generation in radiative recombination.

the order of 10 meV whereas the band gap of semiconductors are on the order of 1 eV. Hence, for electrons and holes to recombine to produce phonons, about 100 phonons must be produced simultaneously. The probability to do so is extremely low and is much smaller than that of radiative recombination. However, crystal imperfections such as point defects or dislocations locally alter the band structure and often provide energy states within the energy gap as shown in Fig. 1-38. If the energy difference between the states is on the order of the phonon energies, it provides a channel for the electron and hole to nonradiatively recombine to produce phonons. This is one of the main failure mechanisms for light-emitting semiconductor devices such as diode lasers. Here, instead of producing light, the electron–hole recombination produces heat.

1-6 PARTICLE TRANSPORT THEORIES

In Sections 1-3 to 1-5, the physics of the three energy carriers in solids—phonons, electrons, and photons—were discussed in detail. Having studied their basic nature, the next step is to investigate their transport within a solid and their interactions with solid microstructures. As we have seen already, all these energy carriers possess a wave nature, and hence their transport can be studied as a problem in wave propagation. Such an approach, however, would require us to keep track of the phase and amplitude of a wave at every scattering process. This is indeed possible when we study the scattering at one or two specific sites or by a well-arranged periodic array of structures. However, when we deal with a large number of scattering events where the scattering occurs randomly in space and time, the wave propagation approach becomes very cumbersome. In such cases, the phase information of the wave invariably becomes insignificant after many scattering events, and what remains as most important for transport studies is the scattering rate or the mean free path of scattering. This lends itself to the particle transport theories in which we pay no attention to the phase but consider phonons, electrons, and photons as particles. As one may expect, this approach breaks down when wave-like phenomena such as interference, diffraction, and tunneling are important. In such cases, the wave propagation approach should be adopted.

A clear distinction between the wave and particle transport approaches can be made based on the time and length scales of the events of interest. Therefore, we will first study these scales in detail. As we shall see, the combination of length and time scales of interest will naturally direct us to the appropriate approach.

1-6-1 Time and Length Scales

In the absence of scattering of energy carriers, energy transport in a solid will occur without any resistance. Scattering of energy carriers poses resistance to energy transport. Therefore, it is necessary to study the scattering process in more detail. Associated with scattering are several characteristic time and length scales. Consider the time scales first.

The smallest time scale is the collision time or duration of collision, τ_c. Collisions are normally considered instantaneous in classical physics. However, this is not entirely true, and there is a finite collision time during wave scattering. This is on the order of the wavelength of the carrier divided by the propagation speed. For electrons at the Fermi energy in a metal, this is about 10^{-15} s (1 fs), whereas for phonons this is about 10^{-13} s (100 fs). The next time scale is the average time between collisions, τ, or the mean free time. For time scales $t \leq \tau$, carriers travel ballistically and the evolution of the system depends strongly on the details of the initial state. Note that τ is not the relaxation time since it takes several collisions to reach equilibrium. Generally, $\tau \gg \tau_c$, although this breaks down in high-electric field transport leading to new transport effects (Ferry, 1991). Collision-induced equilibrium gives rise to the relaxation time, τ_r. Relaxation times are associated with local thermodynamic equilibrium. Since equilibrium is achieved in 5 to 20 collisions, $\tau_r > \tau$. Note that momentum and energy relaxation times of a system can be different. For example, an elastic collision of an electron can change its momentum but not its energy. Hence, energy relaxation times, $\tau_{r\varepsilon}$, are typically longer than momentum relaxation times, τ_{rm}. The last time scale is the diffusion time, τ_d, which is on the order of $\tau_d \approx L^2/\alpha$, where L is the size of the object and α is the thermal diffusivity. This depends on particle speed, v, and mean free time as $\alpha \approx v^2\tau$. Clearly, $\tau_d \approx t_b^2/\tau$ where t_b is the time it takes for the particle to ballistically travel the distance L at speed v. For ballistic transport over a distance L, of course, $\tau \approx \tau_d$.

Note that the diffusion time scale, τ_d, contains a characteristic size of an object, L. This ties in the length scale of the problem. Similarly, the mean free path, ℓ, can be associated with mean free time between collisions, τ, by the relation, $\ell = v\tau$. Note that this is a statistical quantity since collision distances are not fixed. However, the probability p that a particle emerges from a collision and travels a distance x without a collision is related to the mean free path as $p = \exp(-x/\ell)$. Associated with the relaxation time is a length scale, ℓ_r, which is the characteristic size of a volume over which local thermodynamic equilibrium can be defined. Typically $\ell_r > \ell$. Another length scale that is associated with the collision process is the wavelength of the energy carrier. Therefore, the shortest length scale of the phenomenon is the wavelength, λ. The hierarchy of the length scales is generally $\lambda < \ell < \ell_r$.

Depending on time and length scales of interest, different transport laws can be used. Consider first the smallest length and time scales. When L is of the order of wavelength, λ, then wave phenomena such as diffraction, tunneling, and interference are important. For photon transport, wave optics based on Maxwell's equations must be used, whereas for electrons and phonons, quantum transport laws must be developed. When the time scale of interest, t, is of the order collision time scale, τ_c, then again time-dependent wave mechanics must be used. This is typical of high-speed electron transport in quantum structures or ultra-fast femtosecond laser experiments. Now consider the next larger length and time scales, ℓ or ℓ_r, and τ or τ_r. When $L \approx \ell$, ℓ_r and $t \gg \tau$, τ_r, transport is ballistic in nature and local thermodynamic equilibrium cannot be defined. This transport is nonlocal in space. One has to resort to time-averaged statistical particle transport equations. On the other hand, if $L \gg \ell$, ℓ_r and $t \approx \tau$, τ_r, then approximations of local thermodynamic equilibrium can be assumed over space although time-dependent terms cannot be averaged. The nonlocality is in time but not in space. When both $L \approx \ell$, ℓ_r and $t \approx \tau$, τ_r, statistical transport equations in full form should be used and no spatial or temporal averages can be made. Finally, when both $L \gg \ell$, ℓ_r and $t \gg \tau$, τ_r, local thermodynamic equilibrium can be applied over space and time, leading to macroscopic transport laws such as the Fourier law of heat conduction and Ohms law for electrical conduction.

Let us consider the last case first since that is the easiest one and also because it ties the microscopic transport characteristics to the macroscopic world.

1-6-2 Kinetic Theory

Formulation. Consider a plane z, across which particles travel carry mass and kinetic energy. Consider two fictitious planes at $z+\ell_z$ and $z-\ell_z$ on either side of the z-plane as shown in Fig. 1-39. Here ℓ_z is the z-component of the mean free path, ℓ, which makes an angle, θ, from the direction perpendicular to z. On average, the particles moving down from $z+\ell_z$ contain an energy density, u, that is characteristic of the location $u(z+\ell_z)$, whereas those moving up from $z-\ell_z$ have characteristic energy density $u(z-\ell_z)$. If the particles move with a characteristic velocity v, then the net flux of energy in the positive z-direction is

$$q_z = \frac{1}{2} v_z [u(z-\ell_z) - u(z+\ell_z)] \tag{1-118}$$

where v_z is the z-component of the velocity and the factor 1/2 is used because only half of the total number of particles at each location move up from $z-\ell_z$ or down from $z+\ell_z$.

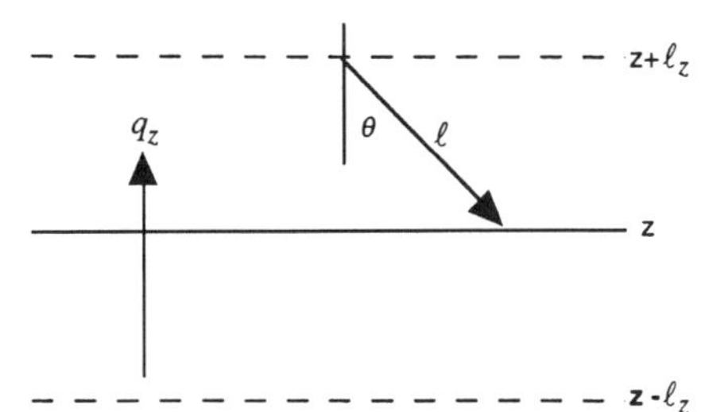

Figure 1-39 Schematic diagram showing energy flux across a z-plane used in kinetic theory.

Using Taylor expansion and keeping only the first-order terms, one gets

$$q_z = -v_z \ell_z \frac{du}{dz} = -(\cos^2\theta) v\ell \frac{du}{dz} \tag{1-119}$$

where it is assumed that $\ell_z = \ell\cos\theta$ and $v_z = v\cos\theta$. Averaging over the whole hemisphere of solid angle 2π, one gets

$$q_z = -v\ell\frac{du}{dz}\left[\frac{1}{2\pi}\int_{\varphi=0}^{2\pi}\int_{\theta=0}^{\pi/2}\cos^2\theta\sin\theta\, d\theta d\varphi\right] = -\frac{1}{3}v\ell\frac{du}{dz} \tag{1-120}$$

where φ is the azimuthal angle, θ is the polar angle, and $\sin\theta d\theta d\varphi$ is the elemental solid angle. Assuming local thermodynamic equilibrium such that u is a function of temperature, one can write the flux as

$$q_z = -\frac{1}{3}v\ell\frac{du}{dT}\frac{dT}{dz} = -\frac{1}{3}Cv\ell\frac{dT}{dz}. \tag{1-121}$$

This is the Fourier law of heat conduction with the thermal conductivity being $k = Cv\ell/3$. Note that we have not made any assumption of the type of energy carrier, and hence this is a universal law for all energy carriers. The only assumption made is that of local thermodynamic equilibrium such that the energy density at any location is a function of the local temperature.

The characteristics of the energy carrier are included in the heat capacity C, velocity v, and the mean free path ℓ. Neglecting photons for now, the thermal conductivity can be written as

$$k = \frac{1}{3}[(Cv\ell)_l + (Cv\ell)_e] \tag{1-122}$$

where the first term is the lattice contribution and the second term comes from electrons. In the case of electrons in a metal, for example, we saw that the heat capacity varies linearly with temperature, $C \propto T$; that the velocity is the Fermi velocity of electrons, which is independent of temperature; and that the mean free path can be calculated from $\ell = v_F\tau$ where τ is the mean free time between collisions, which is discussed in Section 1-4. Although the lattice heat capacity in a metal is much larger than its electronic contribution, the Fermi velocity of electrons (typically 10^6 m/s) is much larger than the speed of sound (about 10^3 m/s). The mean free path of electrons at room temperature is typically on the order of 100 Å, whereas that of phonons is in the range of 50–100 Å. Due to the higher energy carrier speed, the electronic contribution to the thermal conductivity turns out to be more dominant than the lattice contribution. For a semiconductor, however, the electron velocity is not the Fermi velocity but equal to the thermal velocity of the electrons in the conduction band. This can be approximated as $v \approx \sqrt{3k_BT/m}$, which is on the order of 10^5 m/s at room temperature. In addition, the number density of conduction band electrons in a semiconductor is much less than the electron density in metals. Hence, the electronic heat capacity is also lower. Therefore, the lattice contribution to the thermal conductivity dominates over the electronic contribution.

Thermal conductivity of crystalline and amorphous solids. Figure 1-40 plots the thermal conductivities of two metals. The data are the best recommended values based

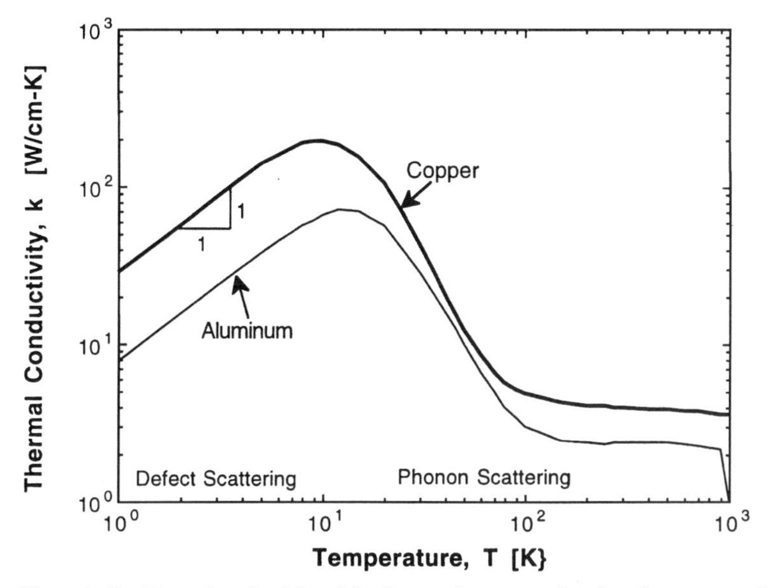

Figure 1-40 Thermal conductivity of aluminum and copper as a function of temperature (Powell et al., 1966). Note that at low temperature the thermal conductivity increases linearly with temperature. In this regime, defect scattering dominates and the mean free path is independent of temperature. The thermal conductivity in this regime depends on the purity of the sample. The linear behavior arises from the linear relation between the electronic heat capacity and temperature. As the temperature is increased, phonon scattering starts to dominate and the mean free path reduces with increasing temperature. To a large extent, the thermal conductivity of a metal is independent of the purity of the sample.

on a combination of experimental and theoretical studies (Powell et al., 1966). At low temperature, the electron mean free path is determined mainly by scattering due to crystal imperfections such as defects, dislocations, grain boundaries, and surfaces. Electron–phonon scattering is frozen out at low temperatures. Since the defect concentration is largely temperature independent, the mean free path is a constant in this range. Therefore, the only temperature dependence in the thermal conductivity arises from the heat capacity, which varies as $C \propto T$. Under these conditions, the thermal conductivity varies linearly with temperature as shown in Fig. 1-40. The value of k, though, is sample-specific since the mean free path depends on the defect density. As the temperature is increased, electron–phonon scattering becomes dominant. The mean free path for such scattering varies as $\ell \propto T^{-n}$ with n usually larger than unity. Hence, the thermal conductivity decreases at higher temperature.

Figure 1-41 shows the thermal conductivity of crystalline diamond samples with different defect concentrations (Berman, 1975). At low temperatures, the thermal conductivities of all the samples are nearly equal and follow the T^3 behavior. The dominant phonon wavelength at low temperatures can be very large as suggested by the relation $\lambda_{dom} = hv_s/k_BT$ (see Section 1-3). Hence, the wavelength can be much larger than crystal imperfections such as point defects, dislocations, and grain boundaries. The mean free path is not limited by the defect scattering but by the size of the crystal. Hence, the mean free path at very low temperatures is temperature independent. Phonon velocity is

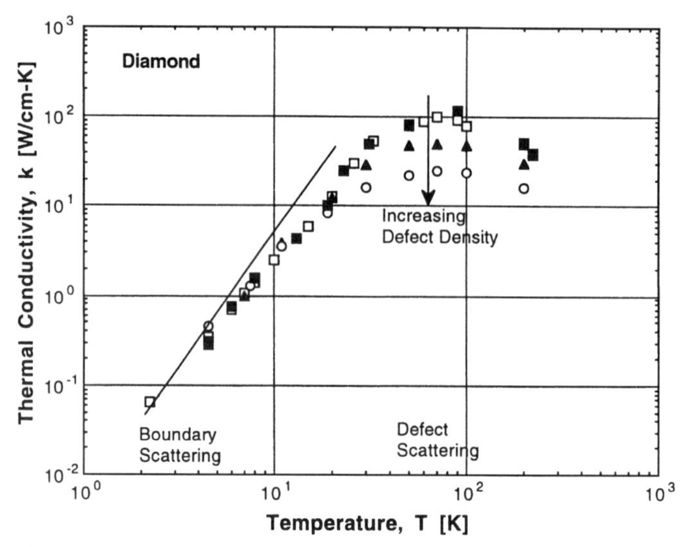

Figure 1-41 Thermal conductivity of diamond as a function of temperature (Berman, 1975). The solid line represents the T^3 behavior.

almost constant and is the speed of sound for long-wavelength acoustic phonons. Note that optical phonons do not contribute to heat conduction since their propagation speed is close to zero. The only temperature dependence comes from the heat capacity, which follows as $C \propto T^3$ at temperatures much lower than the Debye temperature. Hence, the thermal conductivity shows the T^3 behavior. This is often called the *Casimir limit.*

As the temperature is increased, λ_{dom} becomes comparable to the defect sizes, and hence the defect concentration in the crystal determines the thermal conductivity. This is the regime near the peak of the thermal conductivity which occurs when the temperature is on the order of $T \approx \theta_D/10$. In this temperature regime, heat capacity still increases as T^3 as suggested by the data in Fig. 1-16. Although defect scattering is temperature independent, phonon–phonon normal scattering is highly temperature dependent. Although normal scattering does not directly pose any resistance, it distributes the phonon energy to higher frequencies. Because defect scattering is frequency dependent, the effect of N-scattering makes defect scattering somewhat temperature dependent. As the temperature is further increased, the heat capacity becomes a constant and follows the Dulong-Petit law. The phonon density at high frequency and large wavevectors becomes sufficiently high that phonon–phonon Umklapp scattering dominates and determines the phonon mean free path. As the temperature is increased, the mean free path decreases drastically. Hence, the thermal conductivity decreases with increasing temperature. This occurs for temperatures higher than $\theta_D/10$.

Figure 1-42 shows the thermal conductivity for quartz (crystalline SiO_2) and amorphous silica (a-SiO_2) (Cahill and Pohl, 1989). The quartz data follows the T^3 behavior at low temperature, peaks at about 10 K, and then drops with increasing temperature. As discussed before, this is the expected trend for a crystalline solid. However, amorphous silica behaves very differently. The value of the thermal conductivity is much lower than that of the crystalline sample for all values of temperature. In addition, the temperature

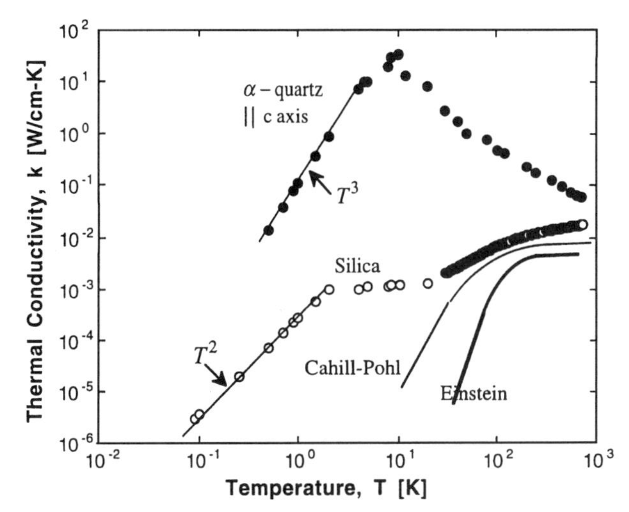

Figure 1-42 Thermal conductivity of quartz and glassy silica as a function of temperature (Cahill and Pohl, 1988). The quartz thermal conductivity exhibits a T^3 behavior at low temperature, a peak at about 10 K, then reduction at higher temperatures. This is typical of a crystalline solid. For amorphous glass the thermal conductivity increases as T^2, plateaus between 1–10 K, and then increases monotonically with temperature. Also plotted are the predictions of the Cahill-Pohl and Einstein models. The Cahill-Pohl model provides accurate predictions for temperatures higher than 50 K but cannot predict the low-temperature behavior. The Einstein model prediction are much lower than the measured values.

dependence of the conductivity is also vastly different. Hence, the model proposed for crystalline solids cannot be applied for such a case. Note that the relation $k = Cv\ell/3$ is still valid although the heat capacity and the mean free path cannot be determined by relations used for crystalline solids.

In 1911, Einstein proposed a model for heat conduction in amorphous solids. In this model, he assumed that all the atoms vibrate as harmonic oscillators at the same frequency ω_E. In addition, he also assumed that a particular oscillator (or atom) is coupled to only first-, second-, and third-nearest neighbors. Hence, the vibrational energy of the oscillator can only be transferred to these atoms. A further assumption was that the phase of these oscillators are uncorrelated and are completely random. Using these assumptions, he derived the thermal conductivity to be

$$k_E = \frac{k_B^2}{\hbar}\frac{\eta^{\frac{1}{3}}}{\pi}\theta_E\frac{x^2e^2}{(e^x-1)^2} \tag{1-123}$$

where η is the number density of oscillators, θ_E is the Einstein temperature, and $x = \theta_E/T$. Unfortunately, the predictions of this theory fall far below those of the measured data as can be observed in Fig. 1-42. The major flaw in this theory is in the assumption that the phase of the neighboring oscillators are uncorrelated. The success of the Debye theory is based on the fact that it considered the coherence of a crystal wave for a distance on the order of a mean free path.

Cahill and Pohl [1988, 1989] recently developed a hybrid model that has the essence of both the localized oscillators of the Einstein model and coherence of the Debye model. In the Cahill-Pohl model, it is assumed that a solid can be divided into localized regions of size $\lambda/2$. These localized regions are assumed to vibrate at frequencies equal to $\omega = 2\pi v_s/\lambda$, where v_s is the speed of sound. Such an assumption is characteristic of the Debye model. The mean free time of each oscillator is assumed to be one half the period of vibration or $\tau = \pi/\omega$. This implies that the mean free path is equal to the size of the region or $\lambda/2$. Using these assumptions, they derived the thermal conductivity to be

$$k_{CP} = \left(\frac{\pi}{6}\right)^{\frac{1}{3}} k_B \eta^{\frac{2}{3}} \sum_i v_{si} \left(\frac{T}{\theta_i}\right)^2 \int_0^{\frac{\theta_i}{T}} \frac{x^2 e^2}{(e^x - 1)^2} dx \qquad (1\text{-}124)$$

where $x = \theta_i/T$ and $\theta_i = v_{si}(\hbar/k_B)(6\pi\eta)^{\frac{1}{3}}$ is a characteristic temperature equivalent to the Debye and Einstein temperatures. The summation is over the two transverse and one longitudinal polarizations for which the speeds of sound can be different. Note that the agreement between the Cahill-Pohl model and the measured thermal conductivity above 50 K is excellent. The validity of the model is further verified by comparing the measured and predicted thermal conductivities of several amorphous solids at 300 K as shown in Table 1-13. Below 50 K, there exists a plateau in the thermal conductivity and a sharp drop at temperatures below 10 K. In the limit $T \ll \theta_i$, the integral in Eq. (124) becomes a constant and the thermal conductivity is expected to vary as $k_{CP} \propto T^2$. Although the trend seems to be correct, the measured values are higher than the predictions by an order of magnitude. This is due to the fact that the Cahill-Pohl model assumes the mean free path to be on the order of the phonon wavelength. At low temperatures ($T < 1$ K), however, the mean free path of the dominant phonons can be larger than the wavelength by a factor of 100.

Although the kinetic theory has been successfully applied to predict the thermal conductivity, it cannot be used under nonequilibrium conditions. For such cases, the Boltzmann transport theory is required.

1-6-3 Boltzmann Transport Theory

General formulation. Local thermodynamic equilibrium in space and time is inherently assumed in the kinetic theory formulation. The length scale that is characteristic of this

Table 1-13 Comparison between the measured thermal conductivity of amorphous solids and the value predicted by the Cahill-Pohl model. Additional data include the number density of atoms, η; the transverse speed of sound, v_{st}; and the longitudinal speed of sound, v_{sl}

Material	η 10^{28} m^{-3}	v_{st} km/s	v_{sl} km/s	Calculated k_{CP} W/m-K	Measured k W/m-K
As_2S_3	3.92	1.44	2.65	0.354	0.246
Ca, KNO_3	6.92	1.73	3.5	0.540	0.490
$CdGeAs_2$	4.11	1.86	3.03	0.431	0.390
Ge	4.41	2.63	4.35	0.621	0.505
Se	3.28	1.06	2.06	0.230	0.140
Si	5.00	4.37	7.36	0.99	1.05
SiO_2	6.63	3.74	5.98	1.04	1.24

volume is ℓ_r, whereas the time scale is τ_r. When either $L \approx \ell, \ell_r$ or $t \approx \tau, \tau_r$ or both, the kinetic theory breaks down because local thermodynamic equilibrium cannot be defined within the system. A more fundamental theory is required. The Boltzmann transport equation is a result of such a theory. Its generality is impressive since macroscopic transport behavior such as Fourier law, Ohm's law, Fick's law, and the hyperbolic heat equation can be derived from this in the macroscale limit. In addition, transport equations such as equation of radiative transfer as well as the set of conservation equations of mass, momentum, and energy can all be derived from the Boltzmann transport equation (BTE). Some of the derivations will be shown here.

In a general form, the BTE can be written as (Ziman, 1960)

$$\frac{\partial f}{\partial t} + \mathbf{v} \cdot \nabla f + \mathbf{F} \cdot \frac{\partial f}{\partial \mathbf{p}} = \left(\frac{\partial f}{\partial t} \right)_{scat} \tag{1-125}$$

where $f(\mathbf{r}, \mathbf{p}, t)$ is the statistical distribution function of an ensemble of particles, which varies with time t, particle position vector $\mathbf{r}$, and momentum vector $\mathbf{p}$. $\mathbf{F}$ is the force applied to the particles. The terms on the left side are called *drift terms*, whereas that on the right is the *scattering term*. The BTE applies to all ensembles of particles—electrons, ions, phonons, photons, gas molecules, etc.—that follow a certain statistical distribution. In the case of electrons under an electric field, for example, the third term on the left side can be written as

$$-\frac{e\mathbf{E}}{\hbar} \cdot \frac{\partial f}{\partial \mathbf{k}} \tag{1-126}$$

where $\mathbf{E}$ is the electric field vector, e is the electron charge, and $\mathbf{k}$ is the electron wave-vector.

The right-hand side of Eq. (1-125) is the rate of change of the distribution due to collisions or scattering. This is the term that restores equilibrium. In its most rigorous form this is very complicated, since collisions transfer particles from one set of $(\mathbf{r}', \mathbf{p}')$ coordinates to another set of $(\mathbf{r}, \mathbf{p})$ coordinates. This can be written as

$$\left(\frac{\partial f}{\partial t} \right)_{scat} = \sum_{\mathbf{p}'} [W(\mathbf{p}, \mathbf{p}') f(\mathbf{p}') - W(\mathbf{p}', \mathbf{p}) f(\mathbf{p})] \tag{1-127}$$

where $W(\mathbf{p}, \mathbf{p}')$ is the scattering rate from state $\mathbf{p}'$ to $\mathbf{p}$. The first term in the summation is from scattering from $\mathbf{p}'$ state to $\mathbf{p}$, and the second term is vice versa. The scattering rates W are often nonlinear functions of $\mathbf{p}$, which makes it quite difficult to solve the BTE. However, a simplification is often made through the relaxation-time approximation:

$$\left(\frac{\partial f}{\partial t} \right)_{scat} = \frac{f_o - f}{\tau(\mathbf{r}, \mathbf{p})} \tag{1-128}$$

where f_o is the equilibrium distribution and $\tau(\mathbf{r}, \mathbf{p})$ is the relaxation time as a function of position and momentum. This approximation linearizes the BTE and implies that if a system is thrown out of equilibrium such that $f - f_o$ is nonzero, then collisions restore equilibrium with the dynamics following an exponential decay $f - f_o \approx \exp(-t/\tau)$. Thus the BTE under the relaxation-time approximation becomes

$$\frac{\partial f}{\partial t} + \mathbf{v} \cdot \nabla f + \mathbf{F} \cdot \frac{\partial f}{\partial \mathbf{p}} = \frac{f_o - f}{\tau(\mathbf{r}, \mathbf{p})}. \tag{1-129}$$

The equilibrium distribution could be of any type: Maxwell-Boltzmann for gas molecules, Fermi-Dirac for electrons, and Bose-Einstein for photons and phonons.

To study energy transport by particles, it is necessary to solve the BTE to determine the distribution function $f(\mathbf{r}, \mathbf{p}, t)$. Once found, the rate of energy flow per unit area or the energy flux can then be written as

$$\mathbf{q}(\mathbf{r}, t) = \sum_{\mathbf{p}} \mathbf{v}(\mathbf{r}, t) f(\mathbf{r}, \mathbf{p}, t) \varepsilon(\mathbf{p}) \tag{1-130a}$$

where $\mathbf{q}(\mathbf{r}, t)$ is the energy flux vector, $\mathbf{v}(\mathbf{r}, t)$ is the velocity vector, and $\varepsilon(\mathbf{p})$ is the particle energy as a function of particle momentum. Note that the units of $f(\mathbf{r}, \mathbf{p}, t)$ is number per unit volume per unit momentum. The summation over momentum space can be changed into an integral over momentum:

$$\mathbf{q}(\mathbf{r}, t) = \int \mathbf{v}(\mathbf{r}, t) f(\mathbf{r}, \mathbf{p}, t) \varepsilon(\mathbf{p}) d^3\mathbf{p}. \tag{1-130b}$$

The integral can be changed to that over energy with the introduction of a density of states $D(\varepsilon)$. The energy flux vector can then be written as

$$\mathbf{q}(\mathbf{r}, t) = \int \mathbf{v}(\mathbf{r}, t) f(\mathbf{r}, \varepsilon, t) \varepsilon D(\varepsilon) d\varepsilon. \tag{1-130c}$$

Fourier and Ohm's laws. Although the solution of the BTE is not trivial, several simplifications can be made. If $t \gg \tau, \tau_r$ is assumed, then the most common simplification is to drop the time-varying term in Eq. (1-127). In addition, if $L \gg \ell, \ell_r$ is assumed, then the gradient term can be approximated as $\nabla f \approx \nabla f_o$ such that in the one-dimensional case, the BTE can be solved to yield

$$f = f_o - \tau v_x \frac{\partial f_o}{\partial x} \tag{1-131}$$

where v_x is the x-component of velocity. This can be called the *quasi-equilibrium* approximation. The only term that contains lack of equilibrium is the scattering term. Local thermodynamic equilibrium is inherently implied by the approximation $df/dx \approx df_o/dx$. However, since the local equilibrium f_o can be defined only over a length scale ℓ_r, the approximation finally boils down to $df/dx \approx \Delta f_o/\ell_r$. This and the time-scale approximations are also made in the kinetic theory, and hence one should expect the same results. Since the equilibrium distribution is a function of temperature, one can express

$$\frac{\partial f_o}{\partial x} = \frac{df_o}{dT}\frac{\partial T}{\partial x}. \tag{1-132}$$

This leads to the energy flux

$$q_x(x) = -\frac{\partial T}{\partial x} \int v_x^2 \tau \frac{df_o}{dT} \varepsilon D(\varepsilon) d\varepsilon. \tag{1-133}$$

The first term containing f_o drops out since the integral over all the directions becomes zero. Equation (1-133) is the Fourier law of heat conduction with the integral being the thermal conductivity, k. If one assumes that the relaxation time and velocity are independent of particle energy, then the integral becomes

$$k = \int v_x^2 \tau \frac{df_o}{dT} \varepsilon D(\varepsilon) d\varepsilon = v_x^2 \tau \int \frac{df_o}{dT} \varepsilon D(\varepsilon) d\varepsilon = \frac{1}{3} C v^2 \tau. \tag{1-134}$$

This is exactly the kinetic theory result $k = Cv\ell/3$. Similar derivations and conclusions can be made for Fick's law.

Ohm's law is characterized by the relation $\mathbf{J} = \sigma \mathbf{E}$, where $\mathbf{J}$ is the current density vector at any point in space, $\mathbf{E}$ is the electric field vector, σ is the electrical conductivity. The Fourier law is the energy analog of the Ohm's law due to the following reasons. The electric field vector, $\mathbf{E}$, can be written as the negative gradient of the electric potential $\mathbf{E} = -\nabla \Phi$ and hence is analogous to the negative gradient of temperature. The energy flux vector, $\mathbf{q}$, is analogous to the current density vector, $\mathbf{J}$, in Ohm's law. Using kinetic theory, it can be shown that the electrical conductivity follows the relation

$$\sigma = \frac{\eta_e e^2 \tau_m}{m} \tag{1-135}$$

where η_e is the density of electrons, e is the electron charge, τ_m is the momentum relaxation time, and m is the electron mass. Using Eq. (1-44) for the heat capacity for electrons, the ratio of thermal conductivity and electrical conductivity can be written as

$$\frac{k}{\sigma T} = \frac{\pi^2}{3} \left(\frac{k_B}{e} \right)^2 = 2.44 \times 10^{-8} \frac{W}{\Omega - K^2} \tag{1-136}$$

where the right-hand side is a constant. This is known as the *Wiedemann-Franz law*, and the constant on the right side is called the *Lorenz number* (Ashcroft and Mermin, 1976). This relation shows the relation between the electrical and thermal conductivity of metals. It must be noted that to derive the Wiedemann-Franz law, it was assumed that the relaxation times for the electrical conductivity and the thermal conductivity were the same. This is not entirely correct; the electrical conductivity is related to the momentum relaxation time, whereas the thermal conductivity is related to the energy relaxation time. However, they are usually close at room temperature or at very low temperatures.

Hyperbolic heat equation. If the Boltzmann transport equation is multiplied by the factor $v_x \varepsilon D(\varepsilon) d\varepsilon$ on both sides and integrated over energy, then the equation transforms into

$$\frac{\partial q_x}{\partial t} + \int v_x^2 \frac{\partial f}{\partial x} \varepsilon D(\varepsilon) d\varepsilon = - \int \frac{f v_x \varepsilon D(\varepsilon) d\varepsilon}{\tau(x, \varepsilon)}. \tag{1-137}$$

The acceleration term is dropped in this equation. Consider the situation that $L \gg \ell$, ℓ_r and $t \approx \tau$, τ_r. Now make the following assumptions: (i) the relaxation time is independent of particle energy and is a constant; and (ii) the quasi-equilibrium assumption is made for the term $\partial f / \partial x = (df_o/dT)(\partial T / \partial x)$. Then Eq. (1-137) becomes

$$\frac{\partial q_x}{\partial t} + \frac{q_x}{\tau} = -\frac{k}{\tau} \frac{\partial T}{\partial x}. \tag{1-138}$$

This is the *Cattaneo equation* which, in combination with the following energy conservation equation,

$$C\frac{\partial T}{\partial t} + \frac{\partial q_x}{\partial x} = 0, \tag{1-139}$$

leads to the hyperbolic heat equation of the form (Joseph and Preziosi, 1989)

$$\tau\frac{\partial^2 T}{\partial t^2} + \frac{\partial T}{\partial t} = \frac{k}{C}\frac{\partial^2 T}{\partial x^2}. \tag{1-140}$$

The solution of Eq. (1-140) is wave-like, suggesting that the temperature field propagates as a wave. The speed of propagation of this wave equal to $\sqrt{k/C\tau}$, which also happens to be the speed of the energy carrier, for example, the speed of sound for phonons. So this model is nonlocal in time but local in space since the temperature represents a spatially localized thermodynamic equilibrium.

It is important to note the assumptions made in deriving the hyperbolic heat equation. The time scales of interest are of the order of the relaxation time, whereas the length scale is much larger than the characteristic size for local thermodynamic equilibrium. The only difference between the derivation of the hyperbolic heat equation and that of the Fourier law is that the transient term $\partial f/\partial t$ is retained for the former. This makes the hyperbolic heat equation nonlocal in time but not in space. The BTE is of course more general and can be used to study nonlocality and nonequilibrium in both space and time (Joshi and Majumdar, 1993).

Mass, momentum, and energy conservation: Hydrodynamic equations. The conservation equations that are encountered in fluid mechanics, heat transfer, and electron transport can be derived as different moments of the BTE. Consider a function, $\phi(\mathbf{p})$, which is a power of the particle momentum $\phi(\mathbf{p}) = \mathbf{p}^n$ where n is an integer $(n = 0, 1, 2, \ldots)$. Its average can be described as

$$\langle\phi(\mathbf{p})\rangle = \frac{1}{\rho}\int \phi(\mathbf{p})f(\mathbf{p})d^3\mathbf{p} \tag{1-141}$$

where ρ is the number density of particles. The BTE is now multiplied by $\phi(\mathbf{p})$ and integrated over momentum. In general form, this gives the *moment equation*

$$\frac{\partial(\rho\langle\phi\rangle)}{\partial t} + \frac{1}{m}\nabla\cdot(\rho\langle\mathbf{p}\phi\rangle) - \rho\mathbf{F}\cdot\left\langle\frac{\partial\phi}{\partial\mathbf{p}}\right\rangle = \rho\sum_{\mathbf{p}'}[\langle W(\mathbf{p},\mathbf{p}')\phi(\mathbf{p}')\rangle - \langle W(\mathbf{p}',\mathbf{p})\phi(\mathbf{p})\rangle]. \tag{1-142}$$

Note that the momentum of each particle can be divided into two components as, $\mathbf{p} = \mathbf{p}_d + \mathbf{p}_r$, where $\mathbf{p}_d$ is the average or drift momentum corresponding to a collective motion of particles in response to an external potential gradient, and $\mathbf{p}_r$ is the random component of the momentum that arises due to thermal motion and is responsible for diffusion. Note that $\langle\mathbf{p}\rangle = \mathbf{p}_d$ since the average of the random component over all the momentum space is zero.

The zeroth moment is when $n = 0$ and $\phi(\mathbf{p})$ is a constant. Using this, one gets the continuity or number conservation equation, which is

$$\frac{\partial\rho}{\partial t} + \nabla\cdot(\rho\mathbf{v}_d) = So - Si \tag{1-143}$$

where $\mathbf{v}_d$ is the drift velocity ($\mathbf{p}_d/m$), So is the source or rate of generation rate of particles and Si is the sink or removal rate of particles. In the case of fluids, there are no sources or sinks, and hence the right side is zero. However, when electrons and holes are considered, the zeroth moment equation can be written for each valley of the electronic structure of a semiconductor or a metal. Intervalley scattering due to electron–electron, electron–hole, electron–photon, or electron–phonon interactions may be responsible for particle exchange between the different valleys and bands. This creates a source or sink in each valley in which case the right side of Eq. (1-143) may be nonzero. However, if all the valleys and bands are considered together, the right side would be zero because charge or mass must be conserved.

The momentum conservation equation is obtained by taking the first moment, $\phi(\mathbf{p}) = \mathbf{p} = m\mathbf{v}$. This yields the following equation:

$$\frac{\partial(\rho\mathbf{p}_d)}{\partial t} + \frac{1}{m}\nabla\cdot(\rho\langle\mathbf{pp}\rangle) - \rho\mathbf{F} = \left(\frac{\partial(\rho\mathbf{p})}{\partial t}\right)_{scat} \tag{1-144}$$

The second term is the average of a tensorial quantity. However, because the average of odd powers of $\mathbf{p}_r$ is zero, we get $\langle\mathbf{pp}\rangle = \mathbf{p}_d\mathbf{p}_d + p_r^2\delta_{ij}$ where δ_{ij} is the unit tensor. The third term of the left side is the what is referred to as a body force term in fluid mechanics. It is perhaps more appropriate to refer to it as a potential gradient term because a thermodynamic force can be written as a gradient of any potential, $\mathbf{F} = -\nabla U$. The potential U is a sum of the gravitational potential G, electrochemical potential, Φ, etc. For electrons in a metal or semiconductor, the force can be due to electric or magnetic fields, which can also be expressed as a gradient of a potential. The right side of Eq. (1-144) is the scattering term. Under the relaxation time approximation, the right side can be assumed to follow

$$\left(\frac{\partial(\rho\mathbf{p})}{\partial t}\right)_{scat} = -\frac{\rho\mathbf{p}}{\tau_m} \tag{1-145}$$

where τ_m is the momentum relaxation time. Therefore, the momentum conservation equation becomes

$$\frac{\partial(\rho\mathbf{p}_d)}{\partial t} + \frac{1}{m}\nabla\cdot(\rho\mathbf{p}_d\mathbf{p}_d) + \frac{1}{m}\nabla\left(\rho p_r^2\right) = -\rho\nabla(G + \Phi + \cdots) - \frac{\rho\mathbf{p}_d}{\tau_m}. \tag{1-146a}$$

The third term on the left side has the form of the kinetic energy of the random particle motion and is representative of the pressure of the particle. Therefore, Eq. (1-146a) can be rewritten in the form

$$\frac{\partial(\rho m\mathbf{v}_d)}{\partial t} + \nabla\cdot(\rho m\mathbf{v}_d\mathbf{v}_d) = -\rho\nabla\left(\frac{P}{\rho} + G + \Phi + \cdots\right) - \frac{\rho m\mathbf{v}_d}{\tau_m} \tag{1-146b}$$

The second term on the left side is often referred to as the advection term. When this negligible, Eq. (1-146b) under zero acceleration reduces to the form

$$\mathbf{v}_d = -\frac{\tau_m}{m}\nabla\left(\frac{P}{\rho} + G + \Phi + \cdots\right). \tag{1-147}$$

In the case of electron transport where $\Phi = eV + k_BT\ln(\rho)$ is the electrochemical potential, one can derive the familiar drift–diffusion equation (Ferry, 1991)

$$\mathbf{J} = \frac{\rho e^2\tau_m}{m}\mathbf{E} + \frac{\rho e\tau_m}{m}\nabla(k_BT\ln\rho). \tag{1-148}$$

Here, the first term is the drift term representing Ohm's law with the electrical conductivity being $\sigma = \rho e^2 \tau_m / m$. The second term is the diffusion term, which gives rise to thermoelectric effects and current flow due to electron concentration gradients. In the case of fluid transport, neglecting the left side of Eq. (1-146b) gives

$$\mathbf{v} = -\frac{\tau_m}{m} \nabla \left(\frac{P}{\rho} \right) \tag{1-149}$$

which is equivalent to the Darcy equation for flows in porous media. It is evident that Eq. (1-144) has the familiar form of the Navier-Stokes equation except for the last term involving collisions. The Navier-Stokes equation can be derived from the BTE using the Chapman-Enskog approximation where the right side of Eq. (1-144) leads to the diffusion term (Vincenti and Kruger, 1977).

The energy conservation equation is obtained from Eq. (1-142) if the second moment is taken $\phi(\mathbf{p}) = p^2$ since energy $\varepsilon = p^2/2m$. This yields the following equation

$$\frac{\partial \xi}{\partial t} + \nabla \cdot \mathbf{J}_\xi = -\rho \mathbf{F} \cdot \mathbf{v}_d + \rho \sum_{\mathbf{p}'} \left[\left\langle W(\mathbf{p}, \mathbf{p}') \frac{p'^2}{2m} \right\rangle - \left\langle W(\mathbf{p}', \mathbf{p}) \frac{p^2}{2m} \right\rangle \right] \tag{1-150}$$

where $\xi = \rho \varepsilon$ is the energy density in J/m^3 and $\mathbf{J}_\xi$ is energy flux vector in W/m^2, which can be expressed in general form as

$$\mathbf{J}_\xi = \mathbf{v}_d \xi + \mathbf{q}. \tag{1-151}$$

The term $\mathbf{v}_d \xi$ is the advection of energy that comes from the drift contribution, and $\mathbf{q}$ is the heat flux vector due to diffusion that arises from the random motion of the particles. This reduces the energy equation to

$$\frac{\partial \xi}{\partial t} + \nabla \cdot (\mathbf{v}_d \xi) = \rho \mathbf{v}_d \cdot \nabla U - \nabla \cdot \mathbf{q} + \left(\frac{\partial \xi}{\partial t} \right)_{So} - \left(\frac{\partial \xi}{\partial t} \right)_{Si} \tag{1-152}$$

where U is the sum of all the potentials discussed earlier. Here, the scattering term from Eq. (1-150) is divided into an energy source and an energy sink term, which will be discussed shortly. The first term on the right side is the work done by a force on the particles and, therefore, must appear in the energy conservation equation. To obtain a relation for $\mathbf{q}$, the next higher moment of the BTE needs to be taken. However, closure is often obtained by assuming the Fourier law $\mathbf{q} = -k \nabla T$. But recalling the fact that Fourier law is derived under the assumption of quasi-equilibrium in both space and time, this may not always be a valid assumption. A higher order relation, which takes into account non-locality in time but quasi-equilibrium in space is the Cattaneo equation for heat flux described in Eq. (1-138).

The energy density ξ of a particle system has contribution from entropic motion as well as from the drift and can be written as

$$\xi = \frac{3}{2} \rho k_B T + \frac{1}{2} \rho m v_d^2. \tag{1-153}$$

Note that the factor 3/2 is valid for particles such as monoatomic gas molecules and electrons, with only three degrees of freedom of motion, each degree possessing an energy of $k_B T/2$. By multiplying the momentum conservation Eq. (1-146) by $\mathbf{v}_d$ and subtracting it out of the energy conservation equation (1-152), the thermal energy conservation equation can be derived as

$$\frac{\partial T}{\partial t} + \mathbf{v}_d \cdot \nabla T + \frac{2}{3} T \nabla \cdot \mathbf{v}_d = \frac{2}{3\rho k_B} \nabla \cdot \mathbf{q} + \left(\frac{\partial T}{\partial t} \right)_{So} - \left(\frac{\partial T}{\partial t} \right)_{Si}. \tag{1-154}$$

Note that the work term $\rho \mathbf{v}_d \cdot \nabla U$ drops out because work increases mechanical energy but does not increase entropy or temperature of a system. Only when this work is dissipated by scattering, the entropy of the system is raised and the temperature increases. The scattering term in Eq. (1-154) can be written as follows (Blotekjaer, 1970)

$$\left(\frac{\partial T}{\partial t} \right)_{So} - \left(\frac{\partial T}{\partial t} \right)_{Si} = -\left(\frac{T - T_o}{\tau_\varepsilon (T, v)} \right) + \frac{m v_d^2}{3 k_B} \left(\frac{2}{\tau_m} - \frac{1}{\tau_\varepsilon} \right) \tag{1-155}$$

where T_o is a reservoir temperature and τ_e is the energy relaxation time. The first term on the right side is simply the energy relaxation term with respect to an equilibrium temperature, T_o. The second term comes from the difference between the momentum and energy relaxation processes. The energy relaxation time is different from the momentum relaxation time because a collision may change the particle momentum but not its energy. Even if both these times were the same, the term would be nonzero. This is the contribution of the kinetic energy of the particle to the temperature rise. Hence, this is the fraction of the work done that is dissipated, resulting in entropy generation and temperature rise. Note that although the work term $\rho \mathbf{v} \cdot \mathbf{F}$ is not present, the term $(2 m v_d^2 / 3 k_B \tau_m)$ in Eq. (1-156) represents the dissipated work that adds thermal energy to the system. If we consider work done on electrons by an external electric field, the electron–phonon interactions eventually dissipate this work and result in energy loss to the phonons. Hence, the reservoir temperature is that of the phonons. The ratio $(\tau_\varepsilon / \tau_m)$ can be called the Prandtl number, *Pr*, of the fluid because the fluid diffusivities are inversely proportional to their respective relaxation times.

Equation of radiative transfer for photons and phonons. Photons and phonons do not follow number conservation as do electrons and molecules. However, they do follow energy conservation. An intensity of photons or phonons can be defined as follows

$$I_k(\mathbf{r}, \mathbf{k}, s, t) = \mathbf{v}(\mathbf{k}, s) f(\mathbf{r}, \mathbf{k}, s, t) \hbar\omega(\mathbf{k}, s) \tag{1-156a}$$

where I_k is the intensity with wavevector $\mathbf{k}$, $\mathbf{v}$ is the velocity at wavevector $\mathbf{k}$, s is the polarization, and $\hbar\omega$ is the energy. The intensity can also be defined in terms of frequency, ω, and angle (θ, ϕ) in polar coordinates corresponding to the direction of vector $\mathbf{k}$ as follows

$$I_\omega(\mathbf{r}, \omega, \theta, \phi, s, t) = \mathbf{v}(\omega, \theta, \phi, s) f(\mathbf{r}, \omega, \theta, \phi, t) \hbar\omega D(\omega, s). \tag{1-156b}$$

If the BTE of Eq. (1-125) is multiplied by the factor $\mathbf{v}(\omega, \theta, \phi, s)\hbar\omega D(\omega, s)$, the following equation is obtained:

$$\frac{\partial I_\omega(\mathbf{r}, \omega, \theta, \phi, s, t)}{\partial t} + \mathbf{v} \cdot \nabla I_\omega(\mathbf{r}, \omega, \theta, \phi, s, t) = \left(\frac{\partial I_\omega}{\partial t} \right)_{scat} \tag{1-157}$$

where

$$\left(\frac{\partial I_\omega}{\partial t}\right)_{scat} = \sum_{(\omega',\theta',\phi',s')} [W(\omega',\theta',\phi',s' \to \omega,\theta,\phi,s)I_{\omega'}(\mathbf{r},\omega',\theta',\phi',s',t)$$
$$- W(\omega,\theta,\phi,s \to \omega',\theta',\phi',s')I_\omega(\mathbf{r},\omega,\theta,\phi,s,t)]$$
$$+ \sum_{(j,\Omega)} [W(j,\Omega \to \omega,\theta,\phi,s)\varepsilon(\mathbf{r},j,\Omega,t)$$
$$- W(\omega,\theta,\phi,s \to j,\Omega)I_\omega(\mathbf{r},\omega,\theta,\phi,s,t)]. \quad (1\text{-}158)$$

Here each W is a scattering rate. It is evident that the scattering term is quite complicated and needs explanation.

Equation (1-157) is the conservation of energy based on the intensity at frequency ω, polarization s, and direction (θ, ϕ). Consider now the first summation in Eq. (1-158). This increases the intensity $I_\omega(\mathbf{r},\omega,\theta,\phi,s,t)$ due to scattering in frequency $\omega' \to \omega$, polarization $s' \to s$, and direction $\theta', \phi' \to \theta, \phi$. The second term is the loss of intensity $I_\omega(\mathbf{r},\omega,\theta,\phi,s,t)$ due to scattering to other frequencies, polarizations, and directions. Note, however, that if photons are considered, then this term represents photon–photon scattering and not between photons and other particles. So this term accounts for scattering among the particle type, either photon–photon or phonon–phonon, respectively. This is often known as the in-scattering term in photon radiative transfer, although scattering usually is considered only in direction $\theta', \phi' \to \theta, \phi$ and not in frequency and polarization. This is because inelastic photon scattering is normally ignored in engineering calculations unless processes such as Raman scattering are involved. For phonon radiative transfer, however, inelastic scattering such as normal and Umklapp processes are very common and must be accounted for in this term. In addition, such phonon–phonon scattering is often between different phonon polarizations—LO, TO, LA, and TA—as allowed by phonon energy and momentum conservation during the collision.

The second summation term in Eq. (1-158) is for increase or decrease in intensity $I_\omega(\mathbf{r},\omega,\theta,\phi,s,t)$ due to interactions with other particles. The particle type is given a tag j, and the phase space defined by momentum and direction is given a tag Ω. For example, an energetic electron in a metal or in the conduction band of a semiconductor can drop in energy by emitting a phonon of a certain polarization (e.g., LO-phonon) due to electron–phonon interactions. Here, the electron is given a tag j and the phonon is given a tag Ω. The frequency, direction and polarization or this phonon is decided by energy and momentum conservation of the scattering process. In photon radiative transfer, this term is often referred to as the blackbody source term. This is true for the particular case of blackbody radiation. However, in a device such as a semiconductor laser or a light-emitting diode, photons are not emitted in a blackbody spectrum but within a certain spectral band that is decided by the semiconductor electronic band structure. Hence, this term is kept as a general emission term in Eq. (1-158). Similarly, there is a loss term when a phonon or a photon is absorbed by another particle and removed from the system.

It is clear that in the most general form as described in Eq. (1-158), the scattering terms pose difficulty for solving. Therefore, the relaxation-time approximation is usually used for convenience in which case the equation of radiative transfer reduces to

$$\frac{\partial I_\omega}{\partial t}+\mathbf{v}\cdot\nabla I_\omega = -\frac{I_\omega}{\tau_s}-\frac{I_\omega}{\tau_a}+\frac{\varepsilon(j,\Omega)}{\tau_e}+\iint_{\omega',\Theta'}\frac{I_{\omega'}(\mathbf{r},\omega',\Theta',t)}{\tau_s}d\omega'\frac{d\Theta'}{4\pi}. \tag{1-159}$$

The first term on the right side is the out-scattering term, with τ_s being the scattering relaxation time; the second term is the photon/phonon absorption (or transfer of energy to other particles such as electrons, or photons to phonons, or phonons to photons, etc.), where τ_a is the absorption time; and the third term is the emission term, with $1/\tau_e$ being the emission rate. Here, energy from other particles is converted and contributed to the intensity I_ω. The last term is the in-scattering term from other frequencies and solid angles, Θ'. In an even simpler form, the equation of radiative transfer can be written as

$$\frac{\partial I_\omega}{\partial t}+\mathbf{v}\cdot\nabla I_\omega = \frac{I_\omega^o - I_\omega}{\tau_s}-\frac{I_\omega}{\tau_a}+\frac{\varepsilon(j,\Omega)}{\tau_e} \tag{1-160}$$

where the in-scattering term is totally ignored but it is assumed that $\omega' \rightarrow \omega$ scattering restores equilibrium that is represented by I_ω^o. This is often the assumption made in phonon radiative transfer, where interfrequency scattering does restore phonon equilibrium.

The equation of radiative transfer will not be solved here since solutions to some approximations of the equation are well known. In photon radiation, it has served as the framework for photon radiative transfer. It is well known (Siegel and Howell, 1992) that in the optically thin or ballistic photon limit, one gets the heat flux as $q = \sigma(T_1^4 - T_2^4)$ from this equation for radiation between two black surfaces. For the case of phonons, this is known as the Casimir limit. In the optically thick or diffusive limit, the equation reduces to $q = -k_p\nabla T$, where k_p is the photon thermal conductivity. The same results can be derived for phonon radiative transfer (Joshi and Majumdar, 1993; Majumdar, 1993).

1-7 NONEQUILIBRIUM ENERGY TRANSFER

The discussion in Section 1-6 concentrated on transport by a single carrier, that is, heat conduction by electrons or phonons, charge transport by electrons, and energy transport by photons. Relatively little attention was paid to energy transfer processes between the energy carriers. For example, Joule heating occurs due to electron–phonon interactions, whereas radiative heating involves photon–electron and photon–phonon interactions. These are examples of what are commonly called heat generation mechanisms.

Traditionally, it is assumed that electrons and phonons within a solid are under local equilibrium such that a heat generation term can be added to the energy conservation equation. For example, Joule heating during electron transport is usually modeled as I^2R, where I is the current and R is the electrical resistance. Such a term is added to the energy conservation equation for the whole solid. Such an equation uses a single temperature T to describe the solid at a point $\mathbf{r}$ and time t. It inherently assumes that there is equilibrium between the electrons and the phonons. However, this is not quite the picture in many cases. The equilibrium between electrons and phonons can be disrupted by several processes. For example, in the presence of a sufficiently high electric field, electrons can be energized and thrown far out of equilibrium from the phonons. Such nonequilibrium conditions can now be achieved in contemporary electronic devices with submicrometer feature sizes. In case of radiative heating in a metal, for example, the electrons can be thrown out of equilibrium from the lattice due to excitations by ultra-short laser pulses that are on the order of 100 fs. Such lasers are now available and are widely used in physics, chemistry, and materials processing. Therefore, it is clear that when modern engineering systems involving transport phenomena become small and fast, the energy dissipation process required by the second law of thermodynamics can take a highly nonequilibrium path. In this section, a close look is taken at microscopic mechanisms of heat generation and dissipation and models are presented to analyze such problems.

1-7-1 Joule Heating in High-Field Electronic Devices

Some simple devices. One of the major goals of the electronics industry is to increase the density of devices on a single chip by reducing the minimum size of features. This has two purposes: (i) to miniaturize and increase the functionality of a single chip, and (ii) to increase the speed of logic operations. By the year 2001, the minimum feature size will reduce to 0.18 μm and the speed and power density will increase significantly. A single chip in the future is likely to contain both power and logic devices. This will lead to high temperature and temperature gradients within a chip. New materials choices based on electrical characteristics also influence the thermal problem. For example, the close proximity of transistors on a high-density silicon (Si) chip requires the use of dielectric material such as silicon dioxide (SiO_2) for electrical insulation between devices. Since the thermal conductivity of SiO_2 is about 100 times lower than Si, it leads to high temperatures and temperature gradients.

Figure 1-43 shows a schematic diagrams of a metal-oxide-semiconductor field-effect transistor (MOSFET) and a metal-semiconductor field-effect transistor (MESFET) (Sze, 1981). The MOSFET is usually made of silicon (Si) and is the workhorse of all logic devices and microprocessors. MESFETs are usually made of III-V materials such as GaAs and are usually used in high-speed communication devices such as microwave receivers and transmitters. GaAs is preferable for such devices since the electron mobility is higher than in Si. This is due because the effective mass of the electron in the conduction band is lower in GaAs. The current-voltage (I-V) characteristics of these devices are also shown. The voltage bias on the gate opens and closes the gate and in effect controls the resistance between the source and the drain. So the drain current is a strong function of the gate voltage. Other high-electron-mobility transistors also operate in a similar fashion except that the electron channel under the gate has different configurations due to clever control and manipulation of material interfaces and properties.

Most of the potential drop between the drain and the source occurs across the gate. So the characteristic electric field in a device is on the order of V_{ds}/L_g where V_{ds} is the drain-to-source voltage and L_g is the gate length. When a voltage bias of about 2 V is applied across a device with a minimum feature size of 0.2 μm, extremely high electric fields (about 10^7 V/m) are generated. The dynamics of an electron can be expressed as $m^*\dot{\mathbf{v}} = -e\mathbf{E}$, where m^* is the effective electron mass, $\dot{\mathbf{v}}$ is the acceleration vector, and $\mathbf{E}$ is the electric field vector. The electron velocity gained between two collisions is equal to $eE\tau/m^*$, where τ is the average time between collisions. When the electric field is very

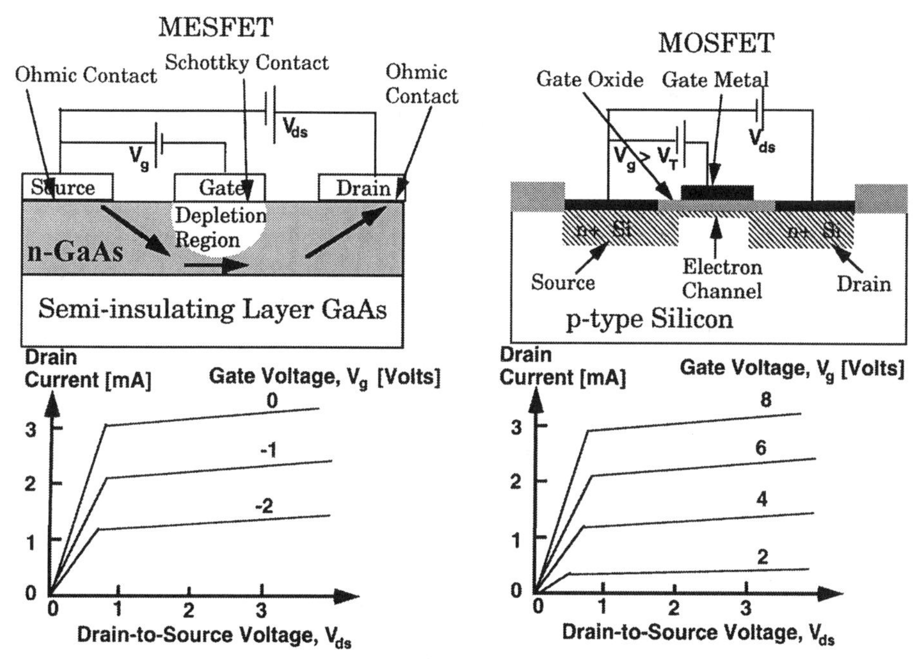

Figure 1-43 Schematic diagrams of a metal-semiconductor field-effect transistor (MESFET) and a metal-oxide-semiconductor field effect transistor (MOSFET). These transistors contain a negative electrode called source, a positive electrode called drain, and the gate in between. The voltage bias on the gate opens and closes the electron channel under the gate. In a MESFET, the metal-semiconductor Schottky contact depletes the region of electrons and thereby forms a narrow channel under the gate. A negative bias on the gate increases the depletion region and reduces the channel size and thereby increases the resistance and reduces the drain current. In a n-channel MOSFET, a positive bias on the gate opens the channel more and increases the current. The devices are usually operated at drain-to-source bias in the 3-5 volt range such that the drain current is in the saturated or flat region of the I-V curves.

high, the velocity and the energy of electron also become very high. Such "hot" electrons are thrown far out of equilibrium with the lattice vibrations. However, the hot electrons collide with the lattice and at some of these collisions, the electron energy is transferred to the lattice to produce phonons. The hot electrons do not always follow Ohm's law, and hence their transport must be studied by the Boltzmann transport equation.

Energy transfer processes. Heat generation occurs by transfer of energy from electrons to phonons. Since Si has two atoms per unit cell, two vibrational modes are present: optical mode and acoustic mode. Similar is the case for GaAs and other III-V materials. As we have seen before, optical phonon energies are higher than those of acoustic phonons. Although electrons interact with both type of phonons, the interactions with optical phonons are restricted to conditions when the electron energy gained from the electric field is higher than the optical phonon energy. So there exists a critical field beyond which electron–optical phonon interactions can occur. In GaAs for instance, the atomic bond is slightly polar and the LO-mode of vibration results in an oscillating dipole which strongly scatters electrons. Hence, electron–LO phonon interaction determines the critical field.

It is important to determine the critical electric field that is necessary to energize electrons sufficiently high so that they emit optical phonons. The energy gained by an electron from the electron field between two collisions is $eEv\tau$. Equating this to the phonon energy $\hbar\omega_{LO}$, we find that the critical field required to emit an optical phonon is

$$E_{cr} = \frac{\hbar\omega_{LO}}{ev\tau} \tag{1-161}$$

LO-phonons in GaAs occur at a single frequency ω_{LO}, and the LO-phonon energy is $\hbar\omega_{LO} = 37$ meV. Femtosecond optical spectroscopy (Elsasser et al., 1991) has revealed that the typical time scale for electron–LO phonon interaction $\tau_{e-LO} \approx 0.1$ ps in GaAs. Therefore, each collision results in transfer of 37 meV between electrons and phonons. Using an electron velocity of 10^5 m/s, the critical electric field in GaAs can be calculated to be about 10^6 V/m.

Figure 1-44 shows the plot of the drift velocity of electrons v_d as a function of the electric field. The drift velocity is related to the current density as $\mathbf{v_d} = \mathbf{J}/e\eta_e$. Under Ohm's law, the drift velocity is related to the electric field as

$$\mathbf{v_d} = \mu\mathbf{E} \quad \text{where } \mu = \frac{e\tau_m}{m^*} \tag{1-162}$$

where μ is called the *electron mobility* and m^* is the *electron effective mass*. In Fig. 1-44, one can clearly see that at low electric fields the electron transport follows Ohmic behavior. Beyond 5×10^5 V/m, the drift velocity saturates and then eventually starts to decrease. The saturation is due to the optical phonon emission. This is because when the electron energy is higher than the LO–phonon energy, the electrons have a very efficient method of transferring energy to the lattice and therefore do not accelerate. The decrease

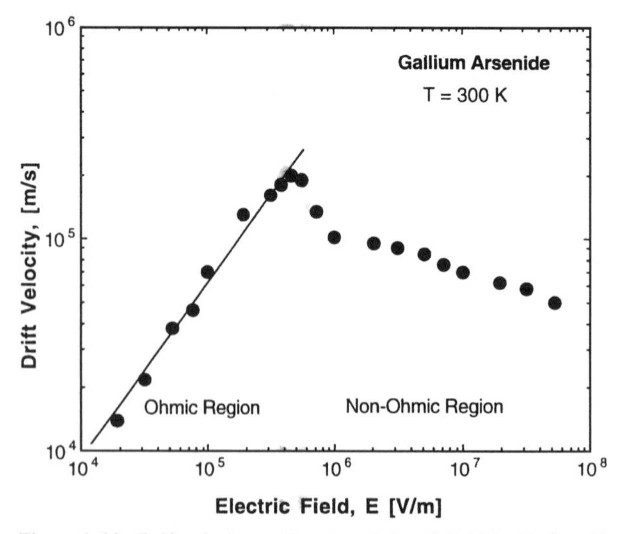

Figure 1-44 Drift velocity as a function of electric field for GaAs at $T = 300$ K (von der Linde et al., 1980). At fields below the critical field, the drift velocity increases linearly with the field, exhibiting Ohmic behavior. For fields higher than the critical field, the drift velocity decreases with electric field due to optical phonon emission and transport through the L and the X valleys where the electron effective mass is higher.

of the drift velocity arises due to intervalley scattering whereby electrons are scattered from the Γ valley to the L and the X valleys. The electron effective mass in these satellite valleys is higher, which results in lower electron mobility.

In Si, the optical phonon energy is about 64 meV and the electron–LO phonon relaxation time is about 0.3 ps. Here the drift velocity does not decrease beyond the critical field since the satellite valleys are separated from the lowest valley by a large energy barrier. However, the saturation occurs due to optical phonon emission.

It is clear that in state-of-the-art submicrometer devices with fields on the order of 10^7 V/m, optical phonons will be generated. Although optical phonons interact with hot electrons, their group velocity is very small and hence they do not conduct any heat. So they eventually decay into acoustic phonons that conduct heat through the device and throughout the package. Therefore, although LO-phonons gain energy from electrons, they must transfer it to acoustic phonons for heat conduction in the solid. Such an energy transfer occurs during scattering of LO-phonons and acoustic phonons, which has a characteristic time scale of $\tau_{LO-A} \approx$ 6–10 ps in GaAs (Ferry, 1991; von der Linde et al., 1980) and about 10 ps in Si (Ferry, 1991). Thus, the electron–LO phonon time scale $\tau_{e-LO} \approx 0.1$ ps is two orders of magnitude faster than τ_{LO-A}. The time scales of electron–phonon and phonon–phonon interactions can be quite different, giving rise to interesting dynamics. Figure 1-45 shows a schematic diagram of the nonequilibrium Joule heating process.

The effect of device temperature on the electrical behavior of the device occurs due to the lattice temperature dependence of the electron scattering rate. When the

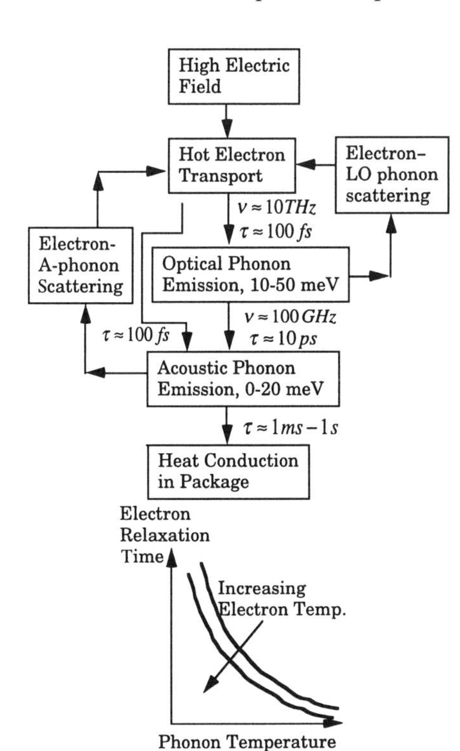

Figure 1-45 A flow chart showing the energy transfer mechanisms during Joule heating in high-electric-field electronic devices. Note that optical phonons will be emitted only when the electric field is higher than the critical field. Otherwise, hot electrons will directly emit acoustic phonons. The number of phonons (or the phonon temperature) influences the electron scattering rate, which in turn changes the device electrical characteristics. The electron scattering rate depends on both electron and phonon temperatures and follows the qualitative trend shown above. The flow chart also shows the typical time scales involved in each process and the energies of phonons.

LO-phonon and acoustic phonon temperatures rise, the electron scattering rate increases, thus increasing the electrical resistance or decreasing the carrier mobility. The coupling of electrical and thermal characteristics suggest that these must be analyzed concurrently.

Governing equations. If the problem is to be solved rigorously, the BTE must be solved for electrons in each valley, optical phonons, and acoustic phonons. The distribution function of each of these depend on six variables, three space variables and three momentum (or energy) variables. The solution to BTE for this complexity becomes very computer intensive, especially due to the fact that the time scales of electron–phonon and phonon–phonon interactions vary by two orders of magnitude. Monte Carlo simulations are sometimes used, although this too is very time consuming. Therefore, researchers have resorted mainly to hydrodynamic equations for modeling electron and phonon transport for practical device simulation.

Based on the mechanism of nonequilibrium Joule heating, the governing equations for charge and energy transport, as derived in Section 1-5, are

$$\nabla^2 V = -\frac{e}{\varepsilon_s}(N_D - \rho); \quad \mathbf{E} = -\nabla V \tag{1-163}$$

$$\frac{\partial \rho}{\partial t} + \nabla \cdot (\rho \mathbf{v}) = 0 \tag{1-164}$$

$$\rho m^* \left(\frac{\partial \mathbf{v}}{\partial t} + \mathbf{v} \cdot \nabla \mathbf{v} \right) = -e\rho \mathbf{E} - \nabla(\rho k_B T_e) - \frac{\rho m^* \mathbf{v}}{\tau_m} \tag{1-165}$$

$$\frac{\partial T_e}{\partial t} + \nabla \cdot (\mathbf{v} T_e) = \frac{1}{3} T_e \nabla \cdot \mathbf{v} + \frac{2}{3nk_B} \nabla \cdot (k_e \nabla T_e) - \frac{T_e - T_{LO}}{\tau_{e-LO}} - \frac{T_e - T_A}{\tau_{e-A}} + \frac{m^* v^2}{3k_B} \left(\frac{3}{\tau_m} - \frac{1}{\tau_{e-LO}} - \frac{1}{\tau_{e-A}} \right) \tag{1-166}$$

$$C_{LO} \frac{\partial T_{LO}}{\partial t} = \frac{3\rho k_B}{2} \left(\frac{T_e - T_{LO}}{\tau_{e-LO}} \right) + \frac{\rho m^* v^2}{2\tau_{e-LO}} - C_{LO} \left(\frac{T_{LO} - T_A}{\tau_{LO-A}} \right) \tag{1-167}$$

$$C_A \frac{\partial T_A}{\partial t} = \nabla \cdot (k_A \nabla T_A) + C_{LO} \left(\frac{T_{LO} - T_A}{\tau_{LO-A}} \right) + \frac{3\rho k_B}{2} \left(\frac{T_e - T_A}{\tau_{e-A}} \right). \tag{1-168}$$

The derivation of these equations is described in detail in Blotekjaer (1970), Fushinobu et al. (1995), Lai and Majumdar (1996), Majumdar et al. (1995), and references therein. Equation (1-163) is the Poisson equation, which satisfies Gauss's law. Here V is the potential, ε_s is the dielectric constant of the medium, N_D is the doping concentration, ρ is the electron concentration, and $\mathbf{E}$ is the electric field vector. Equations (1-165) and (1-166) are the electron momentum and energy conservation equations, which follow the development in Section 1-6. The momentum equation is quite similar to the Navier-Stokes equation of fluid mechanics. By nondimensionalizing this equation, Lai and Majumdar (1996) derived an equivalent electron Reynolds number in terms of device parameters: $Re = eV_{ds}\tau_m^2/(m^* L_g^2)$. For most operating conditions of V_{ds} and L_g and values of m^* and τ_m, Re $\ll 1$, and so the nonlinear convective term can be neglected. In Eq. (1-166), the Fourier law of heat conduction has been assumed for the heat flux, where k_e is the electron thermal conductivity, T_{LO} is the optical phonon temperature, and τ_{e-LO} is the electron energy relaxation time for electron–optical phonon scattering. The right side of Eq. (1-166) contains loss of electron energy to optical phonons and to acoustic phonons, as well as a heat generation term that comes from dissipation of the kinetic energy gained by the electrons. Equation (1-167) represents energy conservation for optical phonons, where C_{LO} is the optical phonon heat capacity. The terms on the right side represent energy gain from electrons and loss to acoustic phonons. Note that there is no heat diffusion term due to negligible group velocity of optical phonons. Equation (1-168) represents energy conservation of acoustic phonons, where C_A is the acoustic phonon heat capacity and T_A is the acoustic phonon temperature, and the right side contains the heat diffusion term (first term on the right side), the term representing energy gain from the optical phonons, as well as that representing energy gain directly from electrons. Note that all the relaxation times used in the energy equations must be the energy relaxation times that involve inelastic scattering.

The numerical solution of these equations for the case of microelectronic devices have been developed and can be found elsewhere (Fushinobu et al., 1995; Lai and Majumdar, 1996; Majumdar et al., 1995). It was found that under steady-state conditions that are realistic for modern devices, the electron temperature is an order of magnitude higher than those of the phonons. This shows the nonequilibrium nature of the Joule heating process. However, the difference between the optical and acoustic phonon temperatures is comparatively much smaller. Hence, under steady-state conditions, it is sufficient to group the optical and acoustic phonon temperatures into one equation called the lattice energy conservation equation. However, it is important to note that this is not valid in high-speed devices. The phonon–phonon scattering time is about 10 ps in most semiconductors at room temperature. This corresponds to a frequency of 100 GHz, which is in the microwave region of the electromagnetic spectrum and is used in wireless communications. Devices that are operated in this regime undergo unsteady heating with a period comparable to the phonon–phonon scattering time. Since electron–optical phonon scattering time is on the order of 100 fs, the electron energy can be efficiently transferred to the optical phonons. However, optical phonons cannot transfer their energy fast enough to the acoustic phonons. Hence, a "phonon bottleneck" is encountered where the optical phonon temperature becomes much higher than the acoustic phonon temperature. This changes the electron scattering rate and thereby changes the electrical performance of the device.

1-7-2 Radiative Heating by Ultra-Short Laser Pulses

Pulsed lasers with pulse widths less than 100 fs and as low as 10 fs are now commercially available. It is important to compare the time scale with some characteristic time scales involved in microscale energy transfer processes. As we have noted earlier, the electron–phonon scattering time scale is on the order of 100 fs in most metals and semiconductors at room temperature. This is also the shortest time it takes for the ions in the solid to undergo one vibrational period. In other words, the highest phonon frequency is about 10^{13} Hz, which corresponds to 100 fs in time period. Such laser pulse widths are much shorter than the phonon–phonon scattering time at room temperature which is on the order of 10 ps.

If we calculate the period of oscillation of an electron wavefunction (i.e., $\tau = \lambda/v$), then using λ of 10 Å and v of 10^6 m/s for a Fermi electron, we get the period to be 1 fs. The period of oscillation for a photon in the visible spectrum is also about 1 fs. Therefore, we see that laser pulses of 10 fs contain at least 10 electromagnetic waves and are longer than the time period for wavefunction of Fermi electrons. Hence, such laser pulses can excite electrons within the duration of a pulse if there are no phonons involved, that is, if the excitation is a direct process. Such pulsed lasers are now widely used to study the electron dynamics in chemical bond formation and dissociation, to investigate carrier relaxation processes in metals and semiconductors, as well as for materials processing with extremely high depth resolution. It is instructive to go through the microscopic energy transfer processes before the governing equations are derived.

Energy transfer processes. The photon energies of such pulsed lasers range from 1–5 eV. Using $k_B T$ as the thermal energy of the electrons, we can see that this corresponds to a temperature level of about 10^4 K. Hence, the electrons that absorb these photons and undergo a direct transition in the Brillouin zone are extremely hot for a very short period of time. Note that the electron–electron scattering rate is typically on the order of 10–50 fs in metals. In semiconductors, this rate depends on the electron concentration in the conduction band but is also typically in this range. Hence, the distribution of electrons is initially highly nonequilibrium with a large number of electrons at the excited state of 1–5 eV due to photon absorption. However, the energy gets quickly redistributed between the electrons such that by about 100 fs the electrons reach the equilibrium Fermi-Dirac distribution. Note, however, that since the electrons do not lose energy to the phonons within this time period, they remain quite energetic. But since equilibrium is established between the electrons, one can define an electron temperature, T_e. During this time, the phonons remain at the original ambient temperature, T_o. So as far as the phonons are concerned, they encounter a hot electron reservoir at T_e. Typically, this is about 10^3–10^4 K. The temperature difference, T_e–T_o, drives energy flow from the electron to the phonons.

The most efficient way for the electrons to lose energy is through emission of optical phonons since their wavevectors span the entire Brillouin zone at the same energy. The electron energy, however, must be larger than the optical phonon energy, which is typically in the 10–50 meV range. In addition, since optical phonon energies are higher than those of acoustic phonons, optical phonon emission is a faster and more efficient way of energy transfer. When polar interactions are involved, such as in polar bonds in GaAs or other III-V or II-VI materials, the coupling to electrons is even stronger. In such cases, LO-phonons are most efficiently emitted by hot electrons. Optical phonons eventually scatter and emit acoustic phonons, which are responsible for lattice heat conduction.

Governing equations. To study rigorously the dynamics and interactions of photons, electrons, and phonons from a particle transport viewpoint, the Boltzmann transport equation for all these individual systems must be solved simultaneously. This is a very challenging task that has not been adopted by most investigators. What people have used are the averaged moments of the Boltzmann equation, which produce the conservation equations (Fushinobu et al., 1996; Qiu and Tien, 1993; van Driel, 1987). It must be noted, however, that the moments always contain less information than the Boltzmann transport equation, and so invariably an assumption needs to be made at some stage.

This is normally called the closure problem. However, these assumptions have been made and they seem to work reasonably well for the cases studied in the past.

The energy conservation equations are as follows:

$$\frac{\partial U_c}{\partial t} = \nabla \cdot (k_c \nabla T_c) - \frac{3\rho k_B}{2}\left(\frac{T_c - T_O}{\tau_{c-O}}\right) - \frac{3\rho k_B}{2}\left(\frac{T_c - T_A}{\tau_{c-A}}\right) + S \quad (1\text{-}169)$$

$$\frac{\partial U_O}{\partial t} = \frac{3\rho k_B}{2}\left(\frac{T_c - T_O}{\tau_{c-O}}\right) - C_O\left(\frac{T_O - T_A}{\tau_{O-A}}\right) \quad (1\text{-}170)$$

$$\frac{\partial U_A}{\partial t} = \nabla \cdot (k_A \nabla T_A) + \frac{3\rho k_B}{2}\left(\frac{T_c - T_A}{\tau_{c-A}}\right) + C_O\left(\frac{T_O - T_A}{\tau_{O-A}}\right). \quad (1\text{-}171)$$

Equation (1-169) is the carrier energy conservation equation, which involves heat conduction by carriers, energy loss to optical phonons that occur at a time scale of τ_{c-O}, energy loss to acoustic phonons at time scale τ_{c-A}, and a source term containing energy increase due to photon absorption. Note that the assumption of Fourier law is not really valid for short time scale studies, but this is an approximation that invokes closure in these hydrodynamic equations. The second equation conserves energy for the optical phonons, where the two terms on the right side represent energy gain from electrons and energy loss to acoustic phonons. The third equation is that for acoustic phonons, where again the Fourier law is used for heat flux. It should be noted that in most of the previous work (Fushinobu et al., 1996; Qiu and Tien, 1993; van Driel, 1987), the lattice has been assumed to be a single thermodynamic system. This implicitly assumes equilibrium between optical and acoustic phonons. However, since the time scales for electron–optical phonon (100 fs) and phonon–phonon (10 ps) interactions are different by two orders of magnitude, it is perhaps not a reasonable assumption. Hence, they have been considered as separate systems in this chapter.

The energy densities in the three systems are

$$U_c = \rho E_g + C_c T_c; \quad U_O = C_O T_O; \quad U_A = C_A T_A \quad (1\text{-}172)$$

where ρE_g is the energy density the electrons gain for excitation across the band gap E_g, and ρ is the density of the electrons. The source term can be written as

$$S = \frac{\alpha(T_c, \rho) 2J\sqrt{\ln 2}}{t_p\sqrt{\pi}}[1 - R(T_{cs}, \rho_s)]\exp\left(-\int_0^y \alpha dz\right)\exp\left[-4\ln 2\left(\frac{t}{t_p}\right)^2\right] \quad (1\text{-}173)$$

where the absorption coefficient $\alpha(T_c, \rho)$ is a function of local carrier temperature and density, J is the laser fluence (J/m^2), t_p is the laser pulse width, and the surface reflectivity R is a function of the surface carrier temperature and density. The carrier density follows the continuity equation

$$\frac{\partial \rho}{\partial t} + \nabla \cdot (\rho \mathbf{v}) = \frac{S_1}{\hbar\omega} - \gamma\rho^3 \quad (1\text{-}174)$$

where S_1 is the absorption source term that corresponds to direct interband transition that excites an electron to the conduction band and creates an electron-hole pair. Here,

$\hbar\omega$ is the photon energy, $\mathbf{v}$ is the carrier velocity, and the last term represents the Auger recombination term. The carrier momentum equation can be written as

$$\rho m^* \left(\frac{\partial \mathbf{v}}{\partial t} \right) = -\nabla(\rho k_B T_e) - \frac{\rho m^* \mathbf{v}}{\tau_m} \tag{1-175}$$

where the last term contains the momentum relaxation time, τ_m.

It is quite evident that the set of conservation equations is nonlinear and highly coupled. They are usually solved numerically, the process of which will not be covered here but can be found in the literature.

1-8 SUMMARY

This chapter provides an introduction to the physics of the three energy carriers—phonons, electrons, and photons—as well as their interactions and transport in solids. The chapter starts by discussing the different types of naturally occurring microstructures such as point defects, dislocations, grain boundaries, as well as fabricated microstructures that include thin films, wires, and quantum dots. The chapter then focuses on the physics of crystal vibrations in solids, introduces the concept of phonons, and develops the theory for heat capacity while discussing the different phonon scattering mechanisms. Next, the chapter looks at the origin of electronic band structure both from chemical bonding and solid-state physics interaction points of view. It develops the free electron theory and discusses the band of common semiconductors. Finally, it shows how to calculate electron concentrations in semiconductors as a function of doping and describes the different electron scattering mechanisms while providing the scattering rates. The chapter then introduces photon interactions in solids, first from a classical microscopic point of view, then from a quantum mechanical approach. The interactions of photons with electrons and phonons in a solid are described in detail. This section also discusses the phenomena of photon emission, which plays a critical role in light emitting devices.

After describing the physics of the three energy carriers, the chapter then studies microscopic transport phenomena in solids. First, the time and length scales that are characteristic of the transport process are identified. Based on these scales, the regimes where continuum or macroscopic theories break down are explained. Having done so, the regimes where microscopic transport phenomena occur in both the wave and particle regimes are also indicated. The monograph does not address wave transport issues such as diffraction, refraction, interference, or tunneling, which occur when the size of an object is on the order of the wavelength of an energy carrier or if the phase information of an energy carrier becomes significant during scattering. Hence, this chapter concentrates only on microscopic particle transport theories.

Kinetic theory is introduced and developed as the initial step toward understanding more complex microscopic transport phenomena. It is used to develop relations for the thermal conductivity which are compared to experimental measurements for a variety of solids. Next, it is shown that if the time or length scale of the phenomena are on the order of those for scattering, kinetic theory cannot be used but instead Boltzmann transport theory should be used. It shows that the Boltzmann transport equation (BTE) is fundamental since it forms the basis for a vast variety of transport laws such as the Fourier law of heat conduction, Ohm's law of electrical conduction and hyperbolic heat conduction equation. In addition, for the ensemble of particles for which the particle number is conserved, such as in molecules, electrons, and holes, the BTE forms the basis for mass, momentum, and energy conservation equations. In cases where particle numbers are not conserved, such as in photons and phonons, the BTE forms the basis for the equation of transfer that span the ballistic to the diffusive transport regimes.

After showing how the BTE forms the fundamental basis for particle transport theories, the monograph describes two case studies—(1) non-equilibrium Joule heating in sub-micrometer transistors; (2) non-equilibrium radiative heating by ultra-short laser pulses. Through these case studies, mechanisms and theories of non-equilibrium energy transfer are introduced.

REFERENCES

Ashcroft, N. W., and N. D. Mermin, *Solid State Physics*, W. B. Saunders, Orlando, FL, 1976.

Berman, R., *Thermal Conduction in Solids*, Oxford University Press, Oxford, 1976.

Berman, R., P. R. W. Hudson, and M. Martinez, "Nitrogen in diamond: Evidence from thermal conductivity," *J. Phys. C*, vol. 9, pp. L430–L434, 1975.

Blotekjaer, K., "Transport equations for electrons in two-valley semiconductors," *IEEE Trans. Electron Dev.*, vol. 17, pp. 38–47, 1970.

Bohren, C. F., and D. R. Huffman, *Absorption and Scattering of Light by Small Particles*, John Wiley, New York, 1983.

Broers, A., "Fabrication limits of electron beam lithography and of UV, X-ray and ion-beam lithographies," *Phil. Trans. A—Phys. Sci. Eng.*, vol. 353, pp. 291–311, 1995.

Cahill, D. G., and R. O. Pohl, "Lattice vibrations and heat transport in crystals and glasses," *Annu. Rev. Phys. Chem.*, vol. 39, pp. 93–121, 1988.

Cahill, D., and R. O. Pohl, "Heat flow and lattice vibrations in glasses," *Solid State Commum.*, vol. 70, pp. 927–930, 1989.

Cahill, D., S. K. Watson, and R. O. Pohl, "Lower limit to the thermal conductivity of disordered crystals," *Phys. Rev. B*, vol. 46, pp. 6131–6140, 1992.

Callaway, J., "Model of lattice thermal conductivity at low temperatures," *Phys. Rev.*, vol. 113, pp. 1046–1051, 1959.

Chelikowsky, J. R., and M. L. Cohen, "Nonlocal pseudopotential calculations for the electronic structure of eleven diamond and zinc-blende semiconductors," *Phys. Rev.*, vol. 112, p. 556, 1976.

Dash, W. C., and R. Newman, "Intrinsic optical absorption in single crystal germanium and silicon at 77 K and 300 K," *Phys. Rev.*, vol. 99, pp. 1151–1155, 1955.

Ehrenreich, H., and H. R. Phillipp, "Optical properties of Ag and Cu," *Phys. Rev.*, vol. 128, pp. 1622–1629, 1962.

Eisberg, R., and R. Resnick, *Quantum Physics of Atoms, Molecules, Solids, Nuclei, and Particles*, 2nd ed., John Wiley, New York, 1985.

Elsasser, T., J. Shah, L. Rota, and L. Lugli, *Phys. Rev. Lett.*, vol. 66, p. 1757, 1991.

Feng, Z. C., and R. Tsu, *Porous Silicon*, World Scientific, River Edge, 1994.

Ferry, D. K., *Semiconductors*, Macmillan, New York, 1991.

Fischetti, M. V., and S. E. Laux, "Monte Carlo analysis of electron transport in small semiconductor devices including band-structure and space-charge effects," *Phys. Rev. B*, vol. 38, pp. 9721–9745, 1988.

Fushinobu, K., A. Majumdar, and K. Hijikata, "Heat generation and transport in submicron semiconductor devices," *J. Heat Transfer*, vol. 117, pp. 25–31, 1995.

Fushinobu, K., L. M. Phinney, and N. C. Tien, "Ultrashort-pulse laser heating of silicon to reduce microstructure adhesion," *Int. J. Heat Mass Transfer*, vol. 39, pp. 3181–3186, 1996.

Gobeli, G. W., and H. Y. Fan, "Semiconductor Research," Second Quarterly Report, Purdue University, 1956; cited in E. J. Johnson, *Semiconductors and Semimetals*, vol. 3, R. K. Willardson and A. C. Beer (eds.), Chap. 6, Academic Press, New York, 1967.

Hull, D., and D. J. Bacon, *Introduction to Dislocations*, 3rd ed., Pergamon Press, New York, 1984.

Israelachvili, J. N., *Intermolecular and Surface Forces*, 2nd ed., Academic Press, San Diego, 1992.

Jacoboni, C., and L. Reggiani, "The Monte Carlo method for the solution of charge transport in semiconductors with applications to covalent materials," *Rev. Mod. Phys.*, vol. 55, pp. 645–705, 1983.

Joseph, D. D., and L. Preziosi, "Heat waves," *Rev. Mod. Phys.*, vol. 61, p. 41, 1989.

Joshi, A. A., and A. Majumdar, "Transient ballistic and diffusive phonon heat transport in thin films," *J. Appl. Phys.*, vol. 74, pp. 31–39, 1993.

Joshi, Y. P., M. D. Tiwari, and G. S. Verma, "Role of four-phonon process in the lattice thermal conductivity of silicon from 300 K to 1300 K," *Phys. Rev. B*, vol. 1, pp. 642–646, 1970.

Kittel, C., *Introduction to Solid State Physics*, 6th ed., John Wiley, New York, 1986.

Klemens, P. G., "Thermal conductivity and lattice vibrational modes," in *Solid State Physics*, vol. 7, F. Seitz and D. Turnbull (eds.), Academic Press, New York, pp. 1–98, 1958.

Lai, J., and A. Majumdar, "Concurrent thermal and electrical modeling of a sub-micrometer silicon devices," *J. Appl. Phys.*, vol. 79, pp. 7353–7361, 1996.

Leon, R., P. M. Petroff, D. Leonard, and S. Fafard, "Spatially resolved visible luminescence of self-assembled semiconductor quantum dots," *Science*, vol. 267, pp. 1966–1968, 1995.

Majumdar, A., "Microscale heat conduction in dielectric thin films," *J. Heat Transfer*, vol. 115, pp. 7–16, 1993.

Majumdar, A., P. I. Oden, J. P. Carrejo, L. A. Nagahara, J. J. Graham, and J. J. Alexander, "Nanometer-scale lithography using the atomic force microscope," *Appl. Phys. Lett.*, vol. 61, pp. 2293–2295, 1992.

Majumdar, A., K. Fushinobu, and K. Hijikata, "Effect of gate voltage on hot electron and hot phonon interaction and transport in sub-micron transistor," *J. Appl. Phys.*, vol. 77, pp. 6686–6694, 1995.

Marrian, C. R. K., *Technology of Proximal Probe Lithography*, SPIE Press, Bellingham, 1993.

Minne, S. C., H. T. Soh, P. Flueckiger, and C. F. Quate, "Fabrication of 0.1-μm metal oxide semiconductor field-effect transistors with the atomic force microscope," *Appl. Phys. Lett.*, vol. 66, pp. 703–705, 1995.

Moss, T. S., and T. D. F. Hawkins, "Infrared absorption of gallium arsenide," *Infrared Phys.*, vol. 1, pp. 111–115, 1962.

Ohring, M., *The Materials Science of Thin Films*, Academic Press, San Diego, 1992.

Powell, R. W., C. Y. Ho, and P. E. Liley, *Thermal Conductivity of Selected Materials*, National Bureau of Standards Reference Data Series 8, Washington, DC, 1966.

Qiu, T. Q., and C. L. Tien, "Heat transfer mechanisms during short-pulse laser heating of metals," *J. Heat Transfer*, vol. 115, pp. 835–841, 1993.

Rayleigh, Lord, *Theory of Sound*, Macmillan, London, 1896.

Siegel, R., and J. R. Howell, *Thermal Radiation Heat Transfer*, 3rd ed., Hemisphere, Washington, 1992.

Stroscio, J. A., and D. M. Eigler, "Atomic and molecular manipulation with the scanning tunneling microscope," *Science*, vol. 254, pp. 1319–1326, 1991.

Sze, S. M., *Physics of Semiconductor Devices*, 2nd ed., Wiley, New York, 1981.

Touloukian, Y. S., and E. H. Buyco, *Thermophysical Properties of Matter*, vol. 5, Plenum, New York, 1970.

van Driel, H. M., "Kinetics of high-density plasmas generated in Si by 1.06- and 0.53-μm picosecond laser pulses," *Phys. Rev. B*, vol. 35, pp. 8166–8176, 1987.

von der Linde, D., J. Kuhl, and H. Klingenburg, "Raman scattering from nonequilibrium LO phonons with picosecond resolution," *Phys. Rev. Lett.*, vol. 44, pp. 1505–1508, 1980.

Vincenti, W. G., and C. H. Kruger, *Introduction to Physical Gas Dynamics*, Robert Krieger Publ., New York, 1977.

Walton, D., "Scattering of phonons by a square-well potential and the effect of colloids on the thermal conductivity. I. Experimental," *Phys. Rev.*, vol. 157, pp. 720–724, 1967.

Walton, D., and E. J. Lee, "Scattering of phonons by a square-well potential and the effect of colloids on the thermal conductivity. II. Theoretical," *Phys. Rev.*, vol. 157, pp. 724–729, 1967.

Waugh, J. L. T., and G. Dolling, "Crystal dynamics of gallium arsenide," *Phys. Rev.*, vol. 132, pp. 2410–2412, 1963.

Worlock, J. M., "Thermal conductivity in sodium chloride crystals containing silver chloride," *Phys. Rev.*, vol. 147, pp. 636–643, 1966.

Ziman, J. M., *Electrons and Phonons*, Oxford University Press, London, 1960.

Ziman, J., *Theory of Solids*, 2nd ed., Cambridge University Press, Cambridge, 1972.

CHAPTER

TWO

HEAT TRANSPORT IN DIELECTRIC THIN FILMS AND AT SOLID–SOLID INTERFACES

David G. Cahill

Department of Materials Science and Engineering,
University of Illinois at Urbana-Champaign,
Urbana, IL 61801

2-1 INTRODUCTION

Heat transport in dielectric thin films plays a vital role in determining the performance of many components and devices used in state-of-the-art engineering systems. For example, control of high-power laser pulses requires antireflection coatings and multilayer dielectric mirrors with low damage thresholds; ceramic thermal barrier coatings extend the operating temperature and therefore improve the efficiency of turbine engines; and superlattice Bragg reflectors in vertical cavity surface emitting lasers (VCSEL) produce a problematic increase in the operating temperature of these optoelectronic devices.

Near room temperature, lattice vibrations are almost solely responsible for energy transport in dielectric materials and most semiconductors. While this fact may at first appear to simplify considerations of energy transport, the wide range of relevant vibrational frequencies and lifetimes often makes a complete understanding of thermal transport troublesome, even for well-characterized bulk materials. Since thin film materials usually are deposited at relatively low temperatures, highly nonequilibrium microstructures and compositions produced by the deposition process compound the intrinsic difficulties of understanding heat transport in dielectric solids.

In addition to the considerations of microstructure and composition, thin films introduce the full complexity of solid–solid interfaces into any complete discussion of heat transport in these materials. Applications of thin films often require layer thicknesses <100 nm. Internal grain boundaries in microcrystalline materials will produce interfaces on even smaller length scales. Since at least some fraction of the phonon spectrum may have mean free paths comparable to or larger than these dimensions, an understanding of *microscale energy transport* is necessary for understanding heat transport in thin films.

In this chapter, we will focus our attention on this less restrictive definition of microscale energy transport: the dimension of the sample or internal structure in the material is less than the mean free path of the excitations responsible for energy transport. The more restrictive regime—that is, that the dimension of the film structure is on the order of the wavelength of the dominant vibrational excitations—to the best of our knowledge, is essentially unexplored for lattice vibrations and will not be discussed here.

Despite the importance of heat transport in thin films and heat transport across solid–solid interfaces for device applications, the measurement of these properties is far from straightforward and has been achieved only in recent years. The analogous transport measurement for thin metals films, the electrical conductivity, is relatively straightforward because electrical current can be completely confined to the film. This simplification is not available in thermal transport measurements: no solid is a perfect thermal insulator, and even vacuum can transport a significant energy flux via blackbody radiation. Thus, we begin our discussion with an analysis of measurement techniques before proceeding to analyze the thermal conductivity of thin films of disordered materials and the effects of interfaces on heat transport by lattice vibrations.

2-2 MEASUREMENT TECHNIQUES

Many methods for the determination of the thermal properties of thin films have been reported in the scientific literature. Because each class of thin film system presents an almost unique set of experimental hurdles to overcome, no single measurement method has yet become universally accepted. In the discussion below, we do not intend a comprehensive review of the literature of experimental methods; our objective is to analyze some of the most useful approaches for characterizing heat transport in thin films—and their possible limitations—using a relatively small selection of successful examples. Goodson and Flik (1994) have also recently reviewed thin film thermal property measurement. We also exclude the characterization of diamond thin films. The extremely high thermal conductivity of diamond coatings ($\Lambda \simeq 2000$ $\mathrm{Wm^{-1}K^{-1}}$) and relatively large thickness (typically $\sim$100 μm) are a special case that has been thoroughly reviewed by Graebner (1993).

2-2-1 Diffusivity Methods

One broadly defined class of methods is based on the measurement of thermal diffusivity D rather than the thermal conductivity Λ. This approach has a long tradition dating back to the work of Ångström (1861), who accurately determined the thermal conductivity of Fe and Cu by periodically switching the temperature of one end of a metal bar between 0°C (ice water) and 100°C (steam) and by measuring the decay of the temperature oscillations along the length of the bar. As long as the heat capacity C per unit volume is known, then a measurement of thermal diffusivity is equivalent to a measurement of $\Lambda : D = \Lambda / C$. Fortunately, when adjusted for porosity, the heat capacity for a given material is not strongly sensitive to microstructure, and for most technologically relevant materials, C is known through measurements of bulk specimens.

To see how the thermal diffusivity is measured in a thin film, consider the solution of the diffusion equation for the one-dimensional (1-D) heat flow geometry shown in

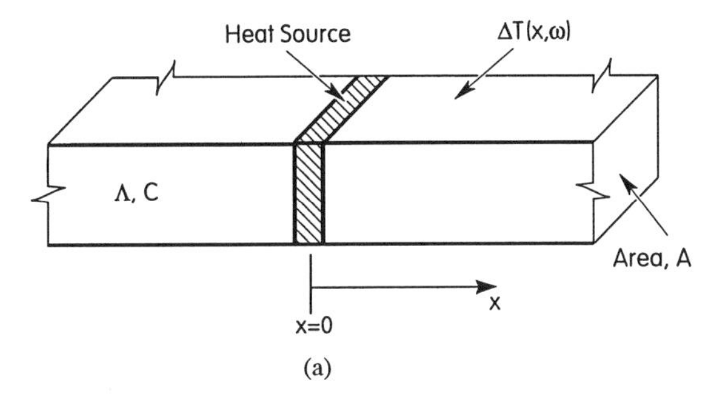

(a)

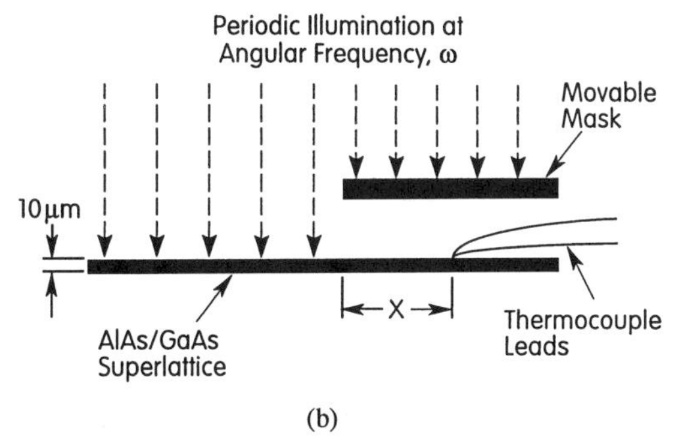

(b)

Figure 2-1 (a) Schematic diagram of 1-D heat flow in a bar of cross-sectional area A and infinite length; x measures the distance from the heat source. (b) Approximation of the 1-D heat flow geometry used by Yao (1987) to measure the in-plane thermal diffusivity of a 10-micron-thick free-standing superlattice of AlAs/GaAs.

Fig. 2-1a. We focus here on frequency domain solutions, that is, on oscillatory heat sources and temperature profiles. For small changes in the temperature $\Delta T \ll T$, the heat diffusion equation is linear, and therefore time domain behavior can be obtained by an integral transform of the frequency domain results. Heat is supplied at a rate P/A, where $P = P_0 \cos(\omega t)$ and P_0 is the amplitude of the power oscillations, t is time, and ω is the angular frequency. In most experimental systems, P is created by optical illumination or joule heating and therefore also includes a static or dc component, which is unimportant for our analysis.

The solution for the amplitude of the temperature oscillations ΔT as a function of distance from the heat source x is straightforward (Carslaw and Jaeger, 1959):

$$\Delta T(x, \omega) = \frac{P}{2A\sqrt{i\omega\Lambda C}} \exp(-q|x|); \quad q^2 = \frac{i\omega}{D}. \tag{2-1}$$

Yao (1987), following Hatta (1985), applied a similar geometry for the measurement of the thermal diffusivity of a 10-micron-thick, free-standing superlattice of AlAs/GaAs (see Fig. 2-1b). In this method, the amplitude of the temperature oscillation, ΔT, is measured as a function of distance x between the edge of a moveable mask and the thermocouple junction used to measure the film temperature. The mask limits the illumination from a halogen lamp to one portion of the sample; Eq. (2-1) must be

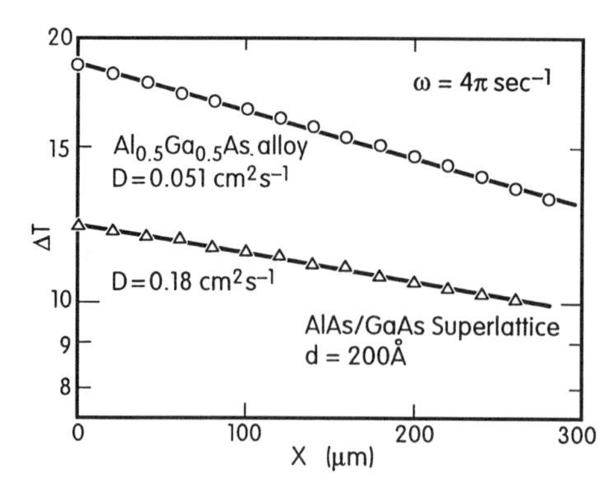

Figure 2-2 Data from Yao (1987) for the temperature oscillations ΔT measured in the geometry described in Fig. 2-1b. Solid lines show the expected exponential dependence of ΔT on x (see Eq. (2-1)).

integrated over the interval x to infinity for a complete comparison with experiment, but the exponential dependence on $-q|x|$ is unchanged.

The data are plotted in Fig. 2-2 to extract D. According to Eq. (2-1), the slope of $\ln(\Delta T)$ versus x is equal to $\sqrt{\omega/(2D)}$. This method is limited to free-standing films or films that have a much larger thermal conductivity Λ_f than that of the substrate Λ_s so that heat conducted by the substrate can be neglected. More precisely, we require $d\Lambda_f \gg t_s\Lambda_f$, where d and t_s are the thicknesses of the film and substrate, respectively.

A measurement of the thermal diffusivity was also used by Chen et al. (1994) to study the through-thickness thermal properties of a VCSEL device structure. The exact structure of the device is complex, but the majority of the thickness is made up of a 70-nm-thick layer of GaAs alternating with a 70-nm-thick short-period AlAs/GaAs superlattice; this pair of layers is repeated $\simeq$50 times to form the Bragg reflectors. Figure 2-3a shows the idealized geometry of 1-D heat flow in a finite-size sample; this

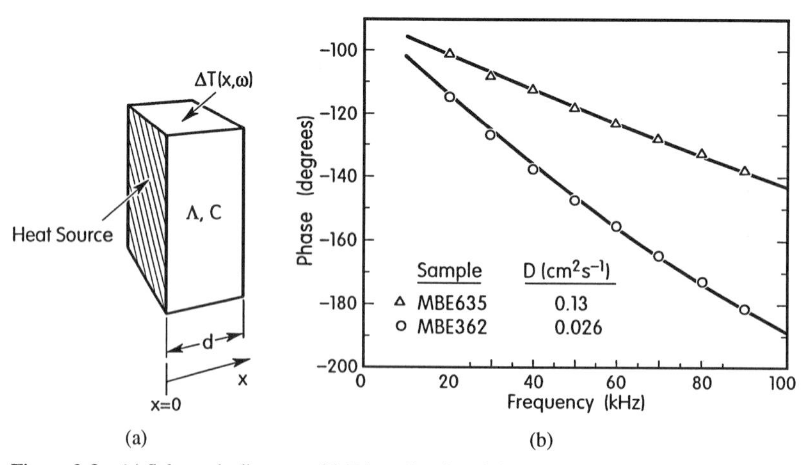

Figure 2-3 (a) Schematic diagram of 1-D heat flow in a finite sized sample. (b) Data from Chen et al. (1994) for the phase of the temperature oscillations at $x = d$ used to measure the through-thickness thermal diffusivity of MBE grown semiconductor layers. MBE635 is a 9.9-μm-thick layer of GaAs grown on a 1.1-μm-thick layer of AlGaAs. MBE362 is a VCSEL device structure, 7.2 μm thick.

geometry is equivalent to the "flash-diffusivity" method that has found wide application in the characterization of bulk, high-temperature materials (Taylor, 1995). Analytical solutions of the heat diffusion equation in this geometry are simplified by the use of Laplace transforms (Carslaw and Jaeger, 1959):

$$\Delta T(x, \omega) = \frac{P}{A\sqrt{i\omega \Lambda C}} \left[\frac{\cosh(qx)}{\tanh(qd)} - \sinh(qx) \right]. \tag{2-2}$$

At $x = d$ (temperature measurement on the opposite side of the film from the heat source), the term in brackets reduces to $\sinh^{-1}(qx)$. Chen et al. (1994) also used optical illumination for the heat source, but they used a metal film resistor as the temperature sensor because the frequency response of a bulk thermocouple junction is far too slow for this application. Data for the phase of the temperature oscillations are plotted in Fig. 2-3b. For the 1-D solution to be valid, the width of the heated area must be large compared to the thermal penetration depth, $\delta = \sqrt{D/w}$. To optimize the accurate measurement of the amplitude and phase of ΔT on the side of the film opposite the heater, the frequency is chosen so that $\delta \sim d$, where d is the film thickness. This condition places a practical limit on the thickness of films that are measurable using this method. For example, if $d \sim 1$ μm and $D = 0.1$ cm^2 sec^{-1}, then $\omega = 10^7$ sec^{-1}, an excitation frequency that places extreme demands on the detection electronics.

The use of metal film temperature sensors deserves further attention. Metal films resistors, patterned by photolithography, provide a flexible method for accurate temperature measurement on micron length scales with excellent frequency response. The resistivity of a pure metal increases almost proportional to T for $T > 50$ K. Impurities, microstructural defects, and interface scattering add a temperature-independent resistivity to the ideal, bulk value. A useful expression for describing the resistivity of a metal film is $\rho = \alpha T^{1+\epsilon} + \rho_0$. ($\epsilon \ll 1$, but this small correction is a useful means of improving the empirical fit of ρ versus T data used for calibrating the metal film thermometer.) The logarithmic derivative $(d \ln \rho)/(d \ln T)$ provides a figure of merit; for most pure metal films near room temperature $(d \ln \rho)/(d \ln T) \approx 1$.

Noncontact methods also have been developed for measuring thin film thermal properties through measurements of the diffusivity $D = \Lambda/C$ or the effusivity $E = \Lambda C$. In photoacoustic methods, the temperature oscillations at the surface of a thin film are monitored by the emissions of acoustic waves produced by the thermal expansion of gas adjacent to the surface. The frequency of the excitation is limited to <20 kHz for atmospheric pressure gases, and low-temperature measurements are enabled by working in a He atmosphere. In practice, extreme care must be taken to insure that light scattered from the surface of the sample does not produce stray acoustic signals from other parts of the sample cell. Because the acoustic signal is a measure of the surface temperature at the location of heating, the photoacoustic response is also more sensitive to variations in the absorption depth than other techniques that use optical illumination for heating but place the temperature measurement at a distance removed from the heat source. A complete discussion of these points has recently been provided by Wood (unpublished).

We end this section with a comment on the validity of the continuum diffusion equation for probing microscale heat transport. Strictly speaking, we can define a local temperature T only by averaging over a length scale that is larger than the mean free paths $l_j(\omega)$ of all phonons. In other words, we require that T varies slowly in the spatial coordinate: $T/(dT/dx) \gg l_j(\omega)$ for all vibrational frequencies ω and modes j. For all phonon scattering mechanisms except diffuse scattering from interfaces, the mean free path l increases with decreasing ω, and so the possibility arises that a significant fraction of the heat flux may be carried by a small fraction of the lattice vibrations with long wavelengths that have extremely large $l_j(\omega)$.

We introduce a practical guideline for when solutions of the diffusion equation will be good approximations, even if these solutions are not rigorously valid. Instead of the criterion $T/(dT/dx) \gg l(\omega)$ for all ω, we relax this condition by requiring that the heat carried by phonons with $l(\omega) > T/(dT/dx)$ is only a small fraction of the total heat flux. Within the Debye model, the spectral distribution of the heat flux is given by $I(\omega)$. Our practical guideline is expressed more formally by defining a "crossover" frequency ω_c using the equation $T/(dT/dx) = l(\omega_c)$ and the condition

$$\int_0^{\omega_c} I(\omega)d\omega \ll \int_0^{\infty} I(\omega)d\omega \, . \tag{2-3}$$

For the measurement schemes described above, $T/(dT/dx) \simeq |1/q|$ and $l(\omega_c) \simeq |1/q|$. The condition expressed by Eq. (2-3) is well in most materials at room temperature as long as $|1/q| > 1\ \mu$m. Measurements on shorter length scales (and equivalently shorter time scales) may encounter strong deviations from the diffusion equation.

2-2-2 Thermal Conductance Methods

In most cases of both scientific and practical interest, a direct measurement of the thermal conductivity Λ is more desirable than a measure of the diffusivity. An idealized geometry for the measurement of a through-thickness thermal conductivity is shown in Fig. 2-4a: the dielectric layer that we wish to characterize is sandwiched between two materials with high thermal conductivity. We define the thermal conductance per unit area of this layer by $G = P/(A\Delta T)$, where P/A is the power per unit area of the heat flux traversing the sample and ΔT is the temperature drop across the layer, $\Delta T = T_1 - T_2$. (We can similarly define the thermal resistance per unit area by $R = 1/G$.) The effective thermal

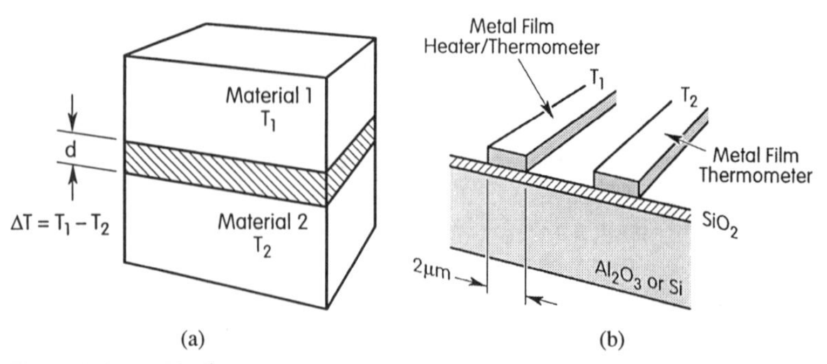

Figure 2-4 (a) Idealized geometry for measuring the thermal conductance produced by a thin film and interfaces. A dielectric film of thickness d is sandwiched between two high thermal conductivity materials. (b) Implementation of the geometry shown in (a) used by Swartz and Pohl (1987) and by Goodson et al. (1994).

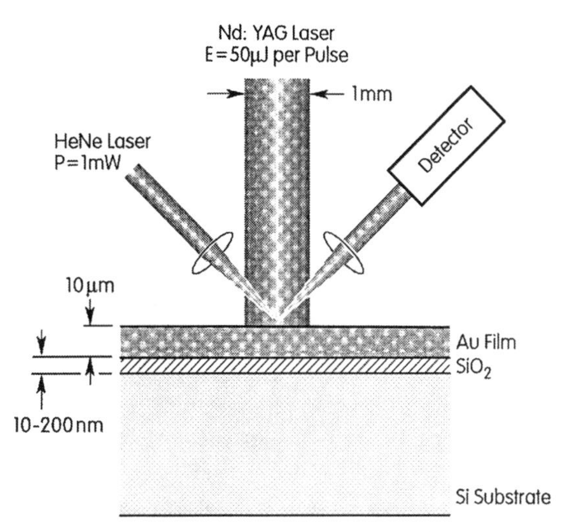

Figure 2-5 Geometry used by Käding et al. (1994) to measure the thermal conductance of thin a-SiO_2. The thermal relaxation time of the Au film, following heating by a short laser pulse, is measured by thermoreflectance (i.e., the temperature dependence of the Au reflectivity).

conductivity of the layer is $\Lambda = Gd$, where d is the layer thickness; we can immediately identify the "microscale" regime: microscale energy transport becomes important when Λ depends on the layer thickness d.

A highly sensitive realization of this measurement geometry was introduced by Swartz and Pohl (1987) and later applied extensively by Goodson et al. (1994) (see Fig. 2-4b). Photolithography is used to produce a parallel arrangement of two narrow metal lines,which are separated by a small distance. One metal line serves as i) material 1 of the idealized geometry shown in Fig. 2-4a; ii) the dc heat source; and iii) the thermometer that measures T_1. The substrate for the film serves as material 2. The temperature of the substrate T_2 is measured by the second metal line. Goodson et al. (1994) applied a third metal line to verify small corrections applied to the results to account for the finite thermal conductivity of the substrate.

Another method that uses essentially the same geometry as shown in Fig. 2-4 but permits a non contact measurement is the thermoreflectance approach used by Käding et al. (1994) (see Fig. 2-5). A relatively thick film of a high-thermal-conductivity metal (typically Au or Al) takes the place of material 1 in Fig. 2-4a. If the heat capacity per unit volume of the thick metal film is C, and if the temperature of the metal film is suddenly and homogeneously increased to $T = T_0 + \Delta T$, then the time evolution of T is given by a simple analogy with a low-pass filter: $T(t) = T_0 + \Delta T \exp(-tG/hC)$, where h is the metal film thickness. A determination of the decay time constant hC/G, combined with the known value of Ch, yields G. In Käding's experiment, the sudden temperature rise in the metal film is produced by a high-energy laser pulse, and the decay of the surface temperature is measured by optical reflectivity; the reflectivity of a metal film is a weak but measurable function of temperature. (The fractional change in reflectivity of Au is $\sim 10^{-6}\ K^{-1}$ at room temperature.) Since only the time constant of the decay is needed to

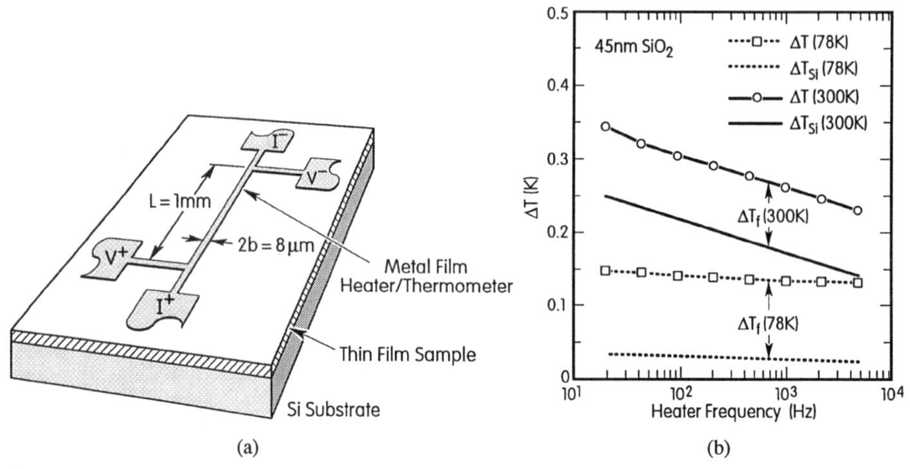

Figure 2-6 (a) Geometry of the 3ω measurement of the thermal conductance of a thin film (Cahill et al., 1994). A single metal line serves as both the heater and thermometer in the experiment. (b) Temperature oscillations of the metal film heat/thermometer as a function of frequency for a 45 nm SiO_2 film deposited on a Si wafer (Lee and Cahill, 1997). The curves marked ΔT_{Si} are the calculated thermal response of the Si substrate.

find G, an absolute calibration of reflectivity versus temperature is not needed. A similar geometry was recently used by Morath et al. (1994) to measure the thermal conductivity of diamond-like carbon and amorphous diamond films approximately 100 nm thick.

A variation of the thermal conductance measurements was recently described by Cahill et al. (1994). Figure 2-6a shows the geometry. A sinusoidal power is produced by joule heating in a narrow metal line at frequency 2ω. As long as the dielectric thin film that we wish to characterize is thermally thin (i.e., the thermal penetration depth in the film is large compared to the film thickness, $\delta = \sqrt{D_f/(2w)} \gg d$), and the thermal conductivity of the film is small compared to the substrate $\Lambda_f \ll \Lambda_s$, the thin film simply adds a frequency-independent ΔT_f to the temperature oscillations at the interface between the dielectric film and the substrate ΔT_s. Furthermore, if the thermal penetration depth in the substrate is small compared to the thickness of the substrate (semi-infinite substrate), ΔT_s can be accurately predicted provided that the heat capacity of the substrate C_s is at least approximately known (the results depends only on $\ln(C_s)$). Λ_s is determined from the same measurement that is used to measure Λ_f. The complete solution to the diffusion equation for this geometry, within these approximations, is given by

$$\Delta T(\omega) = \frac{P}{l\pi\Lambda_s}\int_0^\infty \frac{\sin^2(kb)}{(kb)^2(k^2+q^2)^{1/2}}\,dk + \frac{Pd}{2Lb\Lambda_f}; \quad q^2 = \frac{2i\omega}{D_s}. \tag{2-4}$$

Since $|q|b < 1$, we can approximate this equation by

$$\Delta T(\omega) = \frac{P}{l\pi\Lambda_s}\left[\frac{1}{2}\ln\left(\frac{D_s}{b^2}\right) + \eta - \frac{1}{2}\ln(2\omega) - \frac{i\pi}{4}\right] + \frac{Pd}{2Lb\Lambda_f}, \tag{2-5}$$

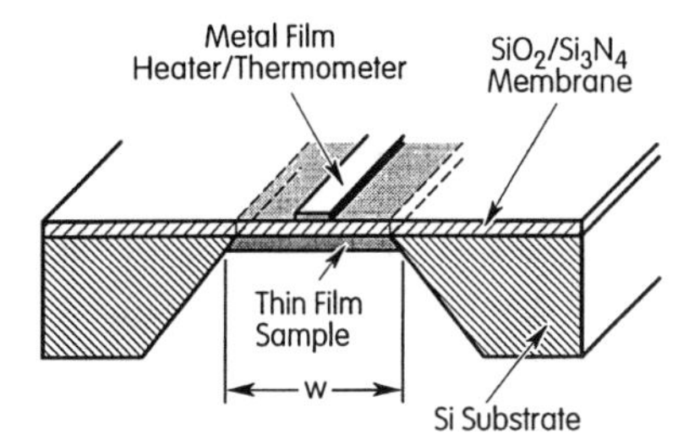

Figure 2-7 In-plane thermal conductivity method as implemented by Schmidt et al. (unpublished). As in Fig. 2-6, a single metal line serves as both the heater and the thermometer for the measurement.

where numerical solution in the limit $|q|b \ll 1$ gives $\eta = 0.92$. Empirically, we have found $\eta = 1.05$ more accurately describes the results of experiments.

The thermal response of an example system, a 45-nm-thick amorphous SiO_2 film deposited on a Si substrate (Lee and Cahill, 1997), is shown in Fig. 2-6b. The temperature oscillations are measured by the same metal line that is the heat source. As in the methods described above, the resistance of the metal line increases with increasing temperature. A power oscillation at frequency 2ω produces a temperature oscillation at 2ω that therefore produces a resistance oscillation at 2ω. This resistance oscillation is probed using a bridge circuit and lock-in amplifier that measures a voltage oscillation at frequency 3ω (Birge and Nagel, 1987). Hence, this measurement technique is often referred to as the 3ω method (Cahill, 1990).

The geometry of Fig. 2-6 was recently modified by Schmidt et al. (unpublished) to enable measurement of the in-plane thermal conductivity of thin film semiconductors and semimetals supported by a SiN_x/SiO_2 membrane (see Fig. 2-7). In one approach, intermediate frequencies are used, $w \gg \sqrt{D/\omega} \gg d$, and heat flow approximates our discussion of 1-D heat flow described above (Fig. 2-1), with the difference that the temperature oscillations are measured at $x = 0$ using the same metal film that serves as the heater. Measurement of ΔT at $x = 0$ yields the product (ΛC) (see Eq. (2-1)), and therefore an independent measurement of C is needed (see Eq. (2-1)) to extract Λ.

2-3 HEAT TRANSPORT IN STRONGLY DISORDERED MATERIALS

To gain an understanding of the thermal conductivity of thin films, we will first analyze the behavior of simple models that have proven useful for bulk materials, and then use these models for insight on the behavior of thin films. The model behavior will be illustrated, when possible, using experimental data for bulk and thin film materials.

2-3-1 Minimum Thermal Conductivity

As described in Chapter 1, the Boltzmann transport equation can be rigorously applied in the analysis of heat transport in most *crystalline* solids because the relaxation time τ satisfies the weak scattering condition $\omega\tau \gg 1$ for all vibrational modes. But most thin film dielectrics are either fully amorphous or microcrystalline with significant atomic scale disorder. Because of these strong deviations from the translational symmetry of a perfect crystal, wave-like solutions (phonons) are no longer well defined. In amorphous and highly disordered crystalline solids, $\omega\tau \sim 1$ for the vast majority of vibrational modes. Only the lowest-energy, longest-wavelength acoustic excitations will have $\omega\tau \gg 1$.

It has long been recognized (see the brief review in Cahill et al. (1990)), however, that the thermal conductivity of an amorphous solid near room temperature is approximately $\Lambda = (1/3)Cvl$ with $l = a$, where C is the specific heat, v is the speed of sound, l is the mean free path of the dominant phonons, and a is the separation between atoms. This fundamental observation is now often referred to as the *minimum thermal conductivity* (Slack, 1979) since no materials have been found where the thermal conductivity near room temperature falls significantly below this value. Consequently, the minimum conductivity is currently thought to be a useful approximation for the lower limit to the conductivity of fully dense solids no matter what the nature or extent of the disorder.

Einstein (1911) theoretically obtained essentially the same equation at the turn of the century. In this effort, which was the first calculation of energy transport by lattice vibrations, Einstein built on his well-known model for the heat capacity of solids. The thermal conductivity of the original Einstein model is zero because no coupling exists between the vibrating atoms, the so-called "Einstein oscillators." By coupling each of the oscillators to the nearest and next-nearest neighbors of a simple cubic lattice, and by making the crucial assumption that the oscillators vibrate with random phases, Einstein found that energy was transported between atoms on a time scale of 1/2 the vibrational period. If we further assume that the heat capacity of the damped oscillators can be approximated by the heat capacity of the undamped oscillators, the thermal conductivity is given by

$$\Lambda_{\text{Eins}} = \frac{k_B^2}{\hbar} \frac{n^{1/3}}{\pi} \Theta_E \frac{x^2 e^x}{(e^x - 1)^2}, \tag{2-6}$$

where Θ_E is the Einstein temperature that characterizes a typical atomic-scale vibrational frequency ω, $\Theta_E = \hbar\omega/k_B$, and $x = (\Theta_E/T)$. Einstein compared his calculation to the thermal conductivity of crystalline KCl and found poor agreement with both the magnitude and temperature dependence of the thermal conductivity. But Einstein did not consider data for amorphous materials, which were available in 1911 and agree quite well with his model.

To remove the uncertainty of choosing the Einstein frequency, Einstein's result can be modified somewhat to include larger oscillating entities than the single atoms considered by Einstein (Cahill et al., 1992). To do this, we borrow from the Debye model of lattice vibrations and divide the sample into regions of size $\lambda/2$ whose frequencies of oscillation are given by the low-frequency speed of sound $\omega = 2\pi v/\lambda$. The lifetime of each oscillator is again assumed to be one-half the period of vibration, that is, $\tau = \pi/\omega$. The thermal conductivity resulting from the random walk of vibrational energy between these localized quantum mechanical oscillators can then be written as the following sum of three Debye integrals:

$$\Lambda_{min} = \left(\frac{\pi}{6}\right)^{1/3} k_B n^{2/3} \sum_j v_j \left(\frac{T}{\Theta_i}\right)^2 \int_0^{\Theta_j/T} \frac{x^3 e^x}{(e^x - 1)^2} dx \ . \tag{2-7}$$

The sum is taken over the three sound modes (two transverse and one longitudinal) with speeds of sound v_j; Θ_j is the cutoff frequency for each polarization expressed in degrees K, $\Theta_j = v_j(\hbar/k_B)(6\pi^2 n)^{1/3}$; and n is the number density of atoms. The physical picture is that of a random walk of energy between localized oscillators of varying sizes and frequencies, and that the dominant energy transport is between nearest neighbors. Equation (2-7) contains no free parameters, for v_j and n are known for most technologically important materials.

In Fig. 2-8a, this calculation is compared to the thermal conductivity of two important oxide thin film materials, SiO_2 (Cahill et al., 1994) and stabilized zirconia. In addition to

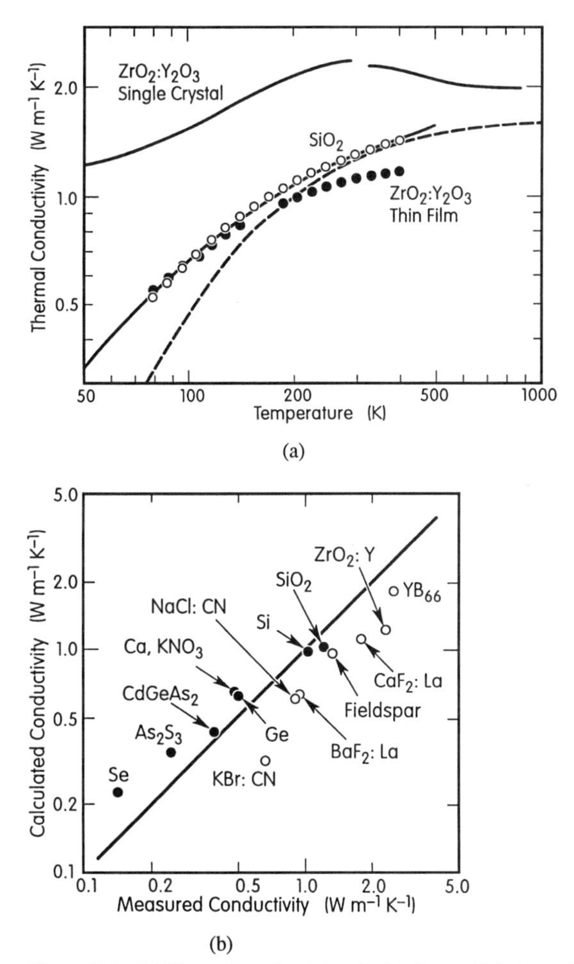

Figure 2-8 (a) Thermal conductivity of a 1-micron-thick thermal oxide SiO_2 film (Cahill et al., 1994) (open circle) and a 5-micron-thick film of stabilized zirconia (filled circles), compared to data for bulk samples (Cahill et al., 1992) and the calculated minimum thermal conductivity Λ_{min}. Λ_{min} is shown for stabilized zirconia; Λ_{min} for a-SiO_2 is almost identical. (b) Comparison of the calculated minimum thermal conductivity and the room-temperature thermal conductivity of a wide variety of amorphous solids (filled circles) and highly disordered crystals (open circles) (Cahill et al., 1992).

its vital importance in the microelectronics industry, SiO_2 is also a common component of optical coatings. Stabilized zirconia is the standard material for thermal barrier coatings in gas turbine engines due to the combination of high melting point, low vapor pressure at elevated temperatures, low thermal conductivity, and good mechanical properties. SiO_2 has an amorphous microstructure, but stabilized zirconia (a solid solution of zirconia ZrO_2 and yttria Y_2O_3) is a highly disordered crystalline solid that contains a large concentration of oxygen vacancies in addition to the random substitution of Y ions on Zr lattice sites. We see that the magnitude and temperature dependence of the conductivities of these thin film materials are close to the predicted Λ_{min}.

The data for thin films are also compared to bulk values. For SiO_2 the agreement between bulk and thin film is almost perfect; phonon scattering at the interfaces of a 1-μ-thick film has no effect on the thermal conductivity. We note that the thermal conductivity of the stabilized zirconia thin film is smaller than the bulk; this behavior is not fully understood at this time, but the reduction is too large to be explained by effects of the interface between substrate and film. Atomic-scale defects and microvoids produced by room-temperature sputter deposition are more likely origins of the reduction in conductivity.

Similar comparisons have been made for a wide variety of materials (see Fig. 2-8b), where the calculated minimum thermal conductivity is compared to the measured thermal conductivity at room temperature (Cahill et al., 1992). Note that this model is a good predictor for heat transport in amorphous solids as well as certain classes of highly disordered crystals, crystals that contain a large concentration of interstitials or vacancies, such as stabilized zirconia.

The minimum thermal conductivity model assumes that long-wavelength phonons with long mean free paths do not contribute significantly to heat transport in a strongly disordered material. We can test this assumption by examining the thermal conductivity of porous Vycor glass (Cahill et al., 1990). Vycor glass is a porous silica glass produced using a borosilicate glass starting material. Spinodal decomposition produces interconnected regions of boron-rich and boron-depleted phases. The boron-rich phase is subsequently dissolved in a basic solution, leaving behind an interconnected network of pores with a uniform pore size of $\sim$10 nm and volume fraction of $\simeq$30%. These pores limit the mean free path of the majority of the lattice vibrations to $\sim$10 nm.

If this additional phonon scattering is unimportant to heat flow in SiO_2 (i.e., long-wavelength phonons with long mean free paths make a negligible contribution to heat transport), then we can use effective medium theory to model the effect of the pores. The conductivity (either electrical or thermal) of a composite structure made of a matrix material and spherical inclusions of a second phase is given by (Landauer, 1952)

$$4\Lambda = (3f_2 - 1)\Lambda_2 + (3f_1 - 1)\Lambda_1 + \left[((3f_2 - 1)\Lambda_2 + (3f_1 - 1)\Lambda_1)^2 + 8\Lambda_1\Lambda_2\right]^{1/2}, \tag{2-8}$$

where Λ_1 is the conductivity of the matrix, Λ_2 is the conductivity of the second phase, and f_1, f_2 are the volume fractions of the matrix and second phase, respectively. For our case (porous Vycor glass), the matrix is fully dense SiO_2 and the second "phase" is voids with zero thermal conductivity. Then, $\Lambda = \Lambda_1(1 - 1.5f_2)$. Figure 2-9 shows

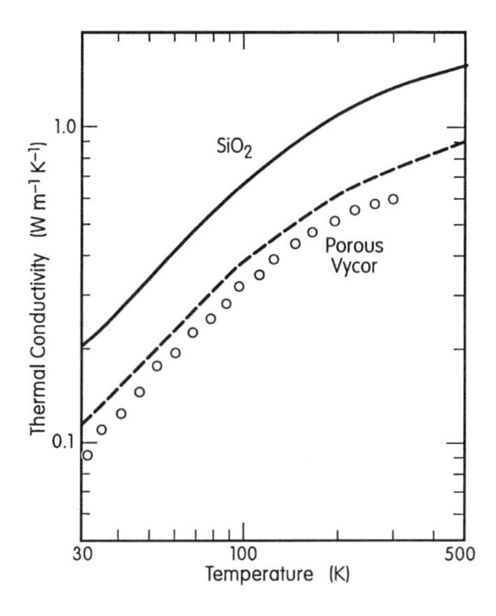

Figure 2-9 Thermal conductivity of a porous Vycor glass (Cahill et al., 1990) compared to the thermal conductivity of bulk SiO_2 film and the prediction of effective medium theory (dashed line). The volume fraction of pores in this sample of Vycor is approximately $f_2 = 0.29$.

a comparison of the experimental results with the predictions of this calculation. The excellent agreement confirms our assumption that long-wavelength phonons make a negligible contribution to heat transport in a-SiO_2.

Recently, Allen and Feldman (1993) and Feldman et al. (1993) have worked to place the concept of a minimum thermal conductivity on a firm theoretical foundation by abandoning the Boltzmann equation, and instead, applying the more general transport equation of Kubo. Starting with a realistic model of the structure of a-Si and descriptions of the atomic potentials, the thermal conductivity of the model structure can be calculated by numerical solution of the Kubo equation. Anharmonicity is ignored. The results of this calculation are compared to the calculated minimum thermal conductivity and data for a 0.5 micron thick film of a-Si:H, with 1 atomic percent H (see Fig. 2-10). Long-wavelength phonons can not be explicitly included in the calculations because of the finite computational cell, but an analysis of the divergence of the mode diffusivity

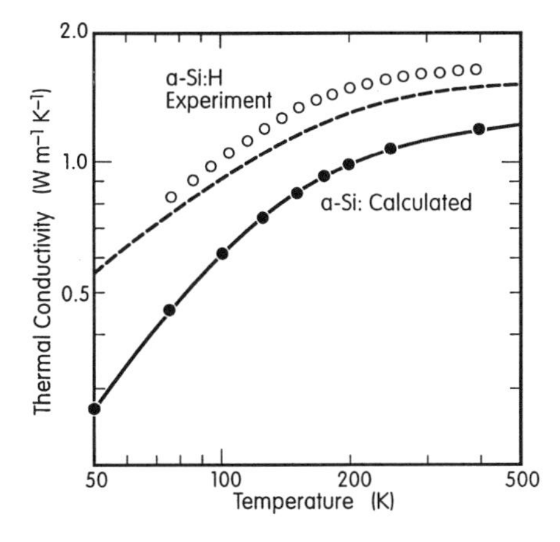

Figure 2-10 Thermal conductivity of a-Si:H with 1 atomic % H (Cahill et al., 1994) compared to the numerical solution of the Kubo transport equation (Feldman et al., 1993). The dashed line shows an extension of the calculation to include heat transport by long-wavelength phonons with a mean free path limited to the film thickness.

(see discussion below) suggests that long wavelength phonons can contribute to heat transport near room temperature. The effect is relatively small, however: $\sim$20% for film thicknesses $\sim$1 μm.

2-3-2 Thermal Conductivity of Amorphous and Microcrystalline Dielectric Thin Films

Even for the highly disordered thin film materials that are the focus of this section, we expect that at sufficiently low frequencies the lattice vibrations are accurately described as phonons with $\omega\tau \gg 1$, since at low frequencies the long wavelength of the lattice vibration can average over the disordered atomic structure. In fact, at low temperatures ($T < 1$ K) in amorphous materials, thermally excited lattice vibrations have $\omega\tau \sim 1000$. In this extreme low-temperature limit, the mean free path is determined by resonant scattering from the so-called tunneling states that are a characteristic feature of the low-temperature properties of highly disordered materials.

To consider the possible influence of low-frequency phonons with long mean free paths on the thermal conductivity of disordered thin film materials near room temperature—and therefore to consider the possible effect of scattering of these phonons by the microstructure and interfaces of thin films—we separate the heat transport by lattice vibrations into two regimes. As in the calculation of the minimum thermal conductivity, we start with a Debye model of the vibrational states and assume that at low vibrational energies the lattice vibrations are wave-like, that is, phonons exist with well-defined wavevector and velocity. In this regime, the kinetic equation is applicable: $\Lambda(\omega) = C(\omega)vl(\omega)/3$, where Λ is the thermal conductivity, C the heat capacity, v the velocity, and l the mean free path. As the energy E of the mode increases, we will further assume that the scattering rate increases due to Rayleigh scattering as ω^4 until $l^{-1} \simeq k/\pi$, where k is the wavevector of the mode.

An example calculation of this type is included in Fig. 2-10. The Rayleigh scattering rate of phonons is given by $\Gamma = A_0\nu^4$ where A_0 is a constant independent of mode and ν is the vibrational frequency of the mode. The computational results of Feldman et al. (1993) suggest that for a-Si, $A_0 \simeq 3 \times 10^{-37}$ sec^3. We also limit the mean free path of the phonons to d, the film thickness. The result of this calculation is equivalent to adding a thermal conductivity of 0.22 W m^{-1} K^{-1} at $T > 50$ K to the predicted minimum thermal conductivity. This value for phonon contribution to the heat transport is only weakly dependent on the film thickness, varying as $\sim d^{0.25}$ but more strongly dependent on the size of A_0; we find that a change in A_0 of a factor of 3 produces a factor of 2 change in the calculated contribution to the thermal conductivity from low-frequency modes.

We do not, however, want to leave the reader with the impression that heat transport in thin dielectric films is completely understood. The thermal conductivity of SiO_2 films produced by state-of-the-art methods for physical vapor deposition of optical coatings—reactive evaporation and magnetron sputtering—are shown in Fig. 2-11. Despite the fact that the density of the sputtered SiO_2 films is nearly identical to bulk SiO_2, the thermal conductivity is reduced to 80% of the bulk value (Cahill and Allen, 1994). Evaporated films are less dense than bulk, but if we model the density deficit as a small volume fraction of roughly spherical voids, the measured density deficit ($\simeq$10%) is far too small to account for the strong reduction in thermal conductivity that is observed.

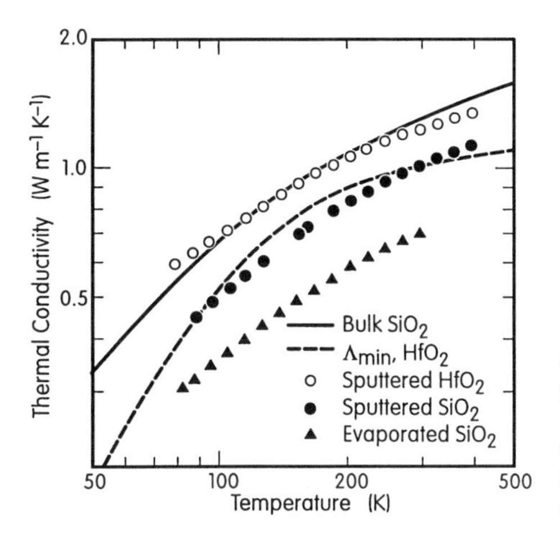

Figure 2-11 Thermal conductivity of sputtered and evaporated SiO_2 thin films (Cahill and Allen, 1994) and sputtered HfO_2 (Lee and Cahill, 1995) compared to the bulk conductivity of SiO_2 and the predicted Λ_{min} for HfO_2; Λ_{min} for a-SiO_2 is similar.

Since data for Vycor show that long-wavelength phonons with mean free paths do not contribute significantly to heat transport in SiO_2 (see the discussion above), we cannot attribute the reduction in thermal conductivity to phonon scattering from interfaces or internal microstructure and must look for an alternative explanation. One possibility, which we believe is relatively unlikely, is that SiO_2 films deposited on substrates held near room temperature have a different atomic scale structure and therefore different vibrational properties than SiO_2 prepared at higher temperatures. A second, more likely possibility is that the density deficit is due to highly anisotropic microvoids that are oriented parallel to the plane of the film. In this proposal, even a small volume fraction of voids could produce a strong reduction in through-thickness heat transport.

We end this section by discussing data for a crystalline film that has heat transport behavior characteristic of an amorphous solid (Lee and Cahill, 1995). Thin films of HfO_2 deposited by room-temperature sputtering are microcrystalline with an average grain size $\sim$100 Å, yet the thermal conductivity again approximates the minimum conductivity value (see Fig. 2-11). The cause of this low thermal conductivity is unclear at this time, but a likely explanation lies in the atomic-scale disorder caused by low-temperature deposition. Vacancies or interstitials created by ion-surface collisions or the incorporation of growth-related defects may produce sufficiently strong phonon scattering that results in glass-like thermal conductivities. Glass-like thermal conductivities are often observed in crystals containing large concentrations of vacancies or interstitials such as stabilized zirconia (see for example Fig. 2-8); in stabilized zirconia every two Y atoms that substitute for two Zr atoms in ZrO_2 add one oxygen vacancy to the lattice.

2-4 SOLID–SOLID INTERFACES

State-of-the-art microelectronic devices contain multiple layers of metallization, dielectrics, and semiconductors. An extreme example of a high density of internal interfaces are the superlattice structures used as the active region and Bragg reflectors of vertical cavity surface emitting lasers. Such a device may contain on the order of 100 interfaces between dissimilar materials in the space of a few microns. Clearly, a detailed understanding of heat transport in such a device requires knowledge of heat transport at solid–solid interfaces.

2-4-1 Diffuse and Acoustic Mismatch Models

One approach to modeling heat transport at a solid–solid interface is the diffuse mismatch model described by Swartz and Pohl (1989). This model is thought to be appropriate for all except the most perfect, epitaxial interfaces between materials. To apply the diffuse mismatch model in its simplest form, we assume i) that phonons incident on an interface are elastically scattered, ii) that the scattered phonon has no other correlation with the incident phonon (i.e., the scattered phonon has no memory of the mode and direction of the incident phonon), and iii) that the solids on both sides of the interface are elastically isotropic.

Since the temperature drop at the interface is small (i.e., any imbalance in the energy flux crossing the interface in the two directions is extremely small compared to the total energy flux crossing the interface), detailed balance can be invoked to greatly simplify the calculations. The transmission coefficient from side 1 to side 2, α_1, is then simply proportional to the ratio of the energy flux incident on the interface from side 2 to the total energy flux incident on both sides of the interface. The energy flux is given by $I_{i,j} = v_{i,j}\hbar\omega N_{i,j}(\omega, T)$, where $v_{i,j}$ is the group velocity on side i of mode j, and $N_{i,j}$ is the product of the density of states and the phonon occupation.

$$\alpha_1 = \frac{\sum_j I_{2,j}(\omega)}{\sum_{i,j} I_{i,j}(\omega)} \tag{2-9}$$

The thermal conductance of the interface per unit area G is calculated using

$$G = \frac{1}{4}\sum_j v_{1,j} \int_0^\infty \alpha_1 \hbar\omega \frac{dN_{1,j}(\omega, T)}{dT} d\omega \,. \tag{2-10}$$

We can further simplify this solution if we use the Debye model for the density of states. (As before, we make a minor extension of the Debye model to treat the transverse and longitudinal modes separately.) Then,

$$N_{i,j}(\omega, T) = \frac{\omega^2}{2\pi^2 v_{i,j}^3[\exp(\hbar\omega/k_B T) - 1)]} \tag{2-11}$$

for $\omega < \omega_{i,j}^{\text{cutoff}}$, and $\omega_{i,j}^{\text{cutoff}} = v_{i,j}(6\pi^2 n_i)^{1/3}$ with n_i the atomic density on side i.

The result of this calculation is shown in Fig. 2-12 with comparison to data by Swartz and Pohl (1987) and by Stoner and Maris (1993) for an Al/Al_2O_3 interface. We see that the model calculation for heat transport between two Debye-like solids is in good agreement with the data at low temperatures but greatly overestimates the thermal conductance of the interface at 300 K. We will return to a discussion of this discrepancy below.

For ideal epitaxial interfaces between similar materials (e.g., AlAs/GaAs interfaces), we might expect that the major assumption of the diffuse mismatch model, namely the random scattering of all vibrational modes by interface disorder, will no longer be valid. Instead, the possibility arises that we may need to consider coherent reflection and transmission of phonons at the interface. In this regime, energy transport across interfaces

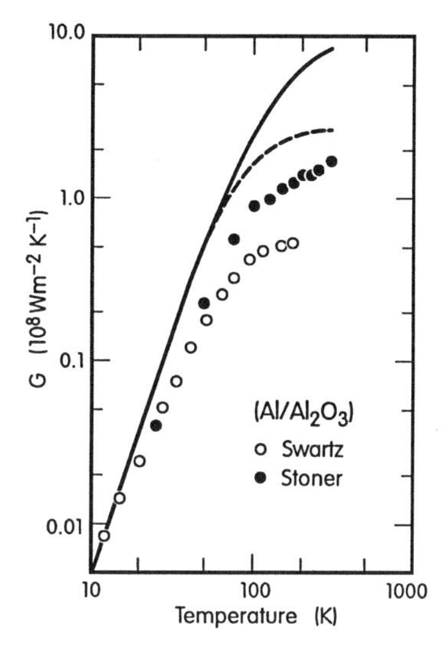

Figure 2-12 Data for thermal conductance of an Al/Al_2O_3 interface as measured by Swartz and Pohl (1987) and by Stoner and Maris (1993). The solid line shows the conductance calculated using the diffuse mismatch model in conjunction with a Debye model for the densities of states (no dispersion). The dashed line shows a microscopic calculation of the acoustic mismatch conductance—the dispersion and vibrational density of states of both materials are modeled using fcc crystal lattices (Stoner and Maris, 1993).

is described by the "acoustic mismatch model" that was developed by Khalatnikov (1952) to explain the small thermal conductance observed between solids and liquid He at $T < 1$ K. Detailed calculations using the acoustic mismatch model are complex, and to the best of our knowledge no experiment has yet been able to distinguish between the acoustic mismatch and diffuse mismatch models for energy transport between solids near room temperature. Some aspects of the acoustic mismatch regime can be incorporated in the diffuse mismatch model by relaxing the requirement that all phonons are randomly scattered at the interface.

As a practical demonstration of the importance of heat transport at solid–solid interfaces, consider a thin amorphous dielectric layer separating a metal film from a semiconductor substrate. As the thickness of the amorphous dielectric layer d decreases, the interfaces between the metal film and the dielectric and between the dielectric and the semiconductor become increasingly important in determining the conductance of the layer. This behavior is illustrated in Fig. 2-13, where we plot the apparent thermal conductivity of thin SiO_2 films deposited by plasma-enhanced chemical vapor deposition for film thickness between 30 nm and 300 nm (Lee and Cahill, 1997). We use the terminology "apparent" thermal conductivity because the measured thermal conductivity is influenced by the finite thermal conductance of the interfaces.

If we assume that the true thermal conductivity of the amorphous film is independent of film thickness, that is, phonons with mean free paths comparable to the film thickness make a negligible contribution to energy transport, we can model the apparent thermal conductivity by

$$\Lambda_a = \frac{G\Lambda/d}{(G + \Lambda/d)}. \tag{2-12}$$

This calculation is compared to the data in Fig. 2-13b to extract a series interface thermal conductance $G = 5 \times 10^7\ \mathrm{W\,m^{-2}\,K^{-1}}$ for the two interfaces at room temperature: $G = G_1G_2/(G_1 + G_2)$ where G_1 and G_2 are the individual interface conductances. As

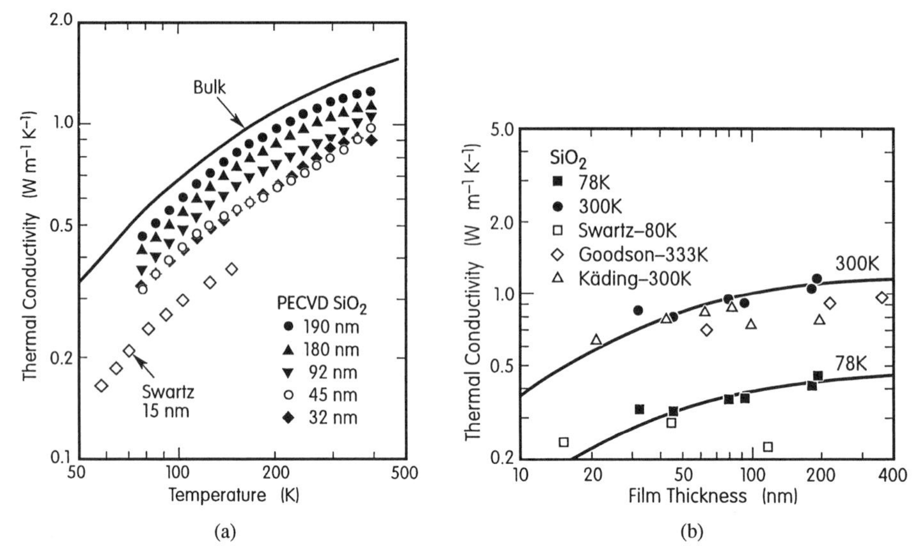

Figure 2-13 (a) Apparent thermal conductivity of thin SiO_2 films as a function of temperature (Lee and Cahill, 1997). (b) Apparent thermal conductivity of SiO_2 films plotted as a function of film thickness. The solid line is a fit to Eq. (2-12) to extract the series interface thermal conductance.

discussed above, this interface conductance is at least an order of magnitude smaller than the conductance calculated with the diffuse mismatch model using the Debye model.

The simplifications of Eqs. (2-9) and (2-11) are apparently unjustified in this case. Heat transport at an interface will be dependent on wavelength, direction, and mode. For example, at small vibrational frequencies we could expect that interface transport will approach the "acoustic-mismatch" behavior. Transport also could be strongly affected by interfacial layers that modify the phonon densities of states and transmission coefficients at the interface. Quantitative comparison with experiment will require that phonon dispersion be considered.

Stoner and Maris (1993) have provided a microscopic calculation of the heat transport between two fcc crystals where the masses of the atoms and the force constants between atoms are varied to model different solid–solid interfaces; the results of this calculation for a Al/Al_2O_3 are also included in Fig. 2-12. While the details of the microscopic calculations are beyond the scope of this chapter, the major difference in the two calculations is the fact that the microscopic model incorporates dispersion of the vibrational modes. In other words, the assumptions of the Debye model, while appropriate and quantitative at low temperatures, may not give reliable calculations near room temperature, particularly for heat transport at interfaces.

2-4-2 Phonon Radiation Transport

Consider again the geometry shown in Fig. 2-4a. If a significant fraction of the heat transport in the dielectric layer is due to phonons with mean free paths comparable to the film thickness, then we cannot treat the two interfaces separately; phonons scattered at one interface can reach the opposite interface without suffering additional scattering. One approach for calculating the heat transport in this geometry is the so-called phonon-radiation transport model (Chen and Tien, 1993; Majumdar, 1993). The terminology refers to the similarity of these model calculations to the solutions for heat transport by black body radiation.

If the dielectric film is sufficiently thin and a nearly perfect crystal, we can ignore phonon scattering within the film; therefore, phonons scatter only at the interfaces between materials. We can apply the diffuse mismatch model to estimate the probability that a phonon striking the interface from the dielectric side crosses the interface and becomes thermalized. In other words, we are assuming that phonons scattering into material 1 or material 2 (see Fig. 2-4a) will subsequently scatter *inelastically* with the distribution of scattered phonon energies determined by the temperature T_1 or T_2, respectively. (For this calculation, we can envision material 1 and material 2 as metals, that is, high-thermal-conductivity materials with large phonon scattering rates that can thermalize lattice vibrations.) Within the assumptions of this model, the thermal conductivity of the dielectric layer is given by (Chen and Tien, 1993)

$$\Lambda_{prt} = nk_B \sum_j v_j \left(\frac{T}{\Theta_j}\right)^3 \int_0^{\Theta_j/T} \frac{x^4 e^x l_{int}}{(e^x - 1)^2} dx, \tag{2-13}$$

$$l_{int} = \frac{3d}{4\left(\alpha_1^{-1} + \alpha_2^{-1} - 1\right)} \tag{2-14}$$

where l_{int} is an effective mean free path for the phonons due to interface scattering, and α_1, α_2 are the transmission coefficients characterizing the two interfaces. For materials with similar speeds of densities of vibrational states, the assumptions of the diffuse mismatch model give $\alpha_i \simeq 0.5$, and therefore $l_{int} \simeq d/4$. Furthermore, with the addition of an intrinsic phonon scattering rate $l_j^{-1}(\omega)$ in the film, the total scattering rate is given by the sum $l_j^{-1}(\omega) + l_{int}^{-1}$ (Chen and Tien, 1993).

The equations for phonon radiation transport *parallel* to the interfaces are more complex; for example, Chen and Tien (1993) have concluded that Matthiessen's rule is not strictly valid in this geometry. But as a starting point for discussion, as suggested by Chen and Tien (1993), the total scattering rate is approximately $l_j^{-1}(\omega) + l_{int}^{-1}$ with $l_{int} = d$.

To make a comparison to experiment, for example, the thermal conductivity of III-V superlattice device structures, we must provide some model for the intrinsic phonon scattering rate in the crystal. Although detailed models have been developed from analysis of phonon scattering mechanisms, we will simply assume that the scattering rates can be modeled by an empirical equation that includes two terms: i) point defect scattering and ii) anharmonic relaxation.

$$l_j^{-1} = A_0 \frac{\omega^4}{v_j} + BT \frac{\omega^2 \exp(-T_u/T)}{v_j^2} \tag{2-15}$$

The three parameters, A_0, B, T_u, are uniquely determined by fitting thermal conductivity data for the bulk crystal: the magnitude of the peak in the thermal conductivity at low-

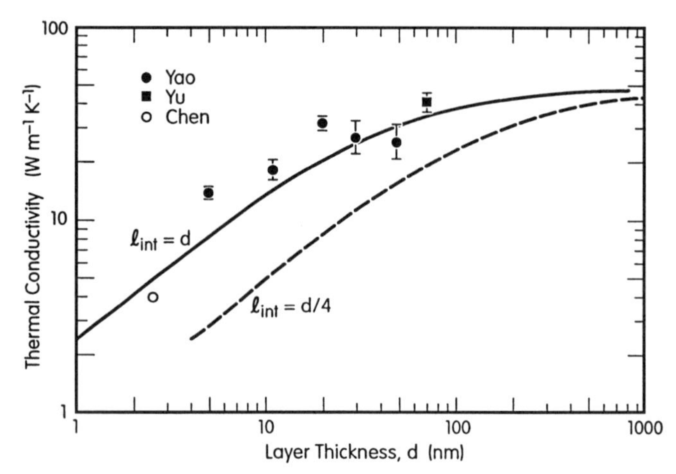

Figure 2-14 Calculated thermal conductivity for GaAs with the addition of an interface scattering rate l_{int}, and comparison to data for perpendicular (Chen et al., 1994) and in-plane thermal conductivity (Yao, 1987; Yu et al., 1995) of III-V superlattices.

temperatures ($T \sim 10$ K) is sensitive to A_0; the thermal conductivity near room temperature is sensitive to B; and T_u can be found to be fitting the temperature at which the low-temperature thermal conductivity is at a maximum. (To perform this fit, the phonon mean free paths are also limited by the dimensions of the crystal used to obtain the bulk thermal conductivity data.) For GaAs we find $A_0 = 6.4 \times 10^{-45}$ sec^3 and $B = 2.5 \times 10^{-6}$ cm K^{-1}. The calculated thermal conductivity of GaAs, using this intrinsic scattering rate in addition to an interface scattering rate of l_{int}, is plotted in Fig. 2-14 for $l_{int} = d/4$ and $l_{int} = d$.

Despite the significance for semiconductor technology, few data exist for heat transport in semiconductor superlattice structures. The available data by Chen et al. (1994), Yao (1987), and Yu et al. (1995) are summarized in Fig. 2-14. Chen et al. (1994) measured the perpendicular transport of a complex VCSEL device structure; the Bragg reflectors include short-period superlattices with an average layer thickness of 1.27 nm. Since the short-period superlattices make up approximately 1/2 of the sample thickness, we (somewhat arbitrarily) plot this data point at $d = 2.5$ nm. Yao (1987) measured a series of superlattice films with transport in the plane of the film. Yu et al. (1995) also measured the in-plane thermal conductivity of an AlAs/GaAs superlattice.

The model calculation, using a fixed interface scattering rate l_{int} and a Debye model for the lattice vibrations, approximates the data for AlAs/GaAs superlattices, but a quantitative description of heat transport in superlattice structures will require more sophisticated treatments that take into account realistic descriptions of the scattering at superlattice interfaces and deviations from the Debye density of states. We note that the dispersion of mode velocities clearly plays an important role in interface heat transport near room temperature (see for example Fig. 2-12). Additional experiments are also clearly needed, particularly experiments where the structural perfection of the interfaces is controlled.

This work was supported by NSF grant no. CTS-9421089.

NOMENCLATURE

A	area
A_0	strength of Rayleigh scattering
a	distance between atoms
B	strength of anharmonic scattering rate
C	heat capacity per unit volume
D	thermal diffusivity
D_s	thermal diffusivity of the substrate
d	film thickness
E	thermal effusivity
f_1	volume fraction of phase 1
G	thermal conductance per unit area
h	thickness of metal film
$\hbar$	Planck's constant divided by 2π
$I(\omega)$	heat flux spectral density
$I_{i,j}(\omega)$	heat flux spectral density of mode j on side i
k	wavevector of a vibrational mode
k_B	Boltzmann's constant
$l(\omega)$	mean free path of phonons with frequency ω
l_j	phonon mean free path of mode j
l_{int}	effective phonon mean free path due to interface scattering
$N_{i,j}$	product of the density of states and phonon occupation of mode j on side i
n	atomic density
P	power
R	thermal resistance per unit area
T	absolute temperature
T_1	absolute temperature of material 1
t	time
t_s	thickness of the substrate
v	sound velocity
v_j	sound velocity of mode j
$v_{i,j}$	sound velocity of mode j on side i
x	position
α_1	phonon transmission coefficient from side 1 to side 2
δ	thermal penetration depth
Λ	thermal conductivity
Λ_{Eins}	Einstein thermal conductivity
Λ_f	film thermal conductivity
Λ_{min}	minimum thermal conductivity
Λ_s	substrate thermal conductivity
Λ_1	thermal conductivity of phase 1
ν	mode frequency
ρ	electrical resistivity
Θ_E	Einstein temperature

Θ_i cutoff frequency for mode i expressed as a temperature
τ relaxation time
ω angular frequency

REFERENCES

Allen, P. B., and J. L. Feldman, "Thermal conductivity of disordered harmonic solids," *Phys. Rev. B*, vol. 48, pp. 12581–12588, 1993.

Ångström, A. J., *Ann. Phys. Chem.*, vol. 114, p. 513, 1861.

Birge, N. O., and S. R. Nagel, "Wide-frequency specific heat spectrometer," *Rev. Sci. Instrum.* vol. 58, pp. 1464–1470, 1987.

Cahill, D. G., "Thermal conductivity measurement from 30–750 K: The 3ω method," *Rev. Sci. Instrum.*, vol. 61, pp. 802–808, 1990.

Cahill, D. G., and T. H. Allen, "Thermal conductivity of sputtered and evaporated SiO_2 and TiO_2 optical coatings," *Appl. Phys. Lett.*, vol. 65, pp. 309–311, 1994.

Cahill, D. G., R. H. Tait, R. B. Stephens, S. K. Watson, and R. O. Pohl, in *Thermal Conductivity*, vol. 21, C. J. Cremers and H. A. Fine (eds.), pp. 3–16, Plenum, New York, 1990.

Cahill, D. G., S. K. Watson, and R. O. Pohl, "Lower limit to the thermal conductivity of disordered crystals," *Phys. Rev. B*, vol. 46, pp. 6131–6140, 1992.

Cahill., D. G., M. Katiyar, and J. R. Abelson, "Thermal conductivity of a-Si:H thin films," *Phys. Rev. B*, vol. 50, pp. 6077–6081, 1994.

Carslaw, H. S., and J. C. Jaeger, *Conduction of Heat in Solids*, Oxford University Press, Oxford, 1959.

Chen, G., and C. L. Tien, "Thermal conductivities of quantum well structures," *J. Thermophys. and Heat Transfer*, vol. 7, pp. 311–318, 1993.

Chen, G., C. L. Tien, X. Wu, and J. S. Smith, "Thermal diffusivity measurement of GaAs/AlGaAs thin-film structures," *J. Heat Transfer*, vol. 116, pp. 325–331, 1994.

Einstein, A., "Elementare Betrachtungen über die thermische Molekularbewegung in festen Körpern," *Ann. Phys.*, vol. 35, p. 679, 1911.

Feldman, J. L., M. D. Kuge, P. B. Allen, and F. Wooten, "Thermal conductivity and localization in glasses: Numerical study of a model of amorphous silicon," *Phys. Rev. B*, vol. 48, pp. 12589–12602, 1993.

Goodson, K. E., and M. I. Flik, "Solid layer thermal conductivity measurement techniques," *Appl. Mechan. Rev.*, vol. 47, pp. 101–112, 1994.

Goodson, K. E., M. I. Flik, L. T. Su, and D. A. Antoniadis, "Prediction and measurement of the thermal conductivity of amorphous dielectric layers," *J. Heat Transfer*, vol. 116, pp. 317–324, 1994.

Graebner, J. E., "Thermal conductivity of CVD diamond: Techniques and results," *Diamond Films Technol.*, vol. 3, pp. 77–130, 1993.

Hatta, I., "Thermal diffusivity measurement of thin films by means of an AC calorimetric method," *Rev. Sci. Instrum.*, vol. 56, pp. 1643–1647, 1985.

Käding, O. W., H. Shurk, and K. E. Goodson, "Thermal conduction in metallized silicon-dioxide layers on silicon," *Appl. Phys. Lett.*, vol. 65, pp. 1629–1631, 1994.

Khalatnikov, I. M., *Zh. Eksp. Teor. Fiz.*, vol. 22, p. 687, 1952.

Landauer, R., "The electrical resistance of binary metallic mixtures," *J. Appl. Phys.*, vol. 23, p. 779, 1952.

Lee, S.-M., and D. G. Cahill, "Thermal conductivity of sputtered oxide films," *Phys. Rev. B*, vol. 52, pp. 253–257, 1995.

Lee, S.-M., and D. G. Cahill, "Heat transport in thin dielectric films," *J. Appl. Phys.*, vol. 81, pp. 2590–2595, 1997.

Majumdar, A., "Microscale heat conduction in dielectric thin films," *J. Heat Transfer*, vol. 115, pp. 7–16, 1993.

Morath, J. C., H. J. Maris, J. J. Cuomo, D. L. Pappas, A. Grill, V. V. Patel, J. P. Doyle, and K. L. Saenger, "Picosecond optical studies of amorphous diamond and diamondlike carbon: Thermal conductivity and longitudinal sound velocity," *J. Appl. Phys.*, vol. 76, pp. 2636–2640, 1994.

Schmidt, R., T. Franke, and P. Häussler, "An improved dynamical method for thermal conductivity and specific heat measurements of thin films in the 100 nm range," unpublished.

Slack, G. A., F. Seitz, and A. G. Turnball (eds.), *Solid State Physics*, vol. 34, p. 1 (see p. 57), Academic Press, New York, 1979.

Stoner, R. J., and H. J. Maris, "Kapitza conductance and heat flow between solids at temperatures from 50 to 300 K," *Phys. Rev. B*, vol. 48, pp. 16373–16387, 1993.

Swartz, E. T., and R. O. Pohl, "Thermal resistance at interfaces," *Appl. Phys. Lett.*, vol. 51, pp. 2200–2202, 1987.

Swartz, E. T., and R. O. Pohl, "Thermal boundary resistance," *Rev. Mod. Phys.*, vol. 61, pp. 605–668, 1989.

Taylor, R., Measurement of thermal properties, in *CRC Handbook of Thermoelectrics Technology*, D. M. Rowe (ed.), CRC Press, Boca Raton, pp. 165–180, 1995.

Wood, J. W., "Thermal properties of arc-evaporated carbon films by photoacoustic calorimetry," Ph.D. Thesis, South Dakota School of Mines and Technology, 1994, unpublished.

Yao, T., "Thermal properties of AlAs/GaAs superlattices," *Appl. Phys. Lett.*, vol. 51, pp. 1798–1800, 1987.

Yu, X. Y., G. Chen, A. Verma, and J. S. Smith, "Temperature dependence of thermophysical properties of GaAs/AlAs Periodic Structure," *Appl. Phys. Lett.*, vol. 67, p. 3554, 1995.

CHAPTER

THREE

MICROSCALE RADIATION PHENOMENA

Jon P. Longtin

Department of Mechanical Engineering, State University of New York at Stony Brook, Stony Brook, NY 11794

Chang-Lin Tien

Department of Mechanical Engineering, University of California at Berkeley, Berkeley, CA 94720

3-1 INTRODUCTION

Thermal energy transport by radiation is, perhaps, the first heat transfer mechanism to be addressed by humankind. Our ancestors sought shelter from the radiant heating of the sun during the day and used fire to cook their food and warm themselves at night. Only recently, at the turn of the century, have significant advances been made in quantifying this mechanism of thermal transport. In fact, the origins of quantum mechanics, which has revolutionized the world with developments such as the transistor, computers, lasers, and nuclear power, started with Max Planck's attempt to predict correctly the thermal radiant emission of heated bodies at different temperatures (Tien and Lienhard, 1979). Since this time, the field of thermal radiation and radiant transfer has grown considerably. Though the basic nature of radiation generation and interaction with matter is well understood for many situations, the overwhelming majority of applications of such theory are for *macroscopic* systems and employ the so-called classical radiation theory.

This chapter addresses the topic of radiation transport on the *microscopic* scale, or microscale. *Microscale* in this sense may refer to microscale *length, time*, or *structure*. Associated with every radiation–matter interaction is a series of (1) characteristic *lengths*, such as the radiation wavelength and specimen dimension; (2) characteristic *times*, such as the duration of the radiation exposure, the lifetime of excited atoms and molecules that constitute the material, and the light propagation time through the material; and (3) fundamental *structural units*, including, for example, atoms, molecules, protein chains, small particles, and micromachined structures. Classical radiation transport makes the following assumptions about the relative magnitude of the characteristic scales in a given problem:

(1) the spatial extent of the material is much greater than the wavelength,
(2) the time duration of the radiation is much larger than the relaxation time of the molecules constituting the material, and

(3) the radiant energy is converted to thermal energy instantaneously (Brewster, 1992; Modest, 1993).

When these characteristic scales become comparable to each other or otherwise deviate from the assumptions on which the classical model is based, the classical model will fail to capture correctly the radiative transport, properties, interaction, and subsequent thermal behavior of the material. Such situations are important for several reasons. First, the deviation from classical theory can be dramatic, differing by orders of magnitude in some cases. Also, major technological advances in the past two decades have resulted in a broad spectrum of devices, technologies, and materials that violate the fundamental assumptions of the classical theory. Examples include ultrashort pulse lasers, contemporary electronic and optoelectronic devices, and novel new materials such as aerogels and nanoparticle materials. Such technology is rapidly being employed in an ever-increasing number of areas. Furthermore, the unique phenomena associated with microscale radiative transport often can be used to advantage. Examples include significantly enhanced thermal insulators that work due to inhibition of radiant and conductive heat transfer; nonequilibrium heating and processing of materials using short-pulse lasers; unique optical properties of extremely small particles, referred to as nanoparticles; and saturable absorption in solids and liquids, which is used to generate extremely short-duration, high-intensity laser pulses.

It is interesting to note that the concept of nonclassical radiative transport is not new. For example, in the field of cryogenics, extremely low temperatures are encountered. At such temperatures, the mean free path (MFP) of the photons responsible for radiative transport can become comparable to the dimension of the material, even at macroscopic length scales (Tien and Cunningham, 1973). Cravalho et al. (1967, 1968) discussed the deviation from classical radiative transport between two dielectric and metal materials separated by a thin gap nearly 30 years ago. More recently, the development of modern microfabrication technology and the ultrashort-pulse laser have dramatically shrunk spatial and temporal dimensions, respectively, resulting in deviations from the classical model for many practical applications of interest. A contemporary review of research in microscale heat transfer, including radiative transfer on the microscale, was presented by Duncan and Peterson (1994).

In the following sections, the impact of microscale length, time, and structure on radiation phenomena is discussed, along with a description of the relevant *radiation* aspects of microscale transport. The reader is referred to Chapter 1, Microscale Energy Transport in Solids, for a comprehensive overview of the basic physics of energy carriers and transport. Aspects of microscale conduction, convection, phase change, and interface transport are addressed in their respective chapters elsewhere in this text.

3-2 PRELIMINARIES

Before proceeding directly to discussions of microscale versus macroscale radiative transport, several basic concepts and definitions are discussed below.

3-2-1 Types of Radiation in Thermal Engineering

There are two primary sources of radiation encountered in contemporary thermal engineering: *thermal radiation*, resulting from a body's temperature (Siegel and Howell,

1992; Modest, 1993), and *laser radiation*, which is generated by stimulated emission in specific materials (Siegman, 1986). Both have unique and useful properties. Thermal radiation results from random transitions from one energy state to another in atoms, molecules, electrons, and ions. Transitions from higher- to lower-energy states, which result in emission of electromagnetic (EM) energy, are termed *radiative relaxations*, and transitions from lower- to higher-energy states due to absorption of EM energy are called *radiative excitations*. Other forms of radiation, including gas plasma discharge, sparks, light from chemical reactions, and fluorescence involve the same basic transition mechanisms, but they will not be discussed here.

Each transition involves one or more *photons*, which are the fundamental quanta of EM energy. Photons and their relationship to other energy carriers are discussed more fully in Chapter 1. Each photon has a wavelength λ and frequency ν that are related to the speed of light c as follows:[1]

$$\lambda\nu = c. \tag{3-1}$$

The energy of a photon E is related to its frequency:

$$E = h\nu = \frac{hc}{\lambda}, \tag{3-2}$$

where h is Planck's constant. The energy of an emitted or absorbed photon E is equal to the energy difference $E_1 - E_0$ between the initial and final energy states in a system undergoing a transition. The *intensity* I of the incident radiation is the energy of N_p photons crossing a unit area per unit time:

$$I = \frac{N_p h\nu}{At} = \frac{N_p hc}{At\lambda}. \tag{3-3}$$

In thermal radiation, the transitions are both thermally driven and random, which results in a statistical distribution of photon energies—and hence wavelengths—in the radiation (Tien and Lienhard, 1979). For an ideal body, or "blackbody," the intensity dependence on wavelength and temperature follows the so-called *Planck distribution* (Brewster, 1992; Siegel and Howell, 1992):

$$I_{b\lambda}(\lambda, T) = \frac{C_1}{\lambda^5[\exp(C_2/\lambda T) - 1]}, \tag{3-4}$$

where $C_1 = 2hc^2 = 11{,}909\ \text{W}\cdot\mu\text{m}^4/\text{cm}^2$; $C_2 = hc/k_b = 14{,}388\ \mu\text{m}\cdot\text{K}$; and T is the temperature in Kelvin. The blackbody distribution for several common temperatures is shown in Fig. 3-1.

Laser radiation, on the other hand, results from stimulated emission from excited atoms and molecules. This radiation is essentially monochromatic, is highly coherent and directional, and can be produced at extremely high intensities and for very short

[1] In this discussion, the radiation is assumed to pass through a vacuum or medium having a refractive index near unity (e.g., $n \approx 1.0$ for air). For a medium with $n > 1$, both the wavelength and speed of light are reduced by a factor of n: $c = c_0/n$, $\lambda = \lambda_0/n$, where c_0 and λ_0 are the free-space (vacuum) light speed and wavelength, respectively. The frequency remains the same, regardless of the material refractive index.

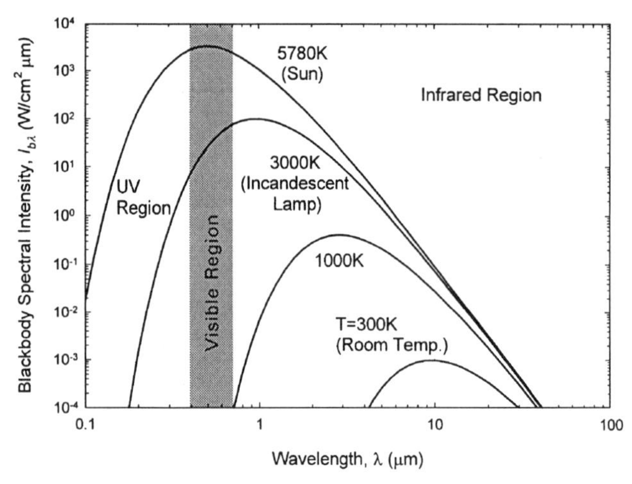

Figure 3-1 Planck blackbody intensity distribution.

durations. Laser radiation is thus a very useful tool in many diverse technologies, including manufacturing, measurement, communication, information storage, biological and medical areas, and consumer products. Although thermal radiation and laser radiation possess distinctly different qualities, both result from the same mechanism of transitions between different energy states.

The process of a substance interacting with radiation can be divided roughly into three steps. First is *absorption*, in which an individual particle (electron, atom, molecule, structure, etc.) absorbs the photon and converts the photon energy to thermal energy. Next, the particle will exchange this energy with its neighbors via collisions with energy carriers and/or other particles. This process is called *thermalization*, and is responsible for transferring the absorbed radiant energy to particles that did not participate directly in the absorption process. Finally, the temperature increase will increase the *emission* of photons, thus increasing the emitted intensity. In the following sections these mechanisms will be discussed in some detail. Specifically, microscale radiative transport at short length, time, and structural scales will be addressed.

3-3 RADIATION PHENOMENA ON THE SPATIAL MICROSCALE

In this section, microscale thermal radiation phenomena associated with small *length* scales is addressed. Tien and Chen (1994) proposed several microscale heat transfer regimes to categorize the nature of the radiation interaction with the medium or system of interest. These regimes are discussed below.

3-3-1 Radiation Length Scales

First, the characteristic lengths associated with the radiation and the matter with which it interacts must be identified. The radiation wavelength λ is of primary importance. For lasers, the radiation is usually highly monochromatic, and the wavelength is dictated

Table 3-1 Characteristic lengths at some common temperatures

Condition	Temperature	Peak wavelength	Coherence length
Near absolute zero	−272°C (1 K)	3 mm	2.2 mm
Liquid nitrogen (LN_2)	−77°C (196 K)	15 μm	11 m
Room temperature	25°C (298 K)	9.7 μm	7.3 μm
Incandescent light bulb	~2700°C (~3000 K)	0.97 μm	0.73 μm
Surface of the sun	5510°C (5783 K)	0.5 μm	0.37 m

by the lasing material (Siegman, 1986). For thermal radiation, however, a spectrum of wavelengths exists, as described by Eq. (3-4). The wavelength of the maximum intensity can be obtained by differentiating Eq. (3-4) with respect to λ and setting the result equal to zero. This yields the familiar *Wien displacement law* (Brewster, 1992; Modest, 1993), which relates the peak intensity wavelength λ_m and the temperature:

$$\lambda_m T \approx 2898\ \mu\text{m}\cdot\text{K}. \tag{3-5}$$

A list of the peak wavelength at common temperatures is shown in Table 3-1.

The second radiation length scale is the coherence length, l_c. This is the distance over which two photons from the *same* light source can generate interference effects when they interact. One unique property of laser light is its extremely large coherence length. The coherence length for the common HeNe laser, for example, typically exceeds several meters (Siegman, 1986). For thermal radiation, however, the coherence length is significantly smaller. For a blackbody in a vacuum, the coherence length is (Metha, 1963; Tien and Chen, 1994):

$$l_c = 0.15ch/k_bT \qquad \text{or} \qquad l_cT \approx 2160\ \mu\text{m}\cdot\text{K}. \tag{3-6}$$

Higher temperatures result in a decreased coherence length. Typical values of l_c for blackbodies are listed in Table 3-1, and the values of l_c and λ_m as a function of temperature are shown in Fig. 3-2. Note that in Eqs. (3-5) and (3-6), T refers to the temperature

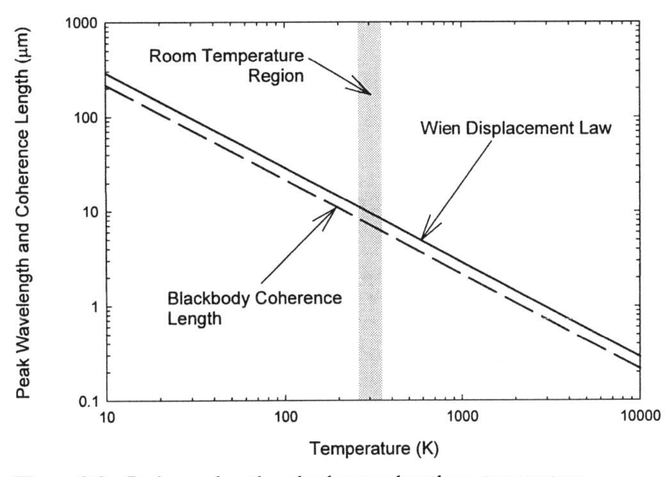

Figure 3-2 Peak wavelength and coherence length vs. temperature.

of the radiation *source* (e.g., the sun) rather than the material it interacts with (e.g., a parking lot).

3-3-2 Medium and Radiation–Medium Length Scales

There is a length scale associated with the radiation–medium interaction, and the medium itself also has several length scales. The *penetration depth* (or *skin depth*), δ, is the distance the radiation penetrates into the material. It is typically specified as the distance at which the intensity decreases to e^{-1} of the value at the surface of the medium. For classical absorption, which obeys Beer's law (Brewster, 1992; Modest, 1993),

$$I(x) = I_0 \exp(-\mu_a x), \tag{3-7}$$

and the absorption depth is simply $\delta = \mu_a^{-1}$. Here μ_a is the absorption coefficient (cm^{-1}), x is the distance into the medium, and I_0 is the incident intensity. The absorption coefficient depends on the wavelength and the material *extinction coefficient, k*. From EM theory (Siegel and Howell, 1992; Modest, 1993),

$$\mu_a = \frac{4\pi k}{\lambda}. \tag{3-8}$$

Absorption depths vary dramatically depending on the material. Metals absorb strongly in the near-infrared, visible, and ultraviolet regions, with typical $\delta \sim 20$–40 nm, while dielectrics like glass and water can have $\delta \sim 1$–10 m or more.[2] For nonlinear or intensity-dependent absorption mechanisms, the absorption profile can deviate significantly from Eq. (3-7), though an effective penetration depth can usually be obtained and the penetration depth concept can still be used.

The medium itself has a characteristic spatial dimension L (e.g., the thickness or diameter). Also, the energy carriers in the medium have characteristic wavelengths and MFPs, as discussed in Chapter 1. In solid metals and semiconductors, both electrons and phonons (quantized lattice vibrations) transport and store thermal energy. For solid dielectrics, only phonons participate in thermal transport and storage. In solids, the bonds between neighboring atoms are very strong, and the constituent atoms remain fixed in their position; the motion (vibration) they experience is only a small perturbation from their nominal position.

For liquids and gases, the situation is somewhat different. With the exception of liquid metals and plasmas, most liquids and gases do not have a significant number of free electrons for radiation absorption and energy transport. The bonding between molecules is considerably weaker, and the molecules undergo significant motion with respect to each other. In this case, the concept of phonons is no longer appropriate and is replaced by a transport mechanism based on collisions between molecules, which can be characterized by a collision rate and a MFP between collisions.

[2] An everyday example is a well-maintained swimming pool, in which one can clearly see the bottom, even three meters below the surface.

3-3-3 First Spatial Microscale Regime

Tien and Chen (1994) define the first microscale regime when[3]

$$L/l_c \precsim O(1). \tag{3-9}$$

Thus, when the characteristic dimension of the medium L becomes comparable to or less than the coherence length l_c, the photons can interact *coherently*, resulting in interference phenomena. As Tien and Chen point out, the optical properties of the material are not affected in this regime. The wave nature of the incident radiation, however, becomes important, so Maxwell's equations must be solved for the electric field in and around the material (Born and Wolf, 1980) to obtain properties such as reflection, transmission, and scattering and absorbing cross sections. Referring to Fig. 3-2, relatively small spatial dimensions are required to see such effects for thermal radiation. Examples include multilayer thin films, collections of fine particles, and micromachined structures. When dealing with laser radiation, however, wave effects can become important at macroscopic length scales due to the extremely long coherence length of laser light.

It is often stated that microscale effects become important when the characteristic material dimension becomes comparable to the radiation wavelength. This is a valid statement, since the ratio of the peak-intensity wavelength in Eq. (3-5) and the coherence length in Eq. (3-6) for thermal radiation is

$$l_c \approx 0.75\lambda_m, \tag{3-10}$$

which indicates that λ is the same order as l_c. However, it must be remembered that $L/l_c \precsim O(1)$ is the true physical criterion for nonclassical, wave interference behavior. Another important point to note is that the discussion and regime definitions in this section refer to a *single* entity, for example, a single-layer thin film, a single particle, or a single fiber. For a *collection* of items, the close proximity of neighboring structures can influence the electric field interacting with the structure of interest. Additional nonclassical phenomena must be considered in this case, as discussed later in Section 3-5.

The exact demarcation between microscale and macroscale effects in the first microscale regime depends on the system at hand. The internal intensity and heating in weakly and strongly absorbing particles was investigated by Tuntomo et al. (1991) and Qiu, Longtin, and Tien (Qiu et al., 1995; Longtin et al., 1995a), respectively. Chen and Tien (1992) and Richter et al. (1993) used the optical coherence theory of light to predict the radiative properties of thin films and delineated the relevant macro- and microscale regimes. Other examples are given in following sections.

3-3-4 Second Spatial Microscale Regime

The second microscale regime is defined as

$$L/\Lambda \precsim O(1) \tag{3-11a}$$

$$\delta/\Lambda \precsim O(1) \tag{3-11b}$$

$$L/\lambda_c \succsim O(1). \tag{3-11c}$$

[3]Note that the notation here differs from that of Tien and Chen (1994).

Equation (3-11a) represents the *size effect*, where the characteristic dimension L is smaller than the carrier mean free path, Λ. Values less than one imply that the majority of carriers strike a boundary rather than each other when they experience a collision, which imposes an upper limit on the maximum MFP and also makes the MFP a function of sample dimension. Equation (3-11b) states that the skin depth is less than the carrier MFP. Since the skin depth δ represents the distance over which the incident radiation intensity decays to e^{-1} of the value at the surface, the magnitude of the electric field changes considerably over this distance. When the skin depth is less than the carrier MFP, carriers near the surface experience a *nonuniform* electric field between collisions, and thus they interact differently with the incident radiation than if the electric field were uniform, as is assumed in the classical model. This phenomenon is called the *anomalous skin effect* (Ziman, 1960) and results in optical properties that differ from the classical case in which $\delta/\Lambda > O(1)$. Armaly and Tien (1970) considered both size effects and anomalous skin effects for optical properties of thin films of gold at cryogenic temperatures,[4] and they found significant deviation in the emissivity when compared to classical predictions.

Equation (3-11c) states that the characteristic material dimension L exceeds the heat carrier wavelength λ_c (see Section 3-4). For the second microscale regime, the optical properties of the material will be altered. Classical models such as the Drude and Lorentz models (Kittel, 1986) can be used by introducing an effective MFP that accounts for size and anomalous skin effects.

3-3-5 Third Spatial Microscale Regime

For the third microscale regime,

$$L/\lambda_c \precsim O(1). \tag{3-12}$$

In this case, the characteristic material dimension is less than the heat carrier wavelength. From the uncertainty principle, every energy carrier has a wavelength or, equivalently, an uncertainty in its position at any given time. For material dimensions less than the carrier wavelength, the wavefunctions that comprise the carrier (i.e., the solutions to the Schrödinger equation) will be perturbed by the presence of the boundaries, resulting in modified electronic energy states and phonon spectra, and the separation of continuous energy levels into discrete states. Quantum mechanics must be used to correctly characterize the transport in such cases. The unique optical properties of a quantum well or a nanoparticle, for example, result from the condition in Eq. (3-12) being satisfied (Yariv, 1989). The technology required to fabricate and analyze such novel structures has only recently developed and, coupled with the tremendous promise that these devices hold, has resulted in vigorous research in the area of quantum-based optical devices.

An interesting application of these various microscale radiation regimes was recently performed for the problem of radiative thermal transport between two metal surfaces separated by a small gap. The current disparity in the experimental data found in the literature

[4]The MFP is a strong function of temperature and increases rapidly with decreasing temperature. At cryogenic temperatures, for example, near 4 K, the MFP of electrons in metals can be over 1000 times that at room temperature.

was well explained by characterizing several different radiation-material interactions using the above microscale radiation regimes (Whale and Cravalho, 1997a,b). Both near and near- and far-field electromagnetic radiation fields as well as the microscopic details of the electrons in the materials were considered in their analysis.

3-4 RADIATION PHENOMENA ON THE TEMPORAL MICROSCALE

In the preceding section it was seen that when the characteristic dimensions of the material approach the inherent microscopic radiation and energy carrier length scales, the nature of the radiation interaction in the material can change. There also exists a characteristic set of *temporal* scales that can give rise to unique, nonclassical behavior in the system at short time scales. These time scales can be extremely fast, ranging from nanoseconds down to femtoseconds (10^{-15} s).

Short-pulse (SP) lasers, which can generate laser pulses as short as several femtoseconds, are by far the most common source of short-time-scale radiant energy. These lasers produce extremely high intensities, often exceeding 10^9–10^{12} W/cm^2 or more. The following discussion will thus focus strongly on SP laser-based radiation interaction with materials. It is important to keep in mind that the following short-time-scale phenomena are *always* present. In many situations, however, these time scales are far faster than that of the incident radiation or the time resolution of the observation, and thus do not result in an observable deviation from the classical model.

The following time scales are important for radiation–material interactions:

1. laser pulse duration, τ_p,
2. thermal diffusion time, τ_d,
3. radiation propagation time, τ_c,
4. relaxation/thermalization time, τ_r.

Note that although the following focuses on pulsed laser radiation, the discussion is valid for all types of short-duration radiation, for example, radiation from a spark discharge or chemical reaction. In such cases, τ_p is taken as the duration of the radiation.

3-4-1 Laser Pulse Propagation and Duration

The *duration* of the pulses emitted from a short-pulse laser is uniquely determined by the laser itself and its associated optics. Pulsed lasers are available over an extremely broad range of pulse durations, ranging from continuous-wave (CW) lasers (infinite τ_p) down to 10^{-14} sec, and over wavelengths ranging from the near infrared (NIR) to the ultraviolet (UV). Another important aspect of the laser pulse is its *temporal intensity distribution.* A laser pulse propagating in a vacuum or nonabsorbing, nonscattering medium is governed by the wave propagation equation:

$$\frac{1}{c}\frac{\partial I}{\partial t} + \frac{\partial I}{\partial x} = 0. \tag{3-13}$$

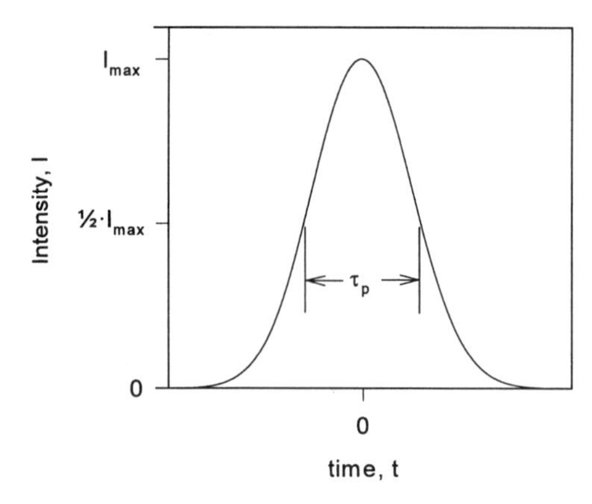

Figure 3-3 Gaussian laser pulse and definition of pulse duration.

The intensity of the pulse varies with time[5]; this variation is known as the temporal pulse shape (Siegman, 1986). The most common pulse shape is the Gaussian profile, shown in Fig. 3-3. The intensity of a Gaussian laser pulse as a function of position x and time t in a medium of refractive index n with no absorption or scattering is

$$I(x,t) = I_m \exp\left(-4\ln 2\left(\frac{t - xn/c}{\tau_p}\right)^2\right). \tag{3-14}$$

The position x is the distance from the point of maximum intensity I_m at $t = 0$. Unlike, for example, the square-wave pulse, the Gaussian laser pulse has no well-defined pulse duration. The most common definition for the pulse duration in this case is the *Full Width at Half Maximum (FWHM)* τ_p, defined as the time the pulse intensity exceeds one-half of the maximum intensity (Fig. 3-3) (Siegman, 1986; Shank and Ippen, 1990). The factor 4 ln 2 is required in Eq. (3-14) to insure that $I(0, \pm\tau_p/2) = I_m/2$, as required by definition.

3-4-2 Thermal Diffusion Time

The characteristic length of the material results in a time scale for conduction:

$$\tau_d \sim \frac{L^2}{\alpha}, \tag{3-15}$$

where α is the thermal diffusivity of the material. This time scale represents the characteristic time for thermal energy to *diffuse* through a material having characteristic length L (Arpaci, 1966). Note that this time scale is based on classical conduction. If microscale length effects, such as those discussed in Chapter 1, become important in the conduction, the appropriate conduction time scale should be used.

3-4-3 Radiation Propagation Time

All EM radiation travels at the speed of light, c_0/n, where c_0 is the speed of light in a vacuum, 3.0×10^{10} cm/s, and n is the refractive index of the medium. Thus, the spatial

[5]The exception is the CW laser, which has a constant, uninterrupted output intensity.

extent of the medium results in a radiation propagation time scale. The time required for radiation to propagate across a medium of characteristic dimension L is

$$\tau_c \sim \frac{Ln}{c_0}. \tag{3-16}$$

The time required for radiation to traverse macroscopic dimensions is usually very short. For example, it takes only 50 ps for light to travel through a 1-cm-thick layer of glass ($n = 1.5$).

Alternatively, a spatial time scale can be determined for radiation with a fixed temporal duration. To determine the spatial extent of a laser pulse of duration τ_p, for example. Eq. (3-16) can be rearranged as

$$l_p \sim \frac{c_0 \tau_p}{n}, \tag{3-17}$$

where l_p is the spatial extent of the laser pulse. For example, for a 1-ps pulse in glass the spatial extent of the pulse is $\sim$300 μm.

3-4-4 Relaxation Time

The material itself also has one or more intrinsic time scales. These time scales reflect the microscopic interaction of the incident radiation with the material. For metals, the radiant energy is absorbed by electrons, which later transfer excess energy to the phonon (lattice) system by electron–phonon collisions (Qiu and Tien, 1993; see also Chapter 1). For very short laser pulses, lattice heating can occur *after* the pulse has been absorbed. The absorption by the electron system is very fast, typically occurring in several femtoseconds. In contrast, electron–phonon thermalization requires a much longer time, on the order of $\sim 10^{-13}$–10^{-12} sec, and this thermalization time is used for τ_r. The absorption and thermalization processes for laser radiation interacting with metals is discussed in detail in Chapter 1.

For dielectrics the number of free electrons at room temperature is insignificant. Nearly all electrons remain bound to their individual molecules at all times. This leads to a different absorption mechanism than that for metals. When an atom or molecule in a dielectric absorbs a photon, transitions to more energetic electronic and vibrational states within the molecule occur. After a characteristic *relaxation time*, the atom or molecule transitions back to its original state, transferring its stored energy to neighboring molecules, and thus heats the material. Under certain circumstances, the atom can re-emit a second photon upon relaxing (called fluorescence). The relaxation time is a function of the material, temperature, and state (solid, liquid, gas) (Tien and Lienhard, 1979; Seilmeier and Kaiser, 1993).

Semiconductors fall in between metals and dielectrics. Semiconductors have small *band gaps*, and thus an incident photon at optical frequencies can provide enough energy to promote an electron to a free-like state similar to that in a metal. The intentional addition of dopants into the semiconductor material further alters the electronic structure of the material, as the dopant atoms can also interact with the incident radiation.

3-4-5 Temporal Microscale Regimes

Of the time scales discussed above, the laser pulse duration is an imposed time scale; the other time scales are all material- or dimension-dependent. Accordingly, the temporal microscale regimes are delineated based on the magnitude of the laser pulse duration compared to the other time scales.

First temporal microscale regime. The first temporal microscale regime is defined as

$$\tau_p \lesssim \tau_d \tag{3-18}$$

or, using Eq. (3.15),

$$\tau_p \lesssim L^2/\alpha. \tag{3-19}$$

Equation (3-18) states that the laser pulse τ_p is comparable or faster than the time required for thermal energy to diffuse through the medium, τ_d. In this case, the thermal energy is deposited in the material before it can be conducted away. As a result, the initial temperature profile will approximate the EM source distribution inside the material (Longtin et al., 1995a). Note that this criterion can hold for macroscopic systems as well as microscopic systems. Due to the quadratic dependence on L in Eq. (3-18), the size of the material will have a pronounced impact on this time scale.

Example: Short-pulse laser heating of absorbing particles. An example where such effects are important is the laser heating of highly absorbing small particles. Longtin et al. (1995a) investigated different heating regimes based on the duration of the laser pulse, the particle size, and the optical and thermophysical properties of a highly absorbing particle in a nonabsorbing medium. Such situations are important in assessing damage thresholds for high-power laser optics (Guenther and McIver, 1989), laser cladding (Komvopoulos and Nagarathnam, 1990), and laser surface cleaning (Kelley and Hovis, 1993; Park et al., 1994). The distribution of radiant energy inside the particle was calculated by Qiu et al. (1995) using rigorous EM theory (Bohren and Huffman, 1983), and is shown in Fig. 3-4 for a spherical platinum particle with a 20 μm diameter exposed to pulsed radiation with a wavelength of 10 μm (CO_2 laser). As can be seen, the internal radiant energy distribution—and hence the heating—is highly nonuniform within the particle. The conduction in the particle was treated classically and solved numerically. For a spherical particle, L is taken as the radius of the particle, a. The thermal diffusivity for platinum is $\sim 5 \times 10^{-5}$ m^2/sec.

Three calculations were performed; for each case, the duration of the laser pulse was changed to alter the relative magnitude of the heating and conduction time scales in Eq. (3-18). The particle is illuminated from the backside in Fig. 3-4, and temperature is plotted along the cross section of the particle just after the laser pulse has passed, i.e., $t = \tau_p$. In Fig. 3-5a, the laser pulse duration is 20 μs, and $\tau_p \alpha / a^2 = 10$, so the laser heating occurs over a time scale much longer than that for conduction. The thermal energy is able to diffuse throughout the particle before the laser pulse has passed, so the temperature distribution is highly uniform, which is markedly different from the radiant source distribution shown in Fig. 3-4.

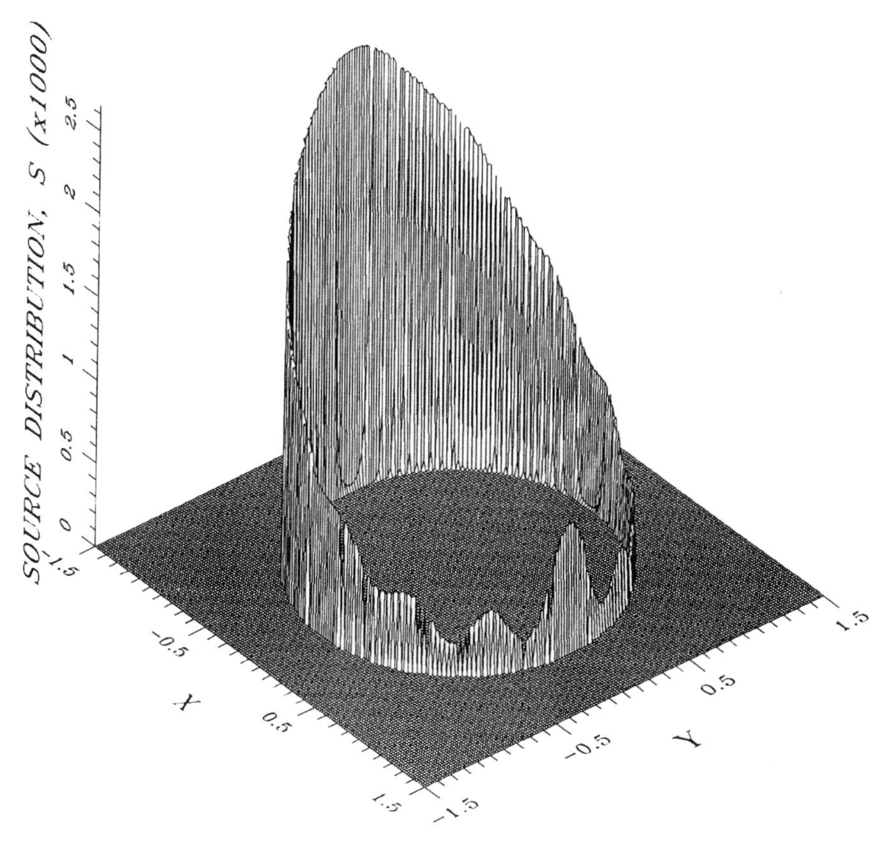

Figure 3-4 Intensity distribution inside Pt particle.

In Fig. 3-5b, the pulse duration has been decreased to 2 μs, and $\tau_p \alpha / a^2 = 1$. In this case, the pulse duration is comparable to the conduction time scale. As can be seen, the temperature distribution is no longer uniform and bears a resemblance to the source distribution. The temperature at the back of the particle is about twice that at the front. In Fig. 3-5c, the pulse duration has been decreased to 200 ns, resulting in $\tau_p \alpha / a^2 = 0.1$. The temperature distribution is highly nonuniform, and both the angular and radial characteristics of the temperature distribution resemble those of the source distribution shown in Fig. 3-4. The peak temperature rise near the back of the particle is over seven times greater than that at the front. As can be seen from this example, the relative time scales for radiation and conduction can have a dramatic impact on the resulting temperature distribution in the material.

Second temporal microscale regime. For the second microscale regime,

$$\tau_p \lesssim \tau_c. \tag{3-20}$$

In such situations, the laser pulse duration is on the order of or shorter than the light propagation time τ_p across the spatial extent of the medium. As a consequence, the

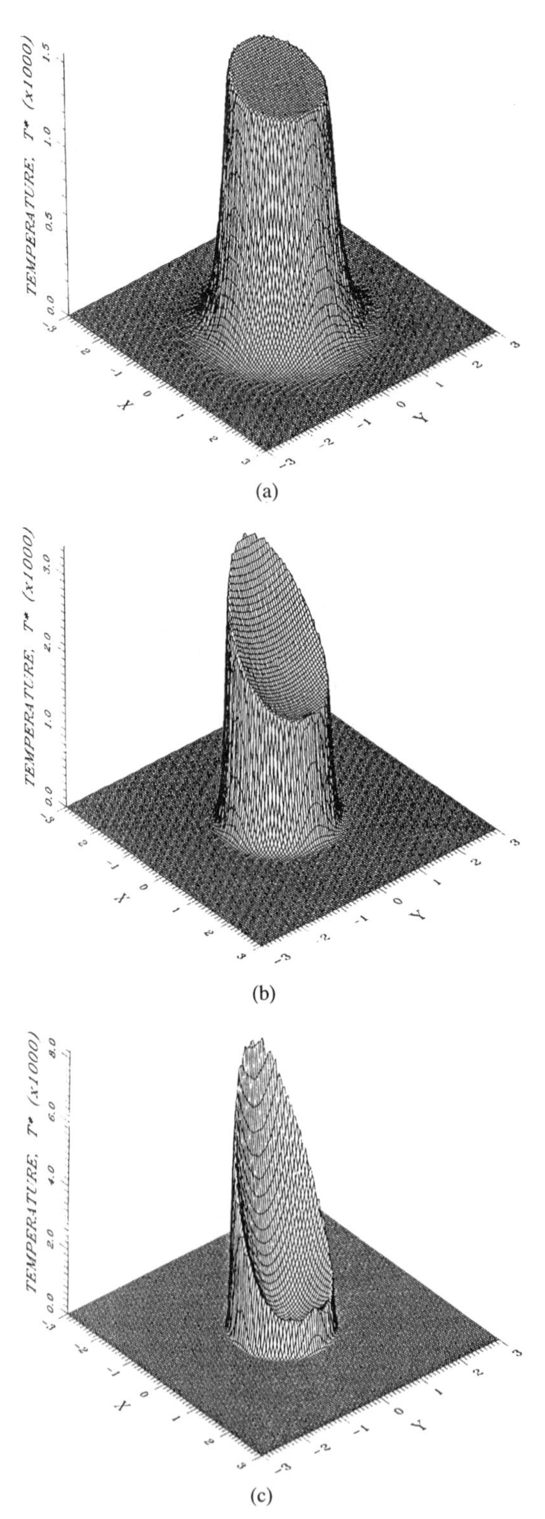

Figure 3-5 Temperature distributions inside particle for different laser pulse durations.

laser pulse intensity will vary as a function of position and must be taken into account if (1) there are intensity-dependent effects exhibited by the medium, such as intensity dependent absorption or scattering, or (2) the short-time-scale nature of the laser pulse propagation is of interest, for example, in diagnostic measurements that rely on the delay time of the laser pulse to obtain information about the medium it passes through (Yamada, 1995a; Kumar, 1995). The propagation of the radiation must be then modeled as a function of position and time.

The degree of complexity required to model the radiation propagation depends on the type of medium and the information required. Three basic possibilities exist. If the medium neither absorbs nor scatters, Eqs. (3-13) and (3-14) can be used to model the pulse propagation. For a medium that absorbs strongly but scatters weakly, Eq. (3-13) can be modified to include the absorption and scattering:

$$\frac{1}{c}\frac{\partial I}{\partial t} + \frac{\partial I}{\partial x} = -(\mu_a + \mu_{sc})I. \tag{3-21}$$

The solution of this equation yields the intensity as a function of position and time. Nonlinear and otherwise intensity-dependent absorption and scattering can be added to the right side of Eq. (3-21) in a similar fashion. The scattering in this case is assumed to be small enough so that re-scattering back into the beam path is insignificant. Laser radiation is also highly unidirectional, making the analysis considerably easier.

If the medium scatters strongly or the intensity is desired *away* from the beam path, then the scattering must be treated rigorously. An equation of radiative transport (ERT) that includes the temporal term and the scattering is recommended (Brewster, 1992; Modest, 1993; Kumar, 1995; Yamada, 1995a). Three common situations for absorption and scattering are illustrated in Fig. 3-6.

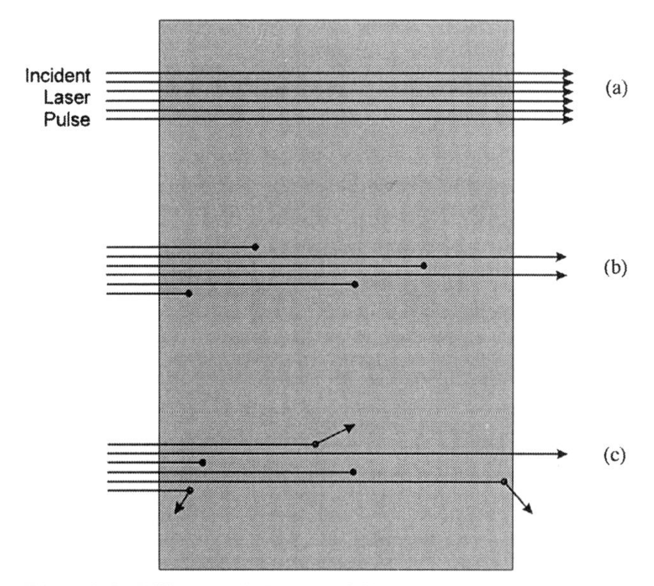

Figure 3-6 Different radiation-material interactions: (a) no absorption or scattering; (b) strong absorption, weak scattering; (c) strong absorption and scattering.

For the one-dimensional case (Kumar, 1995; Yamada, 1995a), the ERT becomes

$$\frac{1}{c}\frac{\partial I}{\partial t} + \xi\frac{\partial I}{\partial x} = -(\mu_a + \mu_{sc})I + S + \frac{\mu_{sc}}{2}\int_{-1}^{1} Ip(\xi' \to \xi)\xi', \tag{3-22}$$

where ξ is the cosine of the angle between the observation and propagation direction, and p is the scattering phase function. The left side of Eq. (3-22) represents propagation of the radiation (both incident and scattered). The first term on the right side represents *linear* absorption and scattering of radiation (hence the negative sign), S represents a source of radiation, and the third term represents radiation scattered into the observation direction. The addition of the scattering term on the right side of Eq. (3-22) makes the solution considerably more difficult. A discussion of various simplifications, solution techniques, and boundary condition issues for strongly scattering media at short time scales has been presented by Kumar (1995) and Yamada (1995a).

Example: Optical imaging using short-pulse lasers. *Optical imaging* in the medical field is a promising technique that takes advantage of the temporal nature of short-pulse propagation in tissues (Yamada, 1995a; Kumar, 1995). The basic idea is that a short laser pulse of known temporal character is directed into a tissue sample. Ultrafast detectors are then used at various points around the tissue body to measure the intensity and time delay of the emerging light. This information, in conjunction with a mathematical model, can be used to noninvasively image structures within the tissue body. Such a measurement is called an *inverse problem* because the results (the time-dependent intensities) are known and, in conjunction with a radiation propagation model, used to determine the geometry and optical properties within the tissue responsible for the measured results. This information can then be used to determine if a tissue is healthy or diseased, or if a foreign object such as a tumor is present. The time-resolved technique has developed only within the past decade or so, as it is only during this time that picosecond laser pulses and detection systems have become readily available.

Living tissues are highly complicated structures and tend to scatter light much more strongly than they absorb (Yamada, 1995a, 1995b). Accordingly, the scattering terms in Eq. (3-22) must be included. Many techniques can be used to process the time-dependent intensity data to obtain an image. Yamada reported results for several techniques including time-gating, the temporally extrapolated absorbance method (TEAM), and a CW technique. The test configuration consisted of two concentric cylinders with different scattering and absorption coefficients. The test configuration and laser beam path are shown in Fig. 3-7. The results for time-gating, TEAM, and CW methods are shown in Fig. 3-8, along with the true image. As can be seen, a fairly quantitative, noncontact measurement of a foreign object in a strongly scattering medium is possible with these techniques.

Third temporal microscale regime. The third microscale time regime involves phenomena that are limited by the relaxation time of the excited electrons, atoms, and/or molecules. The case of free electron absorption and thermalization in metals is discussed in Chapter 1. This section focuses on molecular-based absorption in dielectrics in which the electron remains attached to the molecule. Ultrafast laser interactions with semiconductors are complicated considerably due to the small band gap in these materials and

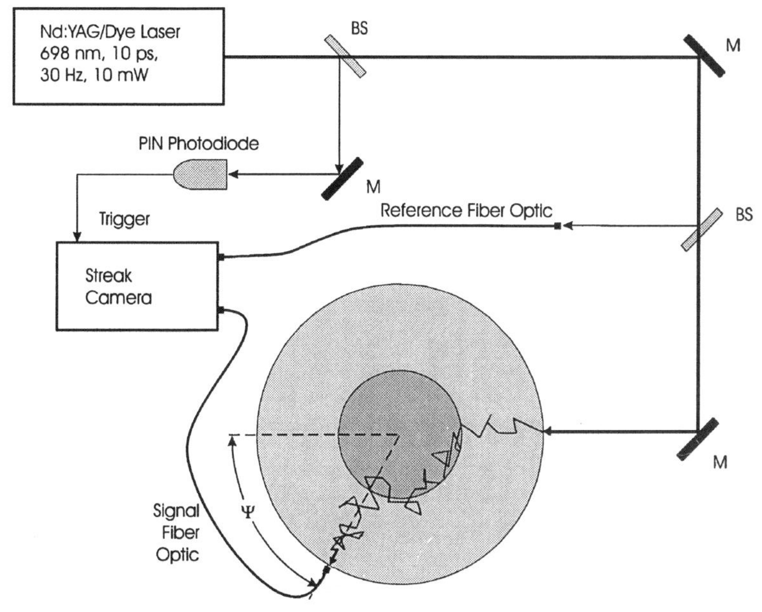

Figure 3-7 Time-resolved optical imaging experiment (Yamada, 1995a).

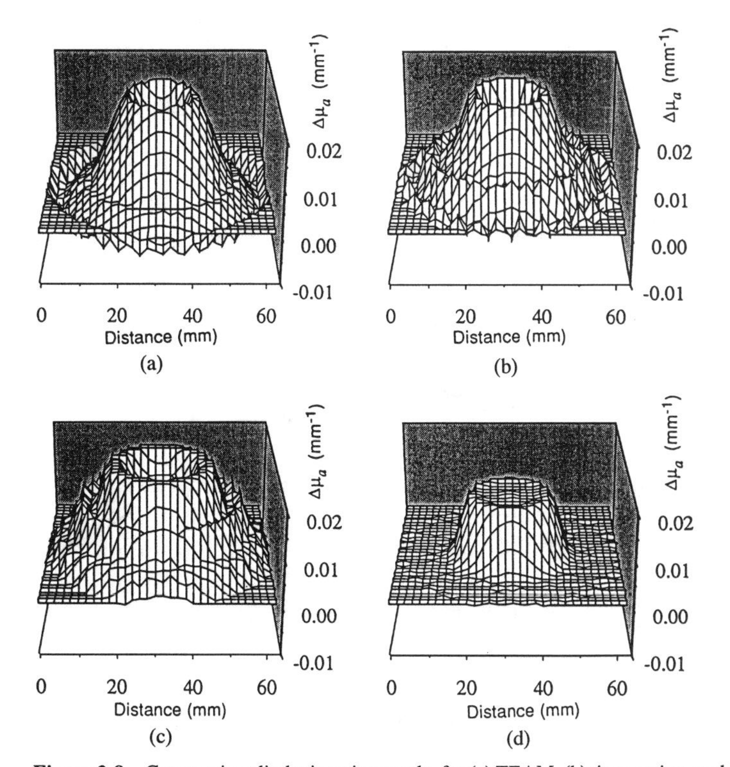

Figure 3-8 Concentric cylinder imaging results for (a) TEAM, (b) time-gating, and (c) CW techniques. True image is shown in (d).

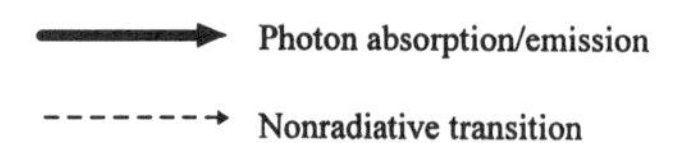

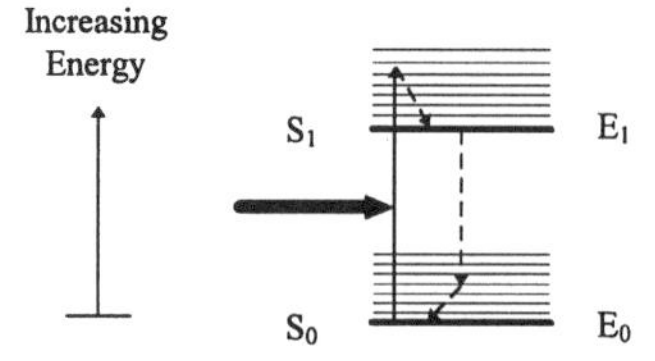

Figure 3-9 Energy states and photon absorption for a molecule.

the addition of dopants. Such interactions are the subject of ongoing research; they will not be discussed here.

In classical absorption, molecules are assumed to relax instantaneously after absorbing radiant energy. As a result, the energy states of the molecules can be well approximated by the equilibrium distribution from statistical mechanics, which is determined solely by the medium temperature (Tien and Lienhard, 1979). This gives rise to a constant classical absorption coefficient μ_a. In reality, molecules have finite relaxation times. Thus, after absorbing a photon, the molecule remains in an excited energy state for a period of time before returning (relaxing) back to its original state, where it exchanges the absorbed energy with its neighbors. If enough molecules are promoted to an excited state before they can relax, the population distribution of the molecules will change dramatically, with many more molecules in excited states than would normally be the case. One common example of such a situation is the operation of the laser itself, which relies on an external optical or electrical source to promote molecules to excited states (Siegman, 1986). The key factor is that the excitation rate exceeds the relaxation rate, yielding the large number of excited molecules required for lasing action.

To understand the radiation interaction with a material, the microscopic nature of the radiation interaction with the molecules constituting the material must be addressed. *Rate equations* often are used to determine the number of molecules in different states as a function of time and intensity. A schematic of the available energy states in a simple two-state molecule is shown in Fig. 3-9. The thick horizontal lines S_0 and S_1 represent, respectively, the ground and first excited states, with energies E_0 and E_1. Associated with each electronic state is a series of unique vibrational states, represented by the thin horizontal lines above each electronic state. Absorption by excited states, triplet transitions, radiative relaxations such as fluorescence and phosphorescence, and stimulated emission (lasing) have not been included in Fig. 3-9.

If a photon incident on a molecule has an energy $h\nu \sim E_1 - E_0$, the molecule can absorb the photon, promoting it from the ground state S_0 to the excited state[6] S_1. After a characteristic relaxation time, τ_r, the molecule relaxes back to the ground state. The

[6]The transition does not *have* to occur between electronic states. Absorption of long-wavelength (low-energy) photons can occur by purely vibrational transitions within the ground state S_0. The CO_2 laser, for example, with a wavelength of 10.6 μm, lases exclusively from vibrational transitions in the CO_2 gas molecules. The discussion here is equally valid for vibration-only transitions.

probability of absorbing a photon is represented by the *effective absorption cross section*, σ (cm^2). The absorption coefficients for molecules in the ground and excited states, σ_{gr} and σ_{ex}, are often dramatically different (Siegman, 1986; Seilmeier and Kaiser, 1993), and both often show a strong dependence on wavelength.

The rate equations represent the combined effects of absorption and relaxation of the excited molecules. For the simple two-level system in Fig. 3-9, the rate equations are (Penzkofer, 1988)

$$\frac{\partial N_0}{\partial t} = \frac{-\sigma_{gr} N_0 I}{h\nu} + \frac{N_1}{\tau_r} \tag{3-23}$$

$$\frac{\partial N_1}{\partial t} = \frac{\sigma_{gr} N_0 I}{h\nu} - \frac{N_1}{\tau_r} \tag{3-24}$$

with the conservation equation

$$C_0 = N_0 + N_1. \tag{3-25}$$

Here N_0 and N_1 represent the number density of molecules (cm^{-3}) in the ground and excited states, respectively, and I is the incident laser intensity. Laser radiation, being essentially monochromatic, is well characterized by a single frequency ν; hence the energy of each photon is $h\nu$, and the number of photons incident on a unit area per unit time is $I/h\nu$. The total photons absorbed is the product of the molecular cross section σ_{gr}, the number of absorbing molecules, N_0, and the photon flux, $I/h\nu$. The reciprocal of the relaxation time $1/\tau_r$ represents the *rate* at which molecules relax per unit time. When multiplied by the number of molecules in the excited state, N_1, this term represents the total number of molecules relaxing per unit time. Equation (3-25) simply states that every interacting molecule must be in either the ground or excited state. These equations thus represent a balance between excitation of molecules due to the incident laser pulse, and relaxation back to the ground state.

At low laser intensities, nearly all of the molecules remain in the electronic ground state.[7] In this case the medium behaves classically, and the classical absorption coefficient μ_a is simply the cross section of the absorbing molecules multiplied by the number density of absorbing molecules in the ground state:

$$\mu_a = \sigma_{gr} N_0 \approx \sigma_{gr} C_0. \tag{3-26}$$

If, however, the laser intensity is high, many molecules will be promoted to an excited state before they can relax to the ground state. Since excited molecules usually absorb significantly less radiation than molecules in the ground state, the absorption in a highly excited medium will be substantially reduced. This effect is called *saturable absorption* and is used routinely in the generation and modification of laser pulses and in diagnostic measurements (Hercher, 1967; Penzkofer, 1988). Referring to Eq. (3-23), the approximate intensity required to keep the same number of molecules in both the ground and

[7] At temperatures below ~1000 K, this is a very good approximation. For temperatures in excess of several thousand Kelvin, the thermal energy of the molecules is sufficient to initiate electronic excitations. Such situations are not considered here.

excited state (i.e., $N_0 \sim N_1$) is $I_s \sim h\nu/\sigma_{gr}\tau_r$. This characteristic intensity is called the *saturation intensity.* For intensities comparable or greater than the saturation intensity, deviation from classical absorption will occur. It should also be noted that if scattering is important, saturable absorption effects can be readily incorporated into the equation of radiative transport, Eq. (3-22).

On an even more fundamental level, the wave nature of the incident light, which is neglected in the rate equations, can be considered. The interaction with the radiation can also be described entirely from a quantum-mechanical perspective. Though still in the developmental stage, such approaches offer the promise of calculating very detailed information about the fundamental laser–material interactions (Kotake and Kuroki, 1993; Kotake, 1995).

Example: Saturable absorption and heating in dielectrics. From the thermal perspective, the altered absorption at high laser intensities will result in a heating profile that deviates from classical behavior. Longtin and Tien (1996a) characterized this deviation for a commercial organic dye, *Q-Switch I*. For the study a Nd:YAG laser with $\lambda = 1064$ nm, $\tau_p = 35$ ps, and a peak intensity of 3×10^9 W/cm^2 was used to irradiate the dye, which had $\tau_r = 11$ ps and $\sigma_{gr} = 6.1 \times 10^{-16}$ cm^2. The saturation intensity $h\nu/\sigma_{gr}\tau$ absorption for this system was $\sim 2.8 \times 10^7$ W/cm^2.

In Fig. 3-10, the calculated temperature as a function of distance into the liquid is shown for both the saturable absorption and the classical models. The laser pulse is incident from the left. For an incident intensity $I_p = 0.1\ I_s = 2.8 \times 10^6$ W/cm^2, the difference between the classical and microscale model is insignificant (i.e., the saturable absorption model reduces to the classical model). Also shown in Fig. 3-10 is the result for

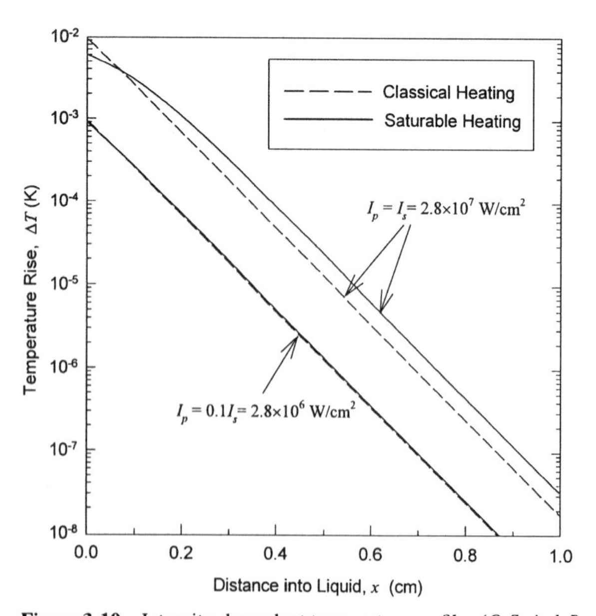

Figure 3-10 Intensity-dependent temperature profiles (*Q-Switch I*).

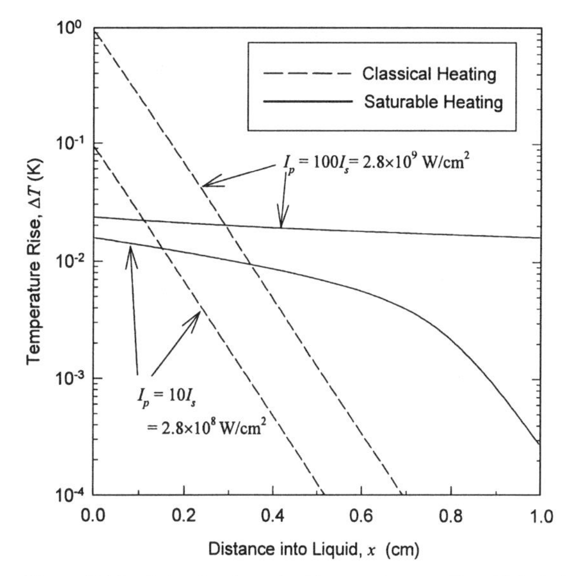

Figure 3-11 Intensity-dependent temperature profiles (*Q-Switch I*).

an incident intensity comparable to the saturation intensity: $I_p = I_s = 2.8 \times 10^7$ W/cm^2. The temperature rise at the front region near $x = 0$ is slightly less than that predicted by the classical model, as a result of saturable absorption. When the liquid saturates, less energy is absorbed, and the heating decreases. As the laser pulse propagates through the liquid, however, enough energy is eventually absorbed to reduce its intensity below I_s, and classical heating resumes for $x > \sim 0.3$ cm.

In Fig. 3-11, the intensity is again increased tenfold: $I_p = 10\ I_s = 2.8 \times 10^8$ W/cm^2. For this case, the temperature at the front of the liquid is nearly an order of magnitude less than that predicted by the classical model. Likewise, the temperature at the rear of the liquid is higher because the laser pulse is not attenuated significantly in passing through the front region of liquid. Around $x \sim 0.8$ cm, there is enough absorption of the pulse to quench saturable heating effects, and the absorption becomes classical. Also shown in Fig. 3-11 are results for an intensity $I_p = 100\ I_s = 2.8 \times 10^9$ W/cm^2. The temperature is nearly flat for the entire path of the laser pulse, since the intensity remains high enough to saturate the entire irradiated region of the liquid. The maximum temperature increase for classical heating, which occurs at $x = 0$, exceeds that for saturable heating by a factor of nearly 50. In general, the temperature predicted by the saturable absorption model is much more uniform than that predicted by the classical model when the saturation intensity is exceeded. Experiments were also performed, with good agreement found with the saturable absorption model (Longtin et al., 1995b).

Other mechanisms. When irradiated with very high laser intensities, the material may ionize and dissociate, resulting in the formation of new species, such as electrons and ions. These particles have their own characteristic lifetimes and can themselves interact with the laser pulse, particularly the electrons that result from ionization. If such phenomena occur, they can be modeled in a fashion similar to the above by accounting for the generation and relaxation of the new species. Also, melting, evaporation, and ablation can occur (Xu et al., 1993; Bennett et al., 1996). Furthermore, these effects can show a *nonlinear* dependence on the incident pulse intensity, for example, during multiphoton ionization (Nikogosyan et al., 1983; Longtin and Tien, 1996b). Such behavior can strongly affect the laser–material interactions.

3-5 RADIATION PHENOMENA ON THE STRUCTURAL MICROSCALE

A third class of microscale radiation phenomena involves radiation interactions with microstructures. *Microstructure* in this sense refers to a *collection* of individual elements, including, for example, particles, thin film layers, molecular clusters, grooves or channels, cells, bubbles, and combinations of these elements. The basic physics of radiation–microstructure interactions do not fundamentally differ from that discussed previously, but the proximity of neighboring elements must be accounted for in terms of both the radiation interaction and the heat transfer.

Two length-scale criteria for microstructures can be established based on the characteristic separation distance l_s between adjacent elements. Note that l_s can be zero, as is the case for multilayer thin films, particle beds, dense bubble nucleation on a solid surface, and micromachined grooves. When either criterion is met, the interaction of the incident radiation with the microstructure will differ from the interaction of radiation with a single, isolated element. The first criterion is

$$l_s \lesssim \lambda, \tag{3-27}$$

which states that the separation distance is smaller than the wavelength of the incident radiation. In this case, the neighboring elements perturb the EM field near the element of interest, rendering the EM solution for an isolated element inapplicable. The second criterion,

$$l_s \lesssim l_c, \tag{3-28}$$

indicates that the separation distance is smaller than the coherence length in Eq. (3-6). In this case, interference effects due to emission from adjacent structures modify the radiation interaction with the microstructure.

Two methods of quantifying radiation interactions with microstructures are possible when Eqs. (3-27) and (3-28) hold. If the microstructure is relatively simple and well defined, the EM equations can be solved rigorously using the exact geometry. Examples include periodic microgrooves and microchannels (Hesketh et al., 1988; Raad and Kumar, 1992), single and multilayer thin films (Wong et al., 1992; Grigoropoulos et al., 1993), insulation made of parallel and woven material (Kumar and White, 1996), and closely packed beds with long-range order (Tien, 1988). For materials whose structure is complicated or whose elements are arranged in a nonuniform, nonrepeating manner, techniques can be applied to obtain averaged values of the radiation interaction behavior. Examples include packed and fluidized beds without long-range order (Tien, 1988), aerogel materials (Richter et al., 1995), and fast-time-scale bubble formation on solid surfaces (Yavas et al., 1994, Kim et al., 1996).

Examples of radiation–microstructure interaction. The interaction of radiation with microstructures is not new; multilayer thin films, for example, have been used for years as mirrors and antireflective coatings on optics. Likewise, a diffraction grating is simply a surface containing hundreds or thousands of fine parallel grooves. Interference in light reflected from the grooves allows these devices to separate white light into its spectral components. The unique *thermal* aspects of microstructure–radiation interactions, on the other hand, have only recently begun to be systematically investigated. As no new physics are required, this section focuses on several specific examples of radiation interaction with microstructures.

Radiation interacting with collections of small particles occurs in many common engineering applications. Examples include combustion processes, packed and fluidized beds, fogs and aerosols, colloidal suspensions (paints, pigments), and packed insulations. If the particles are very far apart from each other, the radiation interaction can be modeled as a superposition of the EM interaction with a single particle. Such a situation is referred to as *independent scattering*. If, however, the particles are close enough to affect the EM fields of neighboring particles, the interaction of the EM radiation with the entire ensemble of particles must be modeled. This condition is called *dependent scattering*. Tien (1988) summarized a variety of different packed and fluidized beds and presented a regime map to determine when the system of particles exhibits independent and dependent scattering. Two parameters were used in his study. The first was the *size parameter* α, which for a spherical particle is defined as the ratio of the circumference to the wavelength:

$$\alpha = \frac{2\pi a}{\lambda}. \tag{3-29}$$

The second parameter was the *particle volume fraction* f_v, which represents the ratio of the volume of solid particles to that of the total volume V. For spherical particles

$$f_v = \frac{4N\pi a^3}{3V}, \tag{3-30}$$

where N is the number of particles with radius a in the volume V. By inspection of Eq. (3-30), it can be seen that the volume fraction is inversely related to the interparticle spacing l_s: a smaller f_v represents a larger interparticle spacing l_s.

The regime map for scattering for several applications of interest is shown in Fig. 3-12 (Modest, 1993). Observe that the division between independent and dependent scattering occurs along the line $C/\lambda = 0.5$ (Brewster and Tien, 1982), where C is the interparticle distance and is related to both α and f_v as follows:

$$\frac{C}{\lambda} = \frac{\alpha}{\pi}\left(\frac{0.905}{f_v^{1/3}} - 1\right). \tag{3-31}$$

As is expected, larger particle volume fractions (small interparticle spacing) result in dependent scattering, while smaller volume fractions allow the independent scattering

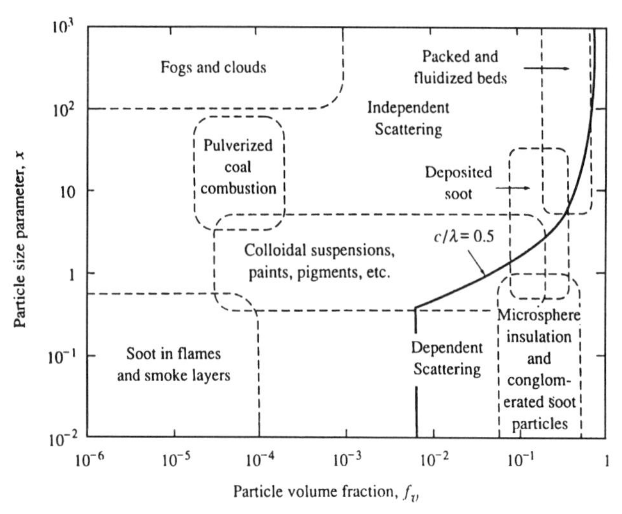

Figure 3-12 Regime map for independent and dependent scattering regimes.

approximation to be made. Also, as α increases with f_v held constant, dependent scattering becomes less pronounced. This occurs because larger particles must have a larger interparticle spacing to maintain the same f_v. It should be mentioned that this regime map reflects *coherent-based* effects only. As pointed out by Tien (1988), perturbation of the near field by adjacent structures depends strongly on the index of refraction and is not accounted for in the graph.

Having considered a single particle and a collection of particles, *aerogels* are an example of the next level of assembly of particles. Aerogels are a unique class of ultra-fine materials that exhibit extremely low volume fractions (Richter et al., 1995). The molecules comprising the aerogel combine to form an intricate, interwoven structure with a very fine pore diameter $\sim$100 nm. Aerogels are extremely light and make exceptionally good thermal insulators because their structure inhibits all three forms of thermal transport: conduction, convection, and radiation. Another unique aspect of aerogel materials is that they can be characterized as a fractal structure over several orders of length scale in the material. Aerogels made of SiO_2 are, due to their porous structure, transparent in the visible region but virtually opaque in the NIR region, as shown in the transmission versus wavelength plot in Fig. 3-13, making them excellent—yet transparent—thermal insulators. Aerogels can ascribe their unique optical behavior to the microstructure of the material, which influences the radiative interaction of the constituent molecules.

Another example of radiation–microstructure interactions is the optical measurement of fast-time-scale bubble nucleation on a laser-irradiated solid surface covered with a liquid (Yavas et al., 1994, Kim et al., 1996). In this case, many tiny bubbles (<1 μm) form on the substrate, and the change in the reflectivity and scattering due to the bubble formation is measured to infer information about their growth dynamics. For dense bubble formations, the scattering is dependent, and an average bubble layer thickness is used (Kim et al., 1996).

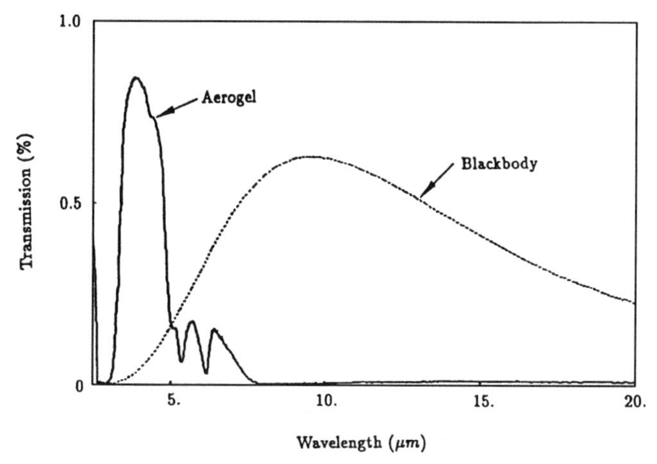

Figure 3-13 Transmission of SiO_2 as a function of wavelength.

Laser interactions with tissues often involve microscale radiation effects, for example, during laser surgery and laser-based diagnostics. Fibrous and particle insulations, both at room temperature and cryogenic temperatures are another important class of applications whose microstructure influences the nature of the radiation interaction (Kumar and White, 1995).

3-6 SUMMARY

The objective of this chapter is to present a quantitative framework for characterizing radiation thermal transport on the microscale and to delineate the important parameters and regimes. Specifically, radiation–material interactions on the *length, time*, and *structural* microscale are discussed. It should be mentioned that the criteria, regimes, and relative importance attached to the material in this chapter are somewhat arbitrary and undergoing constant refinement. This work is by no means meant to be the final word; quite the contrary, it is hoped that research in the microscale thermal transport field will continue to grow and expand, thus requiring more definition, summarization, and categorization.

The field is constantly evolving and research an ongoing endeavor, with many exciting, challenging, and important problems remaining to be addressed. It is hoped that the discussion in this chapter will give the reader an overview of the current state of the art, the important issues that must be addressed, and a set of criteria to determine the appropriate microscale regime for the radiation transport problem at hand.

NOMENCLATURE

A	cross-sectional area [cm^2]
a	particle radius [cm]
C	particle-bed interparticle spacing [cm]

C_0	number density of absorbing molecules [cm^{-3}]
c	radiation propagation speed (speed of light) [cm/s]
E	energy [J]
E_0, E_1	ground, first excited state molecular energy [J]
f_v	solid particle void fraction [·]
h	Planck constant [6.6261×10^{-34} J·s]
I	radiation intensity [W/cm^2]
k	extinction coefficient [·]
k_b	Boltzmann constant [1.3807×10^{-23} J/K]
L	characteristic length dimension [cm]
l_c	coherence length [cm]
l_p	spatial extent of laser pulse [cm]
l_s	separation distance between adjacent microstructures [cm]
N	number of particles in volume V [·]
N_0, N_1	ground, first excited state population number density [cm^{-3}]
n	index of refraction [·]
p	scattering phase function [·]
S	radiation source function [W/cm^3]
S_0, S_1	ground, first excited molecular state [·]
t	time [s]
T	temperature [K]
V	volume [cm^3]
x	distance [cm]

Greek symbols

α	thermal diffusivity [m^2·s], particle size parameter [·]
δ	radiation penetration depth [cm]
Λ	heat carrier mean free path [cm]
λ	wavelength [cm]
λ_c	heat carrier wavelength [cm]
ν	radiation frequency [Hz]
σ	effective molecular cross section for absorption [cm^2]
μ	absorption/scattering coefficient [cm^{-1}]
τ	characteristic time [s]
ξ	cosine of scattering angle [·]

Subscripts

a	absorption
b	blackbody
c	radiation propagation
d	thermal diffusion
ex	first electronic excited state
gr	electronic ground state

m	maximum
p	laser pulse
r	relaxation
s	saturation
sc	scattering
λ	wavelength-dependent

REFERENCES

Armaly, B. F., and C. L. Tien, "Emissivities of thin metallic films at cryogenic temperatures," *Proceedings of the 4th International Heat Transfer Conference*, Paris, vol. 3, p. R1.1, 1970.

Arpaci, V. S., *Conduction Heat Transfer*, Addison-Wesley, Reading, 1996.

Bennett, T. D., D. J. Krajnovich, and C. P. Grigoropoulos, "Separating thermal, electronic, and topographic effects in pulsed laser melting and sputtering of gold," *Phys. Rev. Lett.*, vol. 76, pp. 1659–1662, 1996.

Bohren, C. F., and D. R. Huffman, *Absorption and Scattering of Light by Small Particles*, Wiley, New York, 1983.

Born, M., and E. Wolf, *Principles of Optics*, 6th ed., Pergamon Press, Oxford, 1980.

Brewster, M. Q., *Thermal Radiative Transfer and Properties*, Wiley, New York, 1992.

Brewster, M. Q., and C. L. Tien, "Radiative transfer in packed fluidized-beds: Dependent versus independent scattering," *J. Heat Transfer*, vol. 104, pp. 573–579, 1982.

Chen, G., and C. L. Tien, "Partial coherence theory of thin film radiative properties," *J. Heat Transfer*, vol. 114, pp. 311–318, 1992.

Cravalho, E. G., C. L. Tien, and R. P. Caren, "Effect of small spacings on radiative transfer between two dielectrics," *J. Heat Transfer*, vol. 89C, pp. 351–358, 1967.

Cravalho, E. G., G. A. Domoto, and C. L. Tien, "Measurements of thermal radiation of solids at liquid helium temperatures," in *Progress in Aeronautics and Astronautics*, vol. 21, J. T. Bevans (ed.), pp. 531–542, 1968.

Duncan, A. B., and G. P. Peterson, "Review of microscale heat transfer," *Appl. Mechan. Rev.*, vol. 47, pp. 397–428, 1994.

Grigoropoulos, C. P., H. K. Park, and X. Xu, "Modeling of pulsed laser irradiation of thin silicon layers," *Int. J. Heat Mass Transfer*, vol. 36, pp. 919–924, 1993.

Guenther, A. H., and J. K. McIver, "The pulsed laser damage sensitivity of optical thin films, thermal conductivity," *Laser Particle Beams*, vol. 7, pp. 433–441, 1989.

Hercher, M., "An analysis of saturable absorbers," *Appl. Opt.*, 1967, vol. 6, pp. 947–954, 1967.

Hesketh, P. J., B. Gebhart, and J. N. Zemel, "Measurements of the spectral and directional emission from microgrooved silicon surfaces," *J. Heat Transfer*, vol. 110, pp. 680–686, 1988.

Kelley, J. D., and F. E. Hovis, "A thermal detachment mechanism for particle removal from surfaces by pulsed laser irradiation," *Microelectron. Eng.*, vol. 20, pp. 159–170, 1993.

Kim, D., H. K. Park, and C. P. Grigoropoulos, "Interferometric study on the growth of pulsed-laser-generated submicron bubble layer on a solid surface," *Proc. 31st ASME National Heat Transfer Conference*, vol. 4, HTD vol. 326, pp. 69–77.

Kittel, C., *Introduction to Solid State Physics*, 6th ed., Wiley, New York, 1996.

Komvopoulos, K., and K. Nagarathnam, "Processing and characterization of laser-cladding coating materials," *J. Eng. Mater. Technol.*, vol. 112, pp. 131–143, 1990.

Kotake, S., "Molecular mechanical engineering," *JSME Int. J. Series B-Fluids Therm. Eng.*, vol. 38, pp. 1–7, 1995.

Kotake, S., and M. Kuroki, "Molecular dynamics study of solid melting and vaporization by laser irradiation," *Int. J. Heat Mass Transfer*, vol. 36, pp. 2061–2067, 1993.

Kumar, S., "Radiation transport in participating media at very short time scales," *Symposium on Thermal Science and Engineering in Honor of Chancellor Chang-Lin Tien*, Nov. 14, 1995, Berkeley, CA, pp. 239–245, 1995.

Kumar, S., and S. M. White, "Dependent scattering properties of woven fibrous insulations for normal incidence," *J. Heat Transfer*, vol. 117, pp. 160–166, 1995.
Longtin, J. P., and C. L. Tien, "Saturable absorption during high-intensity laser heating of liquids," *J. Heat Transfer*, vol. 118, pp. 924–930, 1996a.
Longtin, J. P., and C. L. Tien, "Efficient laser heating of transparent liquids using multiphoton absorption," *Int. J. Heat Mass Transfer*, vol. 40, pp. 951–959, 1996b.
Longtin, J. P., T. Q. Qiu, and C. L. Tien, "Pulsed laser heating of highly absorbing particles," *J. Heat Transfer*, vol. 117, pp. 785–788, 1995a.
Longtin, J. P., C. L. Tien, M. M. Kilgo, and R. E. Russo, "Temperature measurement during high-intensity laser–liquid interactions," *Exp. Heat Transfer*, vol. 8, pp. 241–255, 1995b.
Metha, C. L., "Coherence-time and effective bandwidth of blackbody radiation," *Nuovo Cimento*, vol. 21, pp. 401–408, 1963.
Modest, M. F., *Radiative Heat Transfer*, McGraw-Hill, New York, 1993.
Nikogosyan, D. N., A. A. Oraevsky, and V. I. Rupasov, "Two-photon ionization and dissociation of liquid water by powerful laser UV radiation," *Chem. Phys.*, vol. 77, pp. 131–143, 1983.
Park, H. K., C. P. Grigoropoulos, W. P. Leung, and A. C. Tam, "A practical excimer laser-based cleaning tool for removal of surface contaminants," *IEEE Trans. Components, Packaging, Manufacturing Technol., Part A*, vol. 17, pp. 631–643, 1994.
Penzkofer, A., "Passive Q-switching and mode-locking for the generation of nanosecond to femtosecond pulses," *Appl. Phys. B*, vol. 46, pp. 43–60, 1988.
Qiu, T. Q., and C. L. Tien, "Heat transfer mechanisms during short-pulse laser heating of metals," *J. Heat Transfer*, vol. 115, pp. 835–841, 1993.
Qiu, T. Q., J. P. Longtin, and C. L. Tien, "Characteristics of radiation absorption in metallic particles," *J. Heat Transfer*, vol. 117, pp. 340–345, 1995.
Raad, N., and S. Kumar, "Radiation scattering by periodic microgrooved surfaces of conducting materials," *Proceedings of the ASME Winter Annual Meeting: Micromechanical Systems*, DSC vol. 40, pp. 243–258, 1992.
Richter, K., G. Chen, and C. L. Tien, "Partial coherence theory of multilayer thin-film optical properties," *Opt. Eng.*, vol. 32, pp. 1897–1903, 1993.
Richter, K., P. M. Norris, and C. L. Tien, "Aerogels: Applications, structure, and heat transfer phenomena," in *Annual Review of Heat Transfer*, vol. 6, C. L. Tien (ed.), Begell House, New York, Chap. 2, pp. 61–114, 1995.
Seilmeier, A., and W. Kaiser, "Ultrashort intramolecular and intermolecular vibrational energy transfer of polyatomic molecules in liquids," in *Ultrashort Laser Pulses: Generation and Application*, 2nd ed., W. Kaiser (ed.), Springer-Verlag, Berlin, Chap. 7, pp. 279–317, 1993.
Shank, C. V., and E. P. Ippen, "Ultrashort pulse dye lasers," in *Dye Lasers*, 3rd ed., Springer-Verlag, Berlin, Chap. 4, pp. 139–153, 1990.
Siegel, R., and J. R. Howell, *Thermal Radiation Heat Transfer*, 3rd ed., Hemisphere, New York, 1992.
Siegman, A. E., *Lasers*, University Science Books, Mill Valley, CA, 1986.
Tien, C. L., "Thermal radiation in packed and fluidized beds," *J. Heat Transfer*, vol. 110, pp. 1230–1242, 1988.
Tien, C. L., and G. Chen, "Challenges in microscale conductive and radiative heat transfer," *J. Heat Transfer*, vol. 116, pp. 799–807, 1994.
Tien, C. L., and G. R. Cunningham, "Cryogenic insulation heat transfer," *Adv. Heat Transfer*, vol. 9, pp. 349–417, 1973.
Tien, C. L., and J. H. Lienhard, *Statistical Thermodynamics*, Hemisphere, New York, 1979.
Tuntomo, A., C. L. Tien, and S. H. Park, "Internal distribution of radiation absorption in a spherical particle," *J. Heat Transfer*, vol. 113, pp. 407–412, 1991.
Whale, M. D., and E. G. Cravalho, "Analysis of the enhancement of thermal radiation between closely-spaced surfaces due to microscale phenomena," *ASME Proceedings of the 32nd National Heat Transfer Conference: Volume 7, HTD-345*, Baltimore, MD, Aug. 10–12, pp. 41–50, 1997a.
Whale, M. D., and E. G. Cravalho, "Regimes of microscale radiative transfer for exchange of thermal energy between metallic surfaces," *ASME Proceedings of the 32nd National Heat Transfer Conference: Volume 7, HTD-345*, Baltimore, MD, Aug. 10–12, pp. 65–72, 1997b.

Wong, P. Y., C. K. Hess, and I. N. Miaoulis, "Thermal radiation modeling in multilayer thin film structures," *Int. J. Heat Mass Transfer*, vol. 35, pp. 3313–3321, 1992.

Xu, X., S. L. Taylor, H. K. Park, and C. P. Grigoropoulos, "Transient heating and melting transformations in argon-ion laser irradiation of polysilicon films," *J. Appl. Phys.*, vol. 73, pp. 8088–8096, 1993.

Yamada, Y., "Light–tissue interaction and optical imaging in biomedicine," in *Annu. Rev. Heat Transfer*, vol. 6, C. L. Tien (ed.), Begell House, New York, Chap. 1, pp. 1–59, 1995a.

Yamada, Y., "Optical tomography using the temporally extrapolated absorbance method," *Proceedings from the Symposium on Thermal Science and Engineering in Honor of Chancellor Chang-Lin Tien*, Berkeley, CA, Nov. 14, pp. 233–238, 1995b.

Yariv, A., *Quantum Electronics*, 3rd ed., Wiley, New York, 1989.

Yavas, O., P. Leiderer, H. K. Park, C. P. Grigoropoulos, C. C. Poon, W. P. Leung, N. Do, and A. C. Tam, "Optical and acoustic study of nucleation and growth of bubbles at a liquid–solid interface induced by nanosecond-pulsed-laser heating," *Appl. Phys. A (Solids and Surfaces)*, vol. A58, pp. 407–415, 1994.

Ziman, J. M., *Electrons and Phonons*, Oxford University Press, Oxford, 1960.

CHAPTER

FOUR

MELTING AND FREEZING PHENOMENA

R. Stephen Berry

Department of Chemistry and The James Franck Insititute,
The University of Chicago, Chicago, Illinois 60637

4-1 INTRODUCTION

Finite systems, especially small finite systems such as atomic and molecular clusters, exhibit some of the same solid-like and liquid-like behavior of bulk solids and liquids. Because it is possible to treat such systems in some detail, they give us new insights into the phases and phase transitions of bulk matter. However, these systems have many interesting phase-like properties that do not extend to bulk matter, properties that have been recognized but, for the most part, not yet exploited. This discussion will explore the phase-like properties of finite systems in terms that extend simply to bulk matter and are specific to small systems. It will also point out areas where further research is waiting to be done.

The phase-like properties of clusters and nanoscale particles lend themselves to study by simulation and by analytic theory. Until now, there have been a few experimental studies that demonstrate specific phase-like forms of these species (Bartell, 1992; Bartell et al., 1988, 1991b; Buck et al., 1993; Goldstein et al., 1990; Martin et al., 1994; Valente et al., 1984a, 1984b), but very little indeed from the laboratory to reveal the nature of the equilibrium or the transitions between these forms (Bartell et al., 1991a, 1992; Dibble et al., 1992; Ellert et al., 1995). Consequently most of this chapter will treat theoretical methods and especially results and will present almost no experimental evidence. The experimental methods have included electron diffraction, which probes structure directly, and spectroscopic and mass-spectroscopic studies, such as the intensities of mass peaks generated by photo-ionization.

The first fundamental concept that pervades the thermodynamics of small systems and especially the description of their phase-like behavior is the idea that thermodynamic equilibrium of clusters and nanoscale particles must be the equilibrium of an ensemble.

It might be useful occasionally to imagine the state of a single, isolated particle as a kind of equilibrium state, like the state of an isolated rigid rotor or harmonic oscillator, really a stable mechanical state of an isolated dynamical system. However, because clusters and nanoscale particles are complicated dynamical systems, mostly chaotic and ergodic, they do not exhibit periodic behavior and hence lend themselves to descriptions involving some kind of averaging. Because they are presumably ergodic, such a description could emerge from the long-time history of a single system, or, as well, from the instantaneous state of an ensemble of many, many such systems. As in all of statistical thermodynamics, the ensemble of choice depends on the particular problem at hand: microcanonical for a constant-energy system, canonical for a constant-temperature system, grand-canonical for a system at constant chemical potential, isobaric for a system at constant pressure, and so on. In practice, histories of single systems based on computer simulations often yield the information one wants about changes of phase or about the phases of clusters themselves. Such histories come from molecular dynamics simulations, numerical solutions of the equations of Newton's Second Law if the simulations are classical, or the time-dependent Schrödinger (or quantum-mechanical Liouville) equation if the simulations are quantum-mechanical. The alternative approach to simulation is through Monte Carlo methods, which explore the system's phase space but have no dynamics linking one step to the next; these methods are therefore a sort of sampling of an ensemble at an instant, the alternative to a time history and equally valid for any ergodic system. Both have been useful, with molecular dynamics being perhaps a bit more because it gives dynamical as well as thermodynamical information.

"Phase transitions in small systems are gradual, not sharp." This is a commonplace thought that still sometimes appears in discussions of phase changes of clusters and nanoparticles. In one sense this is completely correct (Hill, 1963, 1964), yet there is much precision and a certain kind of sharpness to these changes to be found if we explore their nature in a bit of detail. To begin, we shall refer to the changes of phase exhibited by bulk matter as "phase transitions," and their analogues for finite systems, as closely as they come, as "phase changes," because these are not the same as the phase transitions of bulk systems, although we shall see how we can follow the emergence of bulk transitions from the phase changes of small systems. The phase changes of small systems cannot be classified according to "order" in the Ehrenfest pattern of cataloguing, according to which derivative of the energy or entropy is the lowest to vanish at the point of the transition. Moreover, as we shall see, the Gibbs Phase Rule loses its meaning because the distinction between "phase" and "component" becomes unclear for small systems.

Another distinction between phases of bulk matter and phase-like forms of clusters and nanoparticles is that the latter exhibit many forms that we can call "phases" or phase-like, which do not persist in the limit of very large systems. Table 4-1 lists a menagerie of such forms, with examples or possible examples.

We proceed by first examining the simplest, best-studied forms of clusters, the solid-like and liquid-like forms, and the passages between these forms. This will give information regarding the way a first-order transition occurs, and about limits on metastability and the spinodals of bulk matter. Then we go on to the more exotic phase-like forms of clusters and to the question of coexistence of these forms. Finally, we conclude with some open questions.

Table 4-1 A menagerie of phase-like forms that clusters may exhibit

Phase-like form	Example
solid	any cluster but He_n, at low enough T
soft solid (or "fluxional cluster")	six-particle metal clusters (Sawada et al., 1989); Au_{55} (Sawada et al., 1992).
liquid	Ar_7 (Berry et al., 1988; Briant et al., 1975)
surface-melted	Ar_{55} (Cheng et al., 1991, 1992a; Kunz et al., 1993, 1994; Nauchitel et al., 1980)
core-melted	possibly Ga_n (Kunz et al., 1994)
glassy or amorphous	Ar_n or mixed rare-gas clusters; $(KCl)_n$ (Amini et al., 1979)
"restricted liquid"	Li_8^+ (Jellinek et al., 1994)

4-2 SOLID AND LIQUID CLUSTERS AND THEIR EQUILIBRIA

4-2-1 Solid-Like and Liquid-Like Forms of Clusters

At sufficiently low temperatures or energies, all clusters (with the exception of clusters of helium and possibly some clusters of hydrogen) behave like solids. Their component atoms or molecules undergo small-amplitude vibrations around the equilibrium sites to which they are bound. The vibrations are very nearly harmonic. In simulations, the Lindemann criterion (Lindemann, 1910, 1912)—that the root-mean-square deviation of nearest neighbors be less than about 10%—is satisfied (Berry et al., 1988). Very little diffusion occurs; the mean-square displacement of the particles with time is nearly zero (Berry et al., 1988). The velocity autocorrelation functions have no very low frequency components, meaning that there are no very soft modes of motion in these clusters (Berry et al., 1988). Many clusters exhibit well-ordered geometries, albeit not necessarily geometries consistent with periodic lattices. For example, many kinds of atomic clusters have solid-like structures based on icosahedra. Clusters of rare-gas atoms are generally variations on icosahedral geometries in the global-minimum structures on their potential surfaces, provided they contain no more than a few thousand atoms; larger clusters have close-packed, lattice-based structures. The precise way this change occurs as the number of component atoms increases is not yet understood, but how it happens and at what size depend on the range of the interatomic forces (Doye et al., 1995; Wales et al., 1995). There is sound experimental evidence for such structures (Bartell, 1986; Bartell et al., 1988; Farges et al., 1983, 1986; Raoult et al., 1989; Torchet et al., 1990a, 1990b; Valente et al., 1984a, 1984b). Simulated cold clusters show pair distribution functions with the sharp peaks of successively more distant shells of neighbors, next-nearest neighbors, and so forth, and angular distributions likewise are characteristic of solid-like structures. For example, icosahedral clusters show angular distributions with negligible amplitude at 90°, an angle that simply does not occur for triples of neighbors in that structure (Quirke et al., 1984).

At higher temperatures, many kinds of clusters show liquid-like behavior, at least in simulations (Berry et al., 1988). The particles exhibit linearly increasing mean-square displacements as functions of time, corresponding to well-defined diffusion coefficients, at least until the displacements reach the linear dimension of the cluster. The Lindemann criterion parameter is typically well above 10%. The velocity autocorrelation function has a significant contribution from very low frequency modes, which are the soft modes of a liquid. The pair distribution function and the angular distribution function have the broad form characteristic of liquids. Experimentally, liquid clusters have been identified in a few instances (Bartell et al., 1989). Simulations first revealed such forms (Amini et al., 1979; Briant et al., 1975; Cotterill et al., 1973, 1975; Damgaard Kristensen et al., 1974; Etters et al., 1975, 1977b; Kaelberer et al., 1977; Lee et al., 1973; McGinty, 1973) and implicitly indicated one of the conditions for their existence as recognizable forms of clusters: the liquid had to persist long enough that the system could establish at least vibrational equilibrium in that form. Many of these same simulation studies indicated that clusters of certain sizes (e.g., Ar_7 [the smallest], Ar_{13}, and Ar_{19}, all modeled by pairwise Lennard-Jones potentials) can exhibit well-defined dynamic equilibrium between their solid and liquid forms. The same was confirmed and shown for Ar_{55} soon after (Nauchitel et al., 1980). The nature of the phase change between solid and liquid forms was not yet clear, but some aspects of the equilibrium led to the speculation by Briant and Burton that it might even be first-order. This was puzzling, since dogma had it that first-order transitions are properties of bulk matter and not of small systems.

Some clusters exhibit, in some range of energy or temperature, a floppiness or fluxional character that allows them to explore some limited set of potential minima, yet not become liquid in the sense of exploring their entire potential energy surface. One is the six-particle cluster modeled by the Gupta potential, a system studied by Sawada and Sugano (Sawada, 1987); they have also examined other metal clusters in this manner (Sawada et al., 1992). The tetramers and pentamers of alkali halides exhibit such behavior but also have another kind of nonrigid, phase-like form, namely planar rings, which can pass readily between open rings and "ladders" or rectangles (Luo et al., 1987; Rose et al., 1992). Still another exotic phase-like form of cluster is the very flexible Li_8^+ ion, which has a region of its potential surface in which it is extremely flexible and liquid-like, with the one qualification that one atom, at the center of the cluster, cannot participate in the permutational motions that mix all the other seven and make them liquid-like (Jellinek et al., 1994). We call this kind of system, in which most but not all the component particles are mobile, a "restricted liquid."

4-2-2 Equilibrium of Solid and Liquid Clusters

If the solid-like cluster is treated as an ordinary, near-rigid molecule with small-amplitude vibrations, and the liquid-like cluster is considered a sort of nonrigid, fluxional molecule, then it is possible to construct a quantum-statistical model to rationalize the temperatures at which these two forms may coexist (Berry et al., 1984a, 1984b; Jellinek et al., 1986). This requires postulating that there is at least some temperature at which the two are both locally stable, in the sense that the free energy—if it is expressed in terms of temperature, presumably pressure of volume, and a nonrigidity parameter that serves like an order parameter—has two minima as a function of the nonrigidity parameter at some temperature. This condition, plus the condition of long-enough persistence of each phase, are sufficient to lead to the coexistence of the solid and liquid forms of the cluster. Moreover, such coexistence occurs not along a single curve in the space of two such variables as pressure and temperature; coexistence of solid and liquid clusters, if it occurs, occurs within a *band* of the space of thermodynamic variables of the system (Beck et al., 1987, 1988a, 1988b; Berry et al., 1984a, 1984b, 1988; Davis et al., 1987, 1988).

It is straightforward to see how this happens. To begin, the densities of states of solid and liquid forms can be represented as functions of γ, the nonrigidity or order parameter, at any given temperature. The density of solid-like states is invariably higher at low temperatures, but the density of liquid-like states rises considerably faster with T than that of the solid clusters. This means that the free energy of the solid is lower at low temperatures than that of the liquid and is a monotonically increasing function of the nonrigidity parameter there. The only minimum in $F(T, \gamma)$ occurs for some small value of γ in the solid-like range. However, as the temperature increases, the density of states for high degrees of nonrigidity (i.e., for large γ) tends to lower the free energy for nonrigid forms of the cluster, relative to more rigid forms; the free energy $F(T, \gamma)$ becomes less strongly monotonic and, eventually at a sufficiently high temperature T_f, develops a point of zero slope, that is, $[\partial F(T, \gamma)/\partial\gamma]_T = 0$ at some value of γ. At temperatures above T_f, $F(T, \gamma)$ has two minima, one in the solid-like range and one in the liquid-like range of γ. But as the temperature continues to increase and the density of states of the liquid-like form becomes larger and larger with respect to that of the solid-like form, the curve of $F(T, \gamma)$ as a function of γ continues to tip, more and more, toward the liquid-like side until the system reaches a temperature T_m at which the minimum in $F(T, \gamma)$ near the solid-like end of the scale turns into just a flat spot. At temperatures above T_m, $F(T, \gamma)$ has only a single minimum, and that is in the liquid-like region. This is illustrated in Fig. 4-1.

This argument implies that between T_f at the low end and T_m at the high end, the solid and liquid forms of the cluster may coexist, all at the single pressure for which the curves were constructed. Hence at each pressure there must be a range of temperature within which the free energy has two minima, therefore two locally stable forms, therefore

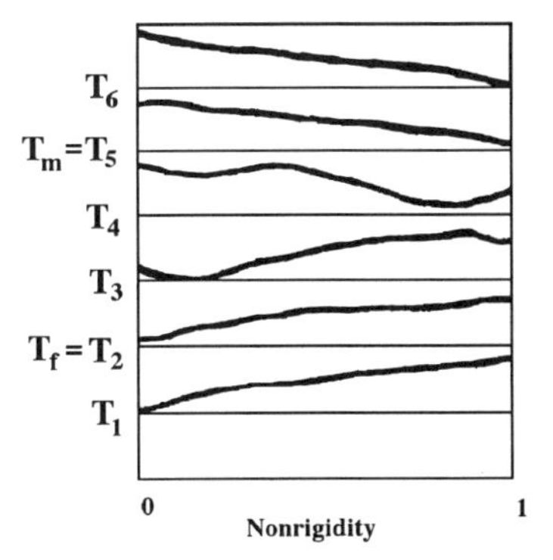

Figure 4-1 The free energy $F(T, \gamma)$ as a function of the nonrigidity parameter, for six temperatures, increasing from T_1 through T_6. Below T_2, $F(T, \gamma)$ has only one minimum, near $\gamma = 0$, in the solid-like end of the scale. At $T_2 = T_f$, $F(T, \gamma)$ develops a point of zero slope near the nonrigid limit (i.e., near $\gamma = 1$). Between $T_2 = T_f$ and $T_3 = T_m$, $F(T, \gamma)$ has two minima—that for lower γ corresponding to a locally stable solid-like form and that for the higher γ—to a locally stable liquid. At $T_3 = T_m$, the free energy has only one minimum and one other point of zero slope and zero second derivative, as a function of γ. Above $T_3 = T_m$, the free energy has only one minimum, corresponding to the one stable form.

two coexisting forms. The relative amounts of these two forms are fixed by a chemical equilibrium constant, $K_{eq} = \exp[-\Delta F(T, \gamma)/kT]$. The argument implies that K_{eq} should have two discontinuities, one at T_f and another at T_m, for each pressure. However, the relative amounts of solid and liquid, fixed as the ratio K_{eq}, vary smoothly between these limits, with the ratio of liquid to solid increasing with T (Berry et al., 1984a, 1984b, 1988). The analytic argument is indeed supported by simulations, both constant-energy and constant-temperature, and both molecular dynamics and Monte Carlo (Amar et al., 1986; Beck et al., 1987, 1988a, 1988b; Berry et al., 1988; Blaisten-Barojas et al., 1986, 1987; Davis et al., 1987, 1988; Garzon et al., 1989; Honeycutt et al., 1987; Sawada, 1987; Sawada et al., 1989, 1992). One sees a bimodal distribution of behavior of (short-term) mean kinetic energies (i.e., of mean vibrational temperatures) in isoergic dynamic simulations of the cluster's evolution and a bimodal distribution of total energies and potential energies in corresponding isothermal molecular dynamics simulations, if the conditions of temperature and pressure put the cluster in the range of coexistence. Otherwise, the distributions are unimodal. The ratio of the fraction of the time the cluster spends as a solid to that spent as a liquid is just the equilibrium constant K_{eq}.

The relation between the phase change between solid and liquid clusters and the solid–liquid first-order phase transition of bulk matter becomes clear from this argument. To see this relation, it is easier to think in terms of a transformation of K_{eq}, specifically the equilibrium distribution function $D_{eq} = (K_{eq} - 1)/(K_{eq} + 1)$, because K_{eq} varies from zero to infinity, with the value 1 when the amounts of solid and liquid are equal, whereas the function D_{eq} varies between -1 (when the system is all solid) to $+1$ (when it is all liquid). At low temperatures, below T_f, D_{eq} is a constant -1; at T_f, it shows a discontinuity, and rises to some finite value greater than -1. Between T_f and T_m, D_{eq} rises monotonically and smoothly, presumably going through zero and up to positive values. Then, at T_m, it again becomes discontinuous and rises to its high-temperature limit of $+1$. Hence between these two limiting temperatures, solid and liquid clusters coexist in a canonical ensemble; alternatively, a single cluster, at constant temperature, passes back and forth between solid-like and liquid-like forms, so long as that temperature lies between T_f and T_m.

If the cluster is small, then the discontinuities in D_{eq} are fairly large and the transition from low to high and negative to positive values is gradual. As the cluster gets larger, the magnitude of the discontinuity decreases, the values of D_{eq} remain close to -1 and $+1$ until D_{eq} comes close to the value of T at which $D_{eq} = 0$, where almost all its change in value occurs. In other words, the discontinuities get smaller and smaller and the continuous change of equilibrium constant becomes sharper and sharper. (There is still an open question of whether fluctuations destabilize the undercooled liquid or superheated solid so much that the discontinuities in D_{eq} are observable (Fisher et al., 1985, 1986; Privman et al., 1983). When the cluster reaches macroscopic size, the discontinuities are immeasurably small, and the continuous change of D_{eq} from very near -1 to very near $+1$ takes place so abruptly that it is in effect discontinuous there! That is the picture of how a first-order melting and freezing transition emerges. Figure 4-2 illustrates the behavior of $D_{eq}(T)$ schematically for three sizes of clusters; as drawn, all three clusters have equal free energies for solid and liquid forms at the same temperature T_{eq}, but this is done here only to emphasize the evolution of the shape of D_{eq} with N. In reality, T_{eq} is a function of N.

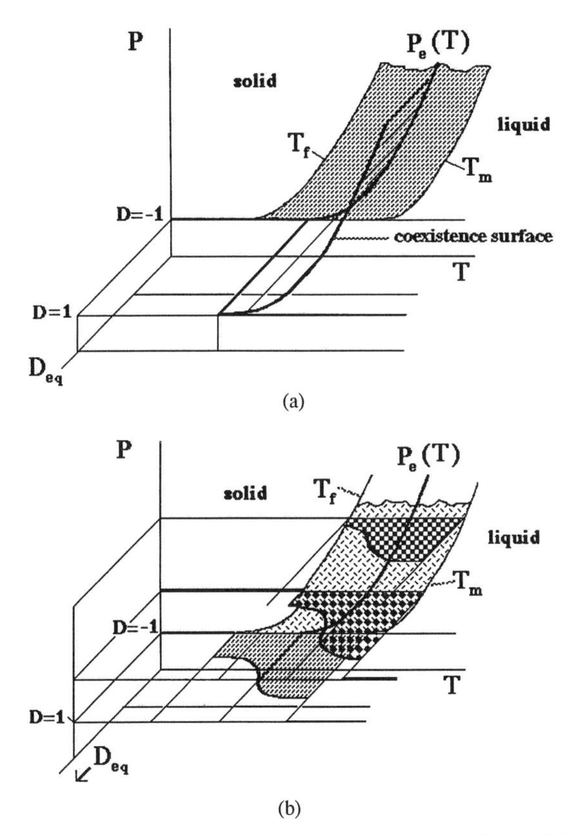

Figure 4-2 Schematic phase diagrams for solid–liquid equilibrium of finite systems: a) a large system, and b) a system of moderate size (e.g., $N \sim 50$ or 100). The three axes correspond to pressure p (vertical), temperature T (horizontal, in the plane of this surface), and D_{eq}, the distribution (projecting horizontally out of the plane of this surface).

A variety of conditions were found in which clusters can coexist in solid and liquid phase-like forms; that is, *sufficient* conditions were discovered and demonstrated, both in simulations (Beck et al., 1987, 1988a, 1988b; Blaisten-Barojas et al., 1986; Briant et al., 1975; Etters et al., 1975, 1977a, 1977b; Honeycutt et al., 1987; Jellinek et al., 1986; Kaelberer et al., 1977; McGinty, 1973) and analytically (Berry et al., 1984a; Wales et al., 1990). These analytic results follow from a model in which the nonrigidity is expressed by the density of defects; their exact nature need not be specified, but the free energy of both solid-like and liquid-like forms must be expressible in power series in this parameter. From this, a statistical theory, quantum or classical, leads to the sufficient conditions that the free energy have two minima within some finite range of temperature. Furthermore, provided the defects either attract each other or lower the vibrational frequencies of the cluster, the interval between the two minima persists as N becomes infinite. In other words, clusters of substances with such defects have solid and liquid forms whose phase changes merge smoothly into conventional first-order phase transitions as the clusters grow to become bulk matter.

Then, more recently, *necessary* conditions were also found for the coexistence of solid and liquid (or any other two dense, phase-like) forms of clusters (Wales et al., 1994), in the sense of the mean temperatures of a microcanonical distribution, or the mean free energies of a canonical distribution having a bimodal distribution. The necessary conditions are three points of inflection in the canonical distribution or in the grand canonical distribution, respectively.

4-2-3 Time Scales and Phases of Small Systems

The possibility to observe solid-like and liquid-like forms of clusters depends, as we have seen, on the individual clusters spending long enough intervals in one phase-like form to establish equilibrium-like properties characteristic of those phases. This requires that the mean time spent in one such phase be long relative not only to the vibrational period of the components; it also must be long relative to the time required for the system to establish well-defined mean-square displacements $\langle d^2(t)\rangle$ of its particles, which means it must establish well-defined diffusion coefficients D since these are directly related: $D = [d\langle d^2(t)\rangle/dt]/6$ in three dimensions. The system must also establish a stable velocity autocorrelation function for a time long enough that the Fourier transform of that autocorrelation function is also stable, which means that the distribution of vibrational frequencies must be stable. Clusters even as small as Ar_7 do exhibit this behavior. However, many other small clusters, such as Ar_{12} and Ar_{14}, do not; they pass between solid-like and liquid-like forms too frequently to establish such properties for purposes, for example, of infrared spectroscopy. In their infrared or Raman spectra, a canonical ensemble of Ar_7, Ar_{13}, or Ar_{19} would look like a collection of solid particles at low temperatures, like a collection of liquid particles at high temperatures (but below a temperature at which the particles would vaporize, of course), and at intermediate temperatures, like a mixture of solid and liquid particles. By contrast, a similar ensemble of clusters of Ar_{12}, Ar_{14}, or Ar_{17} would look like a collection of slush balls at those intermediate temperatures. If the probe were much slower than infrared spectroscopy (for example, if it were a radio-frequency probe), then the observer would see an averaged behavior for all these clusters and would interpret their behavior as slush-like, whatever the size of the cluster. This is because the time scale for residence in solid-like and liquid-like forms for Ar_7, Ar_{13}, and Ar_{19} is of order 50–500 ps, long relative to vibrational line widths but very short compared with times for absorption or emission of a radio-frequency photon. What we mean by "equilibrium" of these clusters depends on the time scale we intend for their description. On a short time scale, they exhibit traditional, distinguishable phases, but on a long time scale, they transform smoothly from solid through slush to liquid.

Because the time scales associated with the dynamics of clusters fall into ranges that we can span with our experimental methods, we become sensitized to these fundamental ideas. They lurk even in the most seemingly elementary concepts of thermodynamics, but they are easy to overlook in contexts where we are not forced to confront them. For example, the notion of a "reversible adiabatic process" would be an oxymoron were it not possible to insulate a system well enough to separate the time scale for its internal thermal equilibration from the much longer time scale for loss of heat to the surroundings (Woods, 1975).

It is important to recognize that time scales are important not only for distinguishing what we mean by a "state" and "equilibrium." It is also important to be aware that what one extracts from a simulation may depend on the time scales of the computation. At the short-time end, it is important to ask whether or not the results one seeks are sensitive to accurate representation of the vibrations. If so, then the time steps must be significantly shorter than the vibrational period. This is an obvious conclusion. However, speeding up a molecular dynamics computation by using long time steps can be dangerously tempting Yet it is at the long end of the time scale that inferences are more likely to be misled. For example, one might wish to infer from isothermal molecular dynamics simulations that the free energy differences between two forms of a cluster are in dynamic equilibrium. If the system is ergodic, then the equilibrium constant is the ratio of the times spent in the two forms. However, the duration of a simulation that would provide a stable value for that ratio is very long indeed, far longer than most current simulations. Moreover, it is important for molecular dynamics to recognize a time scale associated not with the system but with the computation: this is the time scale in which the computation remains mechanically reversible. Typically, this is about 5,000 to 10,000 time steps. Reversing dynamics calculations longer than these typically reveals a loss of significant figures if one tries to recover the initial conditions. This means that long molecular dynamics calculations are not really following reversible, Hamiltonian dynamics, even though they may remain isoergic to within very narrow limits. Rather, they become a kind of slowly randomized on-shell or isoergic walk with strong correlations for short times but none for very long times.

4-2-4 Phase Rule, Maxwell Construction, and Phase Diagrams

The behavior of solid–liquid equilibrium of clusters seems at first sight to contradict the Phase Rule $n = c - p + 2$ for the number of stable phases p, for a system with c components and n degrees of freedom, and the Maxwell construction argument against any second phase existing between the spinodal and the binodal. In a sense this is correct, but for a reason that makes the behavior of clusters fall outside the realm where these two traditional ideas apply. The Maxwell construction argues that between the binodal and the spinodal, only the phase of lower chemical potential μ can be stable *because the free energy* $N\mu$ *of the less stable phase would be enormously higher, effectively infinitely higher, than the free energy of the more stable phase.* Clusters, by contrast with bulk matter, have values of N small enough that the free energy of the less stable form may be not so very much higher than that of the more stable form, so that the ratio $K_{eq} = \exp[-\Delta F(T, \gamma)/kT]$ is still large enough even near the spinodal that both forms may be detected in an ensemble or in the time history of a single system, if it is in the coexistence region.

The Phase Rule implicitly distinguishes phases from components in chemical equilibrium by tacitly supposing that the free energy difference between phases is effectively $\pm\infty$ at all points where it is not zero, while the free energy difference between components may take on any value. Hence if our systems are clusters and not bulk matter, the former supposition does not apply, and the distinction between phases and components is lost—so the Phase Rule is irrelevant to ensembles or time histories of clusters.

We can now understand how clusters and nanoscale systems may exhibit phase-like forms (e.g., crystalline morphologies) that never appear for the corresponding bulk materials. The cluster-specific phases may have local minima in their free energies that are never the lowest free-energy minima, whatever the temperature, pressure or number of particles may be.

Thus far, the discussion has dealt with clusters as systems at only a single pressure. Of course, pressures may change, and with them, the temperature dependence of free energy relationships. The full phase behavior of clusters may be simulated (Cheng et al., 1992b) by using the Nosé method (Nosé, 1984a, 1984b, 1991) to maintain constant temperature and the Anderson method (Andersen, 1980) to keep the simulated system isobaric. The results show how the distribution between solid and liquid forms changes with pressure as well as with temperature. Furthermore, these results give us enough insight to construct phase diagrams for clusters and nanoparticles that are somewhat different from—and extensions of—conventional phase diagrams (Cheng et al., 1992b). In particular, the phase diagram for a cluster requires, in addition to the traditional thermodynamic variables such as pressure p and temperature T, an additional variable that can most usefully be taken as D_{eq}, the distribution function introduced previously. The phase diagram for solid–liquid equilibrium of a bulk system in this representation contains no information beyond what is in a conventional plot of vapor pressure versus temperature; off the curve along which $\mu_{vap} = \mu_{liq}$, the system is either all solid with $D_{eq} = -1$ (at low temperatures) or all liquid with $D_{eq} = +1$ (at high temperatures), and the transition between these two values of D_{eq} is so abrupt that it appears to be a discontinuity. The phase diagram thus consists of a half-plane with $D_{eq} = -1$ and a half-plane with $D_{eq} = +1$, and the two are separated by the curve of the equilibrium vapor pressure.

A small cluster presents a much richer diagram. In the solid–liquid phase diagram of the small cluster, there are indeed discontinuities in D_{eq}, but not at $p(T_{eq})$. These discontinuities occur where D_{eq} is truly discontinuous, along the curves of $p(T_f)$ and $p(T_m)$. These discontinuities separate, respectively, a half-plane where $D_{eq} = -1$ from a curved surface that joins that plane at the lowest fractional values of D_{eq}, and then, that curved surface at the highest fractional values of D_{eq} with the half-plane where $D_{eq} = +1$. The richness of such a diagram is in the curved surface between the two half-planes. Figure 4-2 shows sketches of the two cases just described.

4-2-5 Limits on Metastability of Bulk Phases

This line of reasoning has one important implication for bulk matter, specifically to the spinodal and to the metastable superheated solid and undercooled or supercooled liquid. The logic based on the form of densities of states of solid and liquid forms of matter implied the existence of the limiting temperatures T_f, below which the liquid has no local stability, and T_m, above which the solid has no local stability. Now suppose we cool a bulk liquid and, instead of allowing it to come to thermodynamic equilibrium, keep all of it (in this bulk case, all parts of all the bulk systems in a Gibbsian canonical ensemble) in the vicinity of its local, liquid-like minimum, as an undercooled liquid. This local stability can be maintained at temperatures below T_{eq} just so long as the local minimum in the free energy $F(T, \gamma)$ is there. If that minimum disappears, then so does local stability. Hence the liquid branch of the spinodal exists down only to the temperature

T_f, and the solid branch of the spinodal, up only to the temperature T_m. There are limits to the temperatures, then, beyond which metastable solid and liquid phases cannot exist.

4-3 "SURFACE-MELTED" CLUSTERS AND COEXISTENCE OF MULTIPLE PHASES

Surface melting of rare-gas clusters seemed plausible to Briant and Burton (Briant et al., 1975), and appeared unambiguously in the simulations of larger Lennard-Jones clusters by Nauchitel and Pertsin (Nauchitel et al., 1980). The diagnosis that the surface is liquid and the core is solid comes from standard diagnostics—diffusion coefficient, velocity autocorrelation function, and pair distribution function, for example—with the particles comprising the cluster kept in separate categories, corresponding to the layers of the cluster. Furthermore snapshots of clusters with liquid surfaces show unstructured, amorphous outer layers and ordered, polyhedral or crystalline cores. Typically, in these snapshots a few atoms float in the region just outside the outer layer of the cluster. Lennard-Jones clusters of about 45 or more particles display this behavior within a band of temperature somewhat narrower than the band of liquid-solid coexistence.

The conception of the surface-melted cluster as a polyhedral core with an amorphous, swarm-like liquid coat was the tacit model until animations revealed otherwise (Cheng et al., 1991, 1992a). Animations of such clusters as Ar_{147} (or, more precisely, the model Lennard-Jones cluster "LJ_{147}") show that in the "surface-melted" state, a) the cores are indeed solid-like; b) most of the particles of the outer layer undergo large-amplitude, highly anharmonic motion; and c) a few particles, about 1 in 30 of those in the outer layer, have popped out of the surface and float fairly freely around the cluster's surface. It is these "floaters" that carry the large-amplitude, low-frequency motion and are responsible for the values of the numerical diagnostics that imply that the surface is liquid-like. The other particles that remain in the surface, which are undergoing large-amplitude vibrations, move in a highly collective manner, oscillating around a well-defined polyhedral equilibrium structure! In other words, the surface-melted state, while so distinct and self-organized that it behaves as much like a distinct phase as do the normal solid and liquid forms, is not at all like a conventional liquid coating a conventional solid. This is especially significant in the context of nanoparticles because it implies that the "melted surface" of the cluster in the surface-melted, phase-like form does not automatically provide nuclei for normal, homogeneous melting. This in turn implies that the standard argument, that materials whose liquids wet their solid forms cannot be superheated, is not applicable to clusters, and therefore that clusters may, at least in some cases, be superheated.

Simulations of Lennard-Jones clusters of about 45 or more atoms indeed exhibit bimodal and multimodal distributions, of short-time-average mean kinetic energies or mean vibrational temperatures in isoergic simulations, and of short-time-average or even of instantaneous potential or total energy in isothermal simulations (Kunz et al., 1993, 1994). Examination of the nature of the plateaus in these bimodal and multimodal distributions reveals that they are not all of the same kind. Just above the lowest temperatures, where the distributions are unimodal and the clusters are solid, the bimodal distributions correspond to clusters that spend part of their time as solids and part in the surface-melted phase. At still higher temperatures the distributions become trimodal,

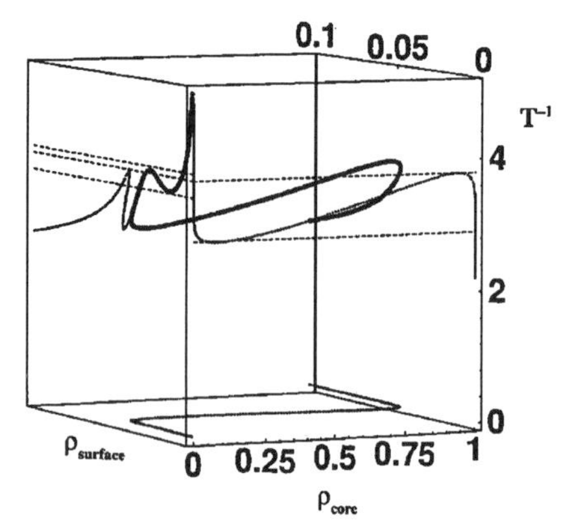

Figure 4-3 A type of phase diagram for a cluster indicating regions of stability and coexistence for three phase-like forms: solid, surface-melted, and liquid. The vertical axis measures T^{-1}, and the two horizontal axes, the densities ρ_{surf} and ρ_{core}, of defects in the surface and core, respectively. The heavy curve is like the curve of stability of a van der Waals gas in a pressure-volume plot; wherever the curve's slope is downward, the system is stable, and where it slopes up, the system fails to satisfy a local stability condition. If one follows this curve from its highest point (lowest T), then each downward branch corresponds to stable phase, whose limits are the points where the curve's direction reverses. Each new stable region along the curve corresponds to a new phase. Thus, in this figure, the solid is stable from the lowest temperatures in the region where there are very few defects in the surface or the core; the surface-melted form is stable only over a finite band of temperature because the curve reverses its direction twice, in the vicinity of very low densities of bulk defects but moderately high densities of surface defects; and the homogeneous liquid is stable in the region of high densities of surface and bulk defects.

then bimodal again, and finally, at the highest temperatures before evaporation dominates the simulations, unimodal. In the trimodal region, the three plateaus correspond to solid, surface-melted, and homogeneously melted phases. The bimodal region above that is the region of dynamic coexistence of surface-melted and liquid phases, and the high-temperature, unimodal region is that of the homogeneous liquid.

We can express the partition function of the system in terms of contributions from the core, the surface, and the floaters, and include the floater–surface and surface–core interactions; these contributions are expressed in terms of parameters reflecting the energy required to produce defects and floaters (Kunz et al., 1993, 1994). With such partition functions in hand, we can search for the limits of the conditions of stability of each phase-like form, just as with the solid–liquid equilibrium. It is convenient to study the results in the form of a kind of phase diagram, such as that in Fig. 4-3. This figure was constructed with parameters that made it correspond to the diagram found from simulations of Ar_{55} and Ar_{147}. It reveals a region of two-phase equilibrium of the solid and surface-melted forms, a region of equilibrium of three phases (solid, surface-melted, and liquid), a region of two-phase equilibrium of surface-melted cluster and solid-like cluster, and a region in which only the liquid is stable.

Not all plausible diagrams of this type have precisely these regions of stability or mutual stability. It is possible to have nonoverlapping regions of bistability, for example (Kunz et al., 1993, 1994). There need not be a tristable region, although the Lennard-Jones system does seem to show one. In fact, with suitable and plausible parameters, this theoretical framework predicts that there could be substances that could have, instead of a surface-melted phase, a core-melted or frozen-surface phase. Naturally one condition for such behavior is that the liquid be denser than the solid. The likely candidate clusters for a frozen-surface phase are those of gallium, indium, and even water. Clusters in this state would be much like ice cubes that have not yet frozen through, but have liquid centers.

4-4 SOME UNSOLVED AND OPEN QUESTIONS

Most of the discussion thus far has dealt with phases and phase changes that are now moderately well understood. However, the study of phase-like forms of clusters and nanoparticles can hardly be said to be complete. We have already pointed out that the role of fluctuations in metastable, superheated, or undercooled systems is still uncertain, at least with regard to whether they would mask the discontinuies of the distribution D_{eq}. There are some others that deserve mention.

One is the question of what the finite-system analogue of a second-order phase transition would be. The true second-order transition of bulk matter has only a single stable minimum with respect to the order parameter, at any temperature. The value of the order parameter at which the free energy is a minimum changes with the system's temperature, and this value moves from a region in which one phase is stable to a region in which the other phase is stable. An analogue of this second-order transition for finite systems might be one which has only a single minimum with respect to the order parameter, for all sizes of clusters. This would be the most obvious formal analogue. But there is another possibility: the system might have two minima for very small or even moderate-size systems, but these minima might converge to a single, common, stable value as N grows very large. This can indeed happen if the defects repel each other, or if they raise the frequencies of the normal modes of the cluster. It may even be possible that both occur, and that there are two kinds of second-order transitions, one emerging from each of the two conditions. This question is under study at the time of this writing (Proykova et al., 1997).

One other open issue deserves mention. It appears from still-unpublished results that clusters may exhibit solid–liquid critical points. The conditions for this to occur are entirely consistent with the finite, nonperiodic nature of the cluster. The stability condition for two phases is, in effect, the condition that the change of entropic contribution to the free energy difference of the phases be matched by the change in energy. If this condition is met at two points, then two phases are in equilibrium. This can be put in terms of two curves that cross at a point. If the two crossing points were to become a single tangent point, that tangent point would be a limit, beyond which the two phases could no longer be distinguished as equilibrium forms of the substance. In short, the point of tangency would be the critical point. It appears at present that there is no logical barrier to the existence of such a point.

It is even possible to argue that there might be solid–liquid critical points for some bulk systems, on the following grounds. The traditional argument that they cannot exist for bulk systems is the symmetry argument, that there can be no continuous transition from the discrete symmetry of the crystalline solid to the continuous translational and rotational symmetry of the amorphous liquid. However, as the temperature of a solid increases, so does the density of its defects, so that establishing the translational symmetry of a hot solid requires averaging over successively longer lengths to establish that average discrete translational symmetry. Likewise, as the pressure increases on a liquid, it becomes more and more ordered. This requires averaging over longer and longer lengths as the pressure increases, in order to establish the continuous translational symmetry of the liquid. If these lengths increase only slowly with N, then the symmetry arguments will be valid. However, if they increase rapidly enough, or if they were even to diverge as temperature or pressure increase, then the symmetry argument would be inapplicable and there could perfectly well be an solid-liquid critical point for a macroscopic system.

4-5 SUMMARY

The phase-like behavior of small clusters leads us not only to a rich variety of new phenomena such as "phases" that do not exist as such in bulk matter, but also to deepened insights into thermodynamics of materials. The solid–liquid equilibrium of an ensemble of clusters reveals, through the banded nature of its coexistence region, the basis of the sharp coexistence curve of bulk solids and liquids, and of the limits of metastability. The coexistence of multiple phase-like forms of clusters and nanoparticles clarifies some of the limits on concepts of "phase" and "component," on the Phase Rule, and on the Maxwell construction normally used to explain the "tie-line" of the solid–liquid phase diagram, with its discontinuities of slope.

Furthermore, the interpretation of phase-like behavior of clusters forces us to reexamine the role of time scales in the thermodynamic interpretation of phenomena when it is possible to make observations on different time scales. Careful consideration of time scales is important not only conceptually but technically as well, especially in the context of molecular dynamics simulations.

Clusters may exhibit coexistence of more than two phases over finite bands of temperature and pressure. This is because, as with solid–liquid equilibria, the various phase-like forms differ from one another by relatively small amounts of free energy, so that detectable amounts of several forms may be present under conditions in which they have nonzero differences in their chemical potentials. Among these forms are not only solid and liquid, but also surface-melted and possibly core-melted (frozen-shell) clusters.

Open questions remain, such as the nature of the finite analogue of second-order phase transitions. Another is the question of how the transition from polyhedral to lattice-based structures occur in those clusters, such as the Lennard-Jones, which are polyhedral for small sizes but close-packed in the bulk crystal. Still another is the question of the structures and phases of molecular clusters, of the extent to which these take on the structures of their bulk counterparts even at small sizes, and of what kinds of phase transitions they may show—clearly an issue closely related to that of the second-order transitions. Finally, this is a subject in which theory and simulation have far outpaced experiment; we can hope that recognition of these ideas will serve as stimuli for new laboratory studies of the phase behavior of clusters and nanoparticles.

ACKNOWLEDGMENTS

The author would like to express his gratitude to his many co-workers whose efforts have helped immeasurably to bring this subject to its present level. The research described here that was carried out at The University of Chicago has been supported by grants from the National Science Foundation.

REFERENCES

Amar, F., and R. S. Berry, "The onset of nonrigid dynamics and the melting transition in Ar_7," *J. Chem. Phys.*, vol. 85, pp. 5943–5954, 1986.

Amini, M., and R. W. Hockney, "Computer simulation of melting and glass formation in a potassium chloride microcrystal," *J. Non-Cryst. Sol.*, vol. 31, pp. 447–452, 1979.

Andersen, H. C., "Molecular dynamics simulations at constant pressure and/or temperature," *J. Chem. Phys.*, vol. 72, pp. 2384–2393, 1980.

Bartell, L. S., "Diffraction studies of clusters generated in supersonic flow," *Chem. Rev.*, vol. 86, pp. 492–505, 1986.

Bartell, L. S., "Inference of cluster phase from considerations of homogeneous nucleation in an evaporative ensemble," *J. Phys. Chem.*, vol. 96, pp. 108–111, 1992.

Bartell, L. S., and J. Chen, "Structure and dynamics of molecular clusters: 2. Melting and freezing of CCl_4 clusters," *J. Phys. Chem.*, vol. 96, pp. 8801–8808, 1992.

Bartell, L. S., and T. S. Dibble, "Electron diffraction studies of the kinetics of phase changes in molecular clusters: Freezing of CCl_4 in supersonic flow," *J. Phys. Chem.*, vol. 95, pp. 1159–1167, 1991a.

Bartell, L. S., and T. S. Dibble, "Kinetics of phase changes in large molecular clusters," *Z. Phys.*, vol. D20, pp. 255–257, 1991b.

Bartell, L. S., L. Harsanyi, and E. J. Valente, "Phases and phase changes of molecular clusters generated in supersonic flow," *J. Phys. Chem.*, vol. 93, pp. 6201–6205, 1989.

Bartell, L. S., L. R. Sharkey, and X. Shi, "Electron diffraction and Monte Carlo studies of liquids: 3. Supercooled benzene," *J. Am. Chem. Soc.*, vol. 110, pp. 7006–7013, 1988.

Beck, T. L., and R. S. Berry, "The interplay of structure and dynamics in the melting of small clusters," *J. Chem. Phys.*, vol. 88, pp. 3910–3922, 1988a.

Beck, T. L., J. Jellinek, and R. S. Berry, "Rare gas clusters: Solids, liquids, slush and magic numbers," *J. Chem. Phys.*, vol. 87, pp. 545–554, 1987.

Beck, T. L., D. M. Leitner, and R. S. Berry, "Melting and phase space transitions in small clusters: Spectral characteristics, dimensions and K-entropy," *J. Chem. Phys.*, vol. 89, pp. 1681–1694, 1988b.

Berry, R. S., T. L. Beck, H. L. Davis, and J. Jellinek, "Solid–liquid phase behavior in microclusters," in *Evolution of Size Effects in Chemical Dynamics, Part 2*, I. Prigogine, and S. A. Rice (eds.), John Wiley and Sons, New York, pp. 75–138, 1988.

Berry, R. S., J. Jellinek, and G. Natanson, "Melting of clusters and melting," *Phys. Rev.*, vol. A30, pp. 919–931, 1984a.

Berry, R. S., J. Jellinek, and G. Natanson, "Unequal freezing and melting temperatures for clusters," *Chem. Phys. Lett.*, vol. 107, p. 227, 1984b.

Blaisten-Barojas, E., and D. Levesque, "Molecular-dynamics simulation of silicon clusters," *Phys. Rev.*, vol. B34, pp. 3910–3916, 1986.

Blaisten-Barojas, E., and D. Levesque, "A molecular dynamics study of silicon clusters," in *Physics and Chemistry of Small Clusters*, P. Jena, B. K. Rao, and S. N. Khanna (eds.), Plenum Press, New York, 1987.

Briant, C. L., and J. J. Burton, "Molecular dynamics study of the structure and thermodynamic properties of argon microclusters," *J. Chem. Phys.*, vol. 63, pp. 2045–2058, 1975.

Buck, U., B. Schmidt, and J. G. Siebers, "Structural transitions and thermally averaged infrared spectra of small methanol clusters," *J. Chem. Phys.*, vol. 99, pp. 9428–9437, 1993.

Cheng, H.-P., and R. S. Berry, "Surface melting and surface diffusion on clusters," *Phys. Rev.*, vol. 206, pp. 241–252, 1991.

Cheng, H.-P., and R. S. Berry, "Surface melting of clusters and implications for bulk matter," *Phys. Rev.*, vol. A45, pp. 7969–7980, 1992a.

Cheng, H.-P., X. Li, R. L. Whetten, and R. S. Berry, "Complete statistical thermodynamics of the cluster solid–liquid transition," *Phys. Rev.*, vol. A46, pp. 791–800, 1992b.

Cotterill, R. M., W. Damgaard Kristensen, J. W. Martin, L. B. Peterson, and E. J. Jensen, "Melting of microcrystals in two dimensions: A molecular dynamics simulation," *Comput. Phys. Commun.*, vol. 5, p. 28, 1973.

Cotterill, R. M. J., "Molecular dynamics studies of melting. III. Spontaneous dislocation generation and the dynamics of melting," *Phil. Mag.*, vol. 32, pp. 1283–1288, 1975.

Damgaard Kristensen, W., E. J. Jensen, and R. M. J. Cotterill, "Thermodynamics of small clusters of atoms: A molecular dynamics simulation," *J. Chem. Phys.*, vol. 60, pp. 4161–4169, 1974.

Davis, H. L., T. L. Beck, P. A. Braier, and R. S. Berry, "Time scale considerations in the characterization of melting and freezing in microclusters," in *The Time Domain in Surface and Structural Dynamics*, G. J. Long and F. Grandjean (eds.), Kluwer Academic Publishers, Boston, pp. 535–549, 1988.

Davis, H. L., J. Jellinek, and R. S. Berry, "Melting and freezing in isothermal Ar_{13} clusters," *J. Chem. Phys.*, vol. 86, pp. 6456–6469, 1987.

Dibble, T. S., and L. S. Bartell, "Electron diffraction studies of the kinetics of phase changes in molecular clusters. 2. Freezing of CH_3Cl_3," *J. Phys. Chem.*, vol. 96, p. 2317, 1992.

Doye, J. P. K., D. J. Wales, and R. S. Berry, "The effect of the range of the potential on the structures of clusters," *J. Chem. Phys.*, vol. 103, pp. 3061–3070, 1995.

Ellert, C., M. Schmidt, C. Schmitt, T. Reiners, and H. Haberland, "Temperature dependence of the optical response of small, open shell sodium clusters," *Phys. Rev. Lett.*, vol. 75, pp. 1731–1734, 1995.

Etters, R. D., R. Danilowicz, and J. Kaelberer, "Metastable states of small rare gas crystallites," *J. Chem. Phys.*, vol. 67, pp. 4145–4148, 1977a.

Etters, R. D., and J. B. Kaelberer, "Thermodynamic properties of small aggregates of rare-gas atoms," *Phys. Rev.*, vol. A11, pp. 1068–1079, 1975.

Etters, R. D., and J. B. Kaelberer, "On the character of the melting transition in small atomic aggregates," *J. Chem. Phys.*, vol. 66, pp. 5112–5116, 1977b.

Farges, J., M. F. deFeraudy, B. Raoult, and G. Torchet, "Noncrystalline structure of argon clusters: I. Polyicosahedral structure of Ar_N clusters, $20 < N < 50$," *J. Chem. Phys.*, vol. 78, pp. 5067–5080, 1983.

Farges, J., M. F. deFeraudy, B. Raoult, and G. Torchet, "Noncrystalline structure of argon clusters. 2. Multilayer icosahedral structure of Ar_N clusters, $50 \leq N \leq 750$," *J. Chem. Phys.*, vol. 84, p. 3491, 1986.

Fisher, M. E., and V. Privman, "First-order transitions breaking O(n) symmetry: Finite-size scaling," *Phys. Rev.*, vol. B32, pp. 447–464, 1985.

Fisher, M. E., and V. Privman, "First-order transitions in spherical models: Finite-size scaling," *Commun. Math. Phys.*, vol. 103, pp. 527–548, 1986.

Garzon, I. L., M. A. Borja, and E. Blaisten-Barojas, "Phenomenological model of melting in Lennard-Jones clusters," *Phys. Rev.*, vol. B40, pp. 4749–4759, 1989.

Goldstein, A. N., V. L. Colvin, and A. P. Alivisatos, "Observation of melting in 30 Å diameter CdS nanocrystals," vol. 206, pp. 271–274, 1990.

Hill, T. L., *The Thermodynamics of Small Systems, Part 1*, W. A. Benjamin, New York, 1963.

Hill, T. L., *The Thermodynamics of Small Systems, Part 2*, W. A. Benjamin, New York, 1964.

Honeycutt, J. D., and H. C. Andersen, "Molecular dynamics study of melting and freezing of small Lennard-Jones clusters," *J. Phys. Chem.*, vol. 90, pp. 4950–4963, 1987.

Jellinek, J., T. L. Beck, and R. S. Berry, "Solid–liquid phase changes in simulated isoenergetic Ar_{13}," *J. Chem. Phys.*, vol. 84, pp. 2783–2794, 1986.

Jellinek, J., V. Bonacic-Koutecky, P. Fantucci, and M. Wiechert, "Ab initio HF SCF study of structure and dynamics of Li_8," *J. Chem. Phys.*, vol. 101, pp. 10092–10100, 1994.

Kaelberer, J. B., and R. D. Etters, "Phase transitions in small clusters of atoms," *J. Chem. Phys.*, vol. 66, pp. 3233–3239, 1977.

Kunz, R. E., and R. S. Berry, "Coexistence of multiple phases in finite systems," *Phys. Rev. Lett.*, vol. 71, pp. 3987–3990, 1993.

Kunz, R. E., and R. S. Berry, "Multiple phase coexistence in finite systems," *Phys. Rev. E*, vol. 49, pp. 1895–1908, 1994.

Lee, J. K., J. A. Barker, and F. F. Abraham, "Theory and Monte Carlo simulation of physical clusters in the imperfect vapor," *J. Chem. Phys.*, vol. 58, pp. 3166–3180, 1973.

Lindemann, F. A., *Phys. Z.*, vol. 11, p. 609, 1910.

Lindemann, F. A., *Engineering*, vol. 94, p. 515, 1912.

Luo, J., U. Landman, and J. Jortner, "Isomerization and melting of small alkalihalide clusters," in *Physics and Chemistry of Small Clusters*, P. Jena, B. K. Rao, and S. N. Khanna (eds.), Plenum Press, New York, p. 201, 1987.

Martin, T. P., U. Näher, H. Schaber, and U. Zimmerman, "Evidence for a size dependent melting of sodium clusters," *J. Chem. Phys.*, vol. 100, p. 2322, 1994.

McGinty, D. J., "Molecular dynamics studies of the properties of small clusters of argon atoms," *J. Chem. Phys.*, vol. 58, p. 4733, 1973.

Nauchitel, V. V., and A. J. Pertsin, "A Monte Carlo study of the structure and thermodynamic behaviour of small Lennard-Jones clusters," *Mol. Phys.*, vol. 40, pp. 1341–1355, 1980.

Nosé, S., "A molecular dynamics method for simulations in the canonical ensemble," *Mol. Phys.*, vol. 52, p. 255, 1984a.

Nosé, S., "A unified formulation of the constant temperature molecular dynamics methods," *J. Chem. Phys.*, vol. 81, pp. 511–519, 1984b.

Nosé, S., "Constant temperature molecular dynamics methods," *Prog. Theor. Phys.*, Suppl. 103, pp. 1–46, 1991.

Privman, V., and M. E. Fisher, "Finite-size effects at first-order transitions," *J. Stat. Phys.*, vol. 33, pp. 385–417, 1983.

Proykova, A., and R. S. Berry, "Analogues in clusters of second-order transitions," *Z. Phys. D*, in press, 1997.

Quirke, N., and P. Sheng, "The melting behavior of small clusters of atoms," *Chem. Phys. Lett.*, vol. 110, pp. 63–66, 1984.

Raoult, B., J. Farges, M. F. DeFeraudy, and G. Torchet, "Comparison between icosahedral, decahedral and crystalline Lennard-Jones models containing 500 to 6000 atoms," *Phil. Mag.*, vol. B60, pp. 881–906, 1989.

Rose, J. P., and R. S. Berry, "Towards elucidating the interplay of structure and dynamics in clusters: Small KCl clusters as models," *J. Chem. Phys.*, vol. 96, pp. 517–538, 1992.

Sawada, S., "Dynamics of transition metal clusters," in *Microclusters*, S. Sugano, Y. Nishina, and S. Ohnishi (eds.), Springer-Verlag, Berlin, pp. 211–217, 1987.

Sawada, S., and S. Sugano, "Dynamics of transition-metal clusters," *Z. Phys.*, vol. D12, p. 189, 1989.

Sawada, S., and S. Sugano, "Structural fluctuations of Au_{55} and Au_{147}: Substrate effect," *Z. Phys.*, vol. D24, pp. 377–384, 1992.

Torchet, G., M. F. deFeraudy, B. Raoult, J. Farges, A. H. Fuchs, and G. S. Pawley, "Cluster model for the monoclinic to cubic transition in SF_6 clusters," *J. Chem. Phys.*, vol. 92, pp. 6768–6774, 1990a.

Torchet, G., J. Farges, M. F. deFeraudy, and B. Raoult, "Electron diffraction studies of clusters produced in a free jet expansion," in *The Chemical Physics of Atomic and Molecular Clusters*, G. Scoles (ed.), North-Holland, Amsterdam, pp. 513–542, 1990b.

Valente, E. J., and L. S. Bartell, "Electron diffraction studies of supersonic jets: VI. Microdrops of benzene," *J. Chem. Phys.*, vol. 80, pp. 1451–1457, 1984a.

Valente, E. J., and L. S. Bartell, "Electron diffraction studies of supersonic jets: VII. Liquid and plastic crystalline carbon tetrachloride," *J. Chem. Phys.*, vol. 80, pp. 1458–1461, 1984b.

Wales, D. J., and R. S. Berry, "Freezing, melting, spinodals and clusters," *J. Chem. Phys.*, vol. 92, pp. 4473–4482, 1990.

Wales, D. J., and R. S. Berry, "Coexistence in Finite Systems," *Phys. Rev. Lett.*, vol. 73, pp. 2875–2878, 1994.

Wales, D. J., and J. P. K. Doye, "Coexistence and phase separation in clusters: From the small to the not-so-small regime," *J. Chem. Phys.*, vol. 103, pp. 3061–3070, 1995.

Woods, L. C., *Thermodynamics of Fluid Systems*, Clarendon, Oxford, 1975.

CHAPTER

FIVE

MOLECULAR CLUSTERS

Susumu Kotake

Department of Mechanical Engineering,
Toyo University,
Kawagoe, Saitama 350, Japan

5-1 INTRODUCTION

Clusters consisting of tens to hundreds of atoms or molecules are a new class of compounds called atomic or molecular clusters. These clusters are expected to have very interesting chemical and physical properties that span the range from molecular to bulk. The number of theoretical and experimental studies of these clusters has grown quickly during the last two decades. The interest in clusters stems from many sources. Small clusters have been found very useful in the study of intermolecular forces as well as physical and chemical dynamics of molecular collisions and energy transfer. Large clusters have been studied for understanding the early stages of nucleation and condensation, catalysis chemistry, and transition phenomena between molecular and condensed bulk materials (Kotake and Glass, 1981). Using newly developed experimental techniques such as laser optics, spectrometry, and cluster beams, as well as powerful, high-speed computers, extensive studies on the physical and chemical features of these clusters have been started in wide ranges of cluster sizes and species.

In addition, these clusters are expected to have many possibilities in engineering applications such as very fine particles and thin films and controlling the process of nucleation and condensation. Work is also started on such applications of the molecular clusters for engineering problems.

The present chapter concerns the formation process of clusters, their molecular structures and physical features, and the control of clustering and condensation processes.

5-2 CLUSTERS AND CLUSTERING

5-2-1 Clusters

Ensembles of atoms or molecules that are bound without any electronic coupling (chemical reaction) are called *atomic clusters* or *molecular clusters* (hereafter, simply "clusters"). These are contrasted with bound atoms or molecules with electronic coupling which are called molecules or polymers as shown in Fig. 5-1. In these clusters, atoms and molecules keep their own atomic and molecular properties, and only the properties as an ensemble of atoms and molecules can be modified or changed. The energy with which the constituting atomic or molecular particles interact electronically to bind together can be evaluated by the potential energy, and the potential energy of cluster n, U_n, is defined as a difference between the energy of the cluster E_n and the sum of constituting particle energies $e_i (i = 1, 2, \ldots, n)$,

$$U_n = E_n - \sum_{i=1}^{n} e_i. \tag{5-1}$$

The potential energy or interaction energy of molecules consists of terms for repulsion, dispersion (attraction), isolated static charges, and induced multiple charges (see Fig. 5-2). Since the partition function of the atomic or molecular configuration at thermal equilibrium can be related to $\exp(-U/kT)$, the clusters of lower potential energy are more favorably realized statistically.

The simplest cluster is the *dimer*, which consists of two atoms or molecules (monomers). Clusters made up of three, four, five, and many atoms or molecules are called *trimers, tetramers, pentamers*, and *highmers*, respectively, and clusters of thousands of atoms or molecules are often called *ultra-fine particles* or *fine particles*. These clusters take a variety of geometrical structures. The variety of the geometrical configuration increases with the number of atoms or molecules in the cluster, that is, the cluster size. In Fig. 5-3, typical configurations of small clusters are shown. Small clusters, in the range of ten to a few hundred atoms or molecules, exhibit noncrystalline structure. At sufficiently low temperatures, these clusters may take some favorable structures influenced by the strong anisotropies in the intermolecular forces with ordered motions of vibration and rotation.

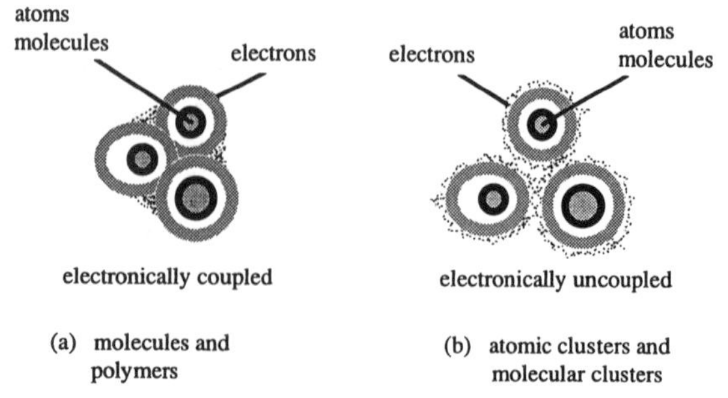

Figure 5-1 Atomic/molecular clusters: atoms or molecules are bound without any electronic coupling but with electronic interactions such as van der Waals forces.

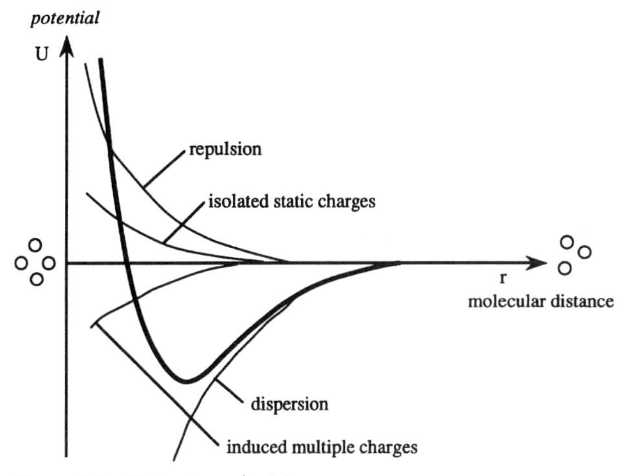

Figure 5-2 Molecular potential.

Depending upon the structure, clusters assume different potential energies. At statistical equilibrium, in which clusters have sufficient relaxation time to equilibrate the intermolecular and intramolecular interactions of atomic and molecular motions, the cluster structure should be the most stable so as to take the lowest potential energy. The most stable structure for a given isomer is that with the most compact configuration. For example, the pyramid-like structure in Fig. 5-3, which has equidistant atoms/molecules, is the most stable tetramer.

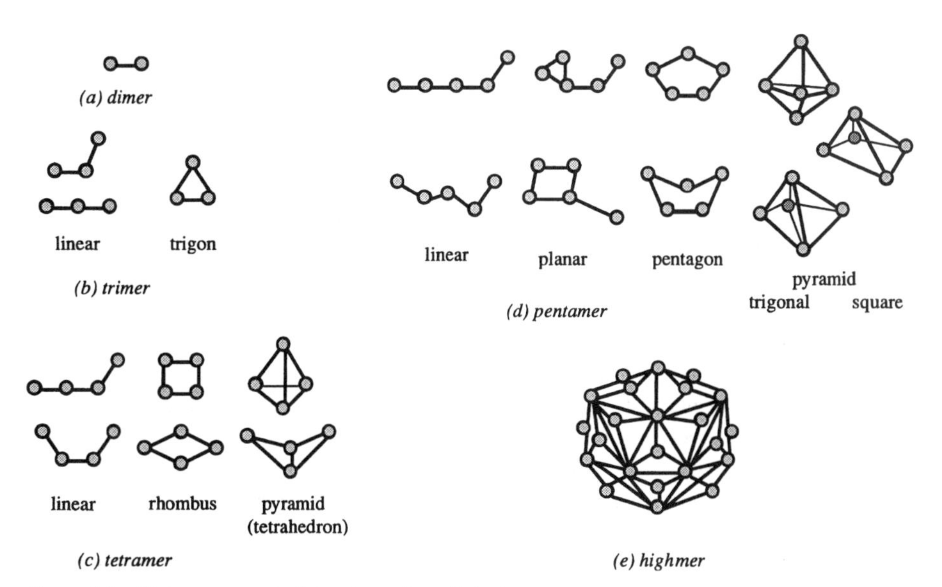

Figure 5-3 Various geometrical structures of small clusters.

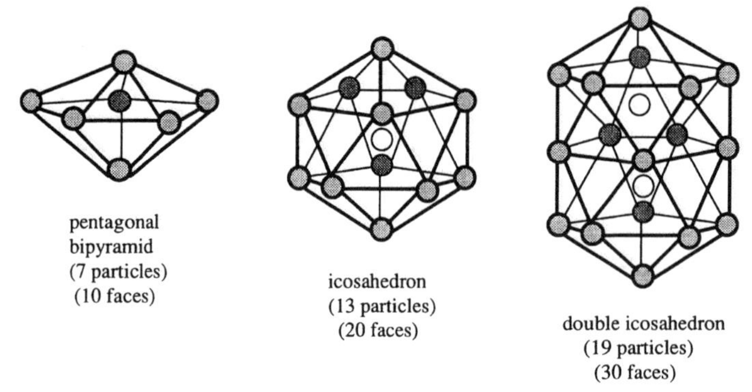

Figure 5-4 Stable structures of rare gas clusters.

The cluster potential energy also depends on the number of atoms or molecules in a cluster exhibiting an irregular size dependence with energy minima at the so-called magic numbers. Clusters of sizes at which the lower potential takes place are more thermodynamically favorable. For clusters of nonpolar molecules such as rare gas atoms, the lower potentials are commonly believed to occur at cluster sizes of 7, 13, 19, . . . , for which the geometrical structure corresponds to pentagonal bipyramid, icosahedron, double icosahedron, etc., respectively (Fig. 5-4). As the cluster size increases, these structures may grow into polyicosahedral, multishell, and then bulk crystalline structures. Much work has been done to study these favorable structures, in clusters ranging from rare gas molecules to complex molecules such as polymers, by employing semi-empirical or ab-initio calculated potentials.

5-2-2 Clustering

Clusters of lowest potential can only be realized at thermal equilibrium. During dynamical processes like phase change, realized clusters are not always of the lowest potential. Because the atomic or molecular bonding without electronic coupling and electronic interaction has no limit to be zero, these clusters are ambiguously defined in their dynamical states such as physical structure and lifetime. Dynamically, there are many possibilities for realization of various structures. To make a practical definition of clusters in dynamic states, a certain threshold of the potential energy or the distance between atoms or molecules is introduced to constitute clusters as shown in Fig. 5-5.

Much effort has been devoted to understanding the processes of cluster formation at the initial stages of nucleation and condensation as well as other dynamic processes of van der Waals molecules. From the standpoint of atomic and molecular collision dynamics, most of the dynamic behavior is similar to that of ordinary chemical reactions. However, these clusters are weakly bound with dissociation energy usually two orders of magnitude smaller than normal chemical bonds. At the beginning of clustering, adsorption of atoms or molecules may be the dominant process, although coagulation and fragmentation of clusters become much more important as the density of clusters increases.

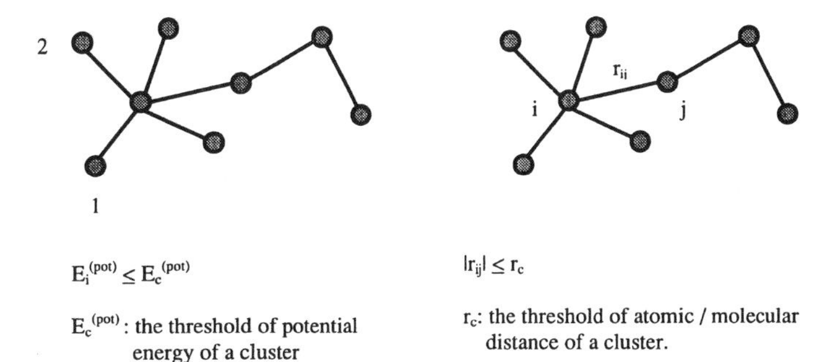

Figure 5-5 Definition of clusters: the threshold of potential energy or distance between atoms or molecules is introduced to define the geometrical structure of a cluster.

Three steps constitute the formation process of clusters (Fig. 5-6), although these steps take place continuously:

(1) Collision between atoms, molecules, or clusters A_m and A_n makes unbound coupling states of these particles $(A_m A_n)^*$,

$$A_m + A_n \rightarrow (A_m A_n)^* \tag{5-2}$$

where the translational, rotational, and vibrational energies, $E_{m+n}^{(trans)}$, $E_{m+n}^{(rot)}$, and $E_{m+n}^{(vib)}$, respectively, are the same as the pre-collision values

$$E_m^{(trans)} + E_n^{(trans)} = E_{(m+n)}^{(trans)}, \qquad E_m^{(rot)} + E_n^{(rot)} = E_{(m+n)}^{(rot)}, \qquad E_m^{(vib)} + E_n^{(vib)} = E_{(m+n)}^{(vib)}. \tag{5-3}$$

(2) Stabilization of the unbound states makes quasi-bound states of cluster $(A_{m+n})^*$,

$$(A_m A_n)^* \rightarrow (A_{m+n})^* \tag{5-4}$$

where the energies are changed due to the stabilization:

$$E_{(m+n)}^{(trans)} \rightarrow E_{m+n}^{(trans)*}, \qquad E_{(m+n)}^{(rot)} \rightarrow E_{m+n}^{(rot)*}, \qquad E_{(m+n)}^{(vib)} \rightarrow E_{m+n}^{(vib)*}. \tag{5-5}$$

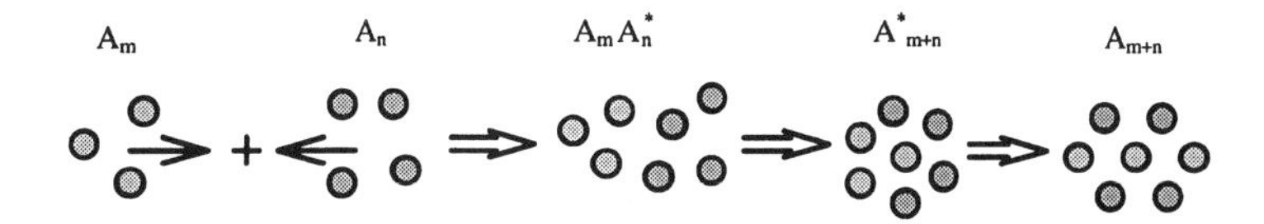

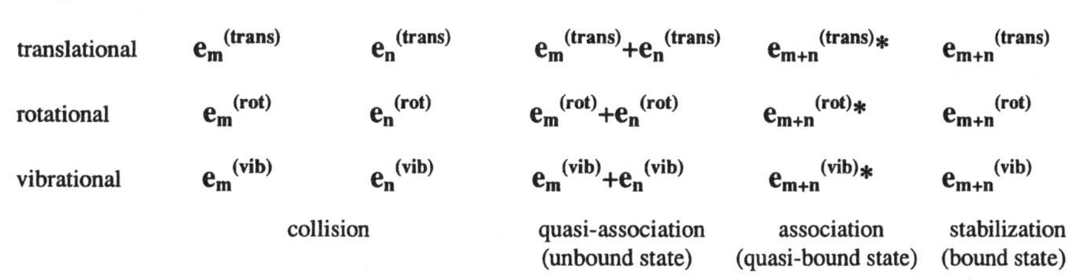

Figure 5-6 Clustering process by collision between atoms, molecules, or clusters; $A_m + A_n \rightarrow A_{m+n}$.

(3) Further stabilization of the quasi-bound states makes bound states of cluster (A_{m+n}),

$$(A_{m+n})^* \rightarrow (A_{m+n}) \tag{5-6}$$

where the energies are further changed or modified due to the stabilization:

$$E_{m+n}^{(trans)*} \rightarrow E_{m+n}^{(trans)}, \qquad E_{m+n}^{(rot)*} \rightarrow E_{m+n}^{(rot)}, \qquad E_{m+n}^{(vib)*} \rightarrow E_{m+n}^{(vib)}. \tag{5-7}$$

The stabilization of the quasi-bound states of the high energy level to the bound state of the lower level requires the extraction of energy from the cluster or the redistribution of energies throughout the cluster. The former may be associated with single or successive collisions of a third-body atom or molecule, and this is the only process (as for dimer formation) that is the starting process of cluster formation.

The statistical reaction processes of these cluster formations usually favor the more exothermic pathway leading to the more stable reaction products. As the initial pre-collision translational energy increases, the process of cluster formation becomes dominated by the collision dynamics, yielding non-statistical products. For example, on the exchange process of the $A+BC$ system ($m_A < m_B$), the mechanism for reaction with the heavier particle (C) to form AC involves three-body interaction of "complex" formation $(ABC)^*$, in which the product AC is randomly scattered. Because of the intermolecular interaction in the $(ABC)^*$ complex, this process shows statistical exothermic reaction behavior. As the collision energy increases, the abstraction of the lighter particle (B) by the incoming one (A) to form AB occurs by a stripping mechanism in which the product (AB) is scattered sharply in the forward direction of the reactants.

The reaction pathway is also affected by the internal energy state of reactants. Due to the lower bond energies of clusters, the rotational barrier to formation is more significant compared to molecular chemical reactions. The rotational barrier $U_{max}^{(J)}$ and the potential minimum $U_{min}^{(J)}$ increase as the rotational state J increases in the potential curve as shown in Fig. 5-7. At some rotational states (J_{max}), the potential becomes strictly repulsive. Clusters having rotational states greater than J_{max} should be dissociative. If the internal energy of the product cluster is less than the rotational barrier but larger than the potential minimum, the product cluster may be rotationally metastable, playing an important role in the clustering process.

After clusters are formed, each collision between clusters and/or monomers leads to dissociation of these clusters as well as association with transferring or redistributing the energies (Fig. 5-8). This dissociation process of a cluster by monomer or cluster collision is greatly enhanced by the excitation of initial vibration and rotation of the cluster.

Thus, the formation process of clusters is a competition between these associations and dissociations. During these processes, clusters never have the physical features of the equilibrium state, possessing a much smaller number of bonds than the most stable structure, that is, the most compact configuration. The most compact configuration takes the greatest number of bonds with the minimum potential energy. The growing clusters have the potential energy larger than the minimum value because of their incompact configuration. At lower temperatures, compact clusters may be more numerous because the surroundings have less kinetic energy, which favors heat transfer

potential
Umn
J rotational state
Jmax
U(J)max
An
en
Am
U(J)min
distance
between Am and An
An
Am
en < U(J)max
An
en = U(J)max
An
en > U(J)max

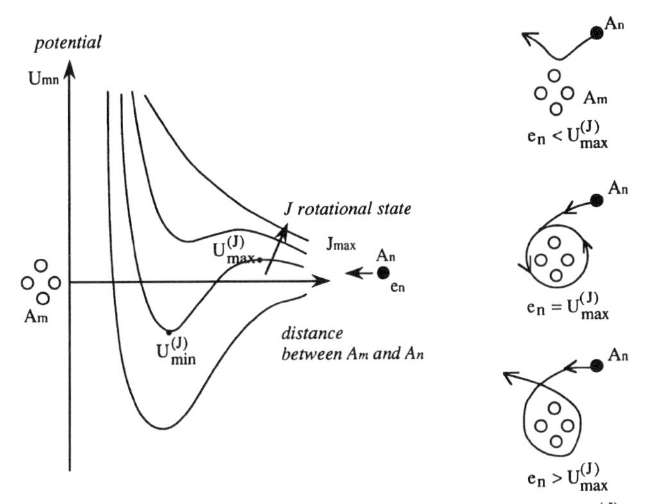

Figure 5-7 Potential barrier for clustering: the rotational barrier ($U_{max}^{(J)}$) and the potential minimum ($U_{min}^{(J)}$) increase as the rotational state (J) increases.

of association rather than dissociation due to destructive collision or excitation to higher energy states.

5-2-3 Clustering Kinetics

These clustering processes can be treated by the molecular dynamics method (MD) with advanced high-speed, large-memory computers. The molecular dynamics method yields information about both static and dynamic behavior of the cluster system by solving the equations of atomic or molecular motions with appropriate potentials.

Denoting the coordinates of clusters as x_i $(i = 1, 2, \ldots, n)$, the equation of molecular dynamics motion can be written as

$$m_i \frac{\partial v_i}{\partial t} = -\frac{\partial U_i}{\partial x_i}, \qquad \frac{\partial x_i}{\partial t} = v_i \quad (i = 1, 2, \ldots, n) \tag{5-8}$$

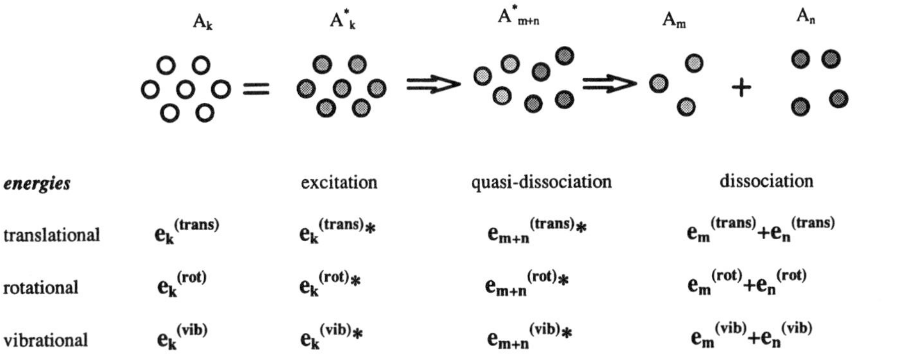

Ak
A*k
A*m+n
Am
An
energies
excitation
quasi-dissociation
dissociation
translational
ek(trans)
ek(trans)*
em+n(trans)*
em(trans)+en(trans)
rotational
ek(rot)
ek(rot)*
em+n(rot)*
em(rot)+en(rot)
vibrational
ek(vib)
ek(vib)*
em+n(vib)*
em(vib)+en(vib)

Figure 5-8 Dissociation of clusters due to excitation of the energy states: $A_k \to A_m + A_n$.

where U_i is the intercluster potential energy, m_i is the mass, and v_i is the velocity of the cluster i. When the cluster i consists of atoms or molecules whose coordinates refer to the center of mass of the cluster, r_i, the molecular dynamics equations are given by

$$m_i \frac{\partial v_{ik}}{\partial t} = -\frac{\partial U_{ik}}{\partial r_{ik}}, \qquad \frac{\partial r_{ik}}{\partial t} = v_{ik} \tag{5-9}$$

where U_{ik} is the intermolecular and intramolecular potential energy, m_i the mass, and v_{ik} the velocity of the kth atom or molecule of the ith cluster.

Solving a set of these dynamic equations for a system consisting of atom–atom, atom–molecule, molecule–molecule, molecule–cluster, or cluster–cluster reactant particles with employing proper intramolecular and intermolecular potentials under an initial condition, the atomic, molecular, and cluster motions can be predicted straightforwardly (Macginty, 1973; Kaelberer and Etters, 1977; Zutrek and Schieve, 1978; Polmeropoulos and Brickmann, 1982; Kelton et al., 1983; Freeman and Doll, 1985; Sano and Kotake, 1989; Ohara and Aihara, 1995; Zhukhovitskii, 1995). It describes the coupling between intracluster (intramolecular) and intercluster (intermolecular) modes, which induces energy transfer between those modes within the clusters as well as between the clusters. The coupling of intracluster modes to intercluster modes is relevant for dissociation of the cluster and energy redistribution in the cluster. As for the electronic states, they usually keep the ground states somewhat modified with intercluster interactions, except for the case of metal, where some of the electrons become delocalized from the ground states.

In these calculations of molecular dynamics, it is of the most importance to predict the intramolecular and intermolecular potentials. These potentials have been estimated empirically or semi-empirically from the statistical data of thermal properties. Simple and typical forms of the potential are composed of repulsive and attractive terms for pairwise interaction between two particles of atom, molecule, or cluster as

$$\begin{aligned} U &= \varepsilon_1 \left(\frac{\sigma_1}{r}\right)^m - \varepsilon_2 \left(\frac{\sigma_2}{r}\right)^n \quad \text{Lennard-Jones potential} \\ &= 4\varepsilon \left\{ \left(\frac{\sigma}{r}\right)^{12} - \left(\frac{\sigma}{r}\right)^6 \right\} \quad (m = 12, n = 6), \end{aligned} \tag{5-10}$$

or

$$\begin{aligned} U &= \varepsilon_1 e^{-\alpha(r-\sigma_1)} - \varepsilon_2 e^{-\alpha(r-\sigma_2)} \\ &= \varepsilon \left\{ e^{-2\alpha(r-\sigma)} - 2e^{-\alpha(r-\sigma)} \right\}. \quad \text{Morse potential} \end{aligned}$$

These potentials in a simple form can give the general and qualitative features of the structures and dynamics of cluster behavior, and have been employed for fundamental studies of molecular dynamics problems.

In order to obtain more quantitative knowledge and better understanding of clusters and clustering behavior, more precise potentials are required. Recent progress in computer speed and memory has made it possible to make ab-initio calculations of the potential by directly solving the Schrödinger equations of the atomic and molecular system with several computational methods. One of the most widely used and highly accurate methods is the MO-LCAO method, where the molecular orbitals are expressed as a linear combination of atomic orbitals, taking into account the configuration interaction self-consistent field (CI SCF) with a large number of orbital basis sets to include the electron interaction (Kim et al., 1992; Zolotoukhina and Kotake, 1993; Rovira et al., 1995). The intercluster (intermolecular) potential can be given as a difference between the total energy of the cluster system and the sum of constituent energies:

$$\begin{aligned} U_n &= E_n - \sum_{i=1}^{n} e_i \\ H_n \Psi_n &= E_n \Psi_n \end{aligned} \tag{5-11}$$

where H_n is the Hamiltonian of the cluster system in Schrödinger equation and Ψ_n is the corresponding wavefunction. At the present state of computer power, however, these methods for large-size clusters of many molecules have limits that stem from the large size of the orbital basis set and sufficient inclusion of electron correlation, although these limits are being rapidly removed with advancing computer capacities.

Averaging molecular dynamics solutions of different molecular states under the same macroscopic initial condition will give the reaction-like kinetics with corresponding reaction rates such as those in chemical kinetics. This kinetic approach to clustering is also obtained with adequate assumptions of the kinetic rate coefficients, k_{mn}. In the process of cluster formation,

$$\begin{aligned} A_m + A_n &\xrightarrow{k_{mn}^{+}} A_{mn} \\ A_{m+n} &\xrightarrow{k_{mn}^{-}} A_m + A_n \end{aligned} \tag{5-12}$$

the reaction rate or formation rate of cluster $[A_{m+n}]^{\circ}$ is given by

$$[A_{m+n}]^{\circ} = k_{mn}^{+}[A_m][A_n] - k_{mn}^{-}[A_{m+n}] \tag{5-13}$$

where $[A]$ is the number density of cluster A. The reaction coefficient k can be given by

$$k_{mn}^{+} = \iint P_{mn}\sigma_{mn}|v_m - v_n| f(v_m) f(v_n) dv_m dv_n \tag{5-14}$$

where v_m is the translational velocity of cluster m, $f(v_m)$ the velocity distribution of cluster m, σ_{mn} is the collision cross section, and P_{mn} is the formation probability of A_{mn} at the collision of A_m and A_n. The formation probability P_{mn} can be predicted analytically with the molecular dynamics calculation or estimated empirically with experimental data. The reverse reaction rate coefficient k_{mn}^{-} can be obtained by the equilibrium condition of the reaction in the same manner as the forward chemical reaction.

The total formation rate of cluster i is then given by

$$\begin{aligned} [A_i]^{\circ} &= \sum_{m=1}^{i-1} k_{mn}^{+}[A_m][A_n]_{(n=i-m)} - \sum_{m=1}^{i-1} k_{mn}^{-}[A_i]_{(n=i-m)} \\ &\quad - \sum_{m=1} k_{mi}^{+}[A_m][A_i] + \sum_{m=1} k_{mn}^{-}[A_m]_{(n=i-m)}. \end{aligned} \tag{5-15}$$

When the number density of monomers is many orders of magnitude larger than the highmer densities, $[A_1] \gg [A_i]\,(i = 2, 3, \ldots)$, only the monomer reactions dominate the clustering process, and the formation rate of cluster i is expressed as

$$[A_i]^\circ = k^+_{(i-1)1}[A_{i-1}][A_1] - k^-_{(i-1)1}[A_i] - k^+_{i1}[A_i][A_1] + k^-_{i1}[A_{i+1}]. \qquad (5\text{-}16)$$

With these equations for calculating the reaction coefficients, the dynamic behavior of clustering processes can be predicted in order to gain insight into the fundamental mechanisms of nucleation and condensation (Kelton et al., 1983; Bauer et al., 1986).

5-2-4 Experimental Studies of Clustering

For experimental studies of clustering, free-jet expansions of gases or effusive vapor flows are commonly used (Rohlfing et al., 1984; Inoue and Kotake, 1989). In the latter, high-temperature vapors of liquid are effused into low-temperature surroundings to be cooled for clusterization. In the former, supersonic free jets of gases or vapors of solid materials are issued from a small aperture nozzle into high vacuum as shown in Fig. 5-9. An isentropic expansion of the gas or vapor is associated with rapid decrease in the temperature. This causes atomic or molecular clustering when the characteristic time of flow becomes less than that of clustering which is on the order of microsecond (Kotake and Glass, 1981).

The characteristic times of clustering can be related to the reaction-rate coefficients of the molecular collision kinetics, being a function of the pressure and temperature as well as the internal states of the flow molecules. Clustering takes much more time as the pressure and temperature decrease, so the clustering process of expanding molecules is essentially frozen downstream of the flow, as shown in Fig. 5-10. The frozen states often occur through a transition region along the jet axis. Upstream of the region, the molecules are of higher pressures and temperatures so that the clustering times are so small in comparison with the flow time that they are essentially at thermodynamic equilibrium. Until the gas reaches the freezing position, the number density of clusters formed increases

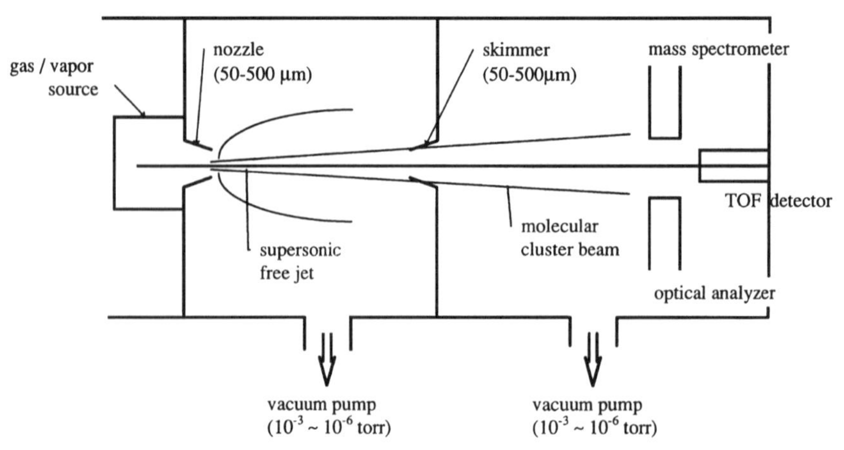

Figure 5-9 Experimental apparatus for cluster studies.

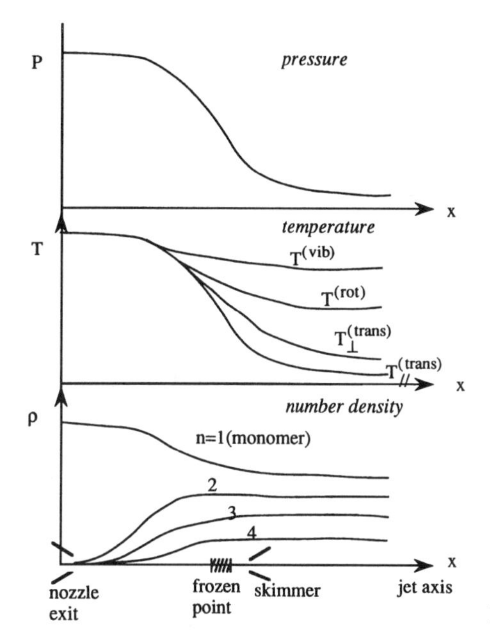

Figure 5-10 Clustering in a supersonic jet.

approximately in correspondence with local thermodynamic equilibrium, and downstream from this position, the number density is kept approximately constant. Around this position, the gas is skimmed with a skimmer into a higher vacuum chamber where the physical and chemical properties of frozen clusters are analyzed with measuring equipment such as mass spectrometers and optical analyzers.

The size or mass of clusters is usually measured with the electric or magnetic behavior of their ionized states, which is related to the cluster size or mass. Studies of electrically neutral clusters involve ionization, followed by analysis and detection of ionized clusters. The ionization process is always associated with dissociation, fragmentation, and destruction of clusters, since the ionization energy is one or two orders greater than the clustering energy. In such cases, much attention has to be paid to the possibility of cluster fragmentation through the ionization. The validity of detected species through ionization of the neutral clusters has been the subject of considerable debate and controversy and is the critical means for obtaining information of neutral clusters. In order to analyze the size distribution of neutral clusters with this method, the dissociation and fragmentation through ionization must be taken into account.

5-3 THERMOPHYSICAL PROPERTIES OF CLUSTERS

Since clusters are composed largely of surface atoms, their chemical and physical properties may well reflect the microscopic details of surface activity. Thus, they are expected to have physical and chemical properties that span the range from molecular to bulk. Questions of interest then arise:

- How do the properties differ from the molecular or bulk ones?
- Over what range of cluster size does the transition between molecular and bulk properties occur?
- Do all physical and chemical properties make transition in the same range of cluster size?
- Do all clusters behave similarly, or do they show significant difference in the transition?

These interesting properties of clusters are molecular structures; electronic, vibrational, and rotational states; physical and chemical reactivities; and thermodynamic properties. For studying these properties, much experimental and computational work has been conducted. First experimental studies used the matrix method in which clusters were formed on a solid matrix, but recent studies employ the supersonic free jet (nozzle beam, or NB) method in which clusters are produced in the process of supersonic expansion of gaseous atoms or molecules issued from a small aperture nozzle into vacuum (Rohlfing et al., 1984; Inoue and Kotake, 1989). Atomic or molecular beams with clusters are then led to the measuring stage after being skimmed and/or collimated to measure the properties with optical, electronic, and magnetic methods. These techniques include mass spectometry, optical spectroscopy, electron spectroscopy, electron diffraction, and electron or photon ionization techniques.

To predict the cluster properties with computers, the Monte Carlo (MC) and molecular dynamics (MD) methods can be also used. In the former (Freeman and Doll, 1985), the molecular structure and motion are changed randomly to obtain the equilibrium state of the cluster. In order to obtain the equilibrium states of a large-size system of a large number of molecules, the Monte Carlo method is more efficient since the molecular dynamics method takes more computer time for all molecules to be relaxed to the equilibrium state, but the MC method gives no information about the dynamical process to the equilibrium state. In contrast, the molecular dynamics method gives the time evolution of the cluster system, which is essential in understanding the kinetic processes.

To calculate stable cluster configurations with the molecular dynamics method, the initial states of the cluster system are relaxed with a constant total energy. This results in different metastable cluster distributions with different kinetic and potential energies, that is, different temperatures. Since the equilibrium state corresponds to the lowest energy for an assigned temperature, the molecular dynamics calculation requires many trials and large calculation times. The Monte Carlo method has the advantage that the temperature remains fixed and the lowest energy state is obtained naturally, but has the disadvantage that time-dependent properties cannot be predicted. Thus both methods are used to complement each other.

5-3-1 Molecular Structure and Internal States

Knowledge of the changing molecular structures and internal states occurring in the transition from gas phase monomers to large clusters is essential for understanding clustering and cluster reaction phenomena. The molecular structures, which determine the internal rotational and vibrational states, are closely related to their electronic structures. There is very little information on electronic structures of large clusters, whereas a great deal of information is available on dimers and small clusters. These electronic structures are mainly studied using ionization features of clusters.

Electronic properties. Properties pertaining to the process of direct ionization, auto-ionization, and fragmentation have been widely studied by various techniques, including vacuum ultraviolet absorption spectroscopy, electron impact method, photo-ionization, photoelectron spectroscopy, photo-ionization resonance spectroscopy, threshold photo-electron spectroscopy, electronic emission spectroscopy, and laser-induced fluorescence. Photoelectron spectroscopic methods are generally superior to others for the investigation of the direct ionization process occurring at energies enough above the initial thresholds. Photo-ionization mass spectrometry is suitable for the measurement of transitions near the first ionization limit, of appearance potentials for dissociative processes, and of auto-ionization peak positions. As with photon spectroscopic techniques with lasers, various methods such as direct single photon ionization, resonant two-photon ionization, photo-dissociation, and radiative repopulation are employed.

Nearly all of the early work on heavier clusters has been matrix isolation studies of metal clusters. These matrix studies of electronic transitions, Raman and resonance Raman effects, and electronic spin resonance can provide knowledge of the electronic structure of clusters. However, these studies are complicated by guest–host and guest–guest interactions on the matrix, and by the difficulty in spectroscopic assignment. The latter arises in assigning spectroscopic features to a specific cluster in the distribution of clusters that are formed on the matrix. Questions always arise pertaining to the identity of the carrier of an observed spectroscopic transition, to the magnitude of the matrix–solute interactions, and to the observed details of rotational and vibrational structure in electronic transition.

For this reason, there has been an increasing amount of work on the electronic states of clusters in the gas phase of nozzle beam. Photo-ionization thresholds as a function of cluster size have been reported. In the studies of photo-ionization of clusters in the gas phase, two problems are to be considered: contamination by fragmentation and true adiabatic threshold values. The contribution of fragmentation to the individual photo-ionization efficiency can be significant and may even be the dominant mechanism for production of lighter clusters. Thus caution must be used in assigning features of a photo-ionization efficiency curve to auto-ionization structure of the neutral molecule of the same mass. For weakly bound clusters, the difference between the ground vibrational states of the neutral and the ions is very small for several tenths of an electron volt above the threshold. Thus, even though ions are formed near the adiabatic ionization threshold via auto-ionization of excited Rydberg states of the neutral, there is no guarantee that the true adiabatic threshold will be observed.

Ionization potentials. The ionization potential is one of the fundamental properties representing molecular electronic states of clusters. The general trend of ionization potential with cluster size is that the potential decreases as the cluster size increases from the atomic ionization potential to the bulk workfunction.

Despite this trend, some clusters—especially pure metal clusters such as in the group Ia (Li, Na, K) and Ib (Cu, Ag, Au)—exhibit odd/even alternation in ionization potential.

In pure metal cluster systems, the effect is currently understood to be a consequence of the odd/even alternation in the occupancy energy of the highest occupied molecular orbital (HOMO). The even-group Ia and Ib clusters are expected to have singlet ground states, whereas the odd clusters have doublet ground states with the unpaired electron in a fairly weakly bound HOMO. In the case of semiconductor clusters, this odd/even alternation in the electronic structure is not observed.

Rare gas clusters do not display a monotonic decrease of photo-ionization with cluster size due to the effects of fragmentation, which largely dominate the spectrum. For rare gas clusters, the effect of fragmentation is the most critical problem for measuring the ionization thresholds for each size cluster. Molecular clusters commonly have photo-ionization spectra with intense structures that rapidly decrease as the cluster size increases. The magnitude of the potential depends significantly on the kind of molecular species or the bond structure of the cluster.

Optical properties. Although molecules in a cluster are bound together by weak van der Waals attractive forces, intermolecular interactions may possibly modify their vibrational and rotational states, resulting in new intermolecular states of vibration and rotation. The former implies modification of infrared absorption, whereas the latter implies modification of far-infrared absorption. As long as the cluster keeps stable internal states, these optical features could be observed experimentally, but for clusters of unstable internal states, the interaction with the light's electromagnetic field brings changes in the cluster structure or dissociation of the cluster due to excitation of the internal state (Shibahara and Kotake, 1994).

Magnetic properties. Since clusters are composed of a small number atoms or molecules, the total spin of the cluster is determined by the individual spin of the constituent particles. If an atom, like aluminum, is an odd-electron system, odd-atom clusters will have half-integer total spin; even-atom clusters will have integer spin. Thus, the magnetic properties of odd-atom and even-atom clusters are very different. Since an observed general trend is that the magnetic moment per atom is reduced as the cluster size increases, these magnetic features could be more remarkable at smaller-size clusters.

5-3-2 Thermodynamic Properties

At very low temperatures, clusters may form crystallites with the molecular axes fixed in some orientation with respect to the crystallite body axes. These crystallites exhibit only vibrational motions without rotational and translational modes. With increasing temperatures, these vibrational molecular fluctuations about this equilibrium orientation increase until, at some temperature (T_o), an orientational order–disorder transition takes place into a hindered rotor, plastic crystallite phase, which is associated with rotational oscillations, adding to vibrational motions as shown in Fig. 5-11. This is evidenced by an abrupt change in the slope of the energy-temperature curve. Until the crystallite melts at some temperature T_m, it is in the plastic crystallite phase, translationally ordered and rotationally disordered, with the molecules becoming nearly free rotors at the melting temperature. The orientational transition is also associated with a structural transition

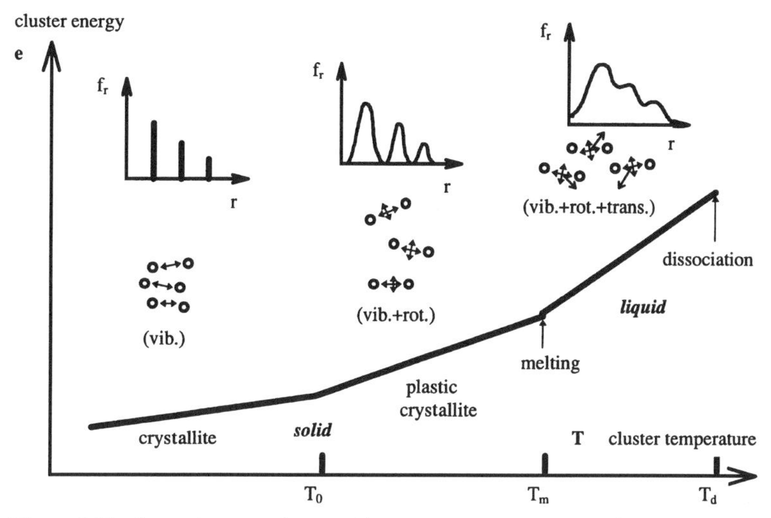

Figure 5-11 Internal energy change with respect to temperature: at low temperatures, only vibrational modes of molecular motion are activated; with temperature increase, rotational and then translational modes are relaxed.

into a configuration nearly identical to those found for clusters of spherical atoms. At the temperature T_m, melting is accompanied by an abrupt change in the radial distribution function for particles and in the slope of the temperature versus energy curve (Fig. 5-11). After melting, the clusters dissociate at the temperature T_d. Beyond this point, thermodynamic properties never stabilize into well-defined values, because particles evaporate from the cluster and the cluster energy decreases toward zero.

Thermodynamic properties such as the internal energy, entropy, and formation free energy can be calculated with the Monte Carlo method or the molecular dynamics method, although it is difficult to confirm such calculations experimentally. Using the Monte Carlo method in the configuration space of the atoms or molecules in the cluster, ensemble averages of thermodynamic variables can be calculated. Generally, the average energy per particle as a function of temperature increases smoothly until some characteristic temperature or cluster size, at which point the slope of the curve (the specific heat) changes abruptly. This change is apparent only when the number of particles in the cluster exceeds a certain value. The bond length distribution function (the probability of a particular bond length) or, alternatively, the radial distribution function (the probability of a particular interparticle distances, f_r) shows very well-resolved, narrow peaks at low temperatures (Fig. 5-11). When the temperature rises, the peaks begin to broaden, which means that a wider range of interparticle separations is being sampled. The maxima of the peaks also shift to larger values. While atoms in a solid usually move with small amplitudes about their equilibrium positions, liquid atoms have the freedom to move with much larger amplitudes within the body of the liquid. This greater freedom is reflected in the nonzero nature of the bond length distribution function between successive peaks. The mean-square fluctuation of the bond lengths increases almost linearly with temperature, until the melting temperature is reached and there is a sudden increase.

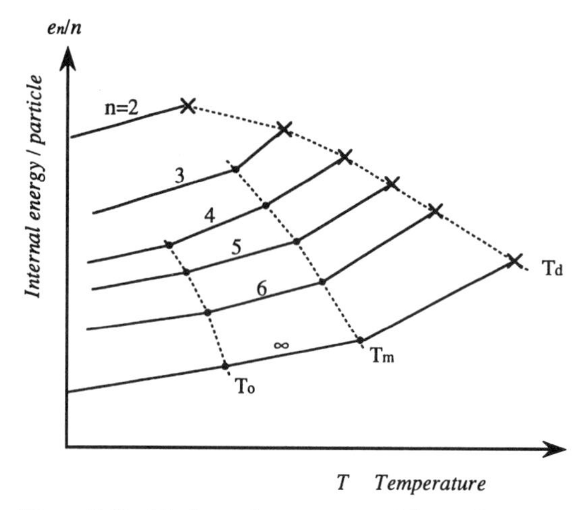

Figure 5-12 The internal energy per particle as a function of temperature. T_o is the temperature of orientational disorder; T_m, the temperature of rotational disorder (melting); T_d, the temperature of translational disorder (dissociation).

Phase-change temperature. Although these features of phase change of clusters depend greatly on the kind of molecular species, the general trends are shown in Fig. 5-12. The temperature of orientational disorder, T_o, the temperature of rotational disorder (melting), T_m, the temperature of translational disorder (dissociation), T_d, and the binding energy are increased with increasing cluster size.

Formation free energy. The free energy of cluster formation usually exhibits local maxima and minima with changing cluster size. These maxima-minima features of the free energy with cluster size are called *magic numbers*. These magic numbers, at which the free energy of cluster formation takes local minima, depend on the kind of molecular species and are observed experimentally to be the most stable clusters. The existence of magic numbers, however, is usually very sensitive to conditions of temperature and pressure, and under some conditions is observed only in the quantum calculations. Much care should be taken when interpreting the magic numbers observed experimentally with the mass spectrometry of ionization, because the ionization process is heavily influenced by the cluster configuration as well as the cluster size.

Specific heat. The specific heat (the derivative of the energy with respect to temperature) can be calculated if the energy is predicted as a function of the temperature. Generally, the specific heat increases rapidly and then gradually approaches the value of the bulk material as the cluster number increases. As the cluster passes through its melting temperature, it deviates considerably from solid-state behavior because its state is much closer to the liquid one. There is a sudden increase in specific heat as the cluster melts.

Reactivities. In neutral cluster reactions, it is difficult to unambiguously assign the reactants and products experimentally. The reactions of cluster ions reduce this ambiguity because it is possible to select a particular cluster size prior to reaction. Thus, almost all experiments on cluster reaction concern those of cluster ions produced in nozzle beams. The clusters are ionized by an electron beam or laser before or after being skimmed. A specific cluster size is mass-selected by a quadrupole or other mass spectrometer, and then passed through a gas cell where reactant gas is introduced. At the end of the gas cell the products and unreacted ions are focused into a second mass spectrometer and mass-analyzed.

Depending on cluster species, clusters show a strong size dependence in their reactivity, varying in their reaction rates with the cluster size n. The onset of bulk behavior of reactivity does not occur gradually but appears to begin abruptly at some cluster size, after showing maxima-minima behavior of reactivity with respect to the cluster size. Some systems of cluster reactants show dissociative or associative reactions, which also depend upon the internal states of reactants as well as the cluster size.

5-4 CONTROL OF CLUSTERING AND CONDENSATION

Nucleation and condensation are the fundamental processes for formation of condensed matter, and material processing, and energy and mass transfer. Control of these processes is the most critical concern in the innovation of new materials with advanced physical and chemical properties. Since the energy associated with these phenomena of nucleation and condensation is two or three orders less than that of chemical reactions (that is, within the range of disturbance fluctuations of the latter), it is not easy to control the processes with sufficient accuracy. Clusters show a variety of static and dynamic features depending on the size and structure with relevant energies distributed in a wide range of thermal levels. This gives great possibilities for the control of nucleation and condensation with these features of clusters and clustering.

In the vicinity of cooled walls or condensates, the existence of clusters depends on the condition of energy transfer between molecules of the surrounding gas as shown in Fig. 5-13 (Hashimoto and Kotake, 1995). Molecules approaching the cooled wall have higher temperatures (energies) than departing molecules that have transferred their energy to the wall. The energy exchange between these oncoming and departing molecules of different temperatures provides a sufficient condition for molecular clustering. The cluster size formed increases closer to the wall, and the thickness of cluster zone depends on the thermal condition of the molecular system and the processes of its energy transfer.

In order for an oncoming molecule to condense onto a cooled wall or onto existing condensate, its translational energy should vanish at the surface of the wall during collision with the surface molecules. In the case of monomers, this vanishing of the translational energy can only happen if the monomer transfers its energy to the surface molecules as shown in Fig. 5-14a ($q \neq 0$). On the contrary, clusters having many internal degrees of freedom such as rotational and vibrational motions can redistribute the internal energies so that the translational energy vanishes on the way to the wall. This redistribution of internal energies of clusters can take place more easily than that for molecules because of small difference of the energy levels. As shown in Fig. 5-14b, the translational energy is changed to the rotational or vibrational energy so that the approaching molecule can stick to the surface molecules. When these energies are redistributed sufficiently, the molecules can condense without any energy transfer to the wall ($q = 0$).

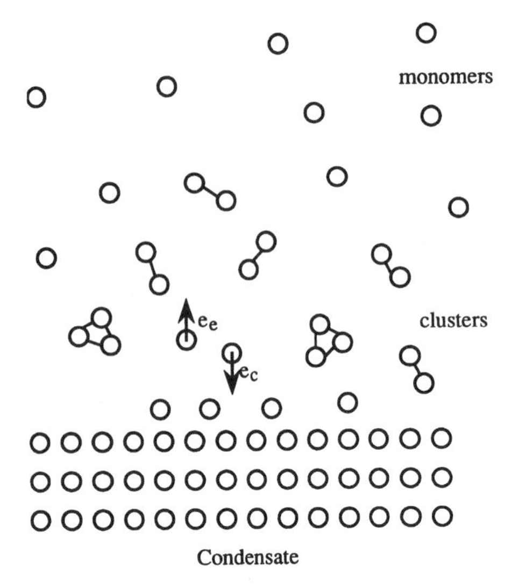

Figure 5-13 Cluster formation in the vicinity of condensate: evaporating molecules have lower energy (e_e) than do condensing molecules (e_c); $e_e < e_c$ makes favorable conditions for cluster formation.

The final molecular configuration of the condensate is determined with rearrangement of surface molecules through surface diffusion and energy transfer. These processes are also influenced by the energy states of sticking molecules, which can be related to the cluster features. These processes can be predicted to some extent with computational methods such as the molecular dynamics calculation with appropriate potentials. The problem is how to obtain the cluster features pertinent to the control of the condensation process. One possible way is the use of a laser technique (Fig. 5-15). As mentioned above, clusters have peculiar absorption spectra in the infrared and far-infrared range of light corresponding to their size and structure. Irradiation by tuned laser light results

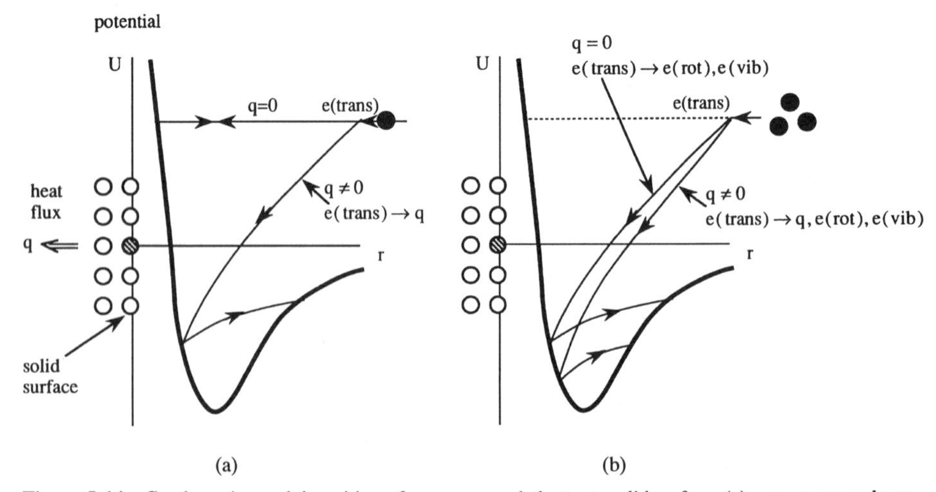

Figure 5-14 Condensation and deposition of monomer and cluster to solid surface; (a) monomer condensation, and (b) cluster condensation.

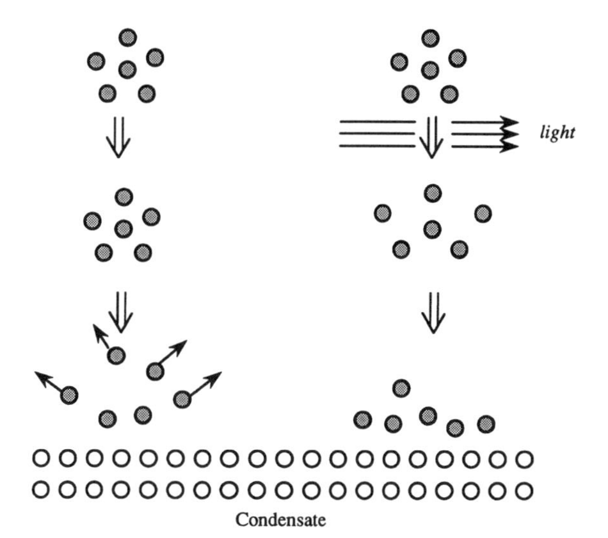

Figure 5-15 Condensation control with laser irradiation, which affects cluster features for condensation.

in formation and dissociation of assigned clusters and then changes in condensation process.

REFERENCES

Bauer, S. H., H.-S. Chiu, and C. F. Wilcox, Jr., "Kinetics of condensation in supersonic expansion (Ar)," *J. Chem. Phys.*, vol. 85, pp. 2029–2037, 1986.

Freeman, D. L., and J. D. Doll, "Quantum Monte Carlo study of the thermodynamic properties of argon clusters," *J. Chem. Phys.*, vol. 82, pp. 462–471, 1985.

Hashimoto, H., and S. Kotake, "In-situ measurement of clustering process near condensate," *Thermal Sci. Eng.*, vol. 3, pp. 37–43, 1995.

Inoue, T., and S. Kotake, "Formation of water clusters in a free molecular jet of binary mixtures," *J. Chem. Phys.*, vol. 91, pp. 162–169, 1989.

Kaelberer, J. B., and R. D. Etters, "Phase transition in small clusters of atoms," *J. Chem. Phys.*, vol. 66, pp. 3233–3239, 1989.

Kelton, K. F., A. L. Greer, and C. V. Thompson, "Transient nucleation in condensed systems," *J. Chem. Phys.*, vol. 79, pp. 6261–6276, 1983.

Kim, K. S., B. J. Mhin, U. S. Choi, and K. Lee, "Ab-initio studies of the water dimer using large basis set: The structure and thermodynamics energies," *J. Chem. Phys.*, vol. 97, pp. 6649–6662, 1992.

Kotake, S., and I. I. Glass, "Flows with nucleation and condensation," *Prog. Aerospace Sci.*, vol. 19, pp. 1–129, 1993.

Mcginty, D. J., "Molecular dynamics studies of the properties of small clusters of argon atoms," *J. Chem. Phys.*, vol. 58, pp. 4733–4742, 1973.

Ohara, T., and T. Aihara, "Molecular dynamics study on phase change and cluster formation," *Thermal Sci. Eng.*, vol. 3, pp. 7–13, 1995.

Polymeropoulos, E. E., and J. Brickmann, "Molecular dynamics study of the formation of argon clusters in the compressed gas," *Chem. Phys. Lett.*, vol. 92, pp. 59–63, 1982.

Rohlfing, E. A., D. M. Cox, and A. Kaldor, "Production and characterization of supersonic carbon cluster beam," *J. Chem. Phys.*, vol. 81, pp. 3322–3330, 1984.

Rovira, M. C. et al., "Ab-initio computation of the potential energy surfaces of water-hydrocarbon complexes:

Minimum energy structures, vibrational frequencies and hydrogen bond energies," *Chem. Phys.*, vol. 200, pp. 319–335, 1995.

Sano, T., and S. Kotake, "Molecular dynamics studies on condensation process of argon," *Prog. Astronaut. Aeronaut.*, vol. 117, pp. 439–446, 1989.

Shibahara, M., and S. Kotake, "Quantum molecular dynamics study of the light absorption of metal clusters," *Trans. JSME*, vol. 60, pp. 9170–177, 1994; *Proc. ASME-JSME Thermal Eng.*, pp. 461–466, 1995.

Zhukhovitskii, D. I., "Molecular dynamics study of cluster evolution in supersaturated vapor," *J. Chem. Phys.*, vol. 103, pp. 9401–9407, 1995.

Zolotoukhina, T. N., and S. Kotake, "Ab-initio study of intermolecular potential for dimers XO-He(X=N,C)," *J. Chem. Phys.*, vol. 99, pp. 2855–2864, 1993.

Zurek, W. H., and W. C. Schieve, "Molecular dynamics study of clustering," *J. Chem. Phys.*, vol. 68, pp. 840–846, 1978.

CHAPTER

SIX

INTERFACIAL FORCES AND PHASE CHANGE IN THIN LIQUID FILMS

Peter C. Wayner, Jr.

Isermann Department of Chemical Engineering, Rensselaer Polytechnic Institute, Troy, NY 12180

6-1 INTRODUCTION

Because of their useful and unique properties, very thin liquid films have innumerable applications in change-of-phase heat transfer. These characteristics are a strong function of the film's thickness, which can range in size from an adsorbed monolayer to relatively thick thin films. The overall resistance to evaporation from a thin film can be small in the thicker limit, where the resistance is primarily due to thermal conduction across the thin film. It can be infinite at the thinner limit, where the intermolecular forces of adsorption and/or capillarity reduce the vapor pressure of the superheated film to that of the surroundings, which is in equilibrium with an assumed bulk fluid sink at saturation. In the contact line region where the liquid, solid, and vapor phases meet, this variation in resistance to phase change can occur (depending on the conditions) over a distance of a few molecular diameters or over a much larger distance. For a conceptual example, we can cite the proposed view that a major portion of the heat flow rate in boiling occurs as a result of evaporation near the liquid–solid–vapor junction. How can such a small but very dynamic region be modeled? Fortunately, it appears that these complex processes can be described using a judicious combination of thermodynamics, interfacial phenomena, and transport phenomena.

Herein, we emphasize the modeling of liquid films in which the average intermolecular force field is affected by the liquid film thickness. Therefore, the thermodynamic internal pressure is a function of the thickness. This functionality occurs because of the varying molecular microstructure near the vapor–liquid and liquid–solid interfaces. For pressures significantly below the critical condition, this thickness range is approximately (a monolayer) $< \delta < 0.1\ \mu\text{m}$. (We note that this is thicker for β films of water.) Since the effects are inversely proportional to the thickness raised to a power (e.g., 3), the most dramatic effects occur at the monolayer scale. Although the region is small and one might also expect a correspondingly small literature, we find that the scientific literature on interfaces is well developed and growing at an explosive rate. This is because interfacial conditions are the important boundary conditions for the physical problem. Here, we apply these scientific insights while keeping in mind the unavoidable differences between the scientific and engineering environments. By necessity, the following material was selected from both new and classical interfacial concepts to fit the particular needs of modeling in change-of-phase heat transfer. The more powerful new techniques based on the Hamaker constant complement the successful classical concepts based on the surface tension. For an engineering perspective, a good example of the practical application of this material is given in Chapter 8 on micro heat pipes. For a much more basic perspective, the reader should keep in mind the fundamental connection between phase change and the electrodynamic nature of interatomic forces.

Since a liquid is deformable, the shape of a liquid film is a function of the three-dimensional intermolecular force field. Therefore, the transport processes in a thin film are a function of the liquid–solid system, temperature, concentration, and the shape, which is a measure of the varying internal pressure (intermolecular force) field. Due to the small scale, a macroscopic model of the transport processes sometimes requires an extension of classical continuum concepts into a thickness realm too thin to be apparently justified. However, due to the dynamic nature of the interface, the continuum model represents a useful average description over time and over a sufficiently large surface area of the thin film. Although roughness can have an additional effect in some cases, roughness greater than that at approximately the molecular level is not addressed herein. There have been experimental evaluations of the applicability of the Kelvin equation to highly curved interfaces. For example, Fisher and Israelachvili (1981) demonstrated the validity of the Kelvin equation as low as 4 nm. Use of the bulk viscosity in extremely small thicknesses is still open to question. For example, conflicting results have been presented by Klein and Kumacheva (1995) and Knudstrup et al. (1995).

As outlined in many recent texts, intermolecular and surfaces forces are the result of the electronic structure of atoms and molecules (e.g., Israelachvili, 1992). At equilibrium, these forces cause the adhesion of one substance to another, the cohesion in bulk liquids, the free energy associated with interfaces and the liquid–vapor phase distribution in a closed system. Since the equilibrium environment is easily disturbed, phase change heat transfer can easily occur. In the vapor–liquid–solid contact line region, all these phenomena are important. The many noncovalent intermolecular interactions can be broadly classified as follows: 1) electrodynamic Lifshitz–van der Waals (LW) forces (comprising the sum of the London dispersion interaction between two apolar molecules or atoms, the Debye induction interaction, and the Keesom orientation interaction); 2) polar forces (hydrogen bonding and Lewis acid-base interactions); and 3) purely electrostatic forces. The London dispersion forces, which are always present, control transport processes in very thin films with apolar systems. These dispersion forces are long-range forces, vary inversely with the film thickness raised to a power, and can be effective over large distances: from approximately 100 nm, at which an extremely small change from the bulk vapor pressure over a thin liquid film occurs, down to interatomic spacing (≈ 0.2 nm), where the vapor pressure of an adsorbed monolayer can be extremely small. As described by the adsorption isotherm, the thickness of an adsorbed film is a function of the surrounding vapor pressure and substrate temperature.

To demonstrate the use of interfacial phenomena in change-of-phase heat transfer in a limited space, the development herein emphasizes pure apolar fluids. In this case, it is possible to start with the frequency-dependent dielectric functions of the participating media and obtain an interfacial heat transfer coefficient using the disjoining pressure concept. This connection between basic interfacial forces and phase change heat transfer was first demonstrated by Derjaguin et al. (1965), who developed a theory of evaporation based on disjoining pressure isotherms. Potash and Wayner (1972) and Wayner et al. (1976) used these ideas to describe the process of evaporation from an extended meniscus. Although polar fluids and mixtures require the use of additional concepts, the heat transfer modeling outlined herein with apolar fluids is also attainable with polar fluids but is currently undeveloped. In order to put both types of fluids in perspective, some introductory material on polar systems is also included. We find that there are ample opportunities for further progress in heat transfer using these polar and apolar interfacial principles, which are available in the scientific literature. However, this progress will depend on the correct use of interfacial physics, which is very sensitive to the chemistry of the system and the local environment.

The material presented in this chapter covers only a very small portion of the field of phase change in thin liquid films. There are numerous topics like condensation, two-phase flow, and stability that are omitted. To obtain additional background information that is related to the current chapter, the following recent books and reviews are suggested: for a further description of intermolecular forces in thin liquid films, see Israelachvili (1992); for a description of aqueous thin films, van Oss (1994); for a general description of phase change phenomena, Carey (1992); and for a review of the dynamics and tendency toward rupture of thin evaporating liquid films, Bankoff (1990).

6-2 THERMODYNAMICS OF THIN FILMS: INTERFACIAL PROPERTIES

6-2-1 Interfacial Free Energy

A molecule in a homogeneous bulk fluid is uniformly surrounded by like molecules with the same average density, whereas a molecule at the liquid–vapor interface is not uniformly surrounded by molecules with the same density. This nonuniformity near the liquid–vapor interface leads to an additional energy: the interfacial free energy per unit area, σ_{lv}. When the liquid–vapor interface is spherical like a vapor bubble, the pressure on the concave vapor side is experimentally found to be greater than that on the liquid side by an amount ΔP, which is called the *capillary pressure*. For an expanding vapor bubble, the differential work causing the expansion of a vapor bubble with radius r is

$$dW = 4\pi r^2(\Delta P)dr. \tag{6-1}$$

This work is stored in the interface as an increase in the *interfacial free energy*:

$$dG = \sigma_{lv} dA. \tag{6-2}$$

Equating these two equations gives Eq. (6-3) below, which can be used to determine the value of σ_{lv} from the experimentally measured value of ΔP:

$$\Delta P = \frac{2\sigma_{lv}}{r}. \tag{6-3}$$

At equilibrium, a superheat ΔT is needed to keep the radius of a particular vapor bubble from changing. Otherwise, the total interfacial free energy $(\sigma_{lv} A)$ would either decrease with vapor condensation or increase with liquid evaporation. For small systems, this pressure and superheat can be large. An approximate equation for ΔT that is frequently used in analyzing boiling heat transfer is (e.g., Carey, 1992)

$$\Delta T = \frac{2\sigma_{lv} T_{sat}}{r\rho_v \Delta h}. \tag{6-4}$$

Example 6-1. To demonstrate the potentially large size of the pressure jump at the liquid–vapor interface for a vapor bubble (radius, $r = 1$ μm) in water ($\sigma_{lv} \approx 59$ mN/m at 373 K), Eq. (6-3) gives $\Delta P = 118$ kPa. To keep this vapor bubble from collapsing, a superheat of approximately $\Delta T = 33$ K is needed. Compared to some of the systems discussed below, this system is not particularly small.

With solid–liquid–vapor systems, additional interfacial effects due to the presence of the solid become important. Figure 6-1a schematically presents a macroscopic drawing of an equilibrium liquid drop with an apparent contact angle, θ, in which the divisions between the phases are *incorrectly* pictured as mathematically sharp. This relatively simple model of a mathematical surface has led to many fruitful concepts. In reality, there is a three-dimensional variation in the molecular structure and density. The real contact angle on the molecular scale, which is believed to have a value different from the conventionally measured macroscopic value, cannot be experimentally viewed because of its small size. Therefore, theoretical pictures have been developed to show the changes in density and contact angle. For example, computer-generated pictures of the contact line region based on molecular-dynamics simulations have been presented by Yang et al. (1992), and mathematically smooth thickness profiles based on one-dimensional continuum models have been developed by Broekhoff and deBoer (1968), Wayner (1982), Brochard-Wyart et al. (1991), and Sharma (1993). A schematic sketch of this view is given in Fig. 6-2. In the one-dimensional continuum model, the film thickness is a function of the effective pressure, $P(\delta)$, and starts with an initial slope of zero at δ_0. A useful initial slope is not apparent in the molecular dynamics model.

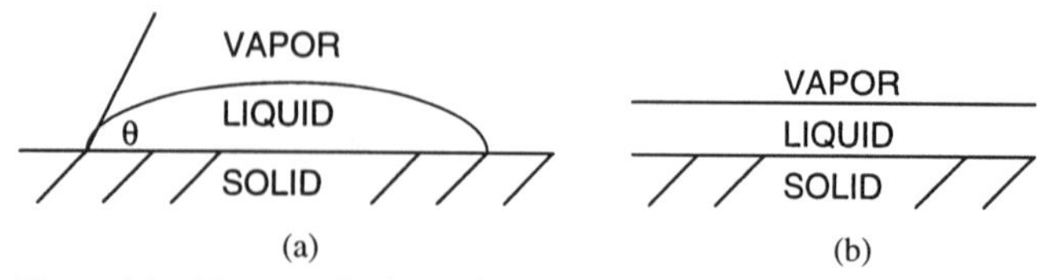

Figure 6-1 Macroscopic views of wetting on a solid substrate: a) partially wetting liquid drop; b) completely wetting liquid film.

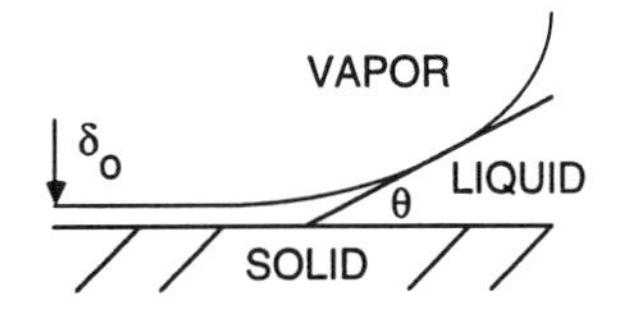

Figure 6-2 Schematic view of partial wetting in the contact line region with a mathematical surface representing the local variation of thickness.

At the very dynamic liquid–vapor interface of a *bulk* liquid, the density varies from the bulk vapor density to the bulk liquid density over a distance on the order of a few nanometers. The normal component of the pressures on both sides of a flat interface are usually taken to be equal, whereas the capillary pressure jump discussed above models a curved interface. However, there is also a cohesive pressure due to intermolecular forces which must be included in the models and which can change by a few thousand atmospheres over the thickness of the interface. Fortunately, this concept of a thick interface with a varying density is not new. For example, in 1893, van der Waals [see Rowlinson (1979)] analyzed the extremely large density gradient located at the liquid–vapor interface. Based on kinetic theory, the vanishing net interfacial heat flux at equilibrium is the difference between two equal, opposite, very large values of evaporation and condensation. There are also excess free energies at the liquid–vapor interface due to the density gradient and at the liquid–solid interface due to both density and composition gradients. All of these concepts are functions of the temperature, pressure, and composition.

To overcome the theoretical difficulties associated with this complex picture, various useful models have been developed. Gibbs (e.g., 1961) developed the very successful surface formalism of a two-dimensional dividing surface: the Gibbs convention. At times, we find this convention and the resulting interfacial free energy extremely useful. In many other situations, like transport processes in ultra-thin liquid films, we find it more useful to include the thickness of the thin film, which is not easily described within the Gibbs convention. To include thickness, a three-dimensional model is needed. This requires either the surface excess convention or the film excess convention. Both the Gibbs and the surface excess conventions are used herein. The surface excess convention allows the disjoining pressure concept to be introduced, which is an effective pressure change due to intermolecular force changes in the thin film. A good discussion and comparison of the details of these two conventions are given by deFeijter (1988).

When a liquid drop is placed on a substrate, the resulting system can be classified as either completely spreading, partially spreading, or nonspreading. For complete spreading, the final equilibrium shape is one of a thin uniform film covering the substrate (presuming that the vapor pressure in the surroundings is sufficient to suppress evaporation). Diagrams of the completely spreading and partially spreading cases in a gravitational field are presented in Fig. 6-1. The picture of the completely wetting case conceptually represents, for an extremely thin case, a monolayer or, at the other extreme, a thick film. For the partially wetting case, a deceptively simple quantitative measure of the complex intermolecular force field in the contact line region is the apparent contact angle, θ. The contact angle is defined as the angle between the tangents to the liquid–vapor and liquid–solid interfaces at a point where the film is sufficiently thick so that the transition regions in the interfaces do not overlap. For some cases, like a small drop on a substrate, this definition is still inexact since the observable tangent angle constantly changes. For the nonwetting case, $\theta > 90°$. At this point, it is becoming obvious that mathematical interfaces and apparent contact angles are relatively crude but (at times) useful descriptions of the complex dynamic molecular world at interfaces and contact lines.

Initially, these spreading observations were simply described using the following macroscopic two-dimensional concepts (e.g., see Israelachvili, 1992). The theoretical work of cohesion (or twice the free energy increase per unit interfacial area in a vacuum), $2\sigma_l$, which is the work required to split a column of liquid into two separate columns, W_c), is

$$W_c = 2\sigma_l \tag{6-5}$$

and the work of adhesion, which is the work required to separate a liquid from a solid, W_a, is

$$W_a = \sigma_l + \sigma_s - \sigma_{sl}. \tag{6-6}$$

A useful model for the liquid–solid interfacial free energy is

$$\sigma_{sl} = \sigma_s + \sigma_l - 2\Phi_{12}\sqrt{\sigma_s \sigma_l} \tag{6-7}$$

in which the parameter Φ_{12} is expected to have a value of unity for simple apolar fluids. Schematics of these processes are presented in Fig. 6-3a and 6-3b.

Complete spreading occurs when the spreading coefficient S, which is the difference between the work of adhesion and the work of cohesion, is positive.

$$S = \sigma_s - \sigma_{sl} - \sigma_l > 0. \tag{6-8}$$

The interfacial free energy can be composed of dispersion (London), d, polar (Keesom), p, induced dipole (Debye), i, electrostatic, e, and acid-base interactions, AB:

$$\sigma = \sigma^d + \sigma^p + \sigma^i + \sigma^e + \sigma^{AB} \tag{6-9}$$

The first three terms in this equation are the Lifshitz–van der Waals (LW) interactions. Simple fluids like the alkanes have only LW interactions. Some representative values

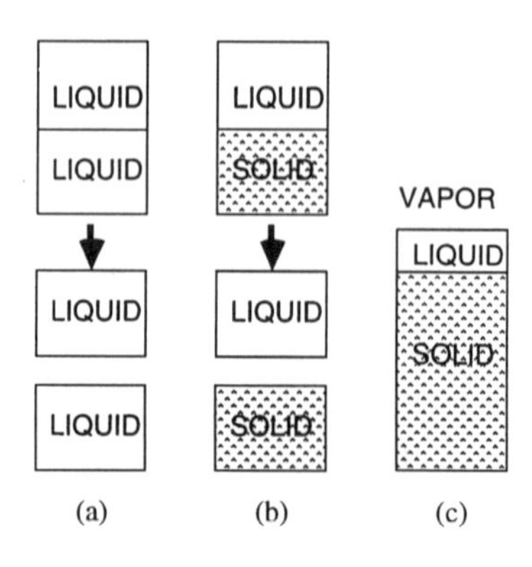

Figure 6-3 a) Illustration of work of cohesion. b) Illustration of work of adhesion. c) Illustration of thin liquid film on solid substrate.

Table 6-1 Surface tension components of various liquids, in mN/m (reprinted from van Oss, 1994, p. 175, by courtesy of Marcel Dekker Inc.)

Liquid	σ^{LW}	σ^{AB}	σ
H_2O	21.8	51	72.8
Glycerol	34	30	64
Ethylene glycol	29	19	48
Formamide	39	19	58
Dimethylsulfoxide	36	8	44

for the terms in Eq. (6-9) for polar fluids are given in Table 6-1. For apolar liquids, $\sigma = \sigma^{LW}$.

To connect these concepts and the Hamaker constant discussed below with experimental observations, we assume (at times) that there is no practical difference between these processes of interfacial formation occurring in a vacuum and an environment saturated with vapor or gas ($\sigma_l \cong \sigma_{lv}, \sigma_s \cong \sigma_{sg}$). However, it is also important to realize that the interfacial free energy values can, in some cases, be substantially different in laboratory air because of the adsorption of foreign vapor molecules like water and hydrocarbons. Experiments on the effect of the environment on the surface tension of stainless steel are discussed below. At liquid–vapor interfaces, impurities may or may not concentrate at the surface and thereby affect the value of the interfacial free energy. A further complication can arise if the environment has a foreign gas that can adsorb on the liquid substrate and change σ_l to σ_{lg}. Therefore, since the experimental determination of the contact angle and surface free energies is not made in a vacuum, we will find that there are many practical complications when we apply these concepts in modeling transport processes. On the other hand, all these additional effects can be experimentally measured and theoretically accounted for. The reader will find it necessary to use, as the situation dictates, both practical and ideal concepts to understand and apply the state-of-the-art in surface transport processes.

The Young-Dupré equation for the contact angle of a liquid on a substrate in its equilibrium vapor environment is (e.g., Wayner, 1982)

$$\sigma_{lv} \cos\theta = \sigma_{sv} - \sigma_{sl}. \tag{6-10}$$

This can also be written as

$$\sigma_{lv} \cos\theta = \sigma_s - \sigma_{sl} - \pi_e \tag{6-11}$$

where $\pi_e = \sigma_s - \sigma_{sv}$ is the equilibrium film pressure which accounts for vapor adsorption on the solid substrate. Bangham (1937) demonstrated that the equilibrium film pressure can be obtained from the adsorption isotherm of the vapor on the substrate, $\Gamma(P_v)$:

$$\pi_e = RT \int_0^{P_s} \Gamma \mathrm{d}\ln P_v \tag{6-12}$$

in which Γ is the number of moles of the adsorbate per unit surface area and P_v is the surrounding vapor pressure. For partially wetting systems with low-surface-energy solid substrates, it has been argued that π_e should be relatively small. On the other hand, high-energy surfaces like metals and oxides have a strong affinity for water and organic vapors. Two additional concepts are the work of wetting, W_w, and spreading, W_s:

$$W_w = \sigma_s - \sigma_{sl}; \qquad W_s = \sigma_s - \sigma_l - \sigma_{sl}. \tag{6-13}$$

6-2-2 Hamaker Constant (Apolar Systems)

To relate the interfacial free energies to the dielectric properties of the materials and to use the surface excess convention for thin films, it is necessary to introduce the Hamaker constant. The Hamaker constant, A_{ii}, which describes the LW interaction between two bulk flat surfaces in a vacuum separated by the van der Waals contact distance, D_s, is, for the liquid columns presented in Fig. 6-3a (Israelachvili, 1992),

$$A_{ll} = 24\pi D_s^2 \sigma_l. \tag{6-14}$$

The Hamaker constant is defined as

$$A = \pi^2 C \rho_1 \rho_2 \tag{6-15}$$

where ρ_1 and ρ_2 are the number of atoms per unit volume in the two volumes and C is a coefficient describing the interaction between atoms. The theoretical value of the separation distance between two semi-infinite slabs when they are in van der Waals contact is approximately 1.65×10^{-10} m. For two bulk solids and a solid–liquid interface, the comparable equations are

$$A_{ss} = 24\pi D_s^2 \sigma_s, \qquad A_{sl} \approx 24\pi D_s^2 \sqrt{\sigma_s \sigma_l}. \tag{6-16}$$

The Hamaker constant between two like bodies is attractive, $A_{ii} > 0$. The procedures by which the Hamaker constant can be obtained from the frequency-dependent dielectric functions of the interacting materials using the DLP theory are addressed below. For now, the following *approximate* equation demonstrates the connection between the Hamaker constant and the refractive index in the visible range, n_o, and the characteristic ultraviolet absorption frequency, ω_{uv} (Israelachvili, 1992):

$$A_{ll} \approx \frac{3\left(n_o^2 - 1\right)^2 \hbar \omega_{uv}}{16\sqrt{2}\left(n_o^2 + 1\right)^{1.5}} \tag{6-17}$$

In order to discuss surface and adhesion energies, Israelachvili (1992) showed that, apart from the bulk cohesive energy, the additional energy of two planar surfaces separated by the distance δ is

$$E = \frac{A}{12\pi}\left(\frac{1}{D_s^2} - \frac{1}{\delta^2}\right). \tag{6-18}$$

The interaction energy is

$$E_i = \frac{-A}{12\pi \delta^2}. \tag{6-19}$$

Equation (6-14) is obtained by taking $E = 2\sigma_l$ at $\delta \to \infty$. The additional potential energy is removed when the two interfaces are brought into contact. The adhesive pressure in contact, F at δ_s, is

$$F = \left(\frac{\partial E}{\partial \delta}\right)_{\delta = D_s} = \frac{A}{6\pi D_s^3} = \frac{4\sigma_l}{D_s} \tag{6-20}$$

Example 6-2. For two planar octane surfaces in van der Waals contact ($D_s = 0.165$ nm, $\sigma_l = 21.6$ mN/m), F is approximately 5,200 atmospheres. Obviously, these intermolecular forces are very large, and large energy gradients can occur over very short separations.

For a thin film on a solid substrate, which is schematically presented in Figure 6-3c, the repulsive force per unit area between the bulk vapor and the bulk solid separated by a liquid film is (e.g., see Truong and Wayner, 1987), for the nonretarded regime,

$$F(\delta) = \frac{A_{slv}}{6\pi\delta^3} = -\Pi, \qquad \delta \lesssim 10 \text{ nm} \tag{6-21}$$

while for a thicker film (retarded regime using a dispersion constant B),

$$F(\delta) = \frac{B_{slv}}{\delta^4} = -\Pi, \qquad \delta \gtrsim 30 \text{ nm.} \tag{6-22}$$

The above equations also give the equivalence between the disjoining pressure concept, Π, and the dispersion force. For completely wetting systems, the vapor is disjoined from the solid by the liquid with a force/area, Π. Calculations of the van der Waals interactions between different materials using the DLP theory showed that A_{slv} and B_{slv} are not constant but, depending on conditions, a weak function of film thickness in their respective regimes. An example of this variation is given in Fig. 6-4, from Truong and Wayner (1987). Keeping this in mind, we can still obtain approximate equations for effective modeling. Caution is advised because two sign conventions are used in the literature. The sign convention used herein makes $A_{slv} < 0$ for completely wetting systems and $A_{slv} > 0$ for partially wetting systems.

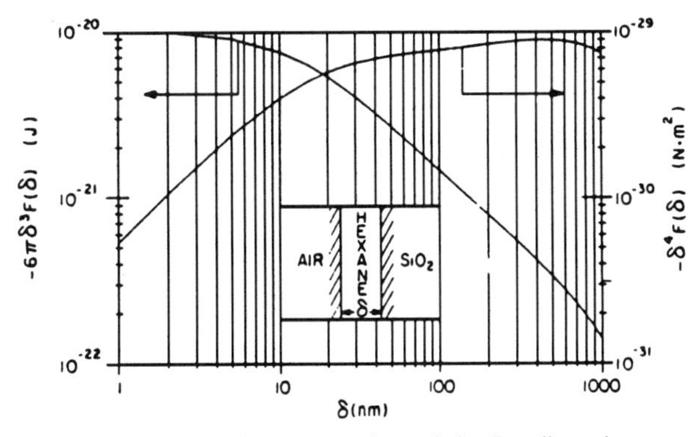

Figure 6-4 Nonretarded $-A_{slv}$ and retarded $-B_{slv}$ dispersion constants as a function of film thickness for hexane wetting silicon dioxide. (From Truong and Wayner, 1987.)

Some of the approximate combining rules for the Hamaker constants are (e.g., Israelachvili, 1992)

$$A_{slv} = A_{sv} + A_{ll} - A_{sl} - A_{lv} \tag{6-23}$$

$$A_{slv} = \left(\sqrt{A_{ss}} - \sqrt{A_{ll}}\right)\left(\sqrt{A_{vv}} - \sqrt{A_{ll}}\right). \tag{6-24}$$

Neglecting cohesion in the vapor gives

$$A_{slv} = A_{ll} - \sqrt{A_{ll}A_{ss}}. \tag{6-25}$$

Using Eqs. (6-14), (6-16), and (6-25) give

$$A_{slv} = 24\pi D_s^2(\sigma_l - \sqrt{\sigma_l \sigma_s}). \tag{6-26}$$

A completely spreading system is one for which $\sigma_s > \sigma_l$ and $A_{slv} < 0$. This means that, for spreading with a completely wetting system, the liquid–solid intermolecular force field causes the liquid to flow along the surface and thereby separate the solid from the bulk gas–vapor mixture. We note that transport can, of course, occur through both the vapor–gas phase and along the surface. It is very difficult to determine which one predominates at the contact line during spreading.

A comparison of experimental surface energies with those calculated on the basis of the Lifshitz theory is presented in Table 6-2. This successful comparison (except for the hydrogen bonding fluids) was obtained using the assumption $\sigma_l \cong \sigma_{lv}$. For comparison

Table 6-2 Comparison of experimental surface energies with those calculated on the basis of the Lifshitz theory (from Israelachvili, 1992)

		Surface energy, σ_{lv} (mJ m^{-2})	
Materials (ε) in order of increasing ε	Theoretical A_{ll} (10^{-20} J)	$A_{ll}/24\pi D_s^2$ ($D_s = 0.165$ nm)	Experimental (20°C)
Liquid helium (1.057)	0.057	0.28	0.12–0.35 (4–1.6 K)
n-Pentane (1.8)	3.75	18.3	16.1
n-Octane (1.9)	4.5	21.9	21.6
Cyclohexane (2.0)	5.2	25.3	25.5
n-Dodecane (2.0)	5.0	24.4	25.4
n-Hexadecane (2.1)	5.2	25.3	27.5
PTFE (2.1)	3.8	18.5	18.3
CCl_4 (2.2)	5.5	26.8	29.7
Benzene (2.3)	5.0	24.4	28.8
Polystyrene (2.6)	6.6	32.1	33
Polyvinyl chloride (3.2)	7.8	38.0	39
Acetone (21)	4.1	20.0	23.7
Ethanol (26)	4.2	20.5	22.8
Methanol (33)	3.6	18	23
Glycol (37)	5.6	28	48
Glycerol (43)	6.7	33	63
Water (80)	3.7	18	73
H_2O_2 (84)	5.4	26	76
Formamide (109)	6.1	30	58

Table 6-3 Nonretarded Hamaker constants A (10^{-20} J) for two identical media interacting across vacuum (air) (from Israelachvili, 1992)

Medium	A (10^{-20} J)
Polystyrene	6.5
Polyvinyl chloride	7.5
PTFE	3.8
Fused quartz	6.3
Mica	10
CaF_2	7
Al_2O_3	14
Fe_3O_4	21
Silicon Carbide	44
Metals (Au, Ag, Cu)	25–40
Liquid He	0.057

with the previous material, additional Hamaker constants for solid materials are given in Table 6-3.

Based on the success demonstrated in Table 6-2, the following equation has also been developed for apolar systems:

$$\sigma_s^d = \frac{1}{4}\sigma_{lv}(1+\cos\theta)^2, \quad \theta > 0 \text{ using dispersion force liquids only and } \pi_e = 0. \tag{6-27}$$

Therefore, the dispersion component of the surface tension of a solid can be determined by measuring the contact angle of an apolar liquid on the solid.

Example 6-3. For a Teflon surface, π_e can be neglected. If the measured contact angle for dodecane ($\sigma_{lv} = 19.8 \times 10^{-3}$ N/m) on Teflon is $\theta = 22°$, Eq. (6-27) gives $\sigma_s^d \cong 18.4 \times 10^{-3}$ N/m.

For completely wetting fluids, the following equation for the interaction between a solid and vapor separated by a thin liquid film can be obtained:

$$\frac{A_{slv}}{12\pi D_s^2} = \sigma_l + \sigma_{sl} - \sigma_s. \tag{6-28}$$

With $\sigma_l \cong \sigma_{lv}$,

$$\sigma_s^d = \frac{1}{\sigma_{lv}}\left[\sigma_{lv} - \frac{A_{slv}}{24\pi D_s^2}\right]^2, \quad \theta = 0 \text{ using dispersion force only liquids.} \tag{6-29}$$

Therefore, the dispersion portion of the interfacial free energy of solids for completely apolar wetting systems can be obtained using the measured value of the Hamaker constant. Since the above equations apply to both wetting and partially wetting systems,

Eqs. (6-27) and (6-29) can be equated to obtain the following equation for the contact angle in partially wetting systems, $A_{slv} > 0$:

$$\cos\theta = 1 - \frac{A_{slv}}{12\pi\sigma_l D_s^2}, \quad \pi_e = 0, \quad \theta > 0. \tag{6-30}$$

There are many additional complications associated with acid–base systems, like water (e.g., see van Oss, 1994).

For example, for some polar systems, we find that

$$\sigma_{sl} = \sigma_s + \sigma_l - 2\sqrt{\sigma_s^d \sigma_l^d} - 2\sqrt{\sigma_s^p \sigma_l^p}. \tag{6-31}$$

Example 6-4. Using Eq. (6-31) for the octane–water system, with $\sigma_l^d = \sigma_l = 21.8$ mN/m for apolar octane and $\sigma_l = 72.75$ mN/m and $\sigma_l^d = 20$ mN/m for polar water, Israelachvili (1992) calculated $\sigma_{ll} = 52.8$ mN/m, which is close to the measured value of $\sigma_{ll} = 50.8$ mN/m. Although good agreement is obtained for many simple hydrocarbon–water systems, the agreement for unsaturated hydrocarbon systems is less.

The reader is reminded that the previously referenced textbooks are a rich source of additional material on surface properties. In the next section, we find it desirable to use the disjoining pressure concept to discuss water.

6-2-3 Polar System (Water/Quartz)

After reviewing the state-of-the-art of contact angles and surface forces, Churaev (1994) concluded that the theoretical prediction of contact angles is still an unsolved problem. This is particularly true for water as demonstrated by the following discussion of the water/quartz system given by Churaev (1994). Note that we are using a sign convention opposite from that used by Churaev. Thin liquid films are asymmetric systems with liquid–solid and liquid–vapor interfaces, and an additional electrostatic disjoining pressure due to the electrical double layer can be present. The three-dimensional disjoining pressure model for a thin liquid film is

$$\Pi(\delta) = \Pi_m(\delta) + \Pi_e(\delta) + \Pi_s(\delta) \tag{6-32}$$

in which the subscripts m, e, and s refer to the molecular (dispersion), electrostatic, and structural components of the disjoining pressure. The structural forces causing the structural component are short range forces at the liquid–solid interface.

Frumkin and Deryagin used the theory of long-range forces to obtain the following relationship between the contact angle and the disjoining pressure:

$$\cos\theta = 1 + \frac{1}{\sigma_{lv}} \int_{\delta_\alpha}^{\infty} \Pi(\delta) d\delta = 1 + \frac{G(\delta_\alpha)}{\sigma_{lv}} \tag{6-33}$$

for $\theta > 0$, $G < 0$ is an excess surface free energy. The characteristic thickness of the adsorbed thin film at $\Pi = 0$ is δ_α. The prediction of the contact angle in polar systems is particularly complicated because each of the components of the disjoining pressure may be either positive or negative, depending on the system.

As an example of a particular application, Churaev (1994) presented the following equation for estimating the contact angle on *hydrophobic* surfaces when all three components of the disjoining pressure are important:

$$\cos\theta = 1 + \frac{1}{\sigma_{lv}}\left\{\frac{-A_{slv}}{12\pi\delta_\alpha^2} - \frac{\varepsilon_0(\Psi_1 - \Psi_2)^2}{8\pi\delta_\alpha} + K\lambda\exp\left(-\frac{\delta_\alpha}{\lambda}\right)\right\} \tag{6-34}$$

where K and λ are parameters, Ψ_1 and Ψ_2 are the surface potentials, δ_α is a film thickness, and ε_0 the permittivity of the fluid. Obviously, the equation for the contact angle and its subsequent use in heat transfer becomes very complicated for polar systems.

Example 6-5. Churaev (1994) used the following numbers with Eq. (6-34) to calculate the contact angle formed by water on a quartz surface hydrophobicized by methylation. Since the dispersion force of the substrate was not significantly affected by the organic monolayer coating formed during methylation, $A_{slv} = -7.2\times10^{-21}$ J for the dispersion term. For the electrostatic component, the potential Ψ_1 of the methylated quartz surface equals −70 mV, whereas Ψ_2 at the liquid–vapor interface equals −40 mV. Using $K = -3\times10^7$ Pa and $\lambda = 2$ nm for the structural component gives $\theta_0 \simeq 75°$. For large contact angle systems with $\delta_\alpha = 0.2$ nm, the hydrophobic effect due to the structural component of the disjoining pressure predominates.

It is interesting to look at the effect of the degree of hydrophilicity on the polymolecular adsorption of water vapor on a quartz surface. In this case, it appears that the short-range forces at the liquid–solid interface can make the film more hydrophilic. However, as discussed by Vigil et al. (1994), even the description of this classical interface is still unresolved. Two experimental isotherms from Gee et al. (1990) are presented in Fig. 6-5. Isotherm 1 was obtained with a completely hydroxylated surface (4.6 OH groups per 100 A^2), whereas Isotherm 2 was obtained using a quartz surface covered with 10% hydroxyl groups. In the former case, the system is completely wetting, whereas, in the latter case, the contact angle was found to be $\theta = 43°$. Although not shown in this figure, bulk film thicknesses are obtained as $P_v \rightarrow P_{sat}$ with the completely wetting system. Hydroxylated surfaces were obtained by using a mild etching solution of 1.5% w/v NH_4HF_2 for 2 hours at room temperature. Gee then removed the hydroxyl

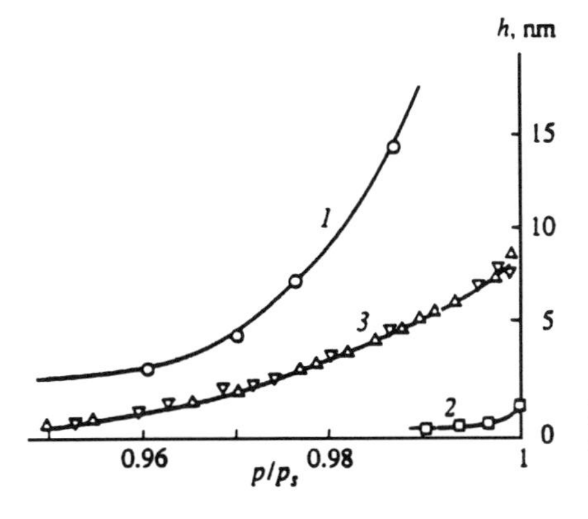

Figure 6-5 Adsorption isotherms of water: reduced vapor pressure of adsorbed film versus film thickness. (From Churaev, 1994.)

groups by heating the surface to 1,323 K (which is higher than necessary) to obtain a dehydroxylated surface. Therefore, the contact angle can be varied by varying the number of hydroxyl groups per unit area. Although this might be difficult to achieve, we wonder if surfaces of varying hydrophilicity could be sufficiently controlled to produce heat transfer systems of engineering importance.

6-2-4 Experimental Determination of the Hamaker Constant

For completely wetting apolar systems, the following equation for the equality of disjoining pressure and the adsorption isotherm can be used to obtain the Hamaker and/or retarded dispersion constants:

$$\Pi = -\frac{RT}{V}\ln\frac{P_v}{P_{sat}}. \tag{6-35}$$

However, experimental control of the system as $P_v \rightarrow P_{sat}$ is problematic since the system is very sensitive to temperature changes of the order of $\Delta T = 10^{-5}$ K (depending on P_v/P_{sat}). For partially wetting apolar systems, the experimentally measured apparent contact angle and Eq. (6-30) can be used to obtain the Hamaker constant. As demonstrated above in Example 6-5 for polar–nonpolar combinations with large contact angles, structural and electrostatic components of the disjoining pressure make the experimental determination of the Hamaker constant more complicated for these systems. In either case, theoretical values can be used to understand the phenomena.

Due to the sensitivity of the system to contamination, DasGupta et al. (1994, 1995) found it advantageous to measure the Hamaker or dispersion constant *in situ* at the start of a change-of-phase heat transfer experiment. For example, the Hamaker or dispersion constant can be experimentally obtained by measuring the equilibrium film thickness profile in the closed square channel presented in Fig. 6-6. This system represents the cross section (at equilibrium) of the constrained vapor bubble thermosyphon (CVBT) concept presented in Fig. 6-7 for a microgravity environment (DasGupta et al., 1995). The CVBT is a larger version of a micro heat pipe. The augmented Young-Laplace equation for a curved thin film in a gravitational field, which is derived in a later section, is

$$\Pi + \sigma_{lv}K = (\rho_l - \rho_v)gh. \tag{6-36}$$

Equilibrium requires that the sum of the dispersion force and the capillary force (chemical potentials per unit volume) equals the hydrostatic force. When the results are for subsequent use in transport processes, this method has advantages relative to classical adsorption studies based on Eq. (6-35). Thicker films can be conveniently studied, and

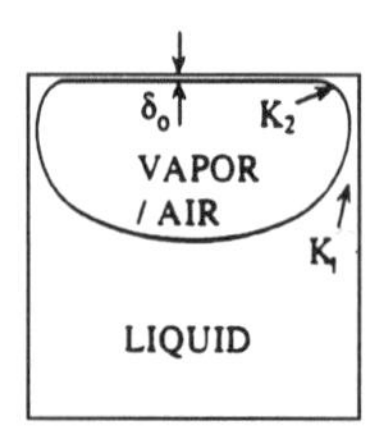

Figure 6-6 Cross section of an extended meniscus in a square channel under the following two gravitational conditions: micrgravitational and earth's gravitational fields. (From DasGupta et al., 1995. Reproduced with permission of the American Institute of Chemical Engineers. Copyright © AIChE 1995. All rights reserved.)

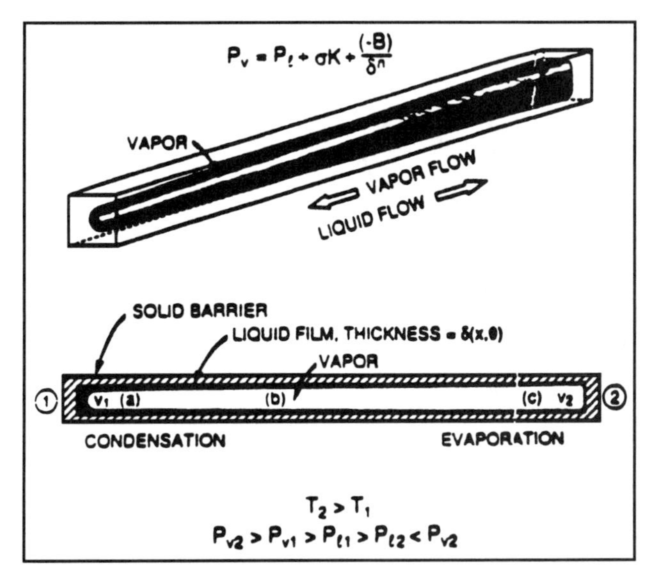

Figure 6-7 Constrained vapor bubble thermosyphon with heat input at End (2) and removal at End (1). (From DasGupta et al., 1995. Reproduced with permission of the American Institute of Chemical Engineers. Copyright © AIChE 1995. All rights reserved.)

a better feel for the connection between the Hamaker or dispersion constant and the pressure in the liquid can be obtained.

For the isothermal nonevaporating case, the liquid pressure will remain the same at a constant gravitational level. Since the variation in hydrostatic head is negligible for a small change in the gravitational level at a particular cross section, the following augmented Young-Laplace equation can be written near the corner both for a point in the thicker portion of the meniscus, K_∞, where the disjoining pressure effects are negligible, and for another point where both effects are present:

$$\sigma K - \frac{B}{\delta^4} = \sigma K_\infty, \quad Q = 0, \quad \delta > 30 \text{ nm}. \tag{6-37}$$

The curvature for a cylindrical interface, K, is (for horizontal x-coordinate)

$$K = \frac{\frac{d^2\delta}{dx^2}}{\left[1 + \left(\frac{d\delta}{dx}\right)^2\right]^{3/2}}. \tag{6-38}$$

Therefore, the Hamaker constant can be obtained by measuring the meniscus profile.

A schematic diagram of the CVBT cell and the experimental setup used by DasGupta et al. (1995) to obtain the dispersion constant is shown in Fig. 6-8. This study was conducted as a precursor to nonisothermal CVBT experiments to determine the dispersion constant, *in situ*, at the start of the experiment. The isothermal results characterize the interfacial force field, and the information is critical to the operation and the heat sink capability of the CVBT. The fused quartz cell was a small cuvette [3 mm × 3 mm in cross section and about 40 mm long], selected to facilitate optical observation and

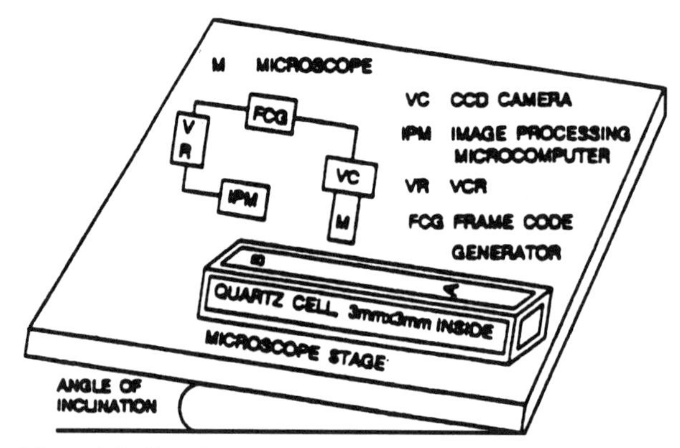

Figure 6-8 Experimental setup from DasGupta et al. (1995). (Reproduced with permission of the American Institute of Chemical Engineers. Copyright © AIChE 1995. All rights reserved.)

measurement of the liquid meniscus. The cell was cleaned after manufacturing and delivered in a closed container ready for spectroscopy. The cell was opened only inside a class-100 clean hood, rinsed repeatedly with pure (99+%, as supplied by Aldrich Chemicals) and filtered test liquid before partially filling it. At the end of each experiment, the cell was heated inside the clean hood at about 300°C for an hour and then rinsed with the new test liquid.

The film thickness profile near the corners of the experimental cell was measured using IAI (image-analyzing interferometry). A detailed description of the IAI techniques and hardware for film thickness profile measurement were presented in Sujanani and Wayner (1991) for a reflecting substrate, and in DasGupta et al. (1995) for a transparent substrate. Earlier, Blake (1975) used a transparent substrate in a related study. The experimental results are presented in Fig. 6-9, from which the Hamaker constant can be determined. Extensions of these procedures can be developed for polar systems.

Example 6-6. Using Eq. (6-35) with $(P_v/P_{sat}) = 0.81$ and $RT/V = 2.1 \times 10^7$ N/m^2 for pentane at room temperature, the value of the disjoining pressure is $4.43 \times$

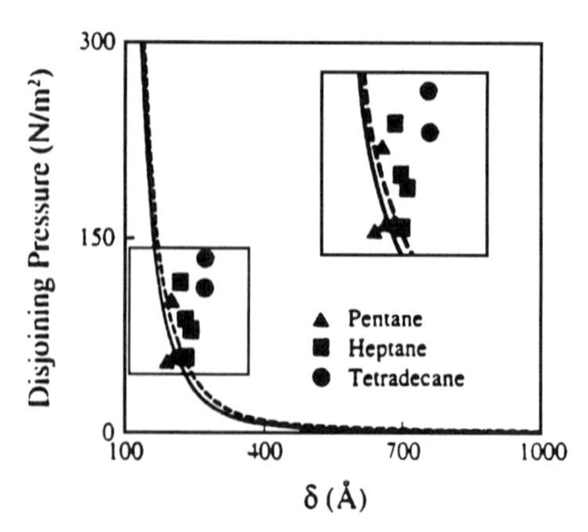

Figure 6-9 DLP theory for an ideal surface versus experiments for the adsorption of pentane and heptane on quartz (solid line) and tetradecane on SFL6 based on UV data only (dashed line). (From DasGupta et al., 1995. Reproduced with permission of the American Institute of Chemical Engineers. Copyright © AIChE 1995. All rights reserved).

10^6 N/m^2. Using $A = -1.28 \times 10^{-20}$ J and Eq. (6-21), the adsorbed film thickness is found to be $\delta = 1.4$ nm.

6-2-5 Engineering Surfaces in Real Environments

The title of this section emphasizes the fact that there is obviously less control of the environment in heat transfer technology or engineering type experiments than there is in thermodynamic theory or scientific type experiments. However, we quickly note that this separation is narrowing in "high technology and/or controlled microscale systems." Increasingly, engineering is using the characteristics of ideal systems to obtain improved devices. Therefore, it is instructive to have a fundamental understanding of the environment, or "contamination." In general, low-surface-energy plastics usually do not have a contamination problem, whereas high-surface-energy substrates, like metals, do. Unless amply demonstrated otherwise, we can presume that most metals are used and/or studied with both a surface layer of oxide and an adsorbed layer of contamination (organic or water). Oxides are contaminated if only by the pervasive presence of water vapor.

In 1922, Harkins and Feldman concluded that practically all liquids should spread on clean metals. Subsequently, Zisman (1964) demonstrated that autophobic liquids self-contaminate the substrate by adsorption, leading to an effective solid surface free energy below that of the autophobic liquid. Autophobic liquids give a partially spreading system where a completely spreading system would be anticipated. Studies and controversies on the wettability of gold are legion. To explain the frequently observed non-wetting of high-energy surfaces by lower-surface-energy fluids, Parsegian et al. (1977) theoretically demonstrated that even a monolayer of hydrocarbon contamination can prevent the spreading of water on gold. As demonstrated below, it takes a large superheat to remove a physically adsorbed organic monolayer.

Metal surfaces with a few monolayers of oxide slightly contaminated with bonded water in a low humidity environment are, of course, much more complicated. A reading of the referenced literature and the discussion of the influence of surface chemistry on the wettability of stainless steel in a recent publication by Mantel and Wightman (1994) gives considerable insight concerning the effects of environment and surface preparation. Using a polar form of Eq. (6-30), Mantel and Wightman (1994) studied the effects that degreasing with solvents, chemical etching, polishing, oxygen and argon plasma treatment, and heating and drying had on the surface level of contamination and "bonded water." The surface energy of the metal was found to decay inversely with the contamination thickness. A good fit between σ_s (the surface energy in mJ/m^2) and δ (the organic thickness layer in nm) was found to be

$$\sigma_s = 11 + 80/(\delta + 0.2). \tag{6-39}$$

The intermolecular distances and type of contamination fixes the various types of interactions. In their experiments, the distance between the oxide/hydroxide film and the test drop of water varied from 0.6 to 3 nm. Using Eq. (6-39), an organic contamination thickness of $\delta = 1$ nm gives $\sigma_s = 77.7$ mJ/m^2. Designing a cleaning procedure to remove all organic containing species from the surface seems unrealistic in some engineering processes. Even with a clean metal surface, the dependence of the dispersive and polar forces on the oxide/hydroxide layer is pronounced. In Mantel and Wightman studies (1994), the as-received oxide thickness on the stainless steel was found to be 3.5–4 nm. A simple cleaning with tetrahydrofuran reduced this value to 3 nm, whereas heating to 200°C leads to partial desorption of water and the elimination of hydroxyl groups and to the formation of a thick ($\delta > 7.5$ nm) layer of oxide. Cleaning the surface with 4M HNO_3 solution increases the thickness of the passive layer by oxidation of chromium. In all cases, the surface can be characterized using contact angle measurements and/or thickness profiles based on the augmented Young-Laplace equation. Since this is not the usual practice in heat transfer studies, these systems are not well characterized, and inconsistent data are found in the literature.

6-2-6 Theoretical Value of the Hamaker Constant From the DLP Theory

There are various inaccuracies associated with the classical Hamaker constant approach (Hamaker, 1937), such as the assumption of pairwise additivity. The problems are avoided in the Dzyaloskinskii Lifshitz-Pitaevskii (DLP) theory, in which the forces are derived in terms of the dielectric properties of the materials (Dzyaloskinskii et al., 1961). This theory enables one to calculate the effect of van der Waals forces (per unit area), $F(\delta)$, on a liquid film, from the temperature and the optical properties of the materials. These procedures are dicussed and related to the material presented in Fig. 6-4 by Truong and Wayner (1987).

6-3 THERMODYNAMICS OF THIN FILMS: MENISCUS PROPERTIES

6-3-1 Kelvin-Clapeyron Equation (Vapor Pressure)

In this section, a combination of the procedures presented by Potash and Wayner (1972), Wayner (1976, 1991a), DasGupta et al. (1994), and Swanson and Peterson (1995) are used to determine the effect of temperature and interfacial phenomena on the vapor pressure of a curved thin liquid film. The process can be described using the schematic quasi-thermodynamic phase diagram for the change of phase process in an evaporating meniscus (see Fig. 6-10) where the Gibbs energy is presented as a function of the pressure (Wayner, 1991b). Udell (1983) and Majumdar and Tien (1990) used the more extensive Gibbs potential diagram in Fig. 6-11 to describe the effects of surface tension

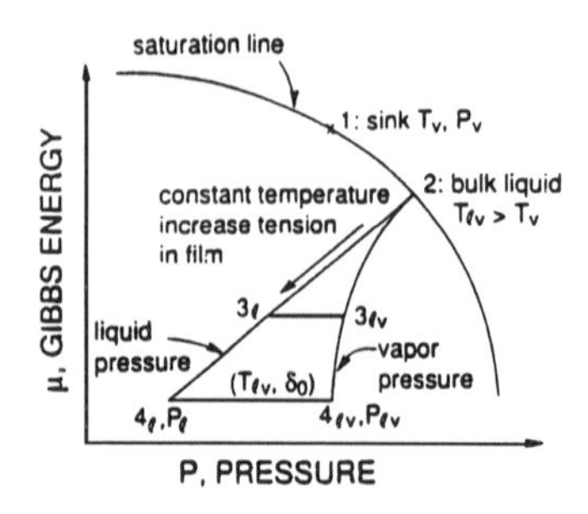

Figure 6-10 Gibbs energy vs. pressure (Wayner, 1991b). Reprinted with permission of IS&T: The Society for Imaging Science and Technology sole copyright owners.

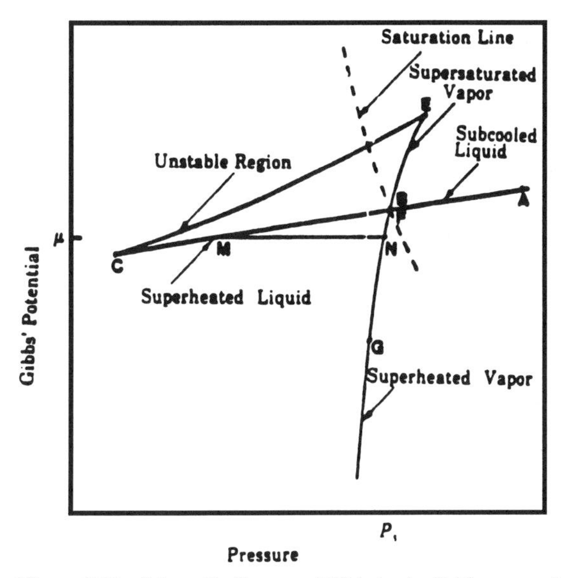

Figure 6-11 Schematic diagram of Gibbs' potential for a pure fluid. (From Majumdar and Tien, 1990.)

on film condensation in a porous media. On the saturation line for a bulk liquid in Fig. 6-10, the vapor pressure increases and the Gibbs energy decreases with an increase in temperature. This is modified in the meniscus under tension. Evaporation can naturally occur from the higher-temperature liquid in the meniscus to a liquid sink kept at a lower temperature. At the base of a relatively large meniscus with a fluid temperature of T_{lv}, the vapor pressure P_{lv} would be close to that of a bulk liquid. If a sink at $T_v < T_{lv}$ with P_v existed, evaporation and vapor flow between locations 2 and 1 would occur. As fluid flows towards the interline along line 2-3_l-4_l, the effective pressure in the liquid, P_l, decreases. The quasi-equilibrium pressure jump across the liquid–vapor interface is represented by a horizontal tie line of constant chemical potential (e.g., 3_l-3_{lv}). As long as the vapor pressure at the liquid–vapor interface, P_{lv}, is greater than the vapor pressure of the sink, P_v, evaporation with vapor flow to the sink occurs.

The Gibbs-Duhem equations for a bulk liquid and a bulk vapor in a gravitational field, $g_x < 0$, are

$$dP_l = s_l dT + \rho_l g_x dx + n_l d\mu_{lg}, \tag{6-40}$$

$$dP_v = s_v dT + \rho_v g_x dx + n_v d\mu_{vg}, \tag{6-41}$$

$$\mu_{ig} = \mu_i - M g_x x. \tag{6-42}$$

Taking the difference between Eqs. (6-41) and (6-42) with $\mu_{lg} = \mu_{vg}$ for a pure system gives

$$d(P_v - P_l) = (s_v - s_l)dT + (\rho_v - \rho_l)g_x\, dx + (n_v - n_l)d\mu_g. \tag{6-43}$$

Combining Eq. (6-43) with $(n_l - n_v) \approx V_l^{-1}$ and the augmented Young-Laplace equation, Eq. (6-44), for the effective pressure in the liquid at the liquid–vapor interface for

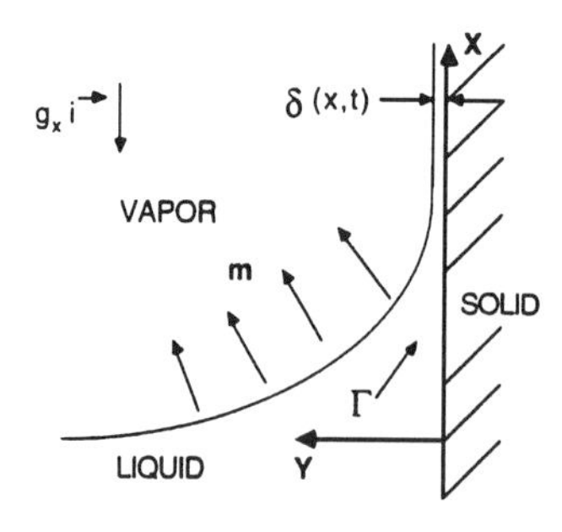

Figure 6-12 Vertical extended meniscus with fluid flow and phase change. (From Wayner, 1992.)

the extended meniscus in Fig. 6-12, with $\Gamma = 0$ and $m = 0$, gives Eq. (6-45) below.

$$P_l = P_v - \Pi - \sigma_{lv} K \tag{6-44}$$

$$d\mu_g = -V_l(d\Pi + \sigma_{lv} dK + K d\sigma_{lv}) - V_l(s_l - s_v)dT - Mg_x dx. \tag{6-45}$$

Using Eqs. (6-45) and (6-42) with $d\mu_i = RT\, d(\ln P_v)$ gives

$$d(\ln P_v) = -\frac{V_l}{RT}(d\Pi + \sigma_{lv} dK + K d\sigma_{lv}) + \frac{V_l \Delta h}{RT^2} dT. \tag{6-46}$$

Integrating Eq. (6-46) between a reference vapor pressure P_{vr}, taken to be the vapor pressure over a bulk liquid ($\Pi = 0$, $K = 0$, $x = 0$, T_v), and a new vapor pressure at the liquid–vapor interface, P_{lv} at Π, K, T_{lv}, gives

$$\ln\left(\frac{P_{lv}}{P_{vr}}\right) = -\frac{V_l}{RT}(\Pi + \sigma_{lv} K + K\Delta\sigma_{lv}) + \frac{V_l \Delta h}{T_{lv} T_v R}(T_{lv} - T_v). \tag{6-47}$$

A linear approximation of the logarithmic term gives Eq. (6-48) for the variation of the vapor pressure of the curved thin film with effective pressure and temperature.

$$P_{lv} - P_{vr} = -\frac{V_l P_{vr}}{RT}(\Pi + \sigma_{lv} K + K\Delta\sigma_{lv}) + \frac{V_l P_{vr} \Delta h}{T_{lv} T_v R}(T_{lv} - T_v) \tag{6-48}$$

It is instructive to use this equilibrium equation to analyze the isothermal effective pressure in the thin film. In Fig. 6-6b, the schematic cross section of a small closed square glass channel partially filled with liquid is presented. In the vapor space *with a vertical x-coordinate*, the change in vapor pressure with vertical distance is given by

$$P_{vx} - P_{vr} = \frac{P_v M}{RT} g_x x. \tag{6-49}$$

Using Eqs. (6-48) and (6-49) with $P_{lv} = P_{vx}$, $\Delta T = 0$, and $\Delta\sigma_{lv} = 0$, we obtain the hydrostatic variation of the effective pressure in an isothermal meniscus with distance:

$$\Pi + \sigma_{lv} K = -\rho_l g_x x. \tag{6-50}$$

In this way, the hydrostatic meaning of disjoining pressure and capillarity is recaptured, and the variation of the effective pressure or tension in the thin liquid film is demonstrated. As discussed previously in connection with Fig. 6-6b, the effective pressure in the liquid film and the Hamaker constant can be obtained by measuring the film profile.

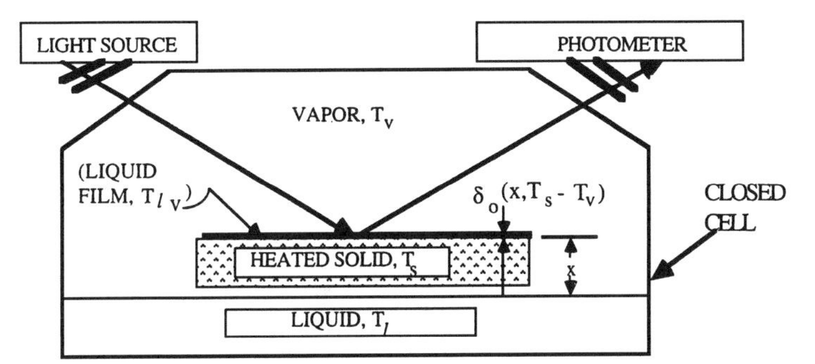

Figure 6-13 Schematic of experimental cell with ellipsometer for use with Eq. (6-54).

Combining Eqs. (6-48) and (6-49) with (on the right side) $P_{vr} \approx P_v$, an average pressure level, and $g = -g_x$ for a vertical film gives for the *nonisothermal* case

$$P_{lv} - P_{vx} = -\frac{V_l P_v}{RT_{lv}}(\Pi + \sigma_{lv}K + K\Delta\sigma_{lv}) + \frac{MgP_v}{RT_v}x + \frac{P_v V_l \Delta h}{RT_{lv}T_v}(T_{lv} - T_v) = 0. \tag{6-51}$$

If $P_{lv} \neq P_{vx}$, Eq. (6-51) can be used along with kinetic theory to calculate the rate of evaporation from (or condensation on) a curved thin film, which is discussed in the next section. For the isothermal equilibrium case, with $T_{lv} = T_v$, $P_{lv} = P_{vx}$, and $\Delta\sigma_{lv} = 0$, this reduces to the identity

$$V_l(\Pi + \sigma_{lv}K) = Mgx. \tag{6-52}$$

Equations (6-47) to (6-51) can be rewritten for constant temperature while retaining the natural logarithm to obtain Eq. (6-53) for a flat film on a vertical substrate.

$$\Pi = \frac{Mgx}{V_l} = -\frac{RT}{V_l}\ln\left(\frac{P_{lv}}{P_{vr}}\right) \tag{6-53}$$

Therefore, the adsorbed flat film thickness is easily related to the static head in the liquid and the variation of the vapor pressure. Experimentally, it is observed that, for a completely wetting system, the film thickness decreases as the height above the pool increases (e.g., Sabisky and Anderson, 1973).

The sensitivity of the equilibrium (to mass transfer, $P_{lv} = P_{vx}$) film thickness, δ_o, to height and temperature difference can be obtained using the experimental design presented in Fig. 6-13 with the following form of Eq. (6-51) for flat film thicknesses:

$$\rho_l g x = -\left\{\frac{\Delta h(T_s - T_v)}{T_{lv}} + \frac{A_{slv}}{6\pi\delta_o^3}\right\}. \tag{6-54}$$

The thickness δ_o is an important characteristic thickness which is used below as the boundary condition at the contact line of a nonequilibrium thin film.

Example 6-7. Use Eq. (6-54) to determine the effect of superheat, ΔT, and hydrostatic head, x, on the equilibrium film thickness for the SiO_2/octane system at 298 K with $A_{slv}/6\pi \approx -3.2 \times 10^{-22}$ J.

ΔT, K	x, mm	δ_0, nm
0	1.7	30
0.001	1.7	7.2
0.01	1.7	3.3
0.1	1.7	1.6
0	12.4 m	1.6
1	1.7	0.7
30	0	0.23

We note that, in this example, $\Delta T \approx 30$ K for a monolayer.

From Eqs. (6-21) and (6-52), it can be seen that the Hamaker constant is negative for a completely spreading system and that the effective pressure or chemical potential per unit volume, $F(\delta) = -\Pi$, in the thin film decreases as the film thickness decreases. Therefore, if a drop of a spreading liquid is placed on a horizontal surface, fluid flows from the *thicker* portion of the ultra-thin film to the *thinner* portion because of this gradient in the disjoining pressure, which is represented by the slope of the thin film. Since gravitational effects are relatively small near the contact line of a horizontal film, the gradient in the effective pressure (F') for flow in a horizontal film can be represented by

$$\frac{dF}{dx} = F' = -\frac{A_{slv}}{2\pi\delta^4}\delta'. \tag{6-55}$$

An example of the variation of chemical potential with film thickness for a thin film is given for a simple spreading case and for a simple finite contact angle case in Fig. 6-14a. In the limit $\delta \to 0$, $\mu \to -\infty$ for both cases, but this is not shown for $\theta > 0$. The variation for a complicated system which includes a fluid-like water on a polar substrate is given in Fig. 6-14b. This represents an adsorbed ultra-thin film in equilibrium at the contact line with a thicker film. A discussion of various isotherms is given by Dzyaloskinskii et al. (1961).

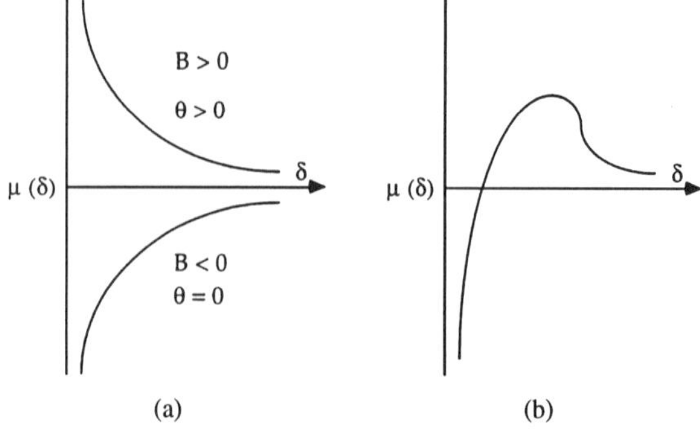

Figure 6-14 Variation of chemical potential with film thickness (Dzyaloskinskii et al., 1961): (a) simple spreading and nonspreading systems; (b) complicated system (e.g., water on a polar substrate).

6-4 QUASI-THERMODYNAMICS OF THIN FILMS: INTERFACIAL MASS FLUX

In 1953, Schrage reviewed the literature and presented and discussed the following equation based on kinetic theory relating the net mass flux of matter crossing a liquid–vapor interface to a jump change in interfacial conditions at the interface:

$$\dot{m} = C_l \left(\frac{M}{2\pi R} \right)^{1/2} \left(\frac{P_{lv}}{T_{lv}^{1/2}} - \frac{P_{vx}}{T_v^{1/2}} \right). \tag{6-56}$$

Herein, we assume that the net mass flux crossing the interface (e.g., evaporation) results from a small vapor pressure drop across an imaginary plane at the interface in which P_{lv} is the quasi-equilibrium vapor pressure of the liquid film at (T_{lv}, K, Π, x) and P_{vx} is the equilibrium vapor pressure of a reference bulk liquid $(K = 0, \Pi = 0, x)$ at a temperature T_v. C_l is the accommodation coefficient, for which the ideal value is 2.0. Neglecting resistances in the bulk vapor space, P_{vx} and T_v can exist at a short distance from the interface, and a resistance to evaporation at the interface can be defined using Eq. (6-56) with $T_i^{1/2} \approx T_{lv}^{1/2} \approx T_v^{1/2}$:

$$\dot{m} = C_1 \left(\frac{M}{2\pi R T_i} \right)^{1/2} (P_{lv} - P_{vx}). \tag{6-57}$$

Combining Eqs. (6-57) and (6-51) gives Eq. (6-58) for the interfacial mass flux:

$$\dot{m} = C_l \left(\frac{M}{2\pi R T_i} \right)^{1/2} \left\{ \frac{P_v M \Delta h_m}{R T_v T_{lv}} (T_{lv} - T_v) - \frac{V_l P_v}{R T_{lv}} (\Pi + \sigma_{lv} K + K \Delta\sigma_{lv}) + \frac{M g P_v}{R T_v} x \right\}. \tag{6-58}$$

There are two interfacial effects that can cause an effective pressure jump at the liquid–vapor interface: capillarity $\sigma_{lv} K$ and disjoining pressure Π. The significance of the disjoining pressure is that the vapor pressure of an adsorbed completely wetting liquid film is reduced by interfacial forces, and therefore a superheated adsorbed liquid film can exist in (vapor pressure) equilibrium with a bulk liquid at a lower temperature. For convenience, we will hereafter emphasize small regions where the gravitational term can be neglected and completely spreading (zero contact angle) fluids with

$$\Pi \approx -\frac{B}{\delta^n} = -\frac{\bar{A}}{\delta^3}. \tag{6-59}$$

The approximation, $n = 3$, in Eq. (6-59) theoretically restricts its use to thicknesses $\delta \lesssim 20$ nm. However, to demonstrate the use of the above in the modeling of more complex real situations below with larger variations in thickness, this restriction will frequently be relaxed. When necessary, the equations can always be expanded or effective values applied for other uses.

The evaporative flux, Eq. (6-58), can be rewritten as

$$\dot{m} = a(T_{lv} - T_v) + b(P_l - P_v) \tag{6-60}$$

with

$$a = C_l\left(\frac{M}{2\pi RT_{lv}}\right)^{1/2}\left(\frac{P_v M\Delta h_m}{RT_v T_{lv}}\right), \qquad b = C_1\left(\frac{M}{2\pi RT_{lv}}\right)^{1/2}\left(\frac{V_l P_v}{RT_{lv}}\right). \tag{6-61}$$

Equation (6-60) states that evaporation is promoted by a superheat and hindered by low effective pressures. We note that, when interfacial effects are important, the effective pressure in the liquid, P_l, is not necessarily equal to the pressure in the vapor or to its normal vapor pressure. For example, the equilibrium vapor pressure of a small droplet is not the same as that of a flat pool of bulk liquid at the same temperature, and the pressure inside the droplet is greater than the surrounding pressure because of surface tension.

The liquid–vapor interface temperature can be related to the substrate temperature through the one-dimensional conduction heat transfer solution for the film:

$$\frac{k}{\delta}(T_s - T_{lv}) = \dot{m}\Delta h_m \tag{6-62}$$

where k is the thermal conductivity of the liquid in the film. Following Moosman and Homsy (1980), Eq. (6-62) may be combined with Eq. (6-60) to eliminate T_{lv} in favor of T_s:

$$\dot{m} = \frac{1}{1 + \frac{a\Delta h_m}{k}\delta}[a(T_s - T_v) + b(P_l - P_v)]. \tag{6-63}$$

This equation demonstrates the effect of the conduction resistance in the film.

6-5 AN EVAPORATING EXTENDED MENISCUS

6-5-1 Microscopic Model

Since gravity has a relatively minor effect on the following material, gravitational effects are neglected hereafter. In Fig. 6-12, the local evaporation rate is linked to the mass flow rate per unit width of the film, Γ, through a material balance:

$$\frac{d\Gamma}{dx} = -\dot{m}. \tag{6-64}$$

Using lubrication theory, the mass flow rate per unit width of the slightly tapered film is related to the pressure gradient in the flow direction by

$$\Gamma(x) = -\frac{\delta^3}{3\nu}\frac{dP_l}{dx}. \tag{6-65}$$

Using Eqs. (6-63)–(6-65), the coupled differential equations are thus Eq. (6-44) and

$$\frac{1}{3\nu}\frac{d}{dx}\left(\delta^3\frac{dP_l}{dx}\right) = \frac{1}{1 + \frac{a\Delta h_m}{k}\delta}[a\Delta T + b(P_l - P_v)] \tag{6-66}$$

where $\Delta T = T_s - T_v$.

The variables are nondimensionalized in the following way. For a strictly flat film, we define $\delta(x) = \delta_0$. Since there is no pressure gradient in such a film, the right side of

Eq. (6-66) must be zero and

$$b(P_l - P_v) = -a\Delta T_o; \qquad \frac{\bar{A}}{\delta_o^3} = -\frac{a}{b}\Delta T_o \tag{6-67}$$

The reference thickness is thus fixed, and the reference pressure is the magnitude of the disjoining pressure of such a film, Π_o, which is obtained from Eq. (6-67). $\bar{A}$ is negative for a spreading film, and $a\Delta T_o = a(T_s - T_v)_o = \dot{m}^{id}$, the ideal evaporative mass flux based on kinetic theory for a given ΔT_o.

The dimensional pressure difference is defined as

$$\phi = (P_l - P_v)/\Pi_o. \tag{6-68}$$

The dimensionless position and film thickness are $\xi = x/l$ and $\eta = \delta/\delta_o$, with the scaling of x given by

$$l = \sqrt{\frac{-\bar{A}}{\nu a \Delta T_o}} \tag{6-69}$$

The dimensionless form of Eqs. (6-66) and (6-44) are

$$\frac{1}{3}\frac{d}{d\xi}\left(\eta^3 \frac{d\phi}{d\xi}\right) = \frac{1}{1+\kappa\eta}(1+\phi) \tag{6-70}$$

$$\phi = -\frac{1}{\eta^3} - \varepsilon\frac{d^2\eta}{d\xi^2} \tag{6-71}$$

where the parameters κ and ε are

$$\kappa = \frac{a\Delta h_m \delta_o}{k}; \qquad \varepsilon = \frac{\sigma\delta_o b\nu}{-\bar{A}}. \tag{6-72}$$

The parameter κ is a measure of the importance of the resistance of the film to thermal conduction. The parameter ε is a measure of the importance of capillary pressure effects relative to disjoining pressure effects. For completely wetting systems, both of these effects reduce the local mass flux to a value less than the ideal mass flux. The right side of Eq. (6-70) is also a dimensionless mass flux, $\dot{M} = \dot{m}/\dot{m}^{id}$ (or, using $\alpha\Delta T_o = \dot{m}\Delta h_m$, an interfacial heat transfer coefficient, α), which can be written as

$$\dot{M} = \frac{[1 - \eta^{-3} - \varepsilon\eta'']}{1+\kappa\eta}. \tag{6-73}$$

Therefore, an interfacial heat transfer coefficient can be obtained from first principles. Considerable insight can be obtained from Eq. (6-73). We find that at the leading edge of an evaporating horizontal flat ($\eta'' = 0$) thin film, $\eta = 1$, the evaporation rate can be zero even through the superheat can be substantial. This condition is required to define a stationary interline for an evaporating thin liquid film with varying thickness that has an equilibrium contact angle equal to zero.

Using four boundary conditions, Eq. (6-70) can be solved numerically to determine the heat transfer characteristics of a stationary, steady-state thin film. Experimental data on the film thickness profile of a steady-state evaporating extended meniscus were

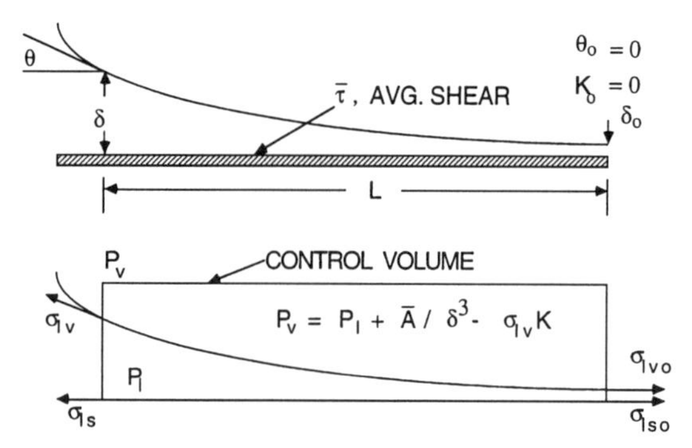

Figure 6-15 Control volume for macroscopic horizontal force balance due to interfacial free energy and shear with $A < 0$. (From Kim and Wayner, 1996.)

analyzed using the above equations and were found to agree with these models (DasGupta et al., 1994). When the right and left sides of Eq. (6-70) are not equal, the above material can be used with the equation of continuity to analyze the transient case.

Example 6-8. Assuming $\kappa\eta = 0$, $\eta'' = 0$, use Eq. (6-84) to determine the value of the dimensionless film thickness where the evaporation rate will be 99% of its substantial ideal value. For these conditions, $\eta = 4.6$, or $\delta = 4.6\,\delta_o$.

6-5-2 Macroscopic Interfacial Force Balance

For a complete description of the evaporating meniscus, we can use both the "macroscopic" (integral) and "microscopic" (differential) viewpoints. A macroscopic horizontal force balance for the completely wetting meniscus presented in Fig. 6-15 gives (Kim and Wayner, 1996)

$$\sigma_{lv}\cos\theta + \sigma_{ls} + \sigma_{lv}K\delta + \frac{-\bar{A}}{\delta^2} + \bar{\tau}L = \sigma_{lvo} + \sigma_{lso} + \sigma_{lvo}K_o\delta_o + \frac{-\bar{A}}{\delta_o^2}. \tag{6-74}$$

In this equation, $\bar{\tau}L$ is the average force due to surface shear stress in the control volume. To obtain Eq. (6-75), Eq. (6-74) can be simplified using the following assumptions: $\sigma_{ls} = \sigma_{lso}$, $\sigma_{lv} = \sigma_{lvo}$, $K_o = 0$, in which the subscript "o" refers to the location of the interline, δ_o. The first two assumptions restrict the contact line thickness to approximately $\delta_o > 1$ nm, and the third assumes that the substrate is smooth.

$$\sigma_{lv}(1 - \cos\theta) = \bar{\tau}L + \sigma_{lv}K\delta + \frac{\bar{A}}{\delta_o^2} - \frac{\bar{A}}{\delta^2} \tag{6-75}$$

We find that the apparent contact angle θ at the thicker end is a function of the downstream viscous losses and interfacial forces.

6-5-3 Experimental Data

Using both microscopic and macroscopic models, Kim and Wayner (1996) analyzed the experimental data obtained in the circular experimental cell presented in Fig. 6-16. The

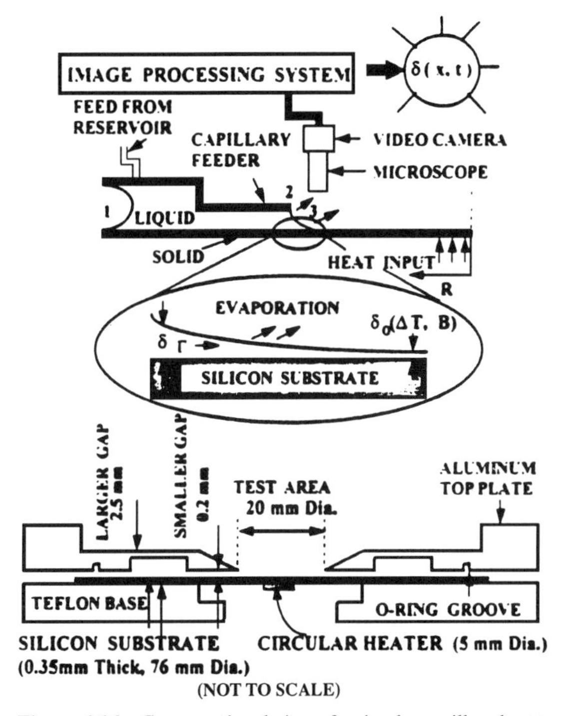

Figure 6-16 Cross-sectional view of a circular capillary heat transfer cell. (From Kim and Wayner, 1996.)

film thickness profiles in the thickness range $\delta < 4\ \mu$m were measured optically using null ellipsometry and image-analyzing interferometry. Using the augmented Young-Laplace equation, Eq. (6-44), the measured film thickness profile gave the pressure field in the evaporating thin liquid film. Using a Kelvin-Clapeyron (KC) model, Eq. (6-70), the pressure and temperature fields gave the evaporative heat flux profile and the local slope (apparent contact angle) as a function of the heat flow rate. The results, which are presented in Figs. 6-17–6-20, demonstrated that there were significant resistances to heat transfer in a small meniscus due to interfacial forces, viscous stresses, and thermal

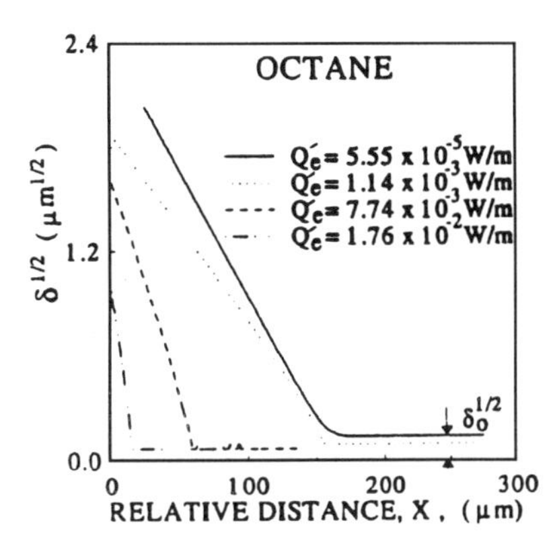

Figure 6-17 Experimental values of $\delta^{1/2}$ versus the relative distance x for octane for various heat flow rates per unit width: $Q'_e = 5.55 \times 10^{-5}$ W/m, $\delta_0 = 1.9 \times 10^{-8}$ m, $\Pi_0 = 1.09 \times 10^{1}$ N/m^2; $Q'_e = 1.14 \times 10^{-3}$ W/m, $\delta_0 = 8.9 \times 10^{-9}$ m, $\Pi_0 = 4.11 \times 10^{1}$ N/m^2; $Q'_e = 7.74 \times 10^{-3}$ W/m, $\delta_0 = 4.5 \times 10^{-9}$ m, $\Pi_0 = 3.20 \times 10^{2}$ N/m^2; $Q'_e = 1.76 \times 10^{-2}$ W/m, $\delta_0 = 4.4 \times 10^{-9}$ m, $\Pi_0 = 8.86 \times 10^{2}$ N/m^2. (From Kim and Wayner, 1996.)

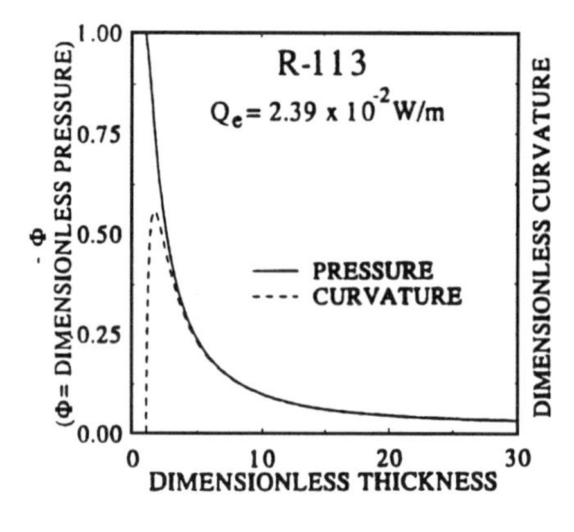

Figure 6-18 Dimensionless pressure and curvature profiles: $\delta_0 = 3.9$ nm, $\Pi_0 = 2.2 \times 10^3$ N/m^2, $Q'_e = 2.39 \times 10^{-2}$ W/m. (From Kim and Wayner, 1996.)

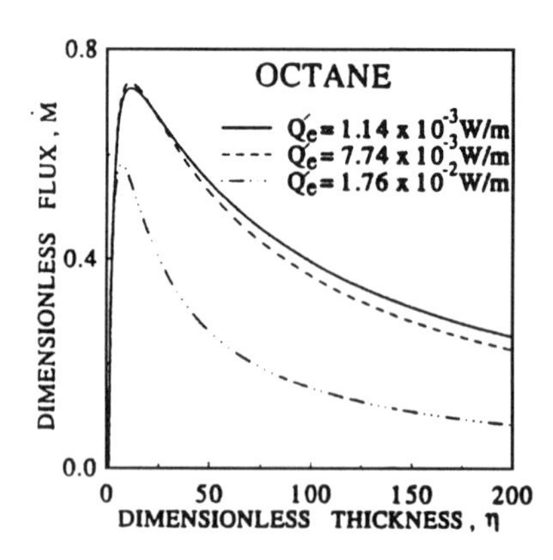

Figure 6-19 Dimensionless flux versus dimensionless thickness: $Q'_e = 1.14 \times 10^{-3}$ W/m, $\delta_0 = 8.9 \times 10^{-9}$ m, $m^{id} = 2.76 \times 10^{-5}$ kg/m^{2}s, $q^{id} = 10$ W/m^2; $Q'_e = 7.74 \times 10^{-3}$ W/m, $\delta_0 = 4.5 \times 10^{-9}$ m, $m^{id} = 6.42 \times 10^{-4}$ kg/m^{2}s, $q^{id} = 223$ W/m^2; $Q'_e = 1.76 \times 10^{-2}$ W/m, $\delta_0 = 4.4 \times 10^{-9}$ m, $m^{id} = 7.42 \times 10^{-3}$ kg/m^{2}s, $q^{id} = 2329$ W/m^2. (From Kim and Wayner, 1996.)

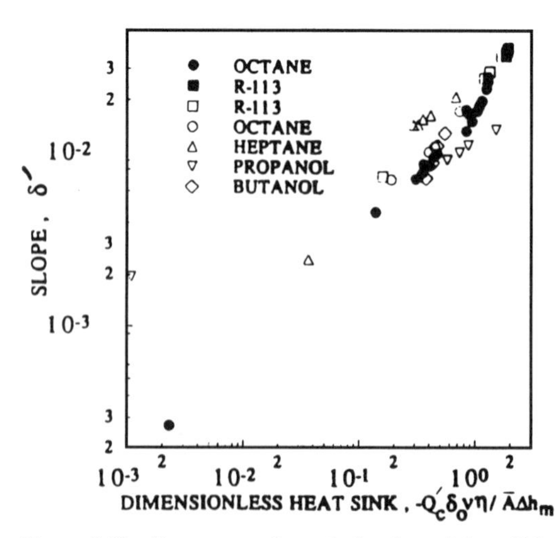

Figure 6-20 Slope versus dimensionless heat sink, $-Q'_c\delta_0\nu\eta/\bar{A}\Delta h_m$, at a film thickness $\delta = 200$ Å. (From Kim and Wayner, 1996.)

conduction in the liquid. Therefore, the ideal heat sink based on kinetic theory and a constant curvature meniscus cannot be attained. The experimental results characterizing the evaporating meniscus were found to agree with the interfacial models described above.

The measured values of $\delta^{1/2}$ versus the relative distance x for octane are given in Fig. 6-17 as a function of the total evaporative heat flow rates per unit width of the meniscus, Q'_e:

$$Q'_e = \int_0^{L_e} q dx \tag{6-76}$$

where L_e represents the length of the region where $0 \leq \delta' \rightarrow \infty$. These results demonstrate that the thickness of the adsorbed film, δ_0, decreased with an increase in Q'_e (which increases with the substrate temperature). In an *isothermal* horizontal system of a spreading liquid on a solid substrate at the exit of a capillary, the curvature should remain constant in the region where dispersion forces can be neglected. So the film profile in this range approximates a parabola, and a plot of $\delta^{1/2}$ versus x is a straight line. For the case where $Q'_e \rightarrow 0$ W/m, $\delta^{1/2}$ versus x is nearly a straight line, which means the system was very close to isothermality. However, for the cases involving evaporation, the lines bend downward, thereby showing the presence of a curvature gradient which causes fluid flow. These results clearly demonstrate that the film thickness profile and curvature are functions of the heat flux. The suction at δ_0 is equal to Π_0 and is a function of the substrate superheat.

In Fig. 6-18, both the dimensionless pressure and curvature are presented as a function of the dimensionless thickness. The dimensional interfacial pressure difference can be obtained by multiplying ϕ by the reference disjoining pressure (Π_0). The data demonstrate that, for a completely wetting system, the curvature starts at zero and becomes a maximum due to a change in the disjoining pressure, and then decreases when it becomes the dominant cause of fluid flow. We also find that, in the thicker region, the curvature approaches a constant value because of the decrease in viscous losses.

In Fig. 6-19, the dimensionless interfacial evaporative flux profiles, Eq. (6-73), are presented for various heat flow rates per unit width of the meniscus. In the adsorbed thin film region, where the film thickness is δ_0, the dimensionless pressure is $\phi = -1$. Therefore, the dimensionless interfacial flux is equal to zero. This result is due to the London–van der Waals force, which keeps the superheated thin film from evaporating. The absolute value of ϕ decreases as the thickness increases. Therefore, the interfacial evaporative flux increases due to the decrease in the effect of interfacial forces. This increase is partially offset by the decline of the interfacial flux resulting from the rise of the conductive resistance, which is due to the increase in the film thickness. Therefore, a maximum is obtained in the flux profile. Additional details concerning the numerical values of the conductive resistances can be found in DasGupta et al. (1994).

The apparent contact angle of the meniscus at the film thickness δ is

$$\theta \mid \delta = \tan^{-1}\left(\frac{d\delta}{dx}\right)_\delta . \tag{6-77}$$

In Fig. 6-20, the slope of the meniscus at $\delta = 200$ Å is presented as a function of a dimensionless contact line heat sink. Using this data, a simple correlation for the slope at $\delta = 200$ Å can be obtained:

$$\delta' = 0.021\left(\frac{-Q_c'\delta\nu}{\bar{A}\Delta h_m}\right)^{0.69}, \qquad \bar{A} < 0,\ \delta = 20\ \text{nm} \tag{6-78}$$

where Q_c' is the value of the evaporative heat flow rate per unit width of the meniscus in the region $\delta_o \leq \delta \leq 20$ nm.

6-6 APPLICATIONS

6-6-1 Microlayer in Nucleate Boiling

Stephan and Hammer (1994) successfully used the above interfacial models to evaluate the influence of meniscus curvature, adhesion forces, interfacial thermal resistance, and wall thermal resistance on nucleate boiling heat transfer for the R114/Cu system. A slightly different evaluation of these models in boiling was presented by Lay and Dhir (1995), and a recent review paper by Sadasivan et al. (1995) discussed the issues in CHF modeling. For the single bubble presented in Fig. 6-21, Stephan and Hammer (1994) calculated the physical details of the boiling heat transfer process presented in Figs. 6-22–6-25.

Their results clearly demonstrated that interfacial phenomena in the micro region have a significant influence on the overall heat transfer process. They concluded that 1) the molecular kinetic resistance at the interface is not negligible; 2) the heat flux reaches a maximum in the micro region, which is about 100 times larger than the burn-out heat flux; 3) due to adhesion forces near the solid, the curvature of the liquid–vapor interface is not constant: 4) the apparent macroscopic contact angle is a result of viscous flow

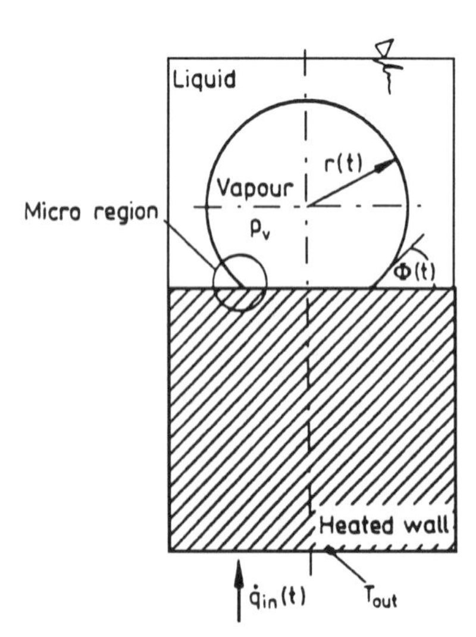

Figure 6-21 Vapor bubble at the liquid–solid interface in boiling. (From Stephan and Hammer, 1994 © Springer-Verlag.)

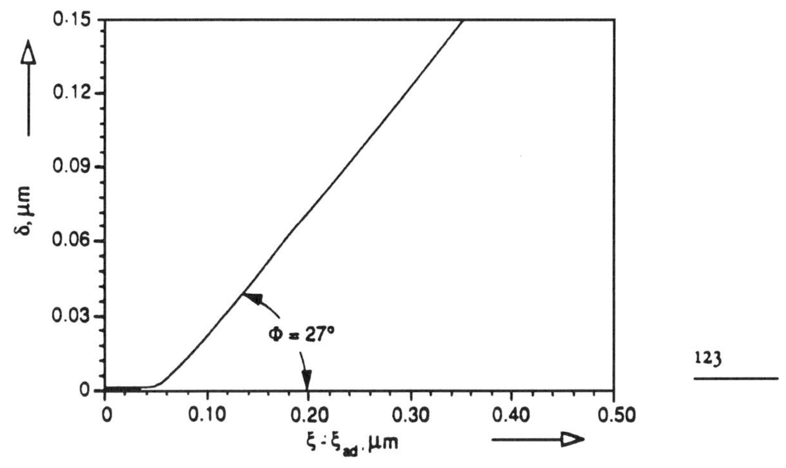

Figure 6-22 Meniscus shape, δ, in the micro region (R114/Cu, $p_v = 2.47$ bar, $T_{out} - T_{sat} = 3.5$ K, $r = 0.125$ mm). (From Stephan and Hammer, 1994 © Springer-Verlag.)

in the micro region; 5) the high evaporation rate leads to a local temperature drop in the heated wall; and 6) the wall temperature is not constant in the micro region. The comparison between the results of the modeling and those of the experiments was quite satisfactory and within the experimental accuracy. In addition to the above conclusions, they demonstrated that the common assumption of an interfacial temperature equal to the saturation temperature of the vapor can lead to a large overprediction of the radial heat transfer coefficient. In an earlier paper, Stephan and Busse (1992) successfully used the same type of modeling to analyze the heat transfer coefficient of grooved evaporator walls.

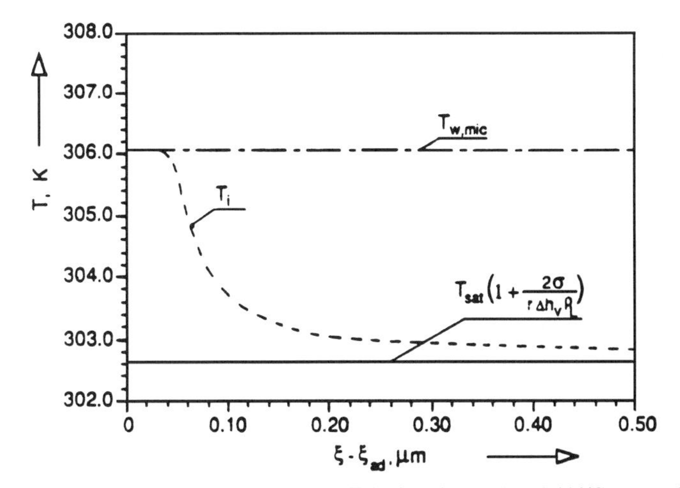

Figure 6-23 Interfacial temperature T_i in the micro region (R114/Cu, $p_v = 2.47$ bar, $T_{out} - T_{sat} = 3.5$ K, $r = 0.125$ mm). (From Stephan and Hammer, 1994 © Springer-Verlag.)

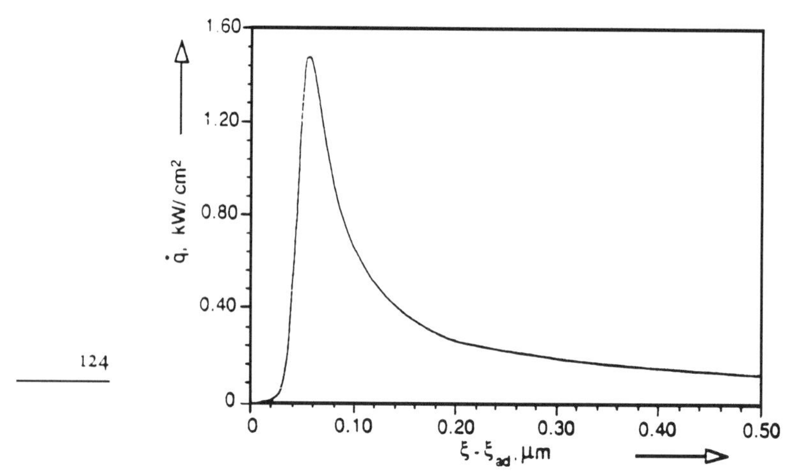

Figure 6-24 Heat transfer coefficient h for different bubble radii (R114/Cu, $p_v = 2.47$ bar, $T_{out} - T_{sat} = 3.5$ K, $r = 0.125$ mm). (From Stephan and Hammer, 1994 © Springer-Verlag.)

6-6-2 Evaporation from Grooves and Capillaries

For a "V"-shaped grooved channel, Xu and Carey (1990) assumed both that flow in the main part of the channel was driven primarily by the capillary pressure difference due to the receding of the meniscus toward the apex of the groove, and that the flow up the groove wall was driven mainly by the disjoining pressure gradient. Comparison of these theoretical results with experimental data suggested that the disjoining pressure may play the central role in evaporation from a microgroove channel. The same concepts

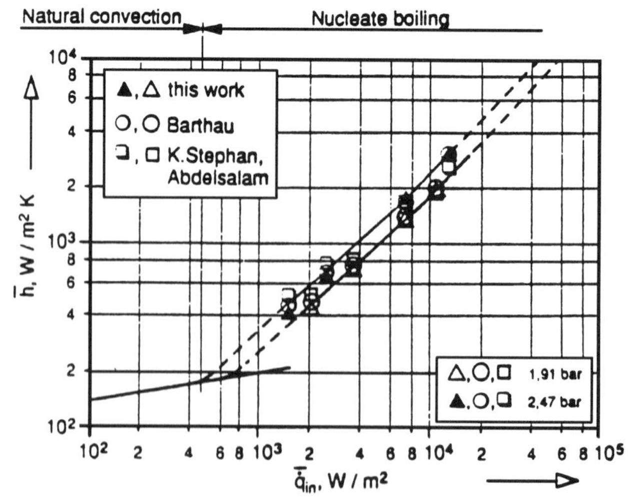

Figure 6-25 Comparison with experiments (Barthau) and correlations (Stephan, Abdelsalam): mean heat transfer coefficient versus mean heat flux (R114/Cu, $p_v = 1.91$ bar and 2.47 bar). (From Stephan and Hammer, 1994 © Springer-Verlag.)

have been used by others to analyze evaporation from grooved structures associated with heat pipes: for example, Kamotani (1978); Holm and Goplen (1979); Khrustalev and Faghri (1995). Holm and Goplen found that a large percent of the evaporative heat transfer occurs in the thin film region. Using a model that included Marangoni effects, a "thermocapillary" limitation for a micro heat pipe was obtained by Swanson and Peterson (1995). Further details concerning the heat pipe applications of these studies are given in Chapter 8 on micro heat pipes.

Swanson and Herdt (1992) evaluated the meniscus evaporation process within a micropore incorporating the three-dimensional Young-Laplace equation, Marangoni convection, LW dispersion forces, and nonequilibrium interfacial conditions. Although most of their numerical results generally followed expectations, the effect of the parameters on the base meniscus profile did not. No change in apparent contact angle was observed with changes in the dimensionless superheat or dispersion number, whereas changes in viscous stresses should have produced one. One possible explanation for this can be found in the macroscopic force balance equation, Eq. (6-75). For a completely wetting fluid, the force at the contact line due to the Hamaker constant has a sign opposite to that of the shear stress. Therefore, changes in these two terms can appear to cancel each other for completely wetting systems, at least at lower fluxes. There was the expected increase in the local mass flux and total mass transfer rate with dispersion number. Hallinan et al. (1994) evaluated the effect of evaporation from the thin film region on the meniscus shape within a micropore. Although the numerical results of Hallinan et al. (1994) also justify the use of the static meniscus curvature as a boundary condition for this particular dynamic case, many additional studies are needed to sort out the microscopic details of the transport processes for various physical situations. The understanding of the evaporating extended meniscus is still rudimentary; much additional theoretical and experimental study is needed to complete the description of this dynamic region and adjudicate the differences.

6-6-3 Rewetting a Hot Substrate (Moving Contact Line)

Equation (6-58) demonstrates that, in an isothermal horizontal ultra-thin film, evaporation or condensation occurs unless $(\Pi + \sigma_{lv} K) = 0$. In addition, the unequal form of Eq. (6-64) demonstrates that motion of the contact line would occur with phase change. Using simple models, the static profile of the contact line region without phase change for the partial wetting case ($\Pi < 0$, $K > 0$) was discussed by Wayner (1982) and for the completely wetting case ($\Pi > 0$, $K < 0$) by Joanny and deGennes (1986).

There have been extensive studies of contact line motion without phase change. These have been extensively reviewed by Kister (1993). In many of these analyses, the infinite stress at the monolayer thickness is mitigated by a slip boundary condition. However, a study of heat transfer naturally leads to the possibility of phase change mitigating the infinite stress at the contact line. For motion due to phase change to occur in an "isothermal," ultra-thin, horizontal film, Eq. (6-58) demonstrates that $(\Pi + \sigma_{lv} K) \neq 0$. In addition, however, we would expect a temperature gradient due to the heat transfer process. The size of the temperature gradient, which is a function of the volatility of the fluid and the heat flux, can be hidden in the small region. In fact, the choice between surface

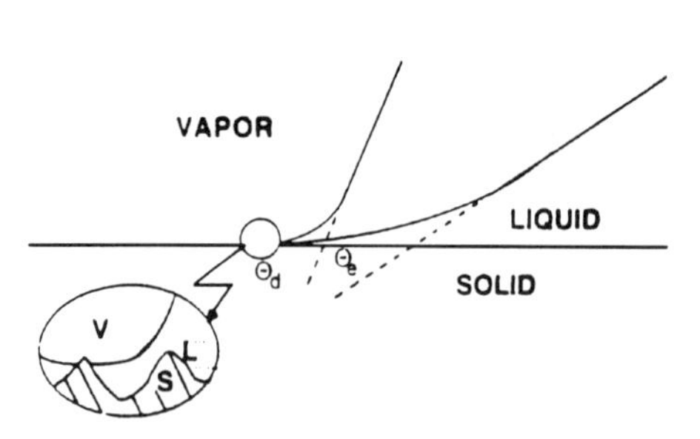

Figure 6-26 Forced change in profile for nonequilibrium system represented by $\theta_e \to \theta_d$ with increase in curvature causing condensation. The resulting phase change, movement, and δ_0 are not shown. (From Wayner, 1994b. Reprinted from *Colloids and Surfaces A, Physicochemical and Engineering Aspects*, vol. 89, by P. C. Wayner, Jr., Mechanical and Thermal Effects in the Forced Spreading of a Liquid Film with Finite Contact Angle, pp. 89–95, © 1994, with kind permission from Elsevier Science-NL Sara Burgerhartstraat 25, 1055 KV Amsterdam, The Netherlands.)

diffusion, slip velocity, or phase change is unresolved. Under nonequilibrium conditions, all three are possible. Contact line motion caused by phase change is discussed next.

Using the above interfacial concepts, Wayner (1994) developed a new physical model for the spreading dynamics of fluids on solid substrates. This model is based on the premise that both interfacial intermolecular forces and temperature control the change-of-phase heat transfer and, therefore, motion in the contact line region. Explicit equations were obtained for the velocity, heat flux, and superheat in the contact line region as a function of a change in the apparent contact angle. To obtain forced flow across the flat surface presented in Fig. 6-26, a forced change in the apparent contact angle, $\theta_e \to \theta_d$, leads to condensation because of a curvature change near the contact line where viscous stresses are very large. Comparison with experimental data demonstrated that the resulting interfacial model of evaporation/condensation not only described the "apparently isothermal" contact line movement in these systems at 20°C but also described the large substrate superheat at the critical heat flux. For the "apparently isothermal" case, it appears that the temperature gradients associated with movement can be localized in a region too small to experimentally study. On the other hand, at the critical heat flux, the turbulence associated with the process leads to the experimental measurement of an average surface temperature. We find that, even though the smallness of the region seems to preclude experimental determination of whether the flow at the contact line occurs by surface diffusion, slip, or phase change, modeling certainly helps our understanding.

A dimensionless equation for the local interfacial superheat in the contact line region, which combines the effects of the change in the apparent contact angle (stress), velocity, and superheat, was found to be

$$\Delta\tau = c + a_1 \ln(1+\gamma) \tag{6-79}$$

in which,

$$\gamma = \frac{a_2 U}{C_1 \cos\theta_{d\eta} \cot\theta_{d\eta}}, \qquad a_2 = \frac{\rho_l}{P_v}\left(\frac{2\pi RT}{M}\right),$$

$$c = \frac{\cos\theta_{e\eta} - \cos\theta_{d\eta}}{\eta - 1}, \qquad a_1 = \frac{\rho_{lM} RT_{lv}\delta_0}{\sigma_{lv}}.$$

This equation is presented in Fig. 6-27 for a given solid–fluid system characterized by $c = 0.05$ with $a_1 = 0.103$ (e.g., octane at 293 K) (Wayner, 1994b). Both the evaporation and condensation processes can be described.

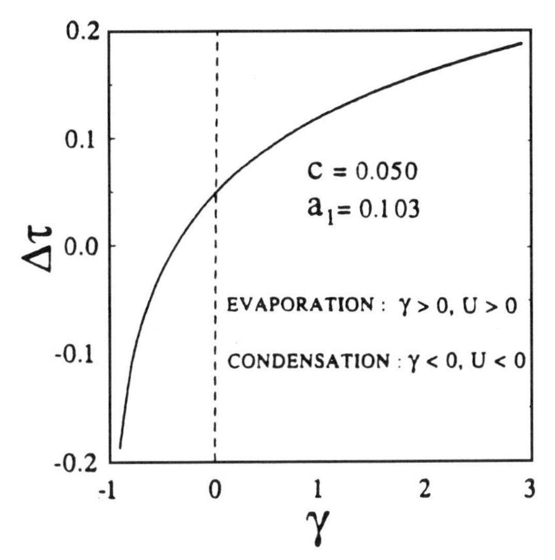

Figure 6-27 Equation (6-79) with $a_1 = 0.103$ for octane (293 K) and $c = 0.05$; $\gamma > 0$ represents evaporation, velocity $U > 0$; $\gamma < 0$ represents condensation. (From Wayner, 1994b. Reprinted from *Colloids and Surfaces A, Physicochemical and Engineering Aspects*, vol. 89, by P. C. Wayner, Jr., Mechanical and Thermal Effects in the Forced Spreading of a Liquid Film with Finite Contact Angle, pp. 89–95, © 1994, with kind permission from Elsevier Science-NL Sara Burgerhartstraat 25, 1055 KV Amsterdam, The Netherlands.)

Assuming $U = 0$ in Eq. (6-79), the following equation is obtained for the dimensionless superheat at the critical heat flux (Wayner, 1994a):

$$c = \frac{\Delta h_m \rho_l \delta_o (T_{lv} - T_v)}{\sigma_{lv} T_v} \tag{6-80}$$

Defining an average change in effective curvature at the contact line as $\Delta K_{eff} = c/\delta_o$ gives the following seeming familiar equation:

$$T_{lv} - T_v = \frac{\sigma_{lv} T_v}{\Delta h_m \rho_1} \Delta K_{eff}. \tag{6-81}$$

It differs from the classical equation for superheat in nucleate boiling since the liquid density appears instead of the vapor density. Using Eq. (6-81) with $c = 0.5$ results in the predictions presented in Table 6-4 for the superheat at the critical heat flux (Wayner, 1994). These results are found to be good estimates of the experimental values found in the literature, which are also presented. Since there are (possibly) large additional effects due to cleanliness, roughness, substrate material, and substrate history, these results are surprising good. In a subsequent paper, Reyes and Wayner (1995) used a different algorithm to obtain better predictions for the experimentally measured superheats on a high-thermal-conductivity substrate, Eq. (6-82). In this case, one experimental data point was needed to characterize the interfacial properties for each system.

$$\frac{\rho_1 \Delta h_m (T_s - T_v)_{\mathrm{CHF}}}{\sigma_{lv}^{1.5} T_v} \cong \mathrm{const} \tag{6-82}$$

Table 6-4 Comparison of theoretical superheat obtained using Eq. (6-81) with experimental data at approximately 1 atm from various references (Wayner, 1994a)

Reference	Experimental K°	Eq. (6.33) K°
H_2/copper Drayer and Timmerhaus (1962)	2	3.4
N_2/copper tube Flynn, Draper and Roos (1962)	10	10
N_2/copper flat (quenching) Peyayopanakul and Westwater (1978)	7	10
H_2O/aluminum flat	28	28
CH_3OH/aluminum flat (quenching) Dhuga and Winterton (1986)	30	23
H_2O/copper flat, $P = 9 \times 10^4$ N/m^2	30	28
CH_3OH/copper flat, $P = 9 \times 10^4$ N/m^2 Wu, Ma, and Li (1982)	27	23
H_2O/copper Gaertner (1965)	44	28
C_5H_{12}/copper, lap E finish	44	52
C_5H_{12}/copper, mirror finish Berenson (1962)	85	52
F113/nickle flat (horizontal) Auracher (1992)	47	45
F113/copper flat (vertical) Liaw and Dhir (1986)	35	45

6-7 SUMMARY

The physical phenomena, theoretical modeling, and experimental results presented above demonstrate that it is possible to start with intermolecular force concepts (as described by either the dielectric properties of the materials or simply the interfacial free energy) and determine the change-of-phase heat transfer characteristics of very thin films from first principles. Generally, these basic principles have many nonisothermal and isothermal applications in engineering. An understanding of interfacial phenomena is important to the proper utilization of small systems in improving heat transfer technology.

ACKNOWLEDGMENT

The development of this chapter was supported in part by the Department of Energy under Grant DE-FG02-89-ER14045. Any opinions, findings, and conclusions or recommendations expressed in this publication are those of the authors and do not necessarily reflect the views of DOE.

NOMENCLATURE

A	Hamaker constant, area
$\bar{A}$	$A/(6\pi)$
a	defined by Eq. (6-61)
B	constant in Eq. (6-22)
b	defined by Eq. (6-61)
C	constant (evaporation coefficient)
c	constant (Eq. (6-79))
D	film thickness
E	surface energy
$F(\delta)$	surface force/area, potential energy/volume
G	excess surface free energy
g	gravitational force per unit mass
h	enthalpy/mass or volume, heat transfer coefficient, Planck's constant, hydrostatic distance
$\hbar$	Planck's constant/2π
K	curvature, parameter
k	Boltzmann's constant, thermal conductivity, absorption coefficient
l	characteristic length defined by Eq. (6-69)
M	molecular weight
m	mass of a molecule, interfacial mass flux
n	index of refraction, molar density
P	pressure
Q	contact line heat sink
q	heat flux
R	gas constant
r	radius of curvature, distance between molecules
S	spreading coefficient
s	entropy per unit volume
T	temperature
U	velocity
V	molar volume
W	work
x	parallel to flow direction
y	axis
α	interfacial heat transfer coefficient
Γ	mass flow rate per unit width of film, adsorbate per unit area
γ	defined in Eq. (6-79)
Δ	difference
δ	liquid film thickness
δ'	slope of liquid–vapor interface
ε	dimensionless group in Eq. (6-72)
η	dimensionless film thickness
θ	angle of inclination, contact angle

κ see Eq. (6-72)
λ parameter
μ chemical potential per molecule or mole, dynamic viscosity
ν kinematic viscosity
ξ dimensionless position
Π disjoining pressure
ρ fluid density, number density of molecules
σ surface free energy per unit area
τ shear stress
ϕ dimensionless pressure difference
Ψ electrostatic surface potential

Subscripts and Superscripts

AB acid-base
d dispersion
e electrostatic
g includes gravity, gas
i interface, average value, induced dipole
id ideal based on superheat only
LW Lifshitz–van der Waals
l liquid
m unit mass, molecular component
o reference
p polar
s solid, structural component
sat saturated
v vapor
x evaluated at x
$'$ derivative

REFERENCES

Bangham, D. H., "The Gibbs adsorption equation and adsorption on solids," *Trans. Faraday Soc.*, vol. 33, pp. 805–811, 1937.

Bankoff, S. G., "Dynamics and stability of thin heated liquid films," *J. Heat Transfer*, vol. 112, pp. 538–546, 1990.

Blake, T. D., "Investigation of equilibrium wetting films of *n*-alkanes on a-alumina," *J. Chem. Soc., Faraday Trans. I*, vol. 71, pp. 192–208, 1975.

Brochard-Wyart, F., J.-M. di Meglio, D. Quéré, and P. G. de Gennes, "Spreading of nonvolatile liquids in a continuum picture," *Langmuir*, vol. 7, pp. 335–338, 1991.

Broekhoff, J. C. P., and J. H. de Boer, "Studies of pore systems in catalysis XIV," *J. Catalysis*, vol. 10, pp. 391–400, 1968.

Carey, V. P., *Liquid–Vapor Phase-Change Phenomena: An Introduction to the Thermophysics of Vaporization and Condensation Processes in Heat Transfer Equipment*, Hemisphere, Washington, 1992.

Churaev, N. V., "Contact angles and surface forces," *Colloid J.*, vol. 56, pp. 631–646, 1994.

DasGupta, S., I. Y. Kim, and P. C. Wayner, Jr., "Use of the Kelvin-Clapeyron equation to model an evaporating microfilm," *J. Heat Transfer*, vol. 116, pp. 1007–1015, 1994.

DasGupta, S., J. L. Plawsky, and P. C. Wayner, Jr., "Interfacial force field characterization in a constrained vapor bubble thermosyphon," *AIChE J.*, vol. 41, pp. 2140–2148, 1995.

Derjaguin, B. V., S. V. Nerpin, and N. V. Churayev, "Effect of film transfer upon evaporation of liquids from capillaries," *Bull. Rilem*, No. 29, pp. 93–98, 1965.

deFeijter, J. A., "Thermodynamics of thin films," in *Thin Liquid Films: Fundamentals and Applications*, I. B. Ivanov (ed.), pp. 1–47, Marcel Dekker, New York, 1988.

Dzyaloskinskii, I. E., E. M. Lifshitz, and L. P. Pitaevskii, "The general theory of van der Waals forces," *Adv. Phys.*, vol. 10, pp. 165–209, 1961.

Fisher, L. R., and J. N. Israelachvili, "Experimental studies on the applicability of the Kelvin equation to highly curved concave menisci," *J. Colloid Interface Sci.*, vol. 80, pp. 528–541, 1981.

Gee, M. L., T. W. Healy, and L. R. White, "Hydrophobicity effects in the condensation of water films on quartz," *J. Colloid Interface Sci.*, vol. 140, pp. 450–465, 1990.

Gibbs, J. W., *The Scientific Papers of J. W. Gibbs: Vol. One, Thermodynamics*, Dover Publications, New York, 1961.

Hamaker, H. C., "The London–van der Waals attraction between spherical particles," *Physica IV*, vol. 4, pp. 1058–1072, 1937.

Harkins, W. D., and A. Feldman, "Films. The spreading of liquids and the spreading coefficient," *J. Am. Chem. Soc.*, vol. 44, pp. 2665–2685, 1922.

Hallinan, K. P., H. C. Chebaro, S. J. Kim, and W. S. Chang, "Evaporation from an extended meniscus for nonisothermal interfacial conditions," *J. Thermophys. Heat Transfer*, vol. 8, pp. 709–716, 1994.

Holm, F. W., and S. P. Goplen, "Heat transfer in the meniscus thin-film transition region," *J. Heat Transfer*, vol. 101, pp. 543–547, 1979.

Israelachvili, J., *Intermolecular and Surface Forces*, 2nd ed., Academic Press, New York, 1992.

Joanny, J. F., and P. G. de Gennes, "Role of long-range forces in heterogeneous nucleation," *J. Colloid Interface Sci.*, vol. 111, pp. 94–101, 1986.

Kamotani, Y., "Evaporator film coefficients of grooved heat pipes," *A Collection of Papers, 3rd International Heat Pipe Conference, Palo Alto, CA*, pp. 128–130, AIAA, 1978.

Khrustalev, D., and A. Faghri, "Heat transfer during evaporation on capillary grooved structures of heat pipes," *J. Heat Transfer*, vol. 117, pp. 740–747, 1995.

Kim, I-Y., and P. C. Wayner, Jr., "The shape of an evaporating completely wetting extended meniscus," *J. Thermophys. Heat Transfer*, vol. 10, pp. 320–325, 1996.

Kistler, S. F., "Hydrodynamics of wetting," in *Wettability*, J. C. Berg, (ed.), Marcel Dekker, New York, 1993.

Klein, J., and E. Kumacheva, "Confinement-induced phase transitions in simple liquids," *Science*, vol. 269, pp. 816–819, 1995.

Knudstrup, T. G., I. A. Bitsanis, and G. B. Westermann-Clark, "Pressure-driven flow experiments in molecularly narrow, straight pores of molecular dimensions in mica," *Langmuir*, vol. 11, pp. 893–897, 1995.

Lay, J. H., and V. K. Dhir, "Shape of a vapor stem during nucleate boiling of saturated liquids," *J. Heat Transfer*, vol. 117, pp. 394–401, 1995.

Mantel, M., and J. P. Wightman, "Influence of the surface chemistry on the wettability of stainless steel," *Surface Interface Anal.*, vol. 21, pp. 595–605, 1994.

Majumdar, A., and C. L. Tien, "Effects of surface tension on film condensation in a porous medium," *J. Heat Transfer*, vol. 112, pp. 751–757, 1990.

Moosman, S., and G. M. Homsy, "Evaporating menisci of wetting fluids," *J. Colloid Interface Sci.*, vol. 73, pp. 212–223, 1980.

Parsegian, V. A., G. H. Weiss, and M. E. Schraeder, "Macroscopic continuum model of influence of hydrocarbon contamination on forces causing wetting of gold by water," *J. Colloid Interface Sci.*, vol. 61, pp. 356–360, 1977.

Potash, M., Jr., and P. C. Wayner, Jr., "Evaporation from a two-dimensional extended meniscus," *Int. J. Heat Mass Transfer*, vol. 15, pp. 1851–1863, 1972.

Reyes, R., and P. C. Wayner, Jr., "An adsorption model for the superheat at the critical heat flux," *J. Heat Transfer*, vol. 117, pp. 779–782, 1995.

Rowlinson, J. S., Translation of J. D. van der Waals, "The thermodynamic theory of capillarity under the hypothesis of a continuous variation of density," *J. Stat. Phys.*, vol. 20, pp. 197–244, 1979.
Sabisky, E. S., and C. H. Anderson, "Verification of the Lifshitz theory of the van der Waals potential using liquid-helium films," *Phys. Rev. A*, vol. 7, pp. 790–806, 1973.
Sadasivan, P., C. Unal, and R. Nelson, "Perspective: Issues in CHF modeling—The need for new experiments," *J. Heat Transfer*, vol. 117, pp. 558–567, 1995.
Stephan, P., and C. A. Busse, "Analysis of the heat transfer coefficient of grooved heat pipe evaporator walls," *Int. J. Heat Mass Transfer*, vol. 35, pp. 383–391, 1992.
Stephan, P., and J. Hammer, "A new model for nucleate boiling heat transfer," *Warme- und Stoffubertragung*, vol. 30, pp. 119–125, 1994.
Sharma, A., "Equilibrium contact angles and film thicknesses in the apolar and polar systems: Role of intermolecular interactions in coexistence of drops with thin films," *Langmuir*, vol. 9, pp. 3580–3586, 1993.
Sujanani, M., and P. C. Wayner, Jr., "Microcomputer-enhanced optical investigation of transport processes with phase change in near equilibrium thin liquid films," *J. Colloid Interface Sci.*, vol. 143, pp. 472–488, 1991.
Swanson, L. W., and G. C. Herdt, "Model of the evaporating meniscus in a capillary tube," *J. Heat Transfer*, vol. 114, pp. 434–441, 1992.
Swanson, L. W., and G. P. Peterson, "The interfacial thermodynamics of micro heat pipes," *J. Heat Transfer*, vol. 117, pp. 195–201, 1995.
Truong, J. G., and P. C. Wayner, Jr., "Effects of capillary and van der Waals dispersion forces on the equilibrium profile of a wetting liquid: Theory and experiment," *J. Chem. Phys.*, vol. 87, pp. 4180–4188, 1987.
Udell, K. S., "Heat transfer in porous media heated from above with evaporation, condensation, and capillary effects," *ASME J. Heat Transfer*, vol. 105, pp. 485–492, 1983.
van Oss, C. J., *Interfacial Forces in Aqueous Media*, Marcel Dekker, New York, 1994.
Vigil, G., Z. Xu, S. Steinberg, and J. Israelachvili, "Interactions of silica surfaces," *J. Colloid Interface Sci.*, vol. 165, pp. 367–385, 1994.
Wayner, P. C., Jr., Y. K. Kao, and L.V. LaCroix, "The interline heat transfer coefficient of an evaporating wetting film," *Int. J. Heat Mass Transfer*, vol. 19, pp. 487–491, 1976.
Wayner, P. C., Jr., "The interfacial profile in the contact line region and the Young-Dupré equation," *J. Colloid Interface Sci.*, vol. 88, pp. 294–295, 1982.
Wayner, P. C., Jr., "The effect of interfacial mass transport on flow in thin liquid films," *Colloids Surfaces*, vol. 52, pp. 71–84, 1991a.
Wayner, P. C., Jr., "Evaporation and stress in a small constrained system," in *Symposium on Coating Technologies for Imaging Materials*, J. Truong (ed.), IS&T, Springfield, VA, pp. 174–181, 1991b.
Wayner, P. C., Jr., "Intermolecular and surface forces with applications in change-of-phase heat transfer," in *Boiling Heat Transfer—Modern Developments and Advances*, R. T. Lahey Jr. (ed.), Elsevier Science, pp. 569–614, 1992.
Wayner, P. C., Jr., "Thermal and mechanical effects in the spreading of a liquid film due to a change in the apparent finite contact angle," *J. Heat Transfer*, vol. 116, pp. 938–945, 1994a.
Wayner, P. C., Jr., "Mechanical and thermal effects in the forced spreading of a liquid film with a finite contact angle," *Colloids Surfaces A*, vol. 89, pp. 89–95, 1994b.
Xu, X., and V. P. Carey, "Film evaporation from a micro-grooved surface—An approximate heat transfer model and its comparison with experimental data," *J. Thermophys. Heat Transfer*, vol. 4, pp. 512–520, 1990.
Yang, J.-X., J. Koplik, and J. R. Banavar, "Terraced spreading of simple liquids on solid surfaces," *Phys. Rev. A*, vol. 46, pp. 7738–7749, 1992.
Zisman, W. A., "Relation of equilibrium contact angle to liquid and solid constitution," in *Contact Angle, Wettability, and Adhesion*, R. F. Gould (ed.), pp. 1–51, American Chemical Society, Washington, DC, 1964.

PART TWO

APPLICATIONS

CHAPTER

SEVEN

THERMAL PHENOMENA IN SEMICONDUCTOR DEVICES AND INTERCONNECTS

Kenneth E. Goodson, Yongho Sungtaek Ju, and Mehdi Asheghi

Department of Mechanical Engineering, Stanford University, Stanford, California 94305

7-1 INTRODUCTION

The performance and reliability of integrated circuits are influenced by the transport of thermal energy in semiconductor devices, interconnects, and passivation. This chapter describes the operating principles of transistors and interconnects with a focus on the physics and practical relevance of thermal phenomena. A review of theories and data for the thermal conductivities of thin layers in circuits quantifies the strong influence of interfaces, impurities, and fabrication details. The hierarchy of transistor simulation tools, which range in complexity from the electron and phonon Boltzmann equations to drift-diffusion theory, are presented with reference to the device technologies where each is appropriate. This chapter also describes transistor and interconnect thermometry techniques, which include scanning probe methods with spatial resolution near 10 nm and optical methods with temporal resolution near 10 ns. The energy transport phenomena documented here are most important for electrostatic discharge (ESD) buffer circuitry, high-power transistors in control systems, high-speed silicon-on-insulator (SOI) and multilayer gallium-arsenide transistors, and multilevel interconnect structures in compact VLSI logic.

The integrated electronic circuit plays a central role in contemporary life and in our capacity for technological advancement. An integrated circuit consists of semiconductor devices, which control the flow of electrons and in some cases generate or detect electromagnetic radiation, and interconnects, which carry the resulting electrical current between devices. We rely on integrated circuit technology for information processing, transportation, communication, defense, health care, household convenience, and entertainment. Some of the more exciting recent advancements of integrated circuit technology are diminishing transportation-related accident rates and emissions through vehicle sensing and control, reducing the size and cost of medical diagnostics and support equipment, and providing inexpensive Internet access through the television.

Energy transport is a by-product of circuit operation. Semiconductor devices and interconnects function through the net transport of electrical charge by electrons. Net charge transport by electrons is accompanied by an increase in their kinetic energy, a large fraction of which is irreversibly transformed into thermal energy by means of electron scattering with other electrons and with the atomic lattice. The resulting thermal energy is transported by electrons and phonons, which are the quanta of lattice vibrational energy. Thermal energy transport does not directly influence the electrical currents flowing in an integrated circuit. But the thermal energy density and its spatial gradients in devices and interconnects, which govern their temperature distributions, can strongly influence both the electrical currents and the circuit lifetime. Since modern circuits consist of multilayer structures with dimensions as small as a few tens of nanometers, many of the relevant energy transport processes occur at micro- and nanoscales.

7-1-1 Integrated Circuit Technologies

Table 7-1 groups integrated circuits into three types, which are distinguished by their applications, figures of merit, and constituent semiconductor devices. Perhaps most familiar are the very-large-scale-integrated (VLSI) circuits found in computers, analog signal processors, video controllers, and random access memory. The sole purpose of these circuits is to manipulate information, a function that does not necessarily require large rates of heat generation. The use of lower operating voltages in VLSI circuits is therefore attractive for reducing the rate of heat generation and power consumption, particularly for portable systems. VLSI circuits achieve high operating frequencies through the use of very compact field-effect and bipolar transistors, in which the electron transit time decreases with decreasing separation between the transistor terminals.

Power integrated circuits provide power amplification and switching in the control systems for a broad variety of products, including vehicles, transmitters, and household appliances. These circuits transmit much electrical current to realize control functions and must often block large voltages. The peak power, voltage, and current handling capabilities are as important as the operating frequency. The transmission of power is fundamentally coupled with the generation of heat, which makes effective heat extraction essential to achieving attractive figures of merit. Many products, such as portable phone transmitters and engine control modules, benefit from continuing miniaturization, which provides strong motivation for reducing the dimensions of power circuits. Miniaturization is in many cases limited by the impact on reliability of the resulting temperature rise in the transistor. Power integrated circuits consist of relatively large field-effect and bipolar transistors, as well as a wealth of hybrid bipolar/field-effect devices.

Optoelectronic circuits generate and detect electromagnetic radiation for applications in lasers, detectors, and navigational systems. Many of these circuits include semiconductor lasers for the generation of visible or near-visible radiation and photodiodes for radiation detection. There is a wealth of thermal phenomena in the photonic semiconductor devices that are used in these circuits, which strongly influence their optical figures of merit (e.g., Agrawal and Dutta, 1993; Chen, 1995, 1996a). Other optoelectronic

Table 7-1 Types of integrated electronic circuits. In addition to the three types described here, there is a large body of research and applications for hybrid technologies, which integrate two or more of the circuit types onto a single chip. One example is *smart-power* circuit technology, which integrates VLSI and power integrated circuits for control systems. Other hybrids use novel bonding technology to integrate semiconductor lasers made from compound materials, such as gallium arsenide, together with VLSI silicon logic

Circuit type	Applications	Figures of merit	Semiconductor devices	Further reading
very-large-scale-integrated (VLSI) circuit	digital and analog processing, memory, video controllers	operating frequency, median lifetime, cost	compact FET, compact bipolar transistor	Veendrick (1992), Shur (1990), Tsividis (1987), Sze (1981), Ismail and Fiez (1994)
power integrated circuit	digital and analog power amplifiers, control systems for vehicles, medical equipment, and appliances	peak voltage, current, and electrical power, median lifetime (often in adverse conditions), cost	large-volume FET, bipolar transistor, and hybrid transistors	Kassakian et al. (1994), Bose (1992), Sze (1981)
optoelectronic circuit	semiconductor laser circuits, vehicle navigational systems	peak optical power, median lifetime, cost	IMPATT diode, laser diode, photodiode, compact FET	Chen (1995, 1996a), Agrawal and Dutta (1993), Shur (1990), Sze (1981)

circuits generate and detect electromagnetic radiation at much longer wavelengths using IMPATT and BARITT diodes for noncontact sensing applications, including velocity detection for smart-vehicle safety systems. The critical figure of merit for optoelectronic circuits is the peak electromagnetic power that can be generated from a circuit of a given volume, as well as the sensitivity and noise equivalent power associated with radiation detection. The maximum working distance of the long-wavelength sensing systems, for example, is governed by the electromagnetic power magnitude that can be reliably generated by their constituent diodes and in most cases is governed by thermal transport.

7-1-2 Relevance of Microscale Thermal Phenomena

Because thermal phenomena are not directly responsible for the electrical functionality of semiconductor devices and interconnects, they often receive only indirect consideration during circuit design. Circuit engineers determine the maximum acceptable circuit temperature through its expected impact on the current-voltage characteristics and reliability of devices and interconnects. The thermal energy transport processes that govern the maximum temperature include macroscopic convection, conduction in the chip packaging, and *microscale* heat generation and transport within and near semiconductor devices and interconnects. Only the last of these three processes falls within the scope of the present chapter. The transport processes occur with timescales and within regions of dimensions varying by orders of magnitude, as indicated by the thermal circuit in Fig. 7-1. Peak temperatures occur in semiconductor devices and interconnects, where heat is generated. The thermal resistances $R_{device\text{-}substrate}$ and $R_{interconnect\text{-}substrate}$ are governed by thermal conduction over a distance between 0.5 and 10 μm, and C_{device} or $C_{interconnect}$ is the effective heat capacity of the region heated significantly above $T_{substrate}$. The timescale for device or interconnect cooling varies between 0.01 and 5 μs. The substrate, which has thickness near 0.5 mm, conducts heat to a chip carrier with timescale comparable to 1 ms. Convection cools the chip carrier over length scales comparable to centimeters and timescales comparable to seconds.

Energy transport within micrometers of semiconductor devices and interconnects is important if $R_{device\text{-}substrate}$ or $R_{interconnect\text{-}substrate}$ is comparable to or larger than the other resistances in the system. In many electronic systems, including most VLSI digital logic, the macroscopic thermal resistances $R_{substrate\text{-}carrier}$ and $R_{carrier\text{-}room}$ dominate and therefore receive much attention (e.g., Aung, 1991; Bar-Cohen and Kraus, 1990; Fletcher, 1990). There are very important exceptions, including silicon-on-insulator (SOI) circuits (e.g., Colinge, 1991; Peters, 1993), whose buried silicon-dioxide layer strongly increases $R_{device\text{-}substrate}$ (e.g., Goodson et al., 1995a). The buried silicon dioxide offers important performance benefits for both VLSI and power circuits because it diminishes the contribution of the substrate to the electrical capacitance of transistors and interconnects and allows complete dielectric isolation of power and logic transistors. Another exception is the multilevel interconnect structures for very compact VLSI circuits, in which the resistance $R_{interconnect\text{-}substrate}$ is augmented by an increased thickness of low-thermal-conductivity passivation between the interconnect and the substrate (Hunter, 1995). The resistance $R_{device\text{-}substrate}$ can be significant in multilayer semiconducting structures, including those based on gallium arsenide, due to the close proximity and

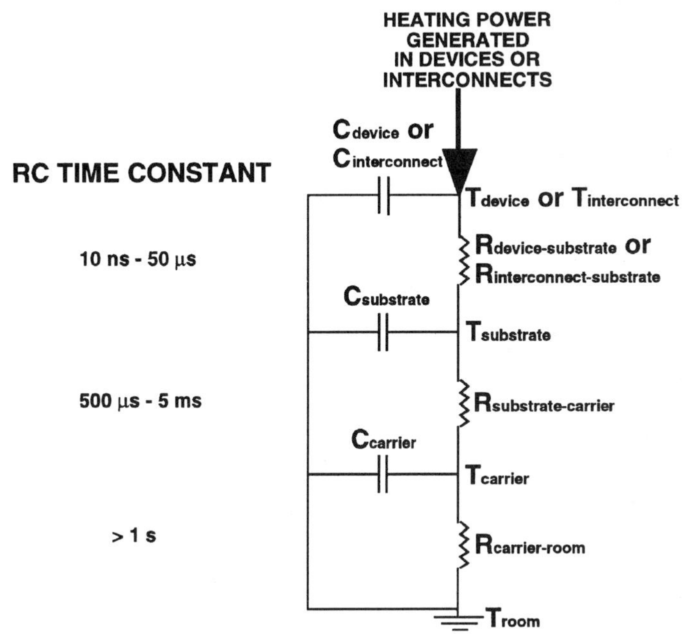

Figure 7-1 An approximate thermal circuit shows the hierarchy of thermal transport phenomena in electronic systems. Microscale thermal energy transport within or near devices and interconnects is dominant if the heating is brief and intense, such as occurs during electrostatic discharge (ESD) and the switching of some power transistors. Microscale thermal energy transport can also be important if either of the steady-state thermal resistances $R_{device\text{-}substrate}$ or $R_{interconnect\text{-}substrate}$ dominates over the macroscale resistances. This can be the case in multilevel interconnect structures, VLSI silicon-on-insulator (SOI) circuits, and in many power devices, particularly those made from silicon-on-insulator or gallium-arsenide substrates.

large number of interfaces in these materials (Chen et al., 1994). This is particularly important for photonic devices and high-frequency power amplifiers, whose figures of merit include the maximum power handling capability.

Figure 7-1 shows that microscale thermal energy transport is extremely important when interconnects and devices are subjected to brief, intense heating pulses. If a large quantity of heat is generated over a period of time less than $R_{device\text{-}substrate}\,C_{device}$ or $R_{interconnect\text{-}substrate}\,C_{interconnect}$, then heat travels only micrometers or less from devices and interconnects during the heating pulse. The energy must therefore be absorbed by a very small volume, which dramatically increases the temperature rise. Such brief heating phenomena are often the result of electrical overstress (EOS), which is a major reliability concern for both power and VLSI circuits. An important example is electrostatic discharge (ESD) (Greason, 1992; Amerasekera et al., 1992; Amerasekera and Verwey, 1992), which results from the contact of chip terminals with charged surfaces during manufacturing and packaging. The resulting current pulse into the terminal is of duration between about 1 and 150 ns and of magnitude as high as 5 A. Figure 7-2 shows optical micrographs of two interconnect structures that failed during ESD characterization experiments. The location of failed regions is influenced by the temperature distribution during the pulse, which is governed by microscale thermal transport and has

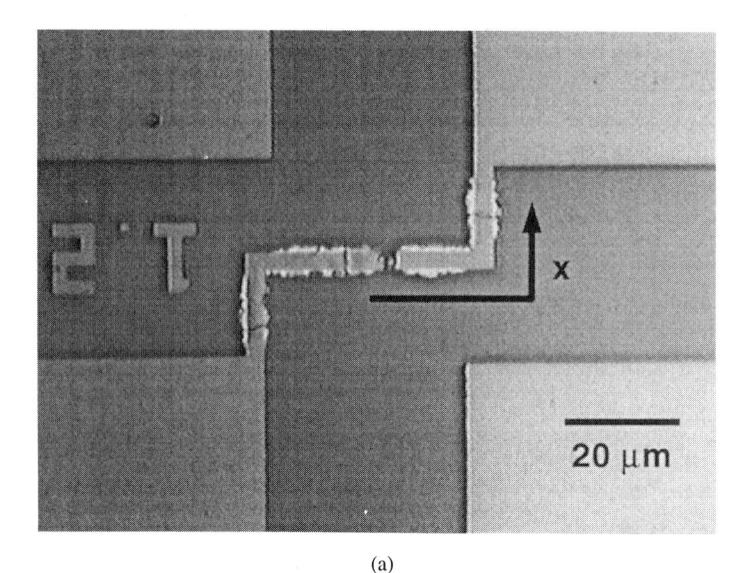

(a)

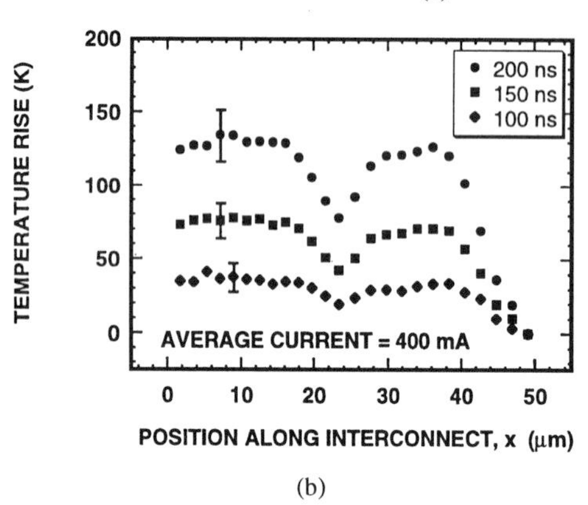

(b)

Figure 7-2 Impact of interconnect self-heating on short-timescale failure. This figure shows optical images (a and c) and transient temperature histories (b and d) of interconnects failed using pulsed currents to simulate electrostatic discharge (ESD). The interconnects were subjected to electrical current pulses of duration 250 ns (a and b) and 2.5 μs (c and d). The temperature distributions are plotted as functions of time after the pulse initiation, exhibiting temperature minima at the interconnect corners. The transient temperature distributions were captured using scanning laser-reflectance thermometry by Ju and Goodson (1996). The interconnect structures were provided by Intel Corporation.

been captured using a rapid thermometry technique described in Section 7-5-2. For pulse durations short compared to the heat diffusion time along the interconnects, the temperature rise and failure regions are distributed more uniformly along the interconnects. Transient heating is common in power circuits, which often switch large voltages or currents with timescales much less than 1 μs. Such circuits may be subjected to large fluctuations in driving voltage due to the transient demands placed on the power source. These fluctuations can have timescales between hundreds of nanoseconds to milliseconds

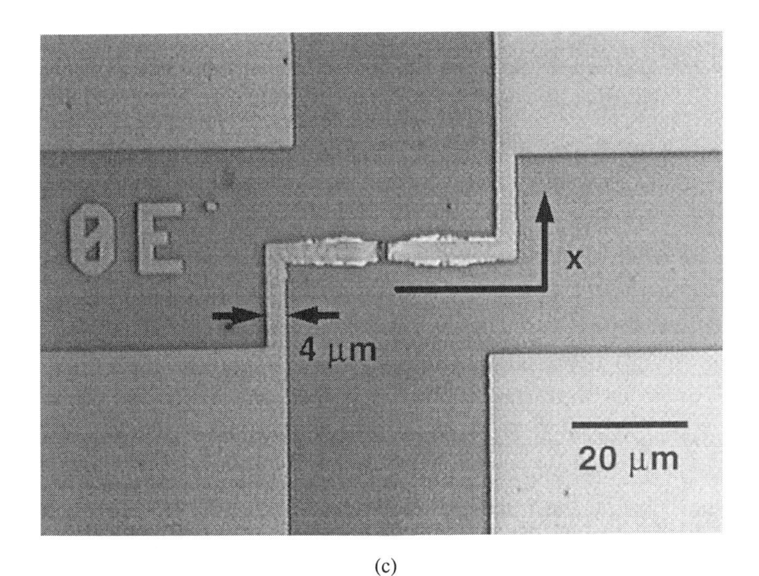

(c)

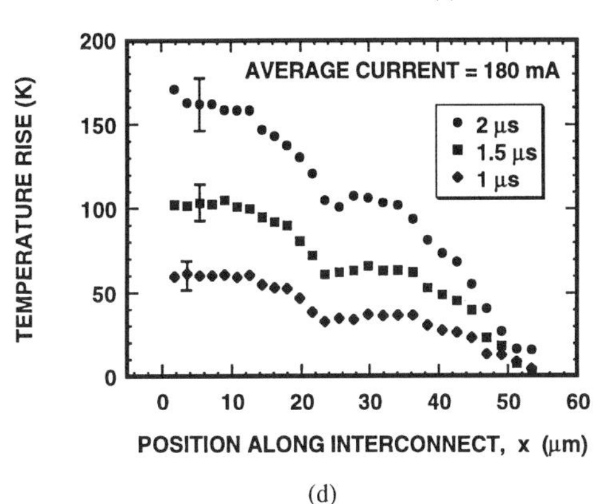

(d)

Figure 7-2 Continued.

and are often accompanied by severe heating. The spectrum of power requirements in an automobile, for example, range from the steady fractions of an ampere drawn across a few volts by microprocessors to the rather extreme needs of discrete Darlington transistors in many sparkplug control modules, which deliver several amperes across several tens of volts at frequencies near 600 Hz. The resulting current and voltage fluctuations are damped with the aid of heat generation in power transistors, which convert available battery power to levels manageable by compact digital logic.

Temperature increases at the circuit level must be carefully distinguished from microscale thermal energy transport. The macroscopic resistances yield a base temperature increase for devices and interconnects, above which the resistances $R_{device\text{-}substrate}$ and $R_{interconnect\text{-}substrate}$ induce micrometer-scale temperature gradients. Most integrated circuit design tools already consider the importance of a base temperature increase on performance and reliability. For example, the influence of a uniform temperature rise on the threshold voltage, electron mobility, and flat-band voltage of field-effect transistors is well documented for circuit modeling (e.g., Su et al., 1994). The impact of uniformly higher junction temperatures on the emitter and collector currents of bipolar transistors is large but well understood, as described in Section 7-2-3. The variation of the interconnect median time to failure resulting from a uniform temperature increase can be predicted using a wealth of data and empirical formulas that consider the grain structure and current density (e.g., Murarka, 1993). This chapter concerns itself with temperature gradients within semiconductor devices and interconnects, which are governed by microscale thermal energy transport.

7-1-3 Objectives

This chapter describes energy transport processes in semiconductor devices and interconnects and identifies the specific circuit technologies whose reliability and performance are most strongly influenced. The chapter also describes the experimental data and the theoretical and experimental tools that can be leveraged to understand these processes. The focus in this work is on thermal phenomena in transistors and interconnects in VLSI and power circuits (the first two types in Table 7-1), rather than those in the photonic devices that have been reviewed elsewhere (Chen, 1995, 1996a). Section 7-2 discusses the basics of charge transport and heat generation in transistors and diodes. Section 7-3 describes experimental data and theory for the thermal properties of insulating, semiconducting, and metallic regions of integrated circuits. Sections 7-4 and 7-5 describe the state of the art for thermal simulation and thermometry of transistors and interconnects.

7-2 BASICS OF CHARGE TRANSPORT AND HEAT GENERATION

This section provides basic information about the electrical operation of and the generation of heat in transistors and interconnects. This section does not attempt to provide a comprehensive description of the operation of these components, since this is available in several of the references cited in Table 7-1. In contrast, this section selects and often qualitatively presents key operating principles in an effort to communicate the necessary background more concisely than other sources.

7-2-1 Near-Equilibrium Transport Relations for Semiconductors

This subsection provides the basic transport relations governing the near-equilibrium operation of transistors. These will be supplemented in Sections 7-4-2 and 7-4-3 with relationships better suited for the far-from-equilibrium transport in many compact and power transistors. Chapter 1 describes the energy states of electrons responsible for charge and energy transport in semiconductors. The energies of these states form two distinct surfaces, called the conduction band and the valence band, when plotted in electron momentum space. The maximum energy in the valence band and the minimum energy in the conduction band are separated by the gap energy, E_g, which at room temperature is about 1.1 and 1.4 eV in silicon and gallium arsenide, respectively. The states in the valence band are mostly occupied near room temperature, and the states in the conduction band are mostly vacant. Electrons in the conduction band can be modeled as free particles with negative charge, which accelerate in the presence of an electric field and transport net quantities of charge and energy. The impact of the crystal lattice is considered through the use of an effective mass, which is different from the free electron mass and is related to the band shape in the vicinity of its minimum energy. The absence of an electron from the valence band results in transport through the net motion of electrons in the presence of an electric field. This can be described using positively charged *holes* with an effective mass governed by the shape of the valence band near its energy maximum. Due to the opposite signs of their electrical charge, electrons and holes accelerate in opposite directions in the presence of an electric field but contribute constructively to the flow of electrical current.

The Fermi energy, E_F, is an upper bound for the energies of occupied electron states in equilibrium at absolute zero temperature. It is also the reference energy for determining the probability that electron states with energy E are occupied in equilibrium at finite temperature, T. The occupation probability is given by the Fermi-Dirac distribution function

$$f_{EQ} = \frac{1}{1 + \exp\left(\frac{E-E_F}{k_B T}\right)} \tag{7-1}$$

where the Boltzmann constant is $k_B = 1.381 \times 10^{-23}$ J K^{-1}. For moderately doped semiconductors, the Fermi level lies between the minimum energy in the conduction band and the maximum energy in the valence band. Since the Fermi level is the chemical potential of electrons, spatial gradients in the Fermi level induce net particle flow.

Equation (7-1) is strictly valid only when the electrons in a semiconductor are in equilibrium, a requirement that is not satisfied in the presence of an electric field. But the electrons in the conduction and valence bands are usually closer to equilibrium when considered independently than is the ensemble. This can be exploited by treating charge transport due to electrons and holes independently, which is achieved by defining separate Fermi levels E_{Fn} and E_{Fp}, temperatures T_n and T_p, and electrical current density vectors $\mathbf{j}_n$ and $\mathbf{j}_p$ for electrons and holes, respectively. The Fermi levels can split even in the presence of modest electric fields. But the electron and hole temperatures are identical and equal to the atomic lattice temperature T_s unless an extremely large electric field or a photonic excitation has induced the electrons and holes to a highly nonequilibrium state. For the near-equilibrium transport described here, it is appropriate to use a single temperature T for the electrons, holes, and phonons. The use of multiple temperatures to describe transport in the presence of very large electric fields is discussed in Section 7-4. The occupation probabilities of the conduction and valence bands are computed using Eq. (7-1) with the separate values of the Fermi level.

Figure 7-3 shows the interaction of energy carriers in semiconductors. The distinction between the acoustic and optical phonons lies mainly in their group velocity, which is comparable to the speed of sound for acoustic phonons and is often negligible for optical phonons. Although both types of phonons scatter with electrons and holes, only the

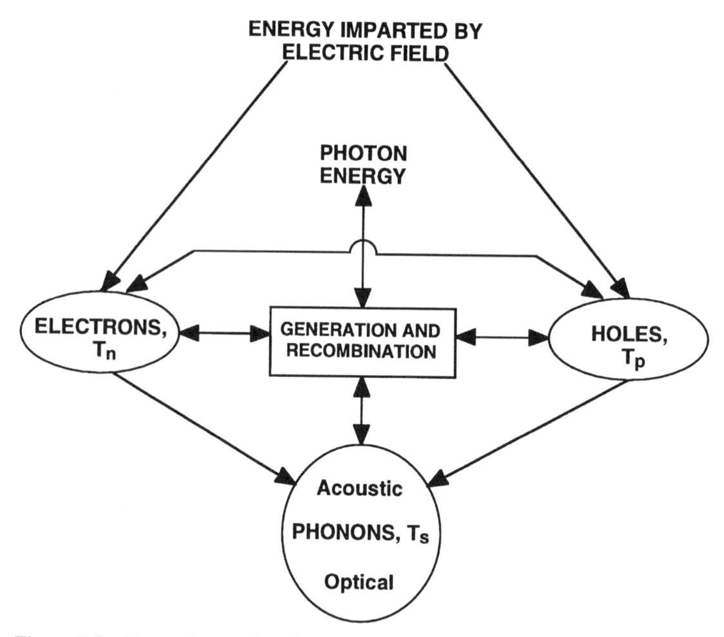

Figure 7-3 Energy-interaction diagram for semiconducting regions of a transistor, adapted from the work of Wachutka (1990). An electric field accelerates electrons and holes, which release energy to the atomic lattice through scattering with optical and acoustic phonons and through nonradiative recombination processes. Electron-hole recombination can also release photons or be induced by photon absorption.

acoustic phonons contribute strongly to thermal conduction. Acoustic phonons dominate thermal conduction in semiconductors, as is described in Section 7-3-2, and are therefore essential for removing the heat generated during the operation of semiconductor devices. The recombination and generation of electron and hole pairs releases and absorbs energy in an amount comparable to the gap energy E_g. The importance of photon generation and absorption depends strongly on the band structure of the semiconductor and is often negligible in VLSI and power integrated circuits.

The electrical current flowing in a semiconductor is modeled using independent components due to electrons and holes. The electron and hole currents are further subdivided into contributions due to an electric field and gradients in their concentration and temperature. The contributions resulting from an electric field, a concentration gradient, and a temperature gradient are called the *drift, diffusion*, and *thermoelectric* electron and hole currents, respectively. When the electric field magnitude is not too large, then electrical current densities are given by drift-diffusion theory:

$$\mathbf{j}_n = \mu_n n\,(\nabla E_{Fn} - qP_n \nabla T) \tag{7-2}$$

$$\mathbf{j}_p = \mu_p p(\nabla E_{Fp} - qP_p \nabla T) \tag{7-3}$$

where n and p are the electron and hole concentrations, respectively, μ_n and μ_p are the low-field electron and hole mobilities, and $q = 1.602 \times 10^{-19}$ C is the charge magnitude of electrons and holes. The total electrical current density is $\mathbf{j}_n + \mathbf{j}_p$. Chapter 1 introduced the mobility as the ratio of the drift velocity and the electric field. Because the absolute velocity of electrons increases with increasing temperature, there is net electron transport away from regions of higher temperature. Because of their negative charge, this net electron transport induces a thermoelectric current in the direction of a positive temperature gradient. Similarly, there is net hole transport away from regions of higher temperature. But in contrast to the case for electrons, the flow of holes induces a negative thermoelectric current in the direction of the temperature gradient due to their positive charge. The electron and hole thermoelectric powers P_n and P_p are therefore negative and positive, respectively, and are derived elsewhere (e.g., Sze, 1981). The drift and diffusion currents are described using a gradient in the Fermi level, which in its simplest form reduces to

$$\nabla E_{Fn} = q\left(\mathbf{E} + \frac{D_n}{\mu_n}\frac{\nabla n}{n}\right) \tag{7-4}$$

$$\nabla E_{Fp} = q\left(\mathbf{E} - \frac{D_p}{\mu_p}\frac{\nabla p}{p}\right) \tag{7-5}$$

where $\mathbf{E}$ is the electric field vector. The electron and hole diffusivities are approximately given by $D_n = \mu_n k_B T/q$ and $D_p = \mu_p k_B T/q$ when the Fermi levels are well outside of the conduction and valence bands, respectively. In the presence of an electric field $\mathbf{E}$, electrons and holes accelerate in opposite directions due to their opposite electrical charge but contribute constructively to the current as shown by Eqs. (7-2)–(7-5). In the absence of a temperature gradient, the electron or hole current density is the sum of a drift current density, which is proportional to the electrical field, and a diffusion current density, which is proportional to the gradient in the electron or hole concentration. In an isothermal semiconductor, an electron or hole current must be accompanied by a gradient in the Fermi level of that carrier.

The electron concentration is an integral over all of the electron states per unit volume in the conduction band of the occupation probability of those states given by Eq. (7-1). When the criterion $(E - E_{Fn})/k_B T \gg 1$ is satisfied, which is frequently the case in the active regions of semiconductor devices, this integral yields an electron concentration per unit volume

$$n = N_c \exp\left[-\frac{(E_c - E_{Fn})}{k_B T}\right] \tag{7-6}$$

where the effective density of electron states at the conduction band edge is

$$N_c = 2\left(\frac{2\pi m_n^* k_B T}{h_P^2}\right)^{3/2} \tag{7-7}$$

and $h_P = 6.625 \times 10^{-34}$ J s is Planck's constant. The effective electron mass m_n^* is defined specifically for calculations of the density of states and considers multiplicity of equivalent minima in the conduction band and anisotropy of the band in wavevector space (e.g., Shur, 1990). The concentration of holes in the valence band can be calculated using a similar approach, yielding

$$p = N_v \exp\left[-\frac{(E_{Fp} - E_v)}{k_B T}\right] \tag{7-8}$$

where the effective density of hole states near the valence band edge is

$$N_v = 2\left(\frac{2\pi m_p^* k_B T}{h_P^2}\right)^{3/2} \tag{7-9}$$

and m_p^* is the effective mass for calculating the density of states of holes. At $T = 300$ K the effective densities of states are $N_c = 2.8 \times 10^{19}$ cm^{-3} and $N_v = 1.04 \times 10^{19}$ cm^{-3} for silicon and $N_c = 4.7 \times 10^{17}$ cm^{-3} and $N_v =$ and 7.0×10^{18} cm^{-3} for gallium arsenide. The concentrations of electrons and holes in a semiconductor are controlled in part by doping. Dopant atoms that can contribute a free electron to the conduction band, such as arsenic and phosphorus in silicon, are called donor impurities and have concentration denoted by N_D. Atoms that can absorb an electron from the valence band, such as boron atoms in silicon, are called acceptor impurities and have concentration denoted by N_A.

To provide a closed set of equations for near-equilibrium modeling, the electric field must be coupled to the electron, hole, and impurity concentrations using the Poisson equation,

$$\nabla \cdot (\varepsilon \mathbf{E}) = q(p - n + N_D - N_A) \tag{7-10}$$

where ε is the absolute permittivity of the semiconductor. Although the impurity concentrations in the Poisson equation denote only those that are ionized, these quantities are very nearly equal to N_D and N_A at room temperature for common impurities in silicon and gallium arsenide. We can also write number-conservation equations for electrons and holes

$$q\frac{\partial n}{\partial t} = \nabla \cdot \mathbf{j}_n + r_g - r_r \tag{7-11}$$

$$q\frac{\partial p}{\partial t} = -\nabla \cdot \mathbf{j}_p + r_g - r_r. \tag{7-12}$$

The balances consider generation and recombination of electron-hole pairs through their rates per unit volume r_g and r_r, respectively. These rates can account for Auger and Shockley-Read-Hall recombination processes and for impact ionization (e.g., Sze, 1981). In equilibrium, when $\mathbf{E} = 0$ and $\mathbf{j}_n = \mathbf{j}_p = 0$, the Fermi level $E_F = E_{Fn} = E_{Fp}$ and electron and hole concentrations are calculated from Eqs. (7-6)–(7-10) alone. For this case, Eq. (7-10) collapses to the charge neutrality condition, $p - n + N_D - N_A = 0$. In the presence of an electric field, the electron and hole Fermi energies, current densities, and concentrations and the electric field distribution can be determined through simultaneous solution of Eqs. (7-2)–(7-12) with appropriate boundary conditions. The electron-hole generation and recombination rates must be determined independently (e.g., Sze, 1981).

An energy equation is required to account for nonisothermal conditions. Figure 7-3 shows that electrons and holes transfer energy to the atomic lattice through scattering and nonradiative recombination. For near-equilibrium transport in semiconducting regions large compared to the free paths of phonons, the energy balance is the conventional heat-diffusion equation

$$C\frac{\partial T}{\partial t} = \nabla \cdot (k\nabla T) + q''' \tag{7-13}$$

where k is the thermal conductivity and C is the heat capacity per unit volume. The thermal conductivities of semiconducting regions in transistors are discussed in Section 7-3-2. Many different expressions can be found in the literature for the volumetric heat generation rate q'''. This dilemma was addressed by Wachutka (1990), who derived

$$q''' = \left[\frac{|\mathbf{j}_n|^2}{qn\mu_n} + \frac{|\mathbf{j}_p|^2}{qp\mu_p}\right] + (r_r - r_g)[E_{Fn} - E_{Fp} + qT(P_p - P_n)] - T[\mathbf{j}_n \cdot \nabla P_n + \mathbf{j}_p \cdot \nabla P_p] - S_R \tag{7-14}$$

Equation (7-14) is consistent with the assumptions of the drift-diffusion theory but is strictly valid only for steady-state simulations. The first term on the right accounts for *resistive* heating due to electron and hole scattering processes and dominates, for example, in field-effect transistors. The second term on the right accounts for heat released due to the net *recombination* of electron-hole pairs and is most important for diodes and bipolar transistors. The third term on the right is the rate of *Thomson heating*, which results from current flow along a gradient in the thermoelectric power. Thomson heating is most important near gradients in the impurity concentration, which influence P_n and P_p. The final term on the right S_R is the net rate of radiation power emission and is most important in direct-bandgap semiconductors, such as gallium arsenide, where electron-hole recombination can stimulate photon emission without participation of a phonon. The radiation term can be neglected for many silicon devices.

7-2-2 Compact Field-Effect Transistors

The building block of most VLSI digital logic circuits is the compact field-effect transistor (FET) shown in Fig. 7-4. The compact FET is also important in a variety of analog circuits, including high-frequency signal processors (e.g., Ismail and Fiez, 1994). Although the FET can be made from many semiconducting materials, including germanium, gallium arsenide, silicon carbide, and even diamond, it is by far most common in silicon. Gallium arsenide is attractive for transistors in the fastest microwave and supercomputer circuits due to the high velocities achieved by electrons and holes in this material. An important lateral dimension of the compact FET is the channel length L_c, which for the best contemporary desktop computers is approaching 0.2 μm as this book goes to press. The lateral dimensions of silicon FETs for digital logic have decreased steadily for three decades, by about one order of magnitude every 15 years (e.g., Sze, 1988). Reducing the lateral dimensions of FETs—in particular the channel length—helps to decrease their electrical capacitance and the electron transit time through the devices. This leads to shorter clock periods and faster computing for digital logic and allows analog circuits to be operated at higher frequencies. Reducing the lateral dimensions of FETs also decreases the substrate area occupied by each transistor, yielding chips that have more transistors and more functionality.

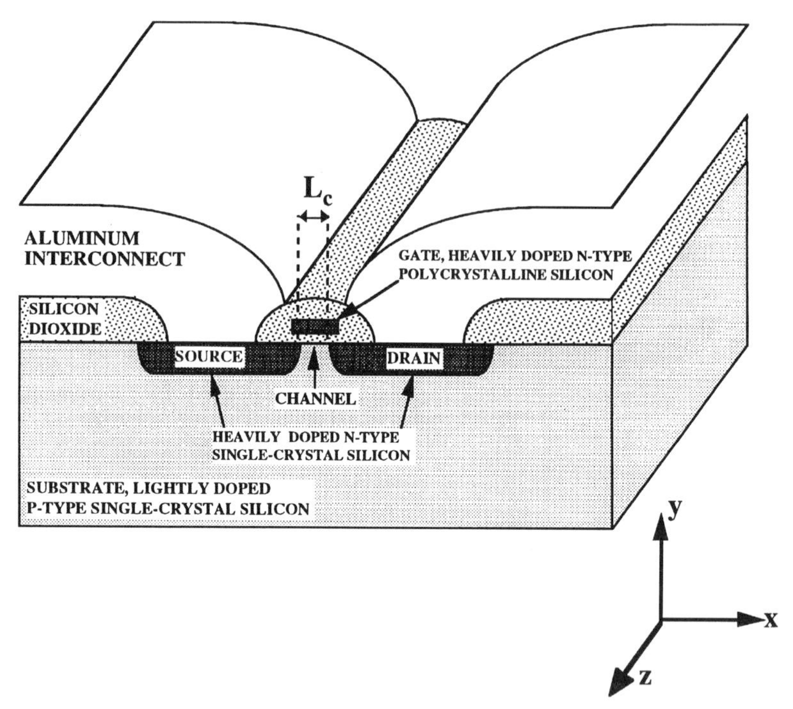

Figure 7-4 Cross-sectional schematic of an n-type field-effect transistor (FET). These devices are also called metal-oxide-semiconductor field-effect transistors (MOSFETs) because the gate can be made from metal. The four terminals of this device are the source and drain for electrons and the gate and substrate. The voltage drop from the drain to the source is V_D, and the voltage drop from the gate to the substrate is V_G. The source and the substrate are often grounded. When the transistor is saturated, resistive heat generation occurs predominantly in the channel region near the drain. The various mechanisms of heat generation are introduced after Eq. (7-14).

A simple and clear introduction to the silicon FET and VLSI digital logic circuits was provided by Veendrick (1992). More detailed descriptions of the FET and the problems associated with its miniaturization were provided by Sze (1981), Tsividis (1987), Yang (1988), and Shur (1990). The following description is for the n-type silicon transistor shown in Fig. 7-4, in which electrons dominate current flow from the drain to the source. The operating principles described here are essentially the same for p-type transistors, in which holes dominate the electrical current, and for transistors made of other semiconducting materials. Some field-effect transistors use no electrical insulator between a metal gate and the channel, but rather exploit the highly nonlinear resistance of the metal–semiconductor contact to prevent gate-channel current leakage. The FET controls electron flow much as a mechanical valve controls the flow of a fluid. The drain-source voltage drop, V_D, is analogous to the pressure drop across the fluid valve, and the gate-substrate voltage drop, V_G, is analogous to the fluid valve position between the extremes of open and closed. When a given potential drop is sustained between the drain and source of the n-type transistor in Fig. 7-4, the resulting electrical current between the drain and the source, I_D, is very small for $V_G = 0$ and increases as V_G approaches a few volts.

The absolute drain-source electrical conductance of an n-type FET, $G_D = I_D/V_D$, is governed by the concentration of free electrons in the semiconductor immediately beneath the gate. For completeness we define also the differential conductance

$$g_D = \left(\frac{\partial I_D}{\partial V_D}\right)_{V_G} \tag{7-15}$$

and the differential transconductance

$$g_T = \left(\frac{\partial I_D}{\partial V_G}\right)_{V_D} \tag{7-16}$$

which are used to modulate signals in analog circuits. The very small drain conductance for $V_G = 0$ is caused by the use of a different impurity type in the channel than is used in the source and drain. The channel of the silicon n-type FET typically has $N_A \sim 10^{17}$ cm^{-3} and $N_D \ll N_A$, which yields far more holes than electrons in the channel, $p \gg n$. In contrast, the source and drain have $N_D \sim 10^{19}$ cm^{-3} and $n \gg p$, which yields no continuous path where the electrons are dominant. The source-channel and channel-drain interfaces serve as p-n junction diodes, one achieves a reverse-bias condition when a voltage is sustained between the source and the drain. Section 7-2-3 shows that in the reverse bias condition of a junction diode, both electrons and holes are depleted from the vicinity of the junction and current flows only through a very large electrical resistance. The source-drain electrical conductance of the transistor is therefore quite small when $V_G = 0$.

A qualitative understanding of the influence of V_G on the electron concentration can be obtained by considering the electrical capacitor formed by the gate, the insulator, and the channel. When the gate voltage increases with respect to the substrate voltage, the gate/insulator and the insulator/channel interfaces become positively and negatively charged, respectively. This increases the concentration of negatively charged electrons in the channel. The gate voltage can be increased such that the electron concentration dominates over that of holes, forming a channel for electron transport and eliminating the two p-n junctions. This condition defines the threshold gate-substrate voltage drop, V_T, above which the gate is said to cause *inversion* in the channel. Inversion is accompanied by depletion of holes from a region surrounding the channel. A drain-source voltage drop V_D induces electron flow from the source to the drain. The resulting drain-source current can be approximately calculated using the analogy to an electrical capacitor, which yields the *charge-control model* (e.g., Shur, 1990). In the simplest form of this model, the drain-source current is

$$I_D = \mu_n C_{ox} \frac{W}{L_c}\left[(V_G - V_T)V_D - \frac{V_D^2}{2}\right], \quad 0 < V_D < V_{D,sat} \tag{7-17}$$

$$I_D = I_{D,sat}, \quad V_D \geq V_{D,sat}. \tag{7-18}$$

The low-electric-field electron mobility is μ_n, the electrical capacitance per unit area of the gate/insulator/channel composite is C_{ox}, and the transistor width is L_c. The

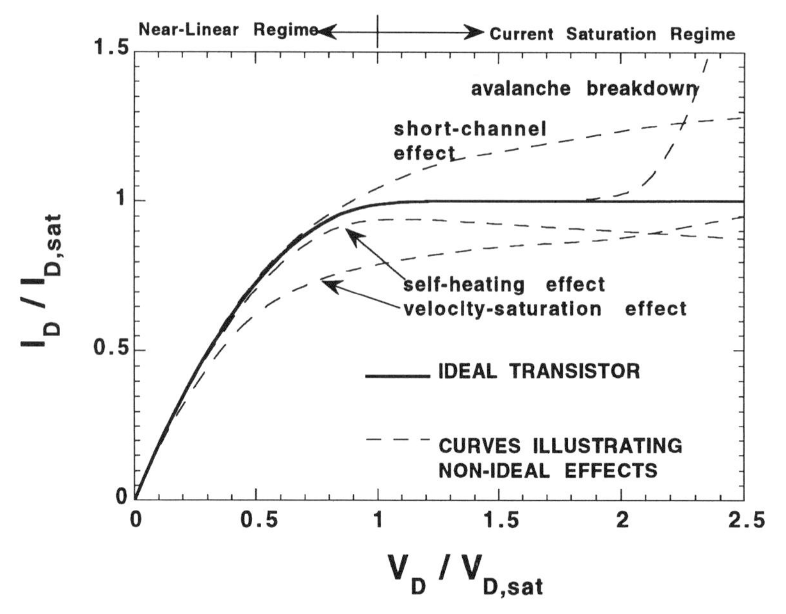

Figure 7-5 Current-voltage characteristics of an n-type FET. The relationship for an ideal transistor is calculated using Eqs. (7-17)–(7-20). Most transistors exhibit a complex combination of the nonideal effects illustrated qualitatively using the dashed curves. The *saturation regime* for the transistor, which describes the condition when the drain-source current varies slowly with changes in the drain-source voltage, must be distinguished from carrier *velocity saturation*. Current saturation results from the elimination of the inversion layer in the vicinity of the drain and occurs even in the absence of carrier velocity saturation.

saturation values of the drain-source voltage and drain-source current both vary with the gate-substrate voltage,

$$V_{D,sat} = V_G - V_T \tag{7-19}$$

$$I_{D,sat} = \mu_n C_{ox} \frac{W}{L_c} \frac{(V_G - V_T)^2}{2}. \tag{7-20}$$

Figure 7-5 plots Eqs. (7-17)–(7-20) in nondimensional form and indicates with dashed curves the deviations from this simple formula, which are discussed below. The absolute drain-source conductance, I_D/V_D, is relatively independent of V_D when $V_D \ll 2(V_G - V_T)$. When this condition is satisfied, I_D varies linearly with V_D and the transistor is said to be in the *linear* regime. When V_D is comparable to V_G, the electron concentration in the channel varies strongly with the lateral coordinate, x, in Fig. 7-4. The voltage across the gate/insulator/channel capacitor is much smaller on the drain side of the device than on the source side, such that the electron concentration is less on the drain side. When the drain voltage is high enough to eliminate inversion on the drain side, the transistor is in the *pinch-off* condition. This reduces the source-drain conductance and causes I_D to vary slowly with V_D. For a transistor operating in this *saturation regime*, the portion of the channel without an inversion layer has very few electrons and therefore dominates both the source-drain electrical resistance and the heat generation in the transistor. The importance of heat generation in this region is evident in Eq. (7-14), which shows that resistive heating increases as the concentration of the dominant charge carrier is reduced. The temperature distribution in a transistor operating in saturation therefore reaches a maximum near the drain side of the channel.

The current-voltage relationship in practical FETs can differ substantially from the idealized relationship predicted in Eqs. (7-17)–(7-20), as indicated qualitatively by the dashed curves in Fig. 7-5. The strong scattering of electrons by optical phonons in the presence of large electric fields causes electron *velocity saturation*, which reduces the carrier mobility and $I_{D,sat}$. For FETs with channel lengths L_c that are less than about 0.5 μm, the simplest charge-control model begins to fail because of the lateral depletion of holes in the vicinity of the heavily doped source and drain regions. This *short-channel effect* decreases the threshold voltage and diminishes the effective channel length, both of which lead to higher drain-source current for a given drain-source voltage. *Avalanche breakdown* can occur in the channel near the drain where the electric fields are largest and results from the formation of electron-hole pairs. Breakdown can inject high-energy electrons into the gate insulator, which permanently depletes the drain region and induces hot-carrier effects on the transistor current-voltage behavior. Another cause of deviation from the solid curve in Fig. 7-5 is the *self-heating effect*. The power deposited in the channel, $P = I_D V_D$, increases with increasing voltage. The temperature rise in the channel also increases, which changes the phonon and electron concentrations. The dominant effect is a reduction in the electron mobility due to an increased scattering rate on phonons, which reduces the drain-source current as indicated in Fig. 7-5. The actual current-voltage behavior of transistors can be simultaneously influenced by many of the effects discussed here.

The relationship between V_G and the source-drain conductance can be interpreted using the shape of the electron energy bands in the channel and Eqs. (7-2) and (7-6). The conduction and valence bands in a semiconductor bend spatially in the presence of an electric field. If this bending occurs in the absence of current flow normal to the channel, which is enforced by the highly resistive gate insulator, then Eq. (7-2) states that for nearly isothermal conditions the Fermi levels are identical and nonvariant in space. This is shown in Fig. 7-6a. The figure also shows that the relative position of the Fermi level between the two bands changes. A positive voltage at the gate of a FET induces a positive field away from the gate within the channel. This bends the conduction band closer to the Fermi level at the insulator/channel interface and is responsible for the inversion of the carrier charge type according to Eq. (7-6). When V_D is appreciable compared to $V_G - V_T$, the voltage in the drain region induces significant electrical current in the direction normal to the insulator and splits the Fermi levels for electrons and holes. This reduces the electron concentration according to Eq. (7-6) and causes the pinch-off condition. Figure 7-6b illustrates the spatial bending of the conduction and valence bands in the channel near the drain. The larger difference between the electron Fermi level and the conduction band-edge energy strongly reduces the electron concentration according to Eq. (7-6).

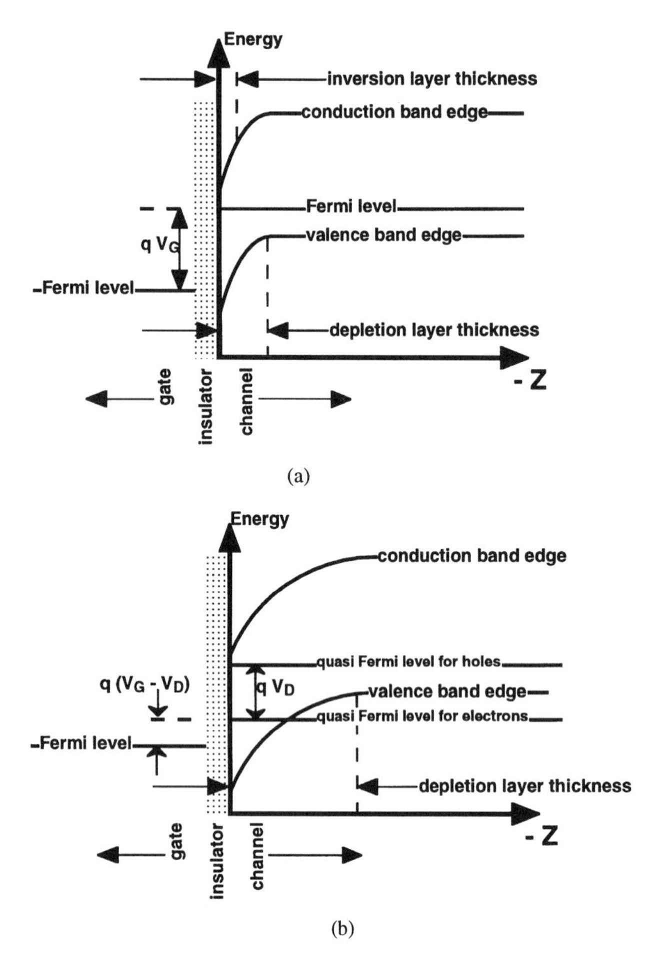

Figure 7-6 Band diagrams in the gate and semiconductor at two sides of the channel in a n-type FET in the saturation regime: a) Source side and b) drain side. The n-type drain forms a p-n junction with the p-type substrate, which induces splitting of the electron and hole Fermi levels in the presence of a voltage V_D.

7-2-3 Compact Diodes and Bipolar Transistors

The bipolar transistor combines two p-n junction diodes, one of which controls the concentration of charge carriers available for transport through the second. First consider the operating conditions of a single p-n junction diode, which are depicted in Fig. 7-7. A diode is the junction of regions doped predominantly with acceptor and donor impurities, respectively. There are large gradients of the concentrations of both electrons and holes near the interface, which induce the diffusion of holes from the p-type (acceptor impurity) region to the n-type (donor impurity) region and diffusion of electrons in the opposite direction. Since the carriers are oppositely charged, these flows both contribute to a positive current from the p-type side to the n-type side. The *external bias voltage* between the p-type side and the n-type side, V_{pn}, is defined to be zero in the absence

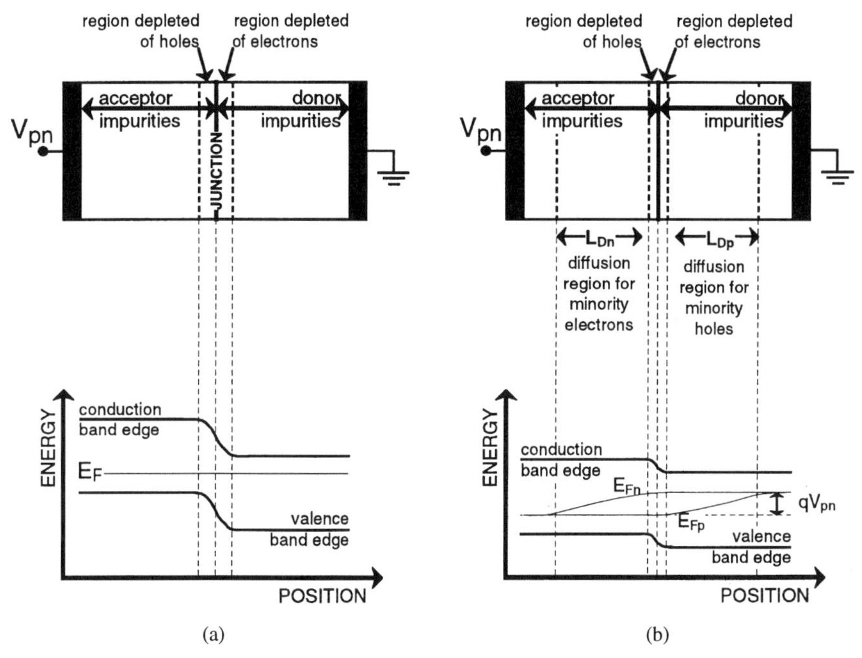

Figure 7-7 Cross-sectional schematics and energy diagrams for an idealized diode structure in the a) zero-bias condition and the b) forward-bias condition. In the forward-bias condition, the depletion regions are compressed and there is a net current flow from the p-type region (acceptor impurities) to the n-type region (donor impurities). The Fermi level splits within the diode to quasi Fermi levels for electrons and holes.

of current flow. In this equilibrium condition, current flow is eliminated by an internal *built-in voltage* V_{bi}, which biases the n-type side positively with respect to the p-type side and is given by

$$qV_{bi} = E_g + k_B T \ln\left(\frac{N_A N_D}{N_c N_v}\right). \tag{7-21}$$

The impurity concentrations N_A and N_D are those in the p-type and n-type sides of the junction, respectively, which are usually less than the effective band-edge densities of electron and hole states. For diodes with at least one side heavily doped, the built-in voltage is comparable to but smaller than the gap energy divided by the electron charge magnitude. The built-in voltage increases with increasing concentration of impurities on either side of the junction. The diffusion of carriers through the junction induces a depletion region on both sides of the p-n junction. In the zero-bias condition, there is no net current flow and the Fermi level does not vary spatially. This can be seen from the energy diagram at the bottom of Fig. 7-7a.

If the diode is forward biased, $V_{pn} > 0$, a current flows from the p-type side to the n-type side. The Shockley equation for the current density through the diode is calculated by examining the diffusion of minority carriers away from the junction outside of the depletion region as they recombine with majority carriers. There is very little electrical field outside of the depletion region, so the diffusion currents in Eqs. (7-2) and

(7-3) dominate. As shown in Fig. 7-7b, the gradients in the minority carrier concentrations prevail over diffusion lengths L_{Dn} and L_{Dp} on each side of the junction, which are large compared to the thicknesses of the depletion regions. The recombination lengths are $L_{Dp} = (D_p \tau_{rp})^{1/2}$ and $L_{Dn} = (D_n \tau_{rn})^{1/2}$, where τ_{rn} and τ_{rp} are the recombination times for electrons and holes. The magnitudes of the hole and electron fluxes follow simple diffusion theory and are approximately $D_p \Delta p / L_{Dp}$ and $D_n \Delta n / L_{Dn}$, respectively, where the diffusion constants are given after Eqs. (7-4) and (7-5). The parameter Δp is the difference in the hole densities at the boundaries of the diffusion region on the n-type side of the junction, and Δn is the difference between the electron densities at the boundaries of the diffusion region on the p-type side. The sum of the electron and hole current densities yields the *Shockley equation* for the total current density magnitude from the p-type side to the n-type side:

$$j = j_s \left[\exp\left(\frac{q V_{pn}}{k_B T} \right) - 1 \right] \tag{7-22}$$

where the saturation current density for reverse-bias conditions is

$$j_s = q N_c N_v \exp\left(\frac{-E_g}{k_B T} \right) \left[\frac{D_p}{L_{Dp} N_D} + \frac{D_n}{L_{Dn} N_A} \right]. \tag{7-23}$$

Equations (7-22) and (7-23) provide a qualitative description of the current-voltage characteristics of diodes, but are not in quantitative agreement with silicon and gallium arsenide diodes at low values of the forward bias. This results because many diodes have dimensions comparable to the electron and hole diffusion lengths. Other problems result from the fact that Eqs. (7-22) and (7-23) neglect the recombination of carriers *within* the depletion region and the often appreciable electric field in the diffusion regions. Diodes subjected to high electric fields experience avalanche breakdown and diodes with very large impurity densities can have strongly modified current-voltage characteristics due to carrier tunneling. In contrast to a FET, in which heat generation is almost entirely resistive, diodes experience significant quantities of heat generation due to the recombination of electron-hole pairs. The resistive and recombination heating is distributed throughout the diffusion regions at rates that diminish with decreasing separation from the junction. Recombination in silicon requires participation of the lattice for momentum conservation, which results in energy transfer to the phonon population. Direct bandgap semiconductors such as gallium arsenide can release the energy in the form of photons, which is the operating principle of semiconductor lasers.

Figure 7-8 is a cross-sectional schematic of a bipolar transistor. The emitter/base and collector/base interfaces are each junction diodes. The diodes form a bipolar transistor when the collector-base and emitter-base junctions are separated by a distance that is less than the minority electron recombination length in the base region. Under normal operation, the emitter-base junction is forward biased and the collector-base junction is reverse biased, that is, the base-emitter voltage drop for the transistor in Fig. 7-8 should satisfy $V_{be} > 0$, and the collector-base voltage drop should satisfy $V_{cb} > 0$. The forward-biased emitter/base junction injects minority electrons into the base. The electrons diffuse towards the collector-base junction and are ultimately swept through it by the potential drop V_{cb}. The current from the emitter to the collector is

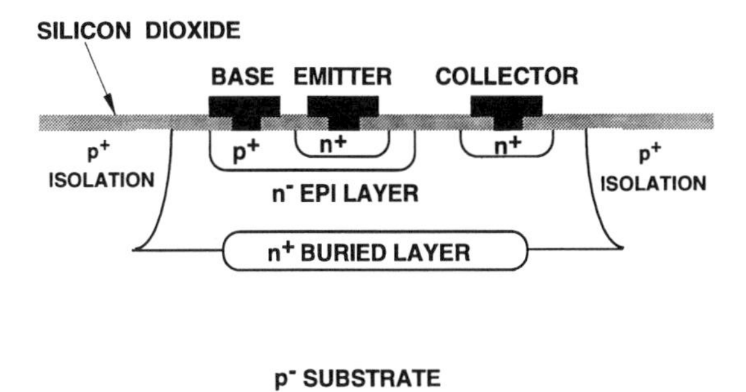

Figure 7-8 Cross section of a bipolar transistor, which integrates two junction diodes. During normal operation, the base-emitter junction is forward biased with $V_{be} > 0$. The voltage difference between the collector and the base is V_{cb}, which during normal operation is reverse biased with $V_{cb} > 0$. Recombination and resistive heat generation occurs in this device in the vicinity of both junctions.

approximately

$$I_e \approx I_c \approx \frac{qD_n A_j N_c N_v}{SN_A} \exp\left(\frac{qV_{be} - E_g}{k_B T}\right) \tag{7-24}$$

where D_n is the electron diffusion constant in the base, A_j is the cross-sectional area of the emitter-base junction normal to Fig. 7-8, S is the separation between the emitter and collector, and N_A is the acceptor concentration in the base. The currents I_e and I_c are those into the emitter and out of the collector, respectively. The first term in the numerator of the argument of the exponential function accounts for the fact that the number density of minority carriers in the base region increases rapidly with the increasing forward bias voltage of the base-emitter junction. The second term accounts for the increasing activation of carriers with increasing temperature. The collector current varies exponentially with the base-emitter voltage, yielding a transconductance

$$g_T = \left(\frac{\partial I_c}{\partial V_{be}}\right)_{V_{cb}} \tag{7-25}$$

which can be much larger than that in a FET. The current I_e is slightly larger than I_c due to holes injected from the base into the emitter. These are responsible for the external current into the base, I_b, which is much smaller than I_c due to the larger impurity concentration of donors in the emitter than acceptors in the base. This allows the transistor to serve as a current amplifier with gain

$$\beta = I_c/I_b \gg 1. \tag{7-26}$$

The common-emitter current gain β is achieved by referencing the input and output potentials of the transistor with respect to the emitter voltage. The heat generation rate in a bipolar transistor is strongly influenced both by the recombination of electron-hole pairs and the scattering of minority carriers as they diffuse within the base.

7-2-4 Power Transistors

The peak power handling capability in a transistor can be limited by the maximum temperature rise that can be sustained without breakdown. For this reason, power transistors usually occupy a much larger volume. This decreases the temperature rise for a given rate of heat generation by increasing the thermal capacitance of the device, which is relevant for the short-timescale heating phenomena discussed in Section 7-1-2. Increasing the volume also increases the effective cooling area for the device, which increases the conductance between the device and the environment and diminishes the temperature rise for heating with longer timescales.

The peak operating current is increased by augmenting the total width of active regions, W, which increases the cross-sectional area through which current may flow. Power analog amplifier FETs, for example, can have channel width/length ratios, W/L_c, greater than 100. The short channel length allows the devices to operate at high frequencies, and the large width increases the current. One method of further increasing W is through the use of multiple interwoven fingers on the surface of a substrate for two potentials of the transistor, such as the source and gate terminals of a FET or the base and emitter terminals of a bipolar transistor. The collector for the bipolar transistor or the drain for the FET is achieved by multiple interwoven fingers or through a contact at the bottom of the substrate. An important disadvantage of power bipolar transistors must be discussed in this context. If one base finger has a larger temperature than the others, then the strong temperature dependence of the collector current relationship in Eq. (7-24) shows that the current can be drawn more strongly through that finger. This further increases its temperature and current leading to thermal instability and the possibility of localized failure. Liu and Bayraktaroglu (1993) and Liou and Bayraktaroglu (1994) discuss this problem for heterojunction bipolar transistors.

Many power transistors achieve high breakdown voltages by increasing the separation between the terminals where current flows in and out of the device. High-voltage field-effect transistors make use of an extended *drift region* between the active inversion layer and the drain to increase the breakdown voltage. In a n-type silicon FET, for example, the drift region is lightly doped with donor impurities and separates the channel region, which is lightly doped with acceptor impurities, from the drain. Increasing the length of the drift region decreases the electric field sustained in this region for a given value of the total drain-source voltage drop, V_D. Since the failure of the transistor due to breakdown results from a large electric field (i.e., the voltage change per unit length), this allows much larger voltage differences to be sustained across the transistor terminals. But extending the drift region also dramatically increases the total electrical resistance of the device when it sustains a large drain-source current. In most high-voltage field-effect transistors, the drift region dominates the total heat generation in the device.

The traditional approach to achieving a long drift region is to place one contact on the back side of a chip, which yields a discrete vertical transistor. Two of the terminals remain at the top surface, such that the transistor effect in the device occurs within fractions of a micrometer of the top surface of the wafer. But the voltage drop is sustained through the entire thickness of the wafer. The disadvantage of these discrete, full-wafer devices is that they cannot be easily integrated into a single chip containing other power devices or logic. An alternative power transistor configuration uses large lateral dimensions to achieve a high breakdown voltage, as in the field-effect transistor shown in Fig. 7-9. This configuration facilitates integration of multiple power devices and even low-power digital logic onto a single silicon chip. The device shown in Fig. 7-9 is made using silicon-on-insulator (SOI) technology (e.g., Colinge, 1991), which electrically isolates

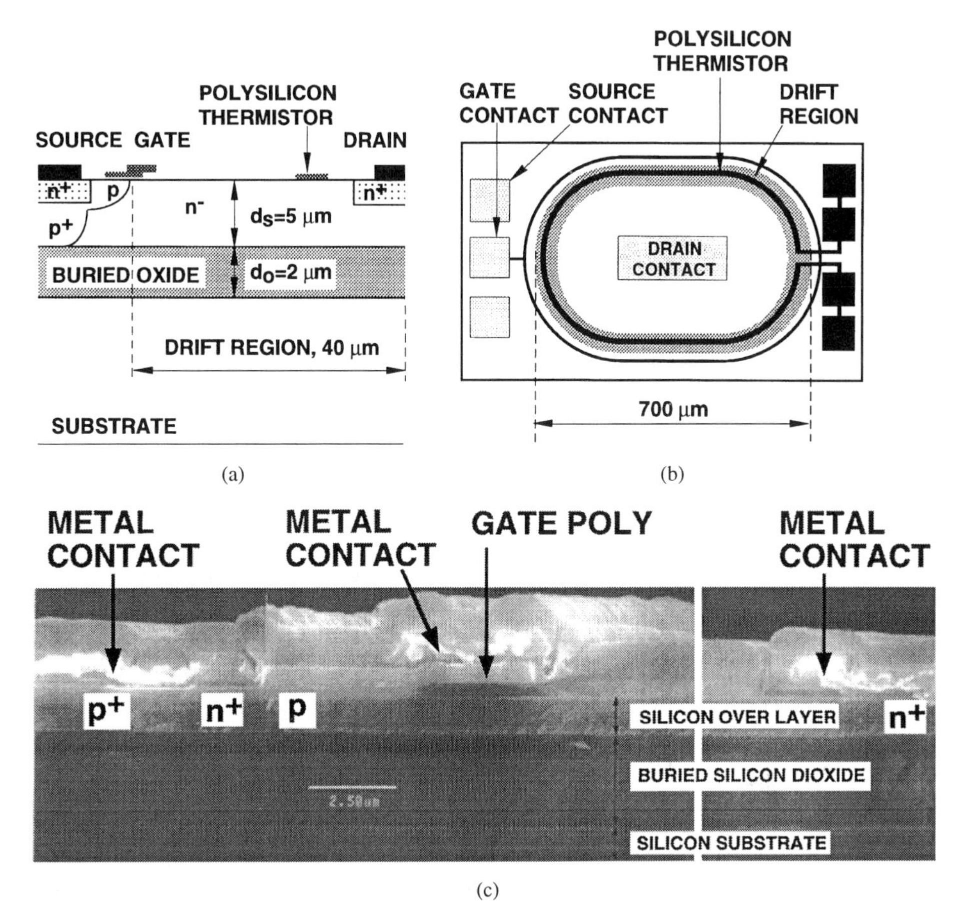

Figure 7-9 Schematics and electron micrograph of the high-power silicon-on-insulator (SOI) field-effect transistor of Leung et al. (1995): a) Schematic cross section, b) schematic top view, c) cross-sectional electron micrograph. The device shown here includes a patterned polysilicon thermistor for thermometry, which is not part of a standard device. The thermometry approach is described in Section 7-5-1. Heat generation in this device is predominantly resistive and within the extended drift region.

silicon active regions from the substrate by means of a buried silicon-dioxide layer. This particular device sustains several Watts of power in the on state (i.e., when $V_G \sim 10$ V) and can block hundreds of volts across the drain-source terminals in the off state. Figure 7-9b shows that the channel traverses a loop in the plane of the substrate, which reduces current leakage around the edges of the device. Figure 7-9c shows an electron micrograph of the cross section of the transistor, which provides information about the actual surface roughness features and step coverage of layers of the transistor.

7-2-5 Interconnects

Interconnects are patterned metal or highly doped semiconducting layers that transmit electrical currents between semiconducting devices. Their design aims to minimize both the temporal delay and the electrical resistance associated with this transmission.

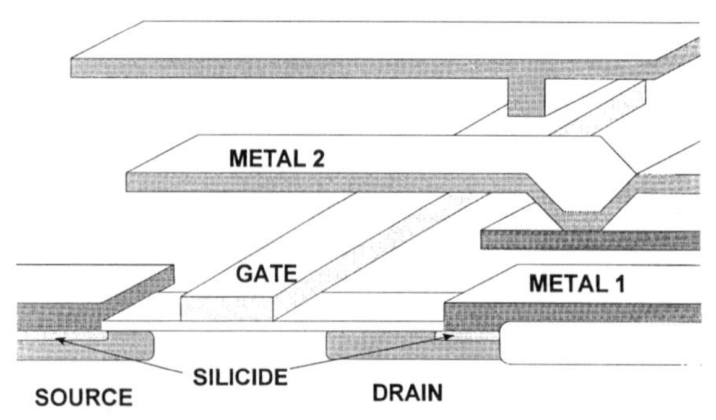

Figure 7-10 Multilayer interconnect structure in a compact VLSI circuit. The interconnects are surrounded by passivation, which is made at present from silicon dioxide or silicon nitride. The vertical interconnects between successive interconnect layers are called *vias*.

Interconnects range in length from several micrometers to several centimeters and are often woven between multiple layers of insulation. Figure 7-10 provides a schematic of multiple interconnection layers in a VLSI circuit in the vicinity of a compact FET. The miniaturization of VLSI circuits necessitates multiple interconnect layers, which lead to increased separation between interconnects and the substrate heat sink. Because the insulation surrounding the interconnects is typically a poor conductor of heat, as discussed in Section 7-3-1, this leads to much higher interconnect temperatures for a given rate of heat generation.

The materials, processing, and reliability of multilevel interconnects were discussed by Murarka (1993) and Wilson et al. (1993). The material choice must balance the need for a low electrical resistance with the need to prevent atomic diffusion in circuits, which occurs near interfaces. The material choice also influences the resistance of interconnects and interconnect-semiconductor contacts to electromigration failure. *Electromigration* is the motion of interconnect atoms in the direction of electron flow, which occurs at a rate that is facilitated by higher temperatures. The gate in silicon circuits is usually made from heavily doped polycrystalline silicon, as shown in Fig. 7-4. Contacts to transistors and higher-level interconnection are typically made from aluminum with a few atomic percent of impurities to impede the flow of vacancies. Diffusion of silicon into aluminum at contacts is usually inhibited by sputtered barrier layers made from silicides or one of several titanium-based compounds. Diffusion barriers are also common between the gate and the gate insulator and between higher-level aluminum interconnects and passivation. Gold is the interconnection of most gallium-arsenide circuits and is alloyed with germanium, zinc, nickel, or germanium in the vicinity of contacts to impede diffusion.

The performance figures of merit for an interconnect are its electrical resistance and the temporal delay associated with sending a current pulse along its length. The electrical resistance is $R_i = (\rho L_i)/(w_i d_i)$, where w_i and d_i are the width and height of the interconnect cross section and L_i is the length. Kinetic theory relates the electrical resistivity

$$\rho = \frac{m}{nq^2\tau_n} = \frac{m}{nq^2}\left(\frac{1}{\tau_{n-s}} + \frac{1}{\tau_{n-i}} + \frac{1}{\tau_{n-gb}}\right) \approx \frac{m}{nq^2}\left(\frac{1}{\tau_{n-s}} + v_F S_i n_i + \frac{v_F}{d_{gb}}\right) \quad (7\text{-}27)$$

to the free-electron effective mass m, charge magnitude q, number density n, and relaxation time τ_n. Equation (7-27) breaks the electron scattering rate, $1/\tau_n$, into components due to electron scattering on phonons, $1/\tau_{n-s}$; localized imperfections such as impurities and vacancies, $1/\tau_{n-i}$; and grain boundaries, $1/\tau_{n-gb}$. Equation (7-27) also shows that the scattering rate due to localized imperfections is proportional to their concentration n_i by means of the scattering cross section S_i. This linear dependence breaks down when the impurities are alloy atoms with concentration comparable to that of the host atoms (e.g., Ziman 1960). Equation (7-27) also shows that the resistivity increases with decreasing characteristic grain dimension, d_{gb}, due to the increasing rate of electron scattering on grain boundaries. The scattering rate increases also with the electron Fermi velocity discussed in Chapter 1, $v_F \sim 10^6$ m s^{-1} for most metals. Equation (7-27) shows that large grains and few impurities lead to improvements in the resistive figure of merit for interconnects. Section 7-3-3 discusses the coupling between the electrical and thermal conductivities in interconnects, which renders the scattering mechanisms discussed here relevant for thermal energy transport.

The temporal delay for a voltage signal to traverse an interconnect is often governed by the electrical capacitor formed by the interconnect and neighboring conducting regions. A semiconductor device must charge this capacitor before it can sustain its steady-state output voltage. For most practical interconnect structures, calculation of the total effective capacitance of an interconnect per unit length requires solution of the three-dimensional Poisson equation. But for the simplest case of a relatively wide interconnect that interacts only with the substrate below, the capacitance is $C_i \approx L_i w_i \varepsilon_{ins}/d_{ins}$, where ε_{ins} and d_{ins} are the absolute permittivity and the thickness of the insulator separating the substrate from the interconnect, respectively. The delay is

$$t_d = R_i C_i = \frac{\rho \varepsilon_{ins}}{d_i d_{ins}} L_i^2 = \frac{R_s \varepsilon_{ins}}{d_{ins}} L_i^2 \tag{7-28}$$

where $R_s = \rho/d_i$ is the sheet resistance of the layer from which the interconnect is patterned. The importance of the permittivity motivates research on novel dielectric materials with low values of this property compared to silicon dioxide, including a variety of organic compounds (e.g., Wilson et al., 1993). For interconnects much shorter than 1 mm, the *capacitive delay* in Eq. (7-28) may be less than the *propagation delay*, which is governed by the finite velocity of a voltage wave in an interconnect. The velocity is governed by the interconnect impedance and is typically within an order of magnitude of the speed of light in a vacuum. The reliability figure of merit for interconnects is the median time to failure, which is limited by electromigration, stress-induced voiding, atomic diffusion, and other phenomena.

7-3 THERMAL TRANSPORT PROPERTIES

A barrier to the precise thermal simulation of integrated circuits is the lack of knowledge of the thermal properties of their constituent layers and interfaces. The thermal conductivities of thin layers can differ substantially from those of bulk materials. Thin layer and interfacial properties are sensitive to the details of the fabrication process of an integrated circuit, which are difficult to reproduce. These problems have motivated much recent experimental and theoretical work on thin-layer and interface thermal properties for integrated circuits. This section summarizes the progress that has been made and provides guidelines for its application in circuit design.

Integrated circuits consist of three different types of patterned layers that are distinguished by their electrical transport properties. *Electrically insulating layers*, such as silicon dioxide, silicon nitride, and polymers, are used to electrically isolate conducting regions and protect them from corrosion. Because of their highly disordered atomic structure, these layers are poor conductors of heat with room-temperature thermal conductivities comparable with or less than 1 W m^{-1} K^{-1}. The active regions in a circuit are made from crystalline *semiconducting layers*, such as silicon, silicon carbide, and gallium arsenide. These layers have relatively high thermal conductivities, of the order of 100 W m^{-1} K^{-1}. *Metal layers*, which serve as interconnects between semiconductor devices, have thermal conductivities in the range from 50–400 W m^{-1} K^{-1}, with the highest values reserved for material with large grains and a few atomic percent or less of impurities.

Techniques for measuring the thermal conductivity in thin layers for integrated circuits were reviewed by Goodson and Flik (1994) and by Cahill (1997). The thermal conductivity measured for layers in integrated circuits is not necessarily a property of the layer material. Thermal boundary resistances and electron and phonon scattering on the layer and grain boundaries cause the apparent conductivities to depend on the direction of heat propagation, even for isotropic materials, on the layer thickness, on the size and orientation of grains, and on the properties of the neighboring materials and interfaces. Measurements yield an *effective* thermal conductivity, which is valid only for a given layer thickness and direction of heat transport. For crystalline and polycrystalline semiconducting and metal layers, which have relatively high thermal conductivities, the in-plane conductivity $k_{a,eff}$ is usually most important because these layers govern lateral heat conduction in multilayer structures. An important exception is for semiconductor devices consisting of multiple layers of semiconducting materials (e.g., gallium-arsenide quantum-well devices), where heat must travel normal to many layers and interfaces. For layers made of materials that are poor thermal conductors, such as the amorphous insulating layers in integrated circuits, the conductivity in the normal direction $k_{n,eff}$ is of greatest importance because these layers dominate the resistance for conduction in that direction.

7-3-1 Electrically-Insulating Layers

The low thermal conductivities of electrically-insulating layers cannot be attributed to a fundamental limitation of dielectric materials. In crystalline or polycrystalline form, electrically-insulating layers have thermal conductivities that approach or even exceed those of metals. Thermal conduction by insulating layers is due to phonon transport, which can be the most effective form of heat conduction in near-perfect crystals. Diamond layers of thickness greater than 10 μm, for example, serve both as electrical insulators and as excellent carriers of heat at room temperature, with thermal conductivities exceeding 1000 W m^{-1} K^{-1} (e.g., Graebner, 1993). The problem with electrically insulating layers in integrated circuits is their disordered atomic structure, which scatters phonons so strongly that it is questionable even to invoke formal transport theory. In the case of silicon dioxide and silicon nitride layers, this disorder is unavoidable using deposition temperatures compatible with the other layers and interfaces in integrated circuits.

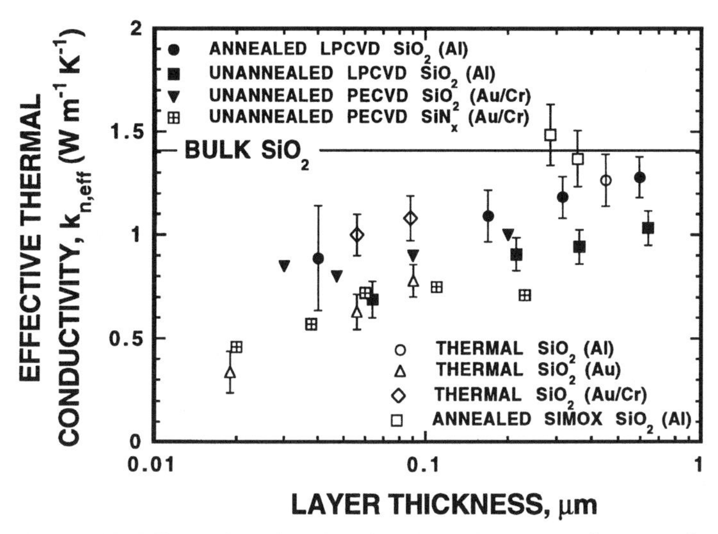

Figure 7-11 Effective thermal conductivities for conduction normal to conventional electrically insulating layers in integrated circuits. The fabrication process and the metallization for each set of data is indicated in the figure. The data for the silicon-nitride layers were measured by Lee and Cahill (1996). The data for the LPCVD silicon-dioxide layers were measured by Goodson et al. (1993a). The data for SIMOX and thermally grown silicon dioxide layers with aluminum coatings were measured by Goodson et al. (1994). Thermally grown silicon-dioxide layers with gold and gold/chromium coatings were measured by Käding et al. (1994).

Figure 7-11 plots $k_{n,eff}$ data as a function of layer thickness for silicon dioxide and silicon nitride layers fabricated using integrated circuit technology. The deposition technology is by far most mature for silicon dioxide, which has been the passivation in silicon-based circuits for several decades (e.g., Adams, 1988). While the data are shown to vary by nearly one order of magnitude, it is helpful to observe that for a given fabrication technique the conductivity decreases with decreasing layer thickness. The data also indicate that at a given layer thickness, the thermal conductivity increases with increasing values of the peak temperature achieved during the fabrication process. The highest conductivities for layers of thickness near 0.4 μm, for example, were measured in silicon dioxide fabricated using Separation by Implantation with Oxygen (SIMOX, e.g., Cellar and White, 1992). The SIMOX process implants oxygen atoms into silicon, which form a disordered oxygen/silicon mixture layer. The disordered layer can be healed into near-stoichiometric silicon dioxide during an anneal near 1500 K for several hours. The next highest conductivities were measured in silicon dioxide grown through thermal oxidation of the silicon wafer, which is performed at temperatures in excess of 1000 K. The dependence of $k_{n,eff}$ on the metallization observed by Käding et al. (1994) can be attributed to the poor adhesion of gold to silicon dioxide in the absence of a chromium layer. The chromium improves physical contact between the materials, which reduces the boundary resistance and increases $k_{n,eff}$. The next highest conductivities were measured in low-pressure chemical vapor deposited (LPCVD) oxides deposited near 900 K. The data show that a subsequent anneal near 1100 K increases the conductivity significantly at a given layer thickness. The lowest observed conductivities are for silicon dioxide fabricated using a plasma-enhanced chemical vapor deposition (PECVD) process, which permits deposition at temperatures as low as 650 K.

When analyzing data for $k_{n,eff}$ of disordered passivation layers, it is often useful to assume that thermal resistances at the layer boundaries, R_{B1} and R_{B2}, are in series with a volume resistance within the layer, d/k_{int}. The total thermal resistance is

$$R_T = \frac{d}{k_{n,eff}} = \frac{d}{k_{int}} + R_{B1} + R_{B2} = \frac{d}{k_{int}} + R_B \tag{7-29}$$

where $R_B = R_{B1} + R_{B2}$ is the sum of the thermal resistances at both boundaries. The Fourier equation, from which Eq. (7-29) is derived, fails for processes in which the scattering of phonons on the layer boundaries is significant when compared with scattering within the layer. This problem has been considered for silicon dioxide by Goodson et al. (1994), who showed that boundary scattering is significant at room temperature only for layers thinner than about 500 Å. The problem of phonon boundary scattering is discussed in the context of diamond layers later in this section. The simplest manner in which to apply Eq. (7-29) is to assume that the layer microstructure and interfacial properties which govern k_{int} and R_B, respectively, depend on the processing conditions, but *not* on the layer thickness. This allows k_{int} and R_B to be extracted from a set of resistance data for layers of varying thicknesses fabricated under similar conditions (e.g., Goodson et al., 1993a). This approach is valid only if the material is relatively homogeneous between its boundaries. Each set of silicon-dioxide data in Fig. 7-11 can be fit with reasonable accuracy using Eq. (7-29) and values with values of k_{int} that depend on the fabrication process and values of R_B comparable to 10^{-8} m^2 K W^{-1}.

Variations in the thermal conductivity k_{int} with the maximum processing temperature can be approximately attributed to changes in the density of the material. Measurements of the room-temperature mass density of LPCVD silicon dioxide layers showed that annealing and rapid cooling in air cause this property to increase (Nagasimi, 1972; Smolinsky and Wendling, 1985). For LPCVD oxides deposited near 650 K, Nagasimi (1972) observed an increase of about 10 percent for a layer annealed at 1273 K. The lower density of the unannealed layers may be due to an Angstrom-scale microstructure different from that of the annealed layers. An alternative hypothesis is that chemical vapor deposition introduces macroscopic pores into the layers, which have characteristic dimensions much larger than the atomic spacing. By assuming pores of dimension greater than 50 Å, which allowed the used of the Fourier equation, the conductivity is approximately (Goodson et al., 1993b)

$$\frac{k_{int}}{k_{bulk}} = 0.457 + 1.10\left(\frac{T_P}{T_r}\right) - 0.557\left(\frac{T_P}{T_r}\right)^2 \tag{7-30}$$

where $T_r = 1305$ K. For annealed layers, T_P is the annealing temperature. For unannealed layers, a very rough approximation can be obtained using the deposition temperature for T_p. While this result qualitatively explains some of the data shown in Fig. 7-11, in particular the dependence of the LPCVD silicon-dioxide conductivities on annealing, the very approximate nature of Eq. (7-30) must be emphasized. For example, the densification process depends strongly on the duration of an anneal, which is not considered. Furthermore, if the densification occurs on an atomic scale, the solution of the heat equation around pores is incorrect. Another problem with the modeling is that silicon dioxide layers fabricated using LPCVD contain up to 4 mass percent of silanol (SiOH), as observed by infrared spectroscopy (Adams, 1988). Impurities account for significant differences in the measured thermal conductivities of bulk fused silicon dioxide. The effect of impurities on the thermal conductivity of LPCVD layers needs to be investigated.

There is a growing interest in the use of unconventional electrical insulators in integrated circuits (e.g., Wilson et al., 1993). Among these are organic materials, such as polymers, which are attractive due to their relatively low permittivities. This minimizes temporal delay according to Eq. (7-28). In bulk form, these materials have thermal conductivities that are lower than those of silicon dioxide passivation. More detailed study is needed for thin films. The impact of the passivation on temperature fields is rendered particularly important by the very large thermal expansion coefficients of polymers, which augment thermally induced mechanical stress in interconnect multilayer systems. Another novel passivation is CVD diamond, which offers benefits due to its high thermal conductivity and high breakdown electric field (Goodson, 1995). The thermal conductivity in diamond layers is strongly nonhomogeneous and anisotropic (Goodson, 1996). For contemporary layers of thickness less than 1 μm, values of $k_{n,eff}$ as large as 100 W m^{-1} K^{-1} have been observed using interconnect structures for the measurements (Goodson et al., 1995b). Figure 7-12 plots the thermal resistance for conduction normal to thin diamond layers as a function of layer thickness. The data

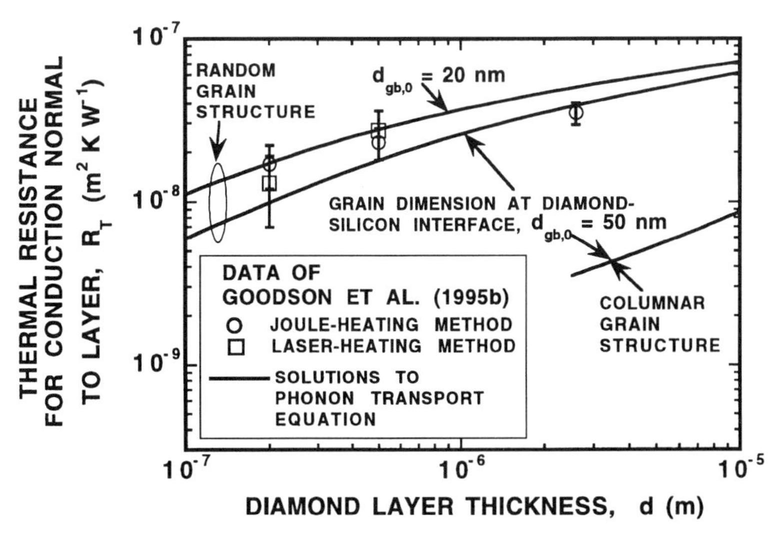

Figure 7-12 Predictions and data for the effective thermal resistance for conduction normal to thin polycrystalline diamond passivation on silicon (Goodson, 1996). Although the data yield thermal conductivities that are considerably less than those in bulk diamond, the resistances are much smaller than those for silicon-dioxide and silicon nitride layers of similar thickness. Diamond passivation offers one possibility for reducing the importance of heat generation in high-density interconnects for compact VLSI circuits. The challenges for realizing this potential lie mainly in the reliable fabrication of composite diamond-silicon-aluminum structures.

are compared with predictions based on the phonon Boltzmann equation to account for scattering of the phonons on the layer boundaries. The higher thermal conductivity owes much to the microstructure of the layers, which is polycrystalline rather than amorphous. The improved microstructure requires relatively high deposition temperatures, which are not compatible with integrated circuit technology after aluminum deposition. Diamond deposition research is making much progress in reducing the processing temperature, which promises to make diamond a more feasible passivation for possible use in IC technology.

7-3-2 Semiconducting Layers

Thermal conduction in semiconductors is dominated by phonon transport even in the presence of large concentrations of electrons or holes. The thermal conductivity in active semiconducting regions in integrated circuits is reduced compared to that of bulk samples by scattering mechanisms not present in the bulk material, such as those depicted in Fig. 7-13. Phonon-boundary scattering is particularly important at low temperatures, where the mean free path would otherwise become arbitrarily large. While phonon-boundary interactions govern phonon transport in any sample at sufficiently low temperature, the reduction is more severe and extends to higher temperatures for confined regions in semiconductor devices. Also important is phonon scattering on imperfections, which can be introduced as part of integrated circuit fabrication processes. Finally, the impurities and additional free carriers in doped semiconducting regions impede heat transport compared to that in bulk intrinsic silicon. There are data available for the thermal conductivity of bulk doped samples (Touloukian et al., 1970), which show a strong reduction for impurity concentrations comparable to those in heavily doped regions of integrated circuits. However, it is not clear if the bulk data are appropriate for active regions in semiconductor devices that are doped using thin-film implantation and diffusion techniques. Due to the highly nonequilibrium situation in the active semiconducting regions, the relative concentrations of impurities and charge carriers can differ vastly from those that prevail in the bulk. Phonon wave confinement may become important when the dominant phonon wavelength is comparable to the layer thickness. This occurs in silicon layers of thickness 500 Å below about 10 K, for example. Confinement reduces

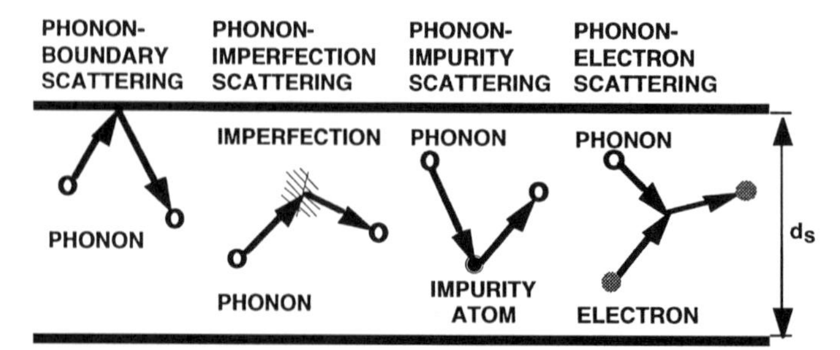

Figure 7-13 Scattering mechanisms that limit phonon conduction in semiconductors. Phonon transport dominates the thermal conductivity of semiconducting materials even in the presence of large concentrations of free electrons and holes.

the states available for transport by phonons of wavelength comparable to or longer than the layer thickness.

Most of the available data for the thermal conductivity of semiconductor layers was measured for samples that were in polycrystalline form. Data for doped polycrystalline silicon layers indicate a reduction of up to 80 percent compared to those available for bulk silicon (Paul et al., 1993, 1994; Von Arx and Baltes, 1992; Von Arx et al., 1995; Mastrangelo and Müller, 1988; Tai et al., 1988). It is likely that the grain boundaries are responsible or related to a large fraction of the reduction. Chen et al. (1994) and Yu et al. (1995) measured transport properties in multilayer gallium arsenide/aluminum-gallium-arsenide superlattice structures at temperatures between about 160 and 325 K, respectively. The room-temperature data for the diffusivity along a superlattice with 1400 Å spatial period are significantly less than those available for bulk gallium arsenide and bulk aluminum arsenide, suggesting strong phonon scattering on the superlattice boundaries. Asheghi et al. (1996) measured the thermal conductivity along crystalline silicon layers in SOI substrates, yielding data that were reduced compared to the conductivity of silicon by orders of magnitude at 15 K. The data agree quantitatively with solutions to the Boltzmann transport equation that account for scattering on the layer boundaries.

The thermal conductivity of semiconductors can be predicted using an approximate solution to the phonon Boltzmann transport equation in the relaxation time approximation together with the Debye model for phonon specific heat. This approach was developed by Callaway (1959) and refined for silicon by Holland (1963), who used a more detailed description of the phonon dispersion relations in this material to better capture the temperature dependence of thermal conductivity. The model of Holland has been adapted to gallium arsenide layers by Chen and Tien (1993), although these calculations would benefit from more extensive bulk data for the determination of the fitting constants. The model of Holland (1963) can be modified to account for the increase in the scattering rate due to the additional imperfections and the small separation between the layer boundaries. The conductivity reduction due to phonon-boundary scattering can be calculated independently for each differential step in the phonon frequency spectrum using the solution to the Boltzmann equation. The model of Holland (1963) is a refinement of the general expression for the phonon thermal conductivity (e.g., Berman, 1976):

$$k_s = \frac{1}{3} v_s^2 \int_0^{\frac{\Theta}{T}} C_s' \tau_s dx_\omega \tag{7-31}$$

where $x_\omega = h_p \omega / (2\pi k_B T)$ is the nondimensional phonon frequency, C_s' is the phonon specific heat per unit volume per unit nondimensional frequency, v_s is the phonon group velocity, Θ is the Debye temperature of the solid, and τ_s is the phonon relaxation time. Holland (1963) separated the conductivity into three integrals of the form of Eq. (7-31) that account for longitudinal, low-frequency transverse, and high-frequency transverse phonons. The three contributions to the conductivity modeled by Holland (1963) differ in the value of the sound velocity and the dependence of the relaxation time on nondimensional frequency and temperature.

Phonon-boundary scattering. To account for phonon scattering on layer boundaries, the relaxation time can be reduced using

$$\tau_s(\omega, T, d_s, p_{sr}) = \tau_{s,NB}(\omega, T) F\left(\frac{d_s}{\Lambda_{s,NB}(\omega, T)}, p_{sr}\right) \tag{7-32}$$

where d_s is a length that quantifies the proximity of the boundaries. For a thin semiconducting layer (e.g., in a silicon-on-insulator (SOI) substrate or a gallium-arsenide multilayer system), d_s is the layer thickness, and the conductivity in Eq. (7-31) is the effective value along the layer. For a semiconducting region in a bulk silicon circuit, which is bounded only on the top by the gate oxide, d_s is the separation from the gate oxide and k_s is the local thermal conductivity in the active region. In this case, the impact of boundary scattering varies strongly with the proximity to the interface. The relaxation time $\tau_{s,NB}(\omega, T)$ is that which would occur in the absence of phonon-boundary scattering and can be determined from measurements and analysis on bulk samples. The boundary scattering reduction ratio F must be determined by solving the phonon Boltzmann equation and depends on the ratio of d_s and the phonon mean free path, $\Lambda_{s,NB}(\omega, T) = v_s \tau_{s,NB}(\omega, T)$, as well as the specular reflection coefficient p_{sr}. If the remaining scattering mechanisms are assumed to be independent, then the scattering rate $1/\tau_{s,NB}$ is the sum of phonon scattering rates on defects and other phonons (Holland, 1963). An exact solution to the Boltzmann equation is available for the determination of $k_{a,eff}$ along layers (Sondheimer, 1952)

$$F(\delta, p_{sr}) = 1 - \frac{3(1 - p_{sr})}{2\delta} \int_1^{\infty} \left(\frac{1}{\xi^3} - \frac{1}{\xi^5}\right) \frac{1 - \exp(-\delta\xi)}{1 - p_{sr}\exp(-\delta\xi)} d\xi. \tag{7-33}$$

The reduced layer thickness is $\delta = d_s/\Lambda_{s,NB}$. The importance of phonon-boundary scattering is reduced by the specular reflection of phonons. Diffuse reflection is caused by the interference of phonon wavepackets reflected by an interface that has a characteristic roughness comparable to or larger than the phonon wavelength. The fraction of phonons reflected diffusely therefore depends strongly on the surface roughness and on the wavelength of the phonons under consideration. The specular reflection coefficient can be approximately estimated from the characteristic dimension of surface roughness, η, and the wavelength λ using (Berman et al., 1955)

$$p_{sr}(\lambda, \eta) = \exp\left(-\frac{16\pi^3\eta^2}{\lambda^2}\right). \tag{7-34}$$

Equation (7-34) treats both incident and scattered waves as if they are normal to the surface, and it has been used in the past (e.g., Ziman 1960) to correlate p_{sr}, η, and λ. A rigorous model that accounts for other angles of incidence and reflection may ultimately prove more appropriate.

Figure 7-14a compares predictions based on Eqs. (7-27)–(7-34) with thermal-conductivity data of Asheghi et al. (1996) for crystalline silicon layers. While data for room temperature are most technically relevant for SOI circuits, a comparison is provided at low temperatures to better illustrate the impact of phonon-boundary scattering. The maximum in the conductivity occurs near 70 K and separates the low-temperature region, where scattering is dominated by imperfections and boundaries, from the high-temperature region, where phonon-phonon scattering dominates. The recommended bulk

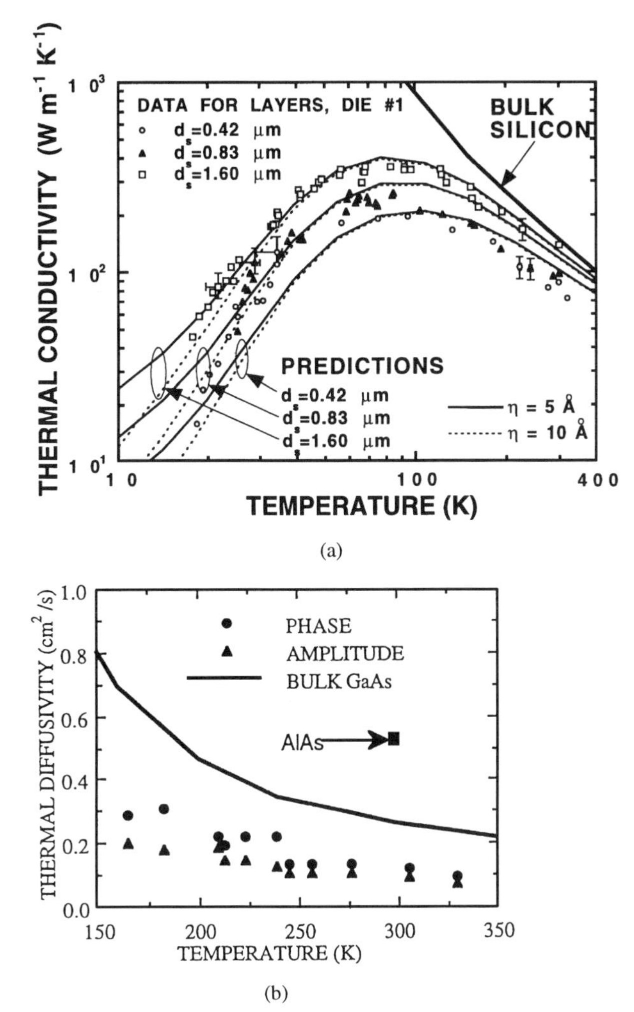

Figure 7-14 Temperature-dependent impact of phonon-boundary scattering on the thermal conductivity along semiconducting layers: a) crystalline silicon layers in SOI substrates (Asheghi et al., 1996). The predictions account for the reduction using the phonon Boltzmann equation and the characteristic dimension of interfacial roughness η. Bulk thermal conductivity data of silicon are provided for reference. b) gallium-arsenide/aluminum-gallium-arsenide superlattice with period 1400 Å (Chen and Yu, 1994). Bulk diffusivity data for aluminum arsenide and gallium arsenide are provided for reference.

conductivity reaches a much higher maximum, 5500 W m^{-1} K^{-1}, near 30 K. The thermal conductivities of silicon device layers are significantly lower than the values recommended for bulk samples due to the much stronger reduction of the phonon mean free path by boundary scattering. The data show clearly that this effect is more prominent for thinner layers. The calculations show that the effect of surface roughness is more pronounced at lower temperatures, where the concentration of the long-wavelength phonons is the greatest. The data agree well with predictions for the 1.6-μm-thick silicon overlayer throughout the entire temperature range. The poorer agreement for the 0.42-μm- and 0.83-μm-thick silicon layers may be due to the higher concentrations of imperfections in these layers. Figure 7-14b shows temperature-dependent thermal-diffusivity data along a gallium-arsenide/aluminum-gallium-arsenide superlattice that has a spatial period of 1400 Å (Yu et al., 1995). The reduction is severe compared to the properties of each of the two constituent materials, suggesting a strong influence of phonon-boundary scattering. Strain-induced crystalline imperfections may also be responsible for some reduction in the diffusivity, although this needs to be investigated.

Phonon scattering on impurities and free electrons. The model described in this section for thin-film thermal conductivities can account for imperfections associated with the processing of semiconducting layers. This is achieved by increasing the scattering rate with terms to account for fabrication-related imperfections. There has been much research on the relationship between the phonon relaxation time and the concentration of dislocations, stacking faults, impurities, and vacancies in crystals (e.g., Berman, 1976; Ziman, 1960). To account for impurities, it is appropriate to use the relationship for the *Rayleigh* scattering of phonon waves,

$$\frac{1}{\tau_{s-i}} = \frac{n_i V^2 \Delta M^2}{4\pi v_s^3 M^2}\omega^4 \tag{7-35}$$

where n_i is the volumetric concentration of the imperfections, M and V are the mass and crystal volume of the host atom, and ΔM is the mass difference introduced by the imperfection compared to the host atom. Figure 7-15 shows predictions and data for the impact of boron and phosphorus impurities on the thermal conductivity of bulk silicon. The data indicate that impurity scattering is important, but clearly a more detailed study is required to establish the validity of Eq. (7-35) and to determine the impact of imperfections introduced during the doping process. This could be particularly important for impurities introduced by implantation, which even after annealing yields large

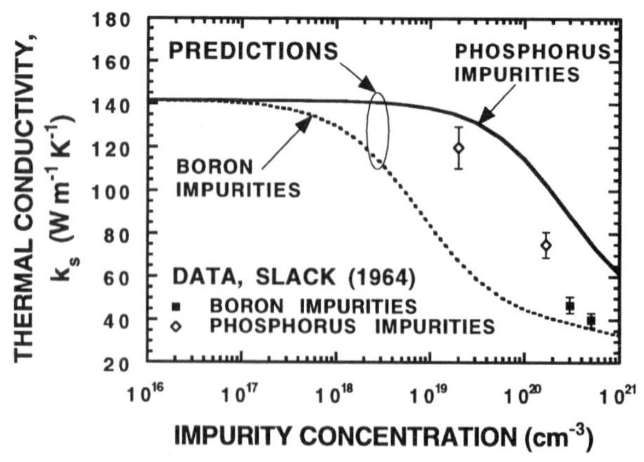

Figure 7-15 Thermal conductivity of silicon as a function of the concentration of boron and phosphorus impurities. The predictions are based on the conductivity model of Holland (1963) with a phonon scattering rate augmented by the expression in Eq. (7-35).

concentrations of dislocations. Silicon samples with identical concentrations of impurities that are introduced using different techniques (e.g., diffusion rather than impure epitaxial growth) may have very different thermal conductivities. Another complication associated with impurities is that they are always accompanied by free electrons and holes. Goodson and Cooper (1995) used perturbation theory to study the influence of phonon-electron scattering on the phonon thermal conductivity. Their model considers electron systems that depart dramatically from equilibrium due to the presence of a large electric field. The predictions suggest that electron scattering can strongly reduce the conductivity for concentrations greater than about 10^{19} cm^{-3}.

7-3-3 Metals

The thermal conductivity in metal layers is dominated by electron transport. The thermal conductivities of metal thin films are rarely measured directly, because they can be predicted with reasonable accuracy from the electrical conductivity by means of the Wiedemann-Franz-Lorenz Law (e.g., Kittel, 1986). This law is based on the kinetic theory of electron transport and assumes that the relaxation times for electron transport of charge and heat are identical. The kinetic expressions for the electron contributions to the thermal and electrical conductivity are $k = C_n v_F^2 \tau_n / 3$ and $\sigma = nq^2\tau/m$, respectively, where v_F is the electron Fermi velocity for the metal, m is the free-electron mass, and $C_n = \gamma T$ is the electron specific heat per unit volume at constant volume. The Sommerfeld parameter γ, the electron number density n, and the electron Fermi velocity are all related by means of the free electron theory. Taking the ratio of the conductivities eliminates the relaxation time and yields a relationship that depends only on temperature,

$$\frac{k}{\sigma} = L_0 T \tag{7-36}$$

where the Lorenz number is a material-independent constant, $L_0 = 2.45 \times 10^{-8}$ W Ω K^{-2}. At room temperature in high-quality aluminum, copper, gold, and tungsten layers, the Lorenz number has been observed experimentally to be 2.20, 2.23, 2.35, and 3.04×10^{-8} W Ω K^{-2}, respectively. The Wiedemann-Franz-Lorenz Law is theoretically sound at temperatures above about fifty percent of the Debye temperature of metals, which is satisfied at room temperature for common metals (e.g., Kittel, 1986). At lower temperatures, scattering on phonons more strongly reduces the electron relaxation time for charge transport than it reduces the electron relaxation time for heat transport, such that the Lorenz number decreases compared to its room-temperature value. This problem disappears at temperatures below about 10 K, where phonon scattering is overwhelmed by scattering on layer boundaries and imperfections in the material. These scattering mechanisms preserve the equality of the electron relaxation times for charge and heat transport, such that Eq. (7-36) is valid at very low temperatures.

Integrated circuits contain many metal layers that are highly impure, such as the silicide and tungsten compounds discussed in Section 7-2-5, and often have small grains. The Wiedemann-Franz-Lorenz Law can be applied with reasonable accuracy to polycrystalline and impure layers because scattering on grain boundaries and localized imperfections yields similar relaxation times for heat and charge transport. However, very impure metals can have conductivities that are augmented by phonon transport and therefore are not predicted by Eq. (7-36). The mean free paths of electrons in pure metals at room temperature is of the order of 10 nm, such that electron-boundary scattering can be expected to be important only in layers thinner than about 100 nm at room temperature. Electron-boundary scattering is relevant only for specialized cryogenic circuits.

7-4 SIMULATION

The electrical engineering of an integrated circuit demands the simulation of charge transport phenomena with characteristic length scales varying by six orders of magnitude. While transistor development must account for electrical currents that vary dramatically over a few tens of nanometers (e.g., normal to the inversion layer of the FET shown in Fig. 7-4), circuit design requires precise modeling of transient electrical currents along interconnects that are sometimes centimeters in length. At every level in this length scale hierarchy, the electrical simulations must account for the temperature fields prevailing in a circuit. Figure 7-1 showed that these temperature fields are governed by thermal transport phenomena spanning a similarly broad range of length scales. This has lead to a wealth of tools for simultaneously simulating electrical and thermal transport for integrated circuits. This section introduces the governing equations of the major techniques and groups the techniques based on their level of detail.

7-4-1 Hierarchy of Governing Equations

Table 7-2 presents the hierarchy of governing equations available for the simulation of semiconductor devices and interconnects. At the top of the hierarchy are the wave equations, which yield wavefunctions for electrons and phonons traveling in semiconducting regions. The use of the Schrödinger wave equation for electrons is standard practice for simulating devices with semiconducting regions of dimensions comparable to or smaller than the electron wavelength, which is the case for multilayer gallium-arsenide/aluminum-gallium-arsenide and silicon/germanium multilayer transistors. Analogous phonon wave simulations are not yet being performed for practical device geometries because the phonon wavelength is only a few Angstroms at room temperature. Electron and phonon transport can be predicted using Boltzmann particle-transport equations, which yield distribution functions for these energy carriers. Table 7-2 does not include the Monte-Carlo technique (e.g., Shur, 1990), which simulates the trajectory of a statistically significant number of particles in momentum and physical space and is equivalent to solving the Boltzmann equation. The Boltzmann equation is a particle number balance with a level of detail that is required only when the scattering rates of electrons or phonons varies significantly within a distance comparable to their respective mean free paths. A more practical approach for many transistor geometries is to solve moments of the Boltzmann equations that enforce conservation of charge, energy, and momentum but do not require knowledge of the particle distribution functions. An even simpler approach is to use the drift-diffusion theory outlined in Section 7-2-1 or to simulate devices as a lumped set of electrical and thermal capacitors and resistors.

Table 7-2 Hierarchy of theoretical approaches for simulating semiconductor devices. All of the equations referenced for electrons have analogous forms for holes

Theory	Transport equations	Independent parameters
Wave	• Electron Schrödinger equation • Elastic continuity or lattice dynamical equations for phonons	Electron and phonon wavefunctions
Kinetic		
1. Detailed electron and phonon number balances	• Electron Boltzmann equation (7-37) or Monte Carlo technique • Phonon Boltzmann equation (7-39)	$\mathbf{E}$, f_n, N, T_n, T_s
2. Averaged number, energy, and momentum balances	• Three electron Boltzmann equation moments, (7-40)–(7-42) • One phonon Boltzmann equation moment, (7-44)	$\mathbf{E}$, n, T_n, $\mathbf{v}_n$, T_s
3. Averaged number and energy balances	• Two electron Boltzmann equation moments, (7-40) and (7-42) • One phonon Boltzmann equation moment, (7-44)	$\mathbf{E}$, n, T_n, T_s
Drift-Diffusion	• Electron drift-diffusion and continuity equations, (7-2) and (7-11) • Heat diffusion equation (7-13) and (7-14)	$\mathbf{E}$, n, T
Lumped-Device	• Temperature-dependent device current-voltage models, e.g., Eqs. (7-17)–(7-20) • Heat diffusion equation, Eq. (7-13)	I, V, T_j

7-4-2 Boltzmann Equations for Electrons and Phonons

Section 7-2-1 presents equations for near-equilibrium transport modeling in semiconductor devices. These equations are not suited for many high-power and compact field effect transistors, where electric fields exceeding 10^6 V m^{-1} cause the electron and hole distribution functions to depart dramatically from those in equilibrium at the lattice temperature. This situation can be accommodated using field-dependent electron and hole mobilities in the drift-diffusion relationships, Eqs. (7-2) and (7-3). However, the local effective mobility in highly nonequilibrium conditions is influenced by the field within a region of dimensions comparable to a few free paths of the carriers. Similarly, the local phonon flux is influenced by scattering phenomena that occur within a few phonon mean free paths. These nonlocal phenomena are very important in compact transistors, where the electric field and the electron and phonon scattering rates can vary dramatically over dimensions less than 0.5 μm. Such transistors cannot be rigorously modeled using the drift-diffusion equations and the macroscopic heat balance, Eq. (7-13). This motivates research on more rigorous transport equations for electrons and phonons. The equations for holes, which are not given here, are analogous to those presented here for electrons.

The transient electron Boltzmann equation is

$$\frac{\partial f_n}{\partial t} + \mathbf{v}_{n,\mathbf{k}} \cdot \nabla_{\mathbf{r}} f_n + \frac{\partial \mathbf{k}}{\partial t} \cdot \nabla_{\mathbf{k}} f_n = \left(\frac{\partial f_n}{\partial t}\right)_{COL} \tag{7-37}$$

where f_n is the electron distribution function among states denoted by electron wavevector $\mathbf{k}$ and position $\mathbf{r}$. The electron velocity is $\mathbf{v}_{n,\mathbf{k}}$. The gradient operators yield derivatives in physical space, described using the subscript $\mathbf{r}$, and wavevector space, described using the subscript $\mathbf{k}$. In the semi-classical model for electron transport, the time rate of change of the wavevector due to an electric field is $2\pi q\mathbf{E}/h_p$. The collision term on the right of Eq. (7-37) can be expressed as an integral over initial and final electron states involving the probabilities for the transitions and the distribution functions. This integral is usually approximated using the equilibrium distribution function in Eq. (7-1) and the relaxation time approximation,

$$\left(\frac{\partial f_n}{\partial t}\right)_{COL} \approx \frac{f_{n,EQ} - f_n}{\tau_n} = (f_{n,EQ} - f_n)\left(\frac{1}{\tau_{n-i}} + \frac{1}{\tau_{n-a}} + \frac{1}{\tau_{n-o}} + \cdots\right). \quad (7\text{-}38)$$

The electron scattering rate $1/\tau_n$ has dominant components due to scattering on impurities, τ_{n-i}; acoustic phonons, τ_{n-a}; and optical phonons, τ_{n-o}. If the temperature of phonons differs from that of electrons, which is the condition in many high-field transistors, then the phonon scattering processes induce relaxation of the electron distribution function to the phonon temperature. This can be described using two separate terms on the right of Eq. (7-38) containing Fermi-Dirac distribution functions referenced with respect to the electron and phonon temperatures. Detailed analysis must distinguish between scattering on nonpolar optical phonons, which are dominant in silicon and germanium, and polar optical phonons, which are important in gallium arsenide and other heteropolar crystals. Scattering on optical phonons grows much more important in high electric fields and is responsible for electron velocity saturation, whose impact on the behavior of a FET was shown in Fig. 7-5. The phonon Boltzmann equation can be similarly written in the relaxation time approximation,

$$\frac{\partial N}{\partial t} + \mathbf{v}_{s,\mathbf{k}} \cdot \nabla_{\mathbf{r}} N = \left(\frac{\partial N}{\partial t}\right)_{COL} \approx \frac{N_{EQ} - N}{\tau_s}$$

$$= (N_{EQ} - N)\left(\frac{1}{\tau_{s-i}} + \frac{1}{\tau_{s-s}} + \frac{1}{\tau_{s-n}} + \frac{1}{\tau_{s-p}} + \cdots\right) \quad (7\text{-}39)$$

where N is the phonon distribution function among states with wavevector $\mathbf{k}$ and position $\mathbf{r}$, $\mathbf{v}_{s,\mathbf{k}}$ is the velocity of phonons with wavevector $\mathbf{k}$, and N_{EQ} is the Bose-Einstein distribution function at the lattice temperature (e.g., Kittel, 1986). The phonon scattering rate $1/\tau_s$ has dominant contributions due to impurities $(1/\tau_{s-i})$, acoustic phonons $(1/\tau_{s-a})$, optical phonons $(1/\tau_{s-o})$, electrons $(1/\tau_{s-n})$, and holes $(1/\tau_{s-p})$. If the electron, hole, and phonon temperatures differ, then the relaxation time approximation on the right of Eq. (7-39) can be expanded into terms that account for the relaxation of phonons to the electron temperature, the hole temperature, and the lattice temperature. The solution of the Boltzmann equations in this case requires simultaneous monitoring of the electron, hole, and phonon temperatures, as well as enforcement of energy conservation for the complete ensemble of energy carriers.

7-4-3 Moments of Boltzmann Equations

Solving the electron, hole, and phonon Boltzmann equations requires a complete description of the distribution functions of these carriers. A more practical approach is to solve conservation equations derived from the Boltzmann equation. These are moments of the transport equations that enforce conservation of charge, momentum, and energy without requiring detailed knowledge of the distribution functions. This is achieved by assuming that electrons and holes each satisfy equilibrium distribution functions, Eqs. (7-6) and (7-8), but with temperatures T_n and T_p that are not necessarily equal to the lattice temperature T_s. For the case of electrons, these moments are often called the hydrodynamic equations (e.g., Bloetekjaer, 1970; Wang, 1985; Majumdar et al., 1995):

$$\frac{\partial n}{\partial t} + \nabla \cdot (n\mathbf{v}_n) = r_g - r_r \tag{7-40}$$

$$\frac{\partial \mathbf{p}_n}{\partial t} + \nabla \cdot (\mathbf{v}_n \mathbf{p}_n) = -qn\mathbf{E} - \nabla(nk_B T_n) - \frac{nm_n^* \mathbf{v}_n}{\tau_{n,M}} \tag{7-41}$$

$$\frac{\partial W_n}{\partial t} + \nabla \cdot (\mathbf{v}_n W_n) = -qn\mathbf{v_n} \cdot \mathbf{E} - \nabla \cdot (\mathbf{v}_n nk_B T_n) + \nabla \cdot (k_n \nabla T_n)$$
$$- \frac{\left(W_n - \frac{3}{2} nk_B T_s\right)}{\tau_{n-s,E}} - r_r \frac{W_n}{n} - r_g E_{imp} \tag{7-42}$$

Equation (7-40) is the number conservation equation for electrons and is equivalent to Eq. (7-11). The average electron velocity vector $\mathbf{v}_n$ is related to the electron current density from Section 7-2-1 through $\mathbf{j}_n = -qn\mathbf{v}_n$. Equations (7-41) and (7-42) are momentum and energy balances, in which the electron momentum and energy densities are $\mathbf{p}_n = m_n^* \mathbf{v}_n$ and W_n, respectively. As written here, the momentum and energy balances assume Maxwellian distribution functions for electrons and holes, which are described using Eqs. (7-6) and (7-8). More detailed moment equations can eliminate this assumption (e.g., Apanovich et al., 1995). The electron effective transport mass m^* differs from the effective mass for density of states calculations introduced in Section 7-2-1. The first two terms on the right of the momentum balance in Eq. (7-41) account for acceleration due to an applied field and a net accumulation of electrons of higher momentum per unit volume due to a gradient in the electron temperature. The final term on the right accounts for momentum loss due to scattering processes and is governed by the electron momentum relaxation time, $\tau_{n,M}$. Terms with similar explanations on the right of the energy balance in Eq. (7-42) are augmented by an energy diffusion term proportional to the electron conductivity, k_n, which can be approximately calculated using the Wiedemann-Franz-Lorenz Law (e.g., Sze, 1981). The energy balance also accounts for the net loss of energy due to electron-hole recombination at the rate r_r and impact ionization at the rate r_g. Impact ionization results from electron-ion collisions which lift an electron to a higher energy state. Each ionization event absorbs energy from the electron gas in the amount E_{imp}. Further expansion of the recombination terms would distinguish between Auger processes, which transmit energy to the electron gas, and processes involving phonon and photon emission (e.g., Sze, 1981; Apanovich et al., 1995).

A large electric field accelerates electrons into states of much higher energy than are occupied in equilibrium at the lattice temperature. Although much of this excess energy is carried by electrons traveling in the direction of the electric field and is described by the average electron velocity vector $\mathbf{v}_n$, electron-electron scattering causes a significant fraction to be distributed without order among the electron wavevectors. This is reflected by the definition for the electron energy density,

$$W_n = n\left(\frac{3}{2}k_B T_n + \frac{1}{2}m_n^*|\mathbf{v}_n|^2\right). \tag{7-43}$$

The first term on the right accounts for electron kinetic energy that is distributed among electron states according to a Maxwellian distribution at the temperature T_n. The second term on the right accounts for excess energy due to net motion of the carriers in a given direction. The electron temperature in the first term is related to the quasi Fermi level for electrons and the electron concentration by means of Eq. (7-6). The hole temperature, T_p, similarly quantifies a directionally averaged departure of the holes from equilibrium at the lattice temperature. The electron and hole temperatures strongly influence the rate of energy transfer to phonons and the phonon thermal conductivity (Goodson and Cooper, 1995).

Since phonons can be created and destroyed, their number is not conserved. Because they can exchange momentum with the crystal lattice through Umklapp scattering processes (e.g., Kittel, 1986), a momentum conservation equation is of little use. Only the energy moment of the phonon Boltzmann equation is relevant. For processes that are not too rapid compared to the phonon relaxation time, the energy moment resembles the macroscopic heat equation,

$$C_s \frac{\partial T_s}{\partial t} = \nabla \cdot (k_s \nabla T_s) + \frac{\left(W_n - \frac{3}{2}nk_B T_s\right)}{\tau_{n-s,E}} + \frac{\left(W_p - \frac{3}{2}nk_B T_s\right)}{\tau_{p-s,E}} + r_r\left(\frac{W_n}{n} + \frac{W_p}{p} + E_g\right) \tag{7-44}$$

where k_s is the phonon thermal conductivity and C_s is the phonon heat capacity per unit volume. The first term on the right accounts for phonon conduction, and the second and third terms account for heat generation due to scattering with electrons and holes. The second and third terms couple Eq. (7-43) with Eq. (7-42) and an analogous equation, not provided here, for the hole energy density W_p. The final term on the right accounts for phonon generation due to the recombination of carriers. This term needs to be reduced if Auger or radiative processes are considered explicitly, because these processes transfer energy to electrons and photons, respectively, rather than to phonons.

The solution of multiple moments of the Boltzmann equations has found broad application for practical bipolar and field-effect transistors. The solution of the charge and energy moments of the electron Boltzmann equation together with the energy moment of the phonon Boltzmann equation is available through commercial software, such as PISCES-2ET (e.g., Yu et al., 1994; Shur, 1990), and has been applied to study both ESD failure in silicon field-effect transistors (Beebe et al., 1994) and the importance of self-heating effects in bipolar and SOI field-effect transistors (e.g., Apanovich et al., 1994, 1995). The analysis of Apanovich et al. (1995) is distinguished by its detailed treatment of heat generation due to electron-hole recombination processes, which are most important for bipolar transistors. These authors compared their predictions for the breakdown collector-base voltage in a submicrometer bipolar transistor with predictions using the simple drift-diffusion theory from Section 7-2-1. While the current-voltage predictions using the two different techniques agree well for low values of the collector-emitter voltage, which corresponds to low electric fields, the breakdown collector-emitter voltage is underpredicted by about 40 percent using drift-diffusion theory. This is attributed to the nonlocal nature of impact ionization, which is better simulated by independently balancing the electron and phonon energies. Apanovich et al. (1995) found that although self-heating of the transistor was important, temperature gradients *within* the transistor were not very important. The temperature rise was dominated by the thermal resistances that were assumed to prevail between the transistor and the environment, which are analogous to the resistances depicted in Fig. 7-1. This indicates that relatively simple simulations using the heat diffusion equation and an isothermal lattice in the channel may be appropriate when coupled with more rigorous simulations of energy generation and transport by the free charge carriers.

Recent work using moments of the electron Boltzmann equation distinguished between the phonon branches in compact silicon and gallium arsenide transistors (Majumdar et al., 1995; Fushinobu et al., 1995; Lai and Majumdar, 1996). These authors used the number, energy, and momentum moments of the electron Boltzmann equation together with separate energy equations for acoustic and optical phonons. Acoustic phonons have appreciable velocity and contribute significantly to net energy transport. In contrast, optical phonons have negligible group velocity and do not contribute significantly to heat transport. But optical phonons can strongly scatter electrons, particularly those in the high energy states occupied in the presence of large electric fields. The relative importance of electron and hole scattering on acoustic and optical phonons depends on the material. For silicon, acoustic scattering dominates at low fields. At high fields in silicon and in general for gallium arsenide, scattering on optical phonons plays a critical role. The extra complexity of an additional energy equation for optical phonons is warranted primarily for transient calculations with heating timescale comparable to the relaxation time for energy transfer between the optical and acoustic phonons. This was estimated to be near 8 ps for gallium arsenide. The results of simulations for a compact silicon FET are provided in Fig. 7-16. Figure 7-16c and 7-16d show the relatively small difference between the predicted temperature rise shape for optical and acoustic phonons in the active region, which is consistent with the data. The independent measurement of optical and acoustic phonon temperatures was achieved using micro-Raman spectroscopy (Ostermeier et al., 1992) and scanning thermal microscopy (Majumdar et al., 1993), techniques that are described in Sections 7-5-2 and 7-5-1, respectively. The impact of lattice heating was found to be more pronounced in gallium-arsenide devices (Fushinobu et al., 1995). The temperature rise was found to be dominated after a few hundred picoseconds by the thermal boundary conditions chosen for the computation domain, similar to the findings of Apanovich et al. (1995).

7-4-4 Drift-Diffusion and Simpler Models

At the next level in the hierarchy are the drift-diffusion equations in Section 7-2-1, which are used together with a single energy balance, Eq. (7-13). This approach has been applied for detailed studies of transistor breakdown (e.g., Amerasekera et al., 1993) and is also available commercially (e.g., Wolbert et al., 1994). The results of Apanovich et al. (1995) place the drift diffusion theory in question for highly nonequilibrium studies. But since the Boltzmann equation moments and the drift diffusion models are both phenomenological, it is not inappropriate to tailor critical parameters, such as the high-field mobility and the impact ionization rate, to account for the nonequilibrium conditions. The weakness of existing transistor simulation capabilities is the large number of unknown microscopic parameters, such as the distinct relaxation times for electron

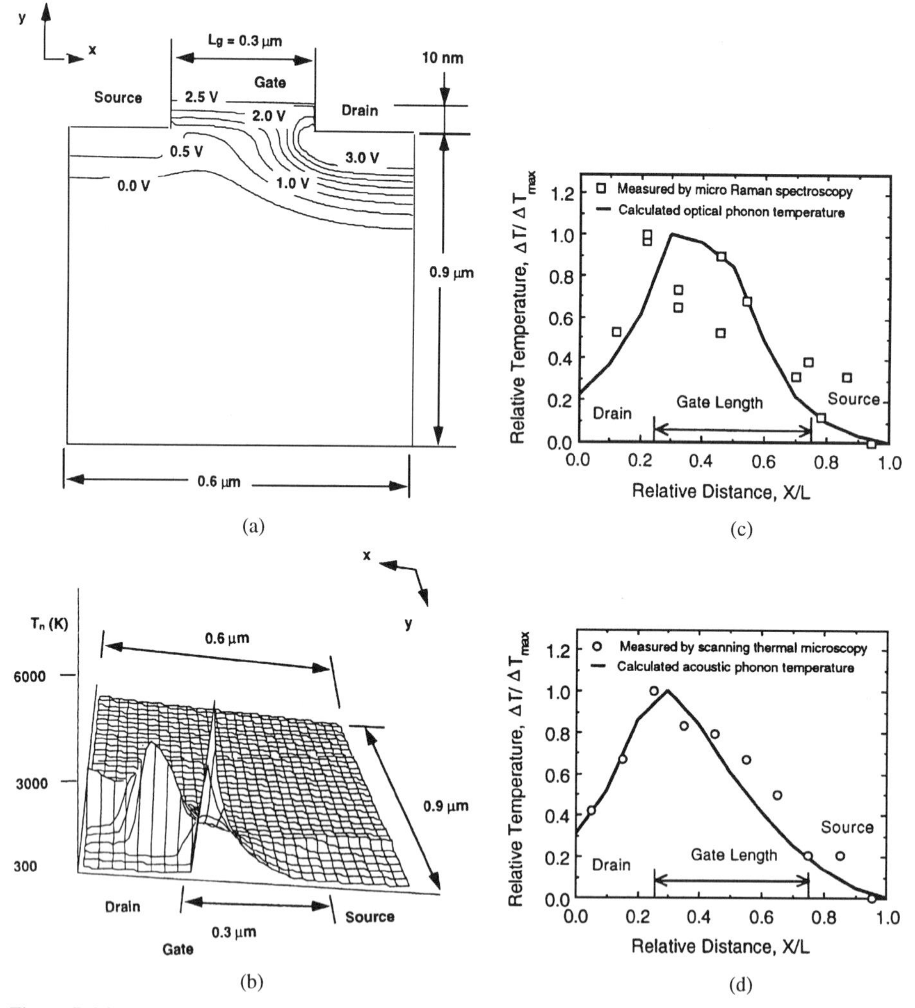

Figure 7-16 Simulation results of Lai and Majumdar (1996) for the distribution of a) electrostatic potential, b) electron temperature, c) optical phonon temperature, and d) acoustic phonon temperature in a compact silicon field-effect transistor. The simulations employed Eqs. (7-40)–(7-42) and distinct energy equations for optical and acoustic phonons. Although the maximum electron temperature approaches 6000 K, the maximum temperature rise of acoustic phonons is about 7 K. Heat generation occurs predominantly near the drain of the transistor, where the pinch-off condition reduces the electron concentration and augments the electric field and resistive heating.

momentum and energy transfer to the lattice. Although a solution using more governing equations may be more rigorous in principle, the practical relevance is often strongly degraded by the uncertainty of the many phenomenological parameters. The last level in the hierarchy in Table 7-2 describes lumped transistor simulations, in which the active regions of the transistor are treated as an isothermal heat source in solutions to Eq. (7-13) (e.g., Goodson and Flik, 1992; Redman-White, 1993). This approach can benefit from the wealth of analytical techniques available for solution of the macroscopic heat diffusion equations, which are documented with sufficient detail in other texts. The resulting temperature rise can be used to modify parameters in current-voltage models for the transistors (e.g., Su et al., 1994). Lumped device models are very useful for circuit-level simulations.

7-5 THERMOMETRY

There has been much progress in recent years on transistor and interconnect thermometry. Much of this progress has used either high-resolution optical diagnostics or the scanning probe microscope, which have improved temporal and spatial resolution, respectively. Thermometry methods are categorized depending on whether they employ electrical or optical signals.

7-5-1 Electrical Methods

Several techniques use electrical-resistance thermometry in patterned bridges. Others use the temperature dependence of the electrical operating characteristics of a semiconductor device. Another electrical method is the use of a thermocouple junction or an electrical-resistance thermometer at the tip of a scanning probe microscope, an approach that is now available commercially and offers outstanding spatial resolution.

Electrical-resistance thermometry. The transistor thermometry methods that lend themselves most readily to precise calibration use patterned electrically conducting bridges made from metals or doped semiconductors. The bridge electrical resistance varies with temperature due to the temperature dependence of electron-phonon scattering rate and, for the case of a semiconducting bridge, the electron concentration. The incorporation of a temperature-sensing resistor within the active region of a semiconductor power diode was suggested by Manduteanu (1988). Another application is interconnect thermometry with the interconnect serving as the thermometer. Maloney and Khurana (1985) and Banerjee et al. (1996) performed transient electrical-resistance thermometry of interconnects subjected to current pulses of duration less than 1 μs. An approach for FET thermometry is to modify the gate such that it can serve as the temperature-sensing bridge, as shown in Fig. 7-17. Estreich (1989) used gate metals of gallium-arsenide metal-semiconductor field-effect transistors (MESFETs) as thermistors to measure their channel temperatures under steady-state conditions. Mautry and Trager (1990) and Goodson et al. (1995a) used the transistor gate as a thermometer to study self-heating of compact FETs made from bulk silicon and SOI wafers, respectively. Figure 7-18 compares steady-state channel temperature rise data measured in compact

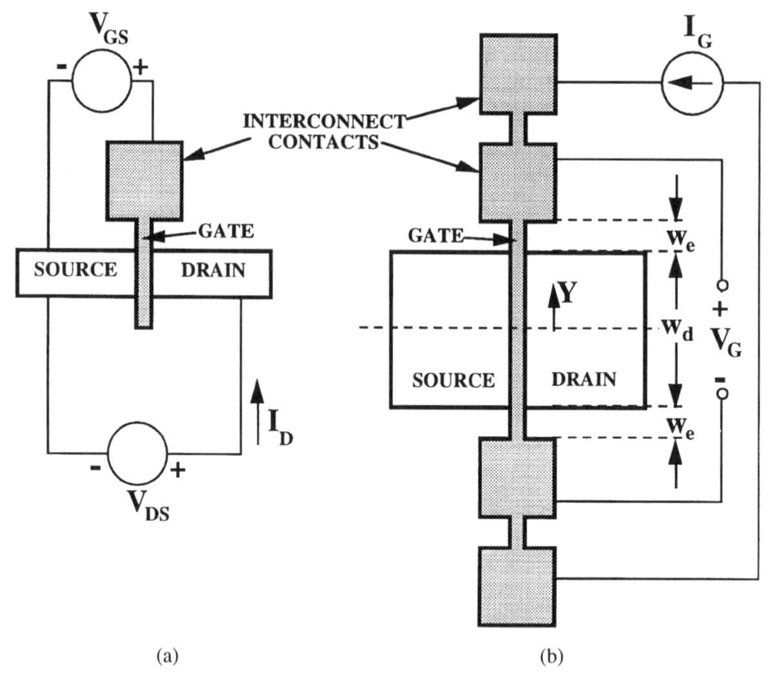

Figure 7-17 Top-view schematics of a) a normal transistor and b) the transistor thermometry structure developed by Goodson et al. (1995a). The gate of the transistor serves as an electrical-resistance thermometer for the temperature in the channel.

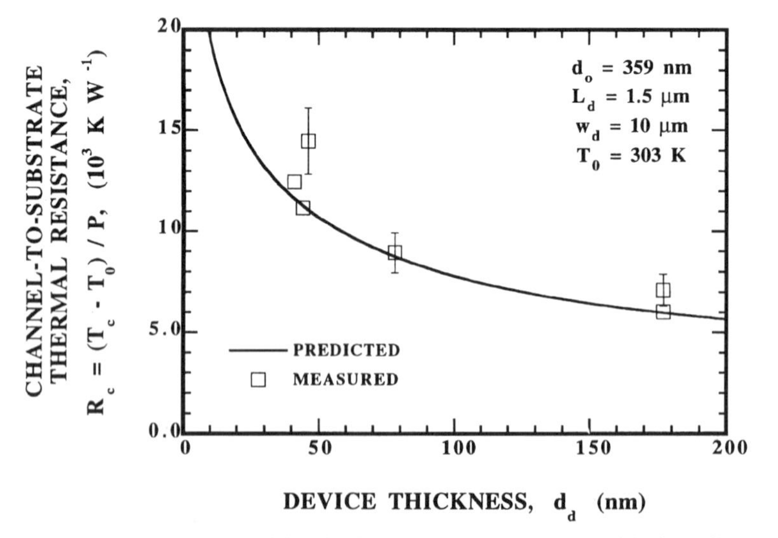

Figure 7-18 Predictions and data for the channel temperature rise in SOI field-effect transistors measured by Goodson et al. (1995a) using the structure shown in Fig. 7-17b. The temperature decreases with increasing thickness of the silicon device layer because this layer is responsible for lateral conduction cooling above the buried silicon dioxide. The predictions are based on an approximate analytical solution to the heat equation in the device and interconnects that models the channel as isothermal (Goodson and Flik, 1992).

SOI FETs with solutions to the heat equation based on the lumped device model described in Section 7-4-4. The predictions and data show that the temperature rise in SOI field-effect transistors varies approximately with the inverse square root of the device layer thickness due to lateral diffusion in this layer. Leung et al. (1995) and Goodson et al. (1996) performed steady-state and transient thermometry, respectively, of the high-power silicon-on-insulator (SOI) shown in Fig. 7-9 using a polysilicon bridge patterned above the drift region. These data showed that the device thickness has a somewhat smaller influence on the temperature rise than in compact FETs due to the large width of the power device, which diminishes the importance of lateral conduction cooling. The temporal resolution that can be achieved using electrical-resistance thermometry is limited by capacitive coupling between the thermometer and the circuit and the reflection of voltage signals within the circuit (Goodson et al., 1996).

The major benefit of using patterned electrical-resistance thermometers is that they can be very precisely calibrated, which facilitates a low experimental uncertainty. But this approach has several important disadvantages: 1) The thermometers provide only the spatially averaged temperature along a patterned bridge at a single location. The spatial resolution can be submicrometer in only one dimension, and no spatial temperature distributions can be captured. 2) Voltage reflections and capacitive coupling in the thermometry circuit limit the temporal resolution to a few hundred nanoseconds. 3) The semiconductor device must usually be modified.

Scanning thermal microscopy (SThM). The scanning tunneling microscope (STM) and atomic force microscope (AFM) have led to a variety of innovations in surface characterization methods, several of which interrogate temperature or thermal properties near a surface. The scanning thermal profiler of Williams and Wickramasinghe (1986) uses heat transfer as a feedback mechanism for the tip-sample separation. Variations in the thermal properties of samples were qualitatively mapped using a heated STM probe by Nonnenmacher and Wickramasinghe (1992), but this approach could not decouple the surface temperature from the separation. Majumdar et al. (1993, 1995) overcame this problem by replacing the cantilever in an AFM with thermocouple wires whose junction formed a scanning probe tip, as shown in Fig. 7-19. This advancement decoupled the tip-surface separation control, which was achieved by monitoring the deflection of the cantilever, from the thermometry signal in the thermocouple circuit. This group has recently developed fabrication technology that yields junctions of dimensions comparable to 100 nm near the tip of a silicon nitride cantilever probe (Luo et al., 1996a).

Figure 7-20 shows topographical and thermometry data for a dual-gate gallium-arsenide metal-semiconductor field-effect-transistor (MESFET) acquired using scanning thermal microscopy. The drain and source contacts are located at the far right and left side of the image. The thermal image indicates that the source and drain metallizations remain cold while the gates are hot. This is consistent with the fact that most of the heat generation in the transistor occurs in the channel beneath the gate. SThM has also been applied to study the failure of semiconductor devices after ESD stressing. Figure 7-21 compares temperature fields measured in an operating silicon FET before and after electrical stressing, showing a local hotspot in the region where destructive breakdown has reduced the electrical resistance. The spatial resolution that can be achieved using

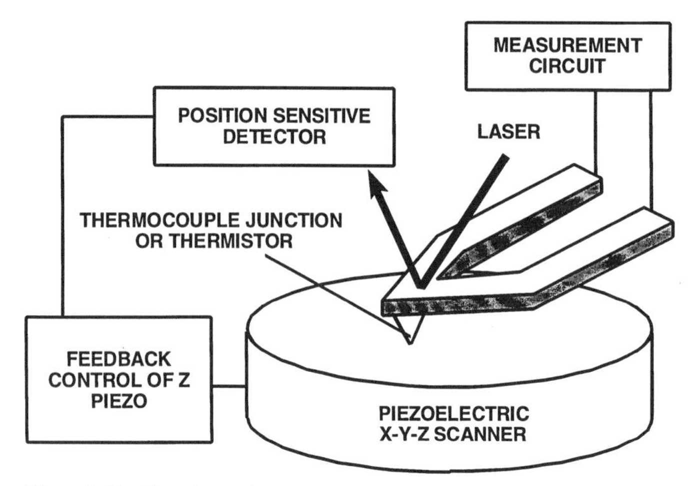

Figure 7-19 Experimental apparatus for high-resolution scanning thermal microscopy using an Atomic Force Microscope (AFM) (after Majumdar et al., 1993). Chromel and alumel leads serve both as the cantilever for force measurements and as legs of the thermocouple circuit. Subsequent research (Luo et al., 1996a) has improved the spatial resolution using junctions fabricated at the tips of conventional silicon-nitride cantilevers.

SThM is near 100 nm, making it better suited for steady-state failure analysis of semiconductor devices than far-field optical techniques. Luo et al. (1996b) captured the temperature distribution inside a cloven vertical-cavity surface-emitting laser using the same technique, as shown in Fig. 7-22. The peak temperature occurs at the intersection of the optical axis and the active region.

SThM provides the best spatial resolution among the existing transistor thermometry techniques and has the important advantage that modification of devices is not essential.

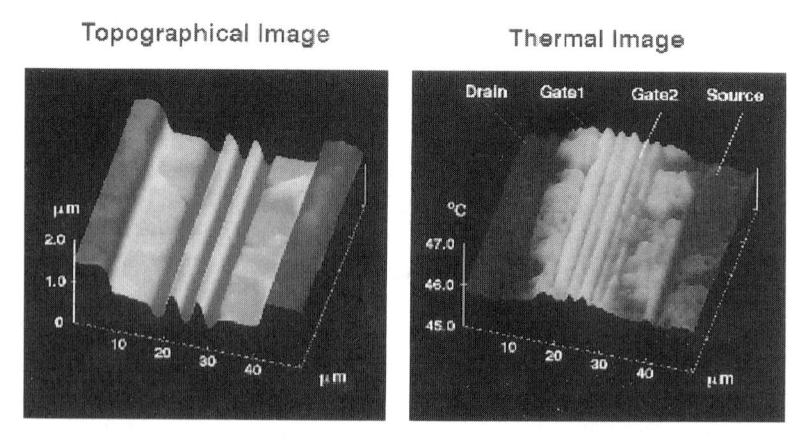

Figure 7-20 Topographical and thermal images of a dual-gate metal-semiconductor field-effect transistor (MESFET) made of gallium arsenide (Majumdar et al., 1995). The gates have a larger temperature than the source and drain regions. The transistor was provided by Motorola Corporation.

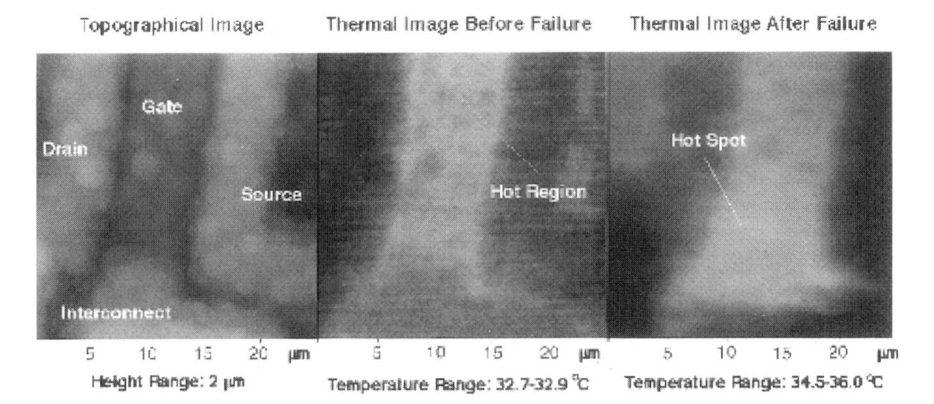

Figure 7-21 Topographical and thermal images of a compact silicon field-effect transistor before and after failure due to an ESD pulse. A hotspot appears after failure due to a short circuit near the gate interconnect. Grains can be resolved by means of the temperature field (Lai et al., 1995). The transistor was provided by Motorola Corporation.

There are important drawbacks with this approach: 1) The temperature drop between the scanning tip and the surface is difficult to determine precisely, such that the magnitude of the temperature rise cannot usually be determined with low uncertainty. 2) The tip-surface thermal conductance is influenced by topographical features, such that the tip-surface temperature drop may vary during a scan. This causes spurious information in the shape of measured temperature profiles. 3) Because this is an electrical measurement, capacitive and reflection effects dramatically complicate measurements performed with timescales less than 1 μs.

Thermometry using temperature-sensitive electrical parameters. Section 7-2 showed that the electrical behavior of a transistor and its component junctions can be strongly influenced by temperature. This makes possible thermometry of a transistor with little or no modification. The sensitivity for thermometry can be strongest for diodes within semiconductor devices, such as the base/emitter diode in a bipolar transistor and the drain/channel or source/channel diodes in silicon field-effect transistors (e.g., Blackburn, 1988; Cain et al., 1992). Zweidinger et al. (1996) proposed a systematic method to extract the thermal resistance for bipolar transistors based on TSEPs. One approach is to first calibrate the TSEP using an external temperature controller. During the calibration step, the rate of heat generation in the device should induce a negligible temperature rise. During the measurement, the transistor is heated using an electrical current of magnitude typical during operation. The device electrical operation is interrupted briefly, and the TSEP is monitored using the lower calibration current. Cain et al. (1992) measured junction temperatures in high-frequency power transistors using the base-emitter voltage as a TSEP, as depicted in Fig. 7-23. Liu and Yuksel (1995) used the turn-on voltage of aluminum-gallium-arsenide/gallium-arsenide bipolar transistors for thermometry. Arnold et al. (1994) used the low-current drain-source voltage across SOI power transistors to determine the transient temperature rise. Another approach to performing TSEP thermometry is to fit the current-voltage characteristics of a device

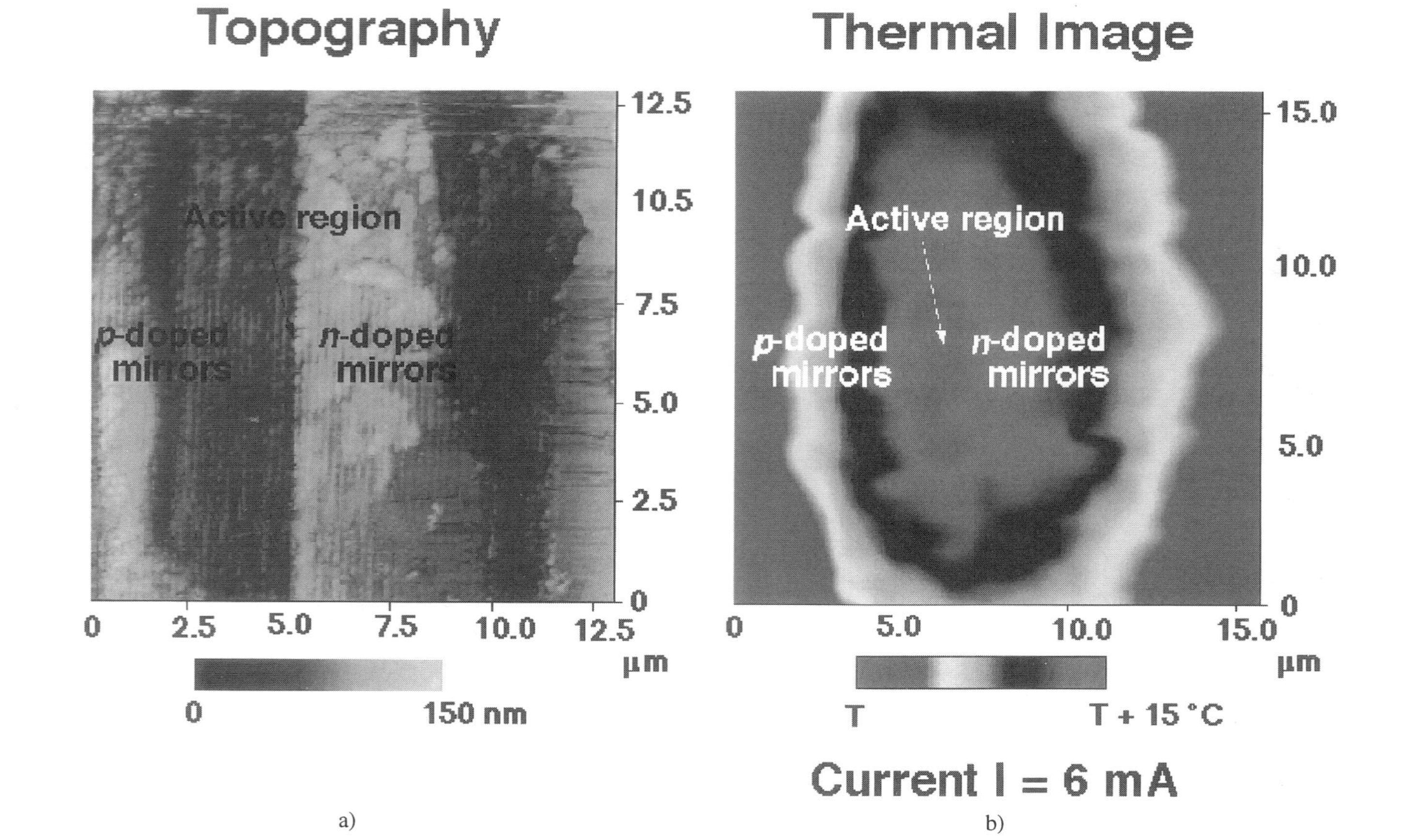

Figure 7-22 a) Topographical and b) thermal images of the cross section of a vertical-cavity surface-emitting laser obtained by scanning thermal microscopy (Luo et al., 1996b). The vertical lines in the topographical image are the alternative layers of 70-nm-thick aluminum-arsenide and 60-nm-thick $Al_{0.16}Ga_{0.48}As$ multilayers for the Bragg-reflecting n and p doped mirrors. The topographical contrast comes from swelling of the aluminum arsenide layers due to oxidation. The thermal image is obtained with a current of 6 mA.

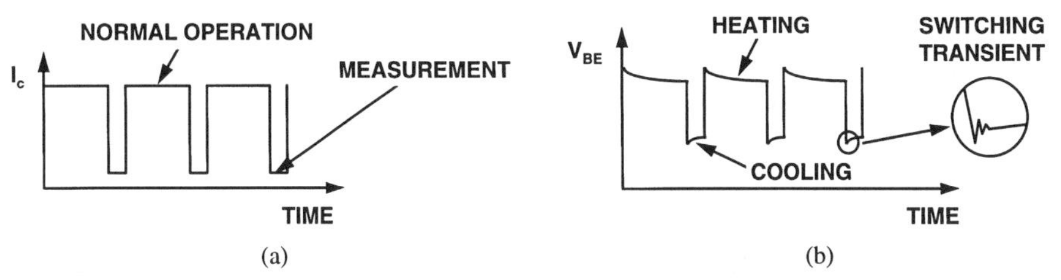

Figure 7-23 Collector current and base-emitter voltage drop during TSEP thermometry of a bipolar transistor (Cain et al., 1992). The normal operation of the transistor is interrupted for thermometry, which results in errors associated with short-timescale cooling.

using temperature-dependent parameters, such as the mobility, threshold voltage, and saturation velocity in a SOI FET (Su et al., 1994).

The major advantage of TSEP thermometry measurement is that it can be performed on fully packaged devices with little or no modification. The temperature is measured directly within the transistor rather than at the surface of a sample, which is the case for SThM and many of the optical methods. This can be important when there is much passivation separating active regions from the surface. But there are major drawbacks associated with transient TSEP thermometry: 1) The low-power calibration condition cannot be established in the transistor with arbitrary speed, due to electrical capacitive effects and voltage reflections in the circuit. This results in temporal delay between the termination of the high-power heating and the thermometry that can be comparable to or larger than 1 μs. The device cools significantly during this period, particularly if the resistance $R_{device\text{-}substrate}$ in Fig. 7-1 is large and the delay is longer than the characteristic timescale $R_{device\text{-}substrate}\, C_{device}$. The error due to this cooling and any approximate compensations in the data extraction can be quite large. Another complication with TSEP thermometry is the impact of a nonisothermal temperature distribution in the device on the measured electrical parameter, which is typically calibrated under isothermal conditions.

Noise thermometry. The rate of current flow through an electrical resistor experiences statistical fluctuations due to scattering with phonons, which induces thermal voltage noise. The thermal voltage noise associated with a resistance R is given by $(4\, k_B T R \Delta f)^{1/2}$, where Δf is the frequency bandwidth under consideration. Measurement of the thermal noise voltage of a resistor can thus yield its temperature. Bunyan et al. (1992) fabricated a doped silicon resistor normal to the direction of current flow in a SOI FET and within the plane of the substrate. The noise along this resistor was monitored during operation of the transistor, yielding data for the steady-state channel temperature.

7-5-2 Optical Methods

There is a wealth of thermometry techniques that exploit the temperature dependence of optical properties near the surface of a semiconductor device or interconnect. Optical methods use radiation interaction with either the surface of a device or interconnect, in some cases probed through transparent passivation, or with a specially deposited thin layer on the surface. Optical methods achieve higher temporal resolution by not contributing to the thermal mass or the electrical capacitance of the system. Some optical methods offer the advantage of requiring no physical contact with the sample, which eliminates the possibility of spurious influences of a probe on the temperature field in the device or interconnect.

Laser-reflectance thermometry. Laser-reflectance thermometry detects temperature changes near a surface using the dependence of the optical reflectance on temperature (e.g., Miklos and Lorincz, 1988). The reflectance is measured using a probe laser. For a small temperature change ΔT, the relative change in surface reflectance R_{opt} can be expressed as $\Delta R_{opt}/R_{opt} = C_{TR}\Delta T$, where C_{TR} is the thermoreflectance coefficient. The thermoreflectance coefficient has magnitude comparable to 10^{-5} or 10^{-4} K^{-1} for metals and semiconductors depending on the wavelength. The calibration of the thermoreflectance coefficient at silicon surfaces has been discussed by Qiu et al. (1993). The calibration is complicated by heat diffusion and photon interference if there is overlying passivation (e.g., Ju and Goodson, 1996). The spatial resolution of conventional far-field laser-reflectance thermometry is limited by diffraction of the probing laser beam to a dimension comparable to the wavelength. Initial progress on *near-field* laser-reflectance thermometry of interconnects using a scanning optical fiber promises spatial resolution well below 100 nm (Goodson and Asheghi, 1996). The temporal resolution is limited to the order of the relaxation times associated with the scattering processes governing the reflectance and the establishment of temperature near the surface. In practice, the temporal resolution for thermometry is usually limited by electrical noise in the detection circuit.

Figure 7-24 shows an apparatus for performing far-field scanning laser-reflectance thermometry on integrated circuits. This technique is most readily applied to metal

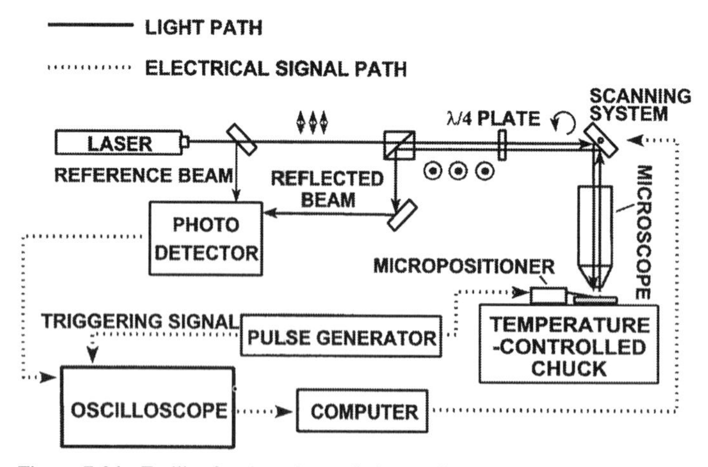

Figure 7-24 Facility for short-timescale laser-reflectance thermometry (Ju et al., 1996; Ju and Goodson, 1996), which yielded the thermometry data in Fig. 7-2. Radiation from a laser diode is coupled into an optical microscope and focused onto the surface. The reflected radiation is decoupled and monitored using a fast detector. The focus location of the laser is scanned over the sample surface using two galvanometrically controlled rotating mirrors. While the spatial precision of the scanning system is about 250 nm, the spatial resolution of the thermometry is limited by diffraction of the far-field radiation to slightly better than 1 μm.

interconnect and via structures, whose temperature fields have been captured in the steady state (Claeys et al., 1993) and in the presence of submicrosecond current pulses (Ju and Goodson, 1996). Because of the small radiation penetration depth in metals, laser-reflectance thermometry at metal surfaces yields a precisely defined location for temperature measurement and minimizes the interaction of the probe beam with the electrical currents flowing in a circuit. Figures 7-2b and 7-2d depict temperature fields in interconnect structures subjected to brief electrical stressing, which yielded the closely coupled distribution of failure regions in Figs. 7-2a and 7-2c. The use of visible or near-visible radiation allows temperature distributions to be captured through silicon dioxide and other transparent passivation, although this complicates the calibration for transient measurements (Ju and Goodson, 1996). Laser-reflectance thermometry has also been applied on semiconductor devices. Epperlein (1993) studied the impact of temperature fields on the failure of semiconductor laser mirrors and Ju et al. (1996) captured the transient evolution of temperature fields in the drift region of a SOI power transistor, as shown in Fig. 7-25. The thermometry of the transistor improved over previous TSEP thermometry in that device (Arnold et al., 1994), which determined the spatially averaged temperature field in the drift region.

The primary benefits of laser-reflectance thermometry are the abilities to interrogate very brief phenomena, such as ESD failure of interconnects and devices, and to capture temperature distributions through passivation transparent at the wavelength of the probe laser. The disadvantages are: 1) the spatial resolution is limited to the order of 500 nm for far-field techniques and 50 nm for near-field techniques; 2) thermometry of active semiconducting regions must take care not to influence the operation of the transistor through photon-induced generation of electron-hole pairs; and 3) calibration is complicated by the presence of different materials, topography, and passivation at

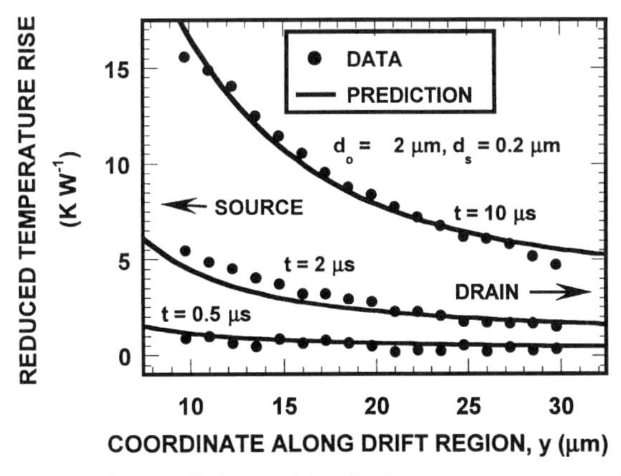

Figure 7-25 Predictions and data for the transient temperature distribution above the drift region of a high-power silicon-on-insulator transistor (Ju et al., 1996). The data were obtained using the scanning laser-reflectance thermometry facility in Fig. 7-24. The temperature increases with decreasing separation from the source side of the drift region due to the spatially varying impurity concentration. The impurity concentration is tailored to reduce the maximum electric field in the device, which makes possible the highest possible breakdown voltage (Leung et al., 1995).

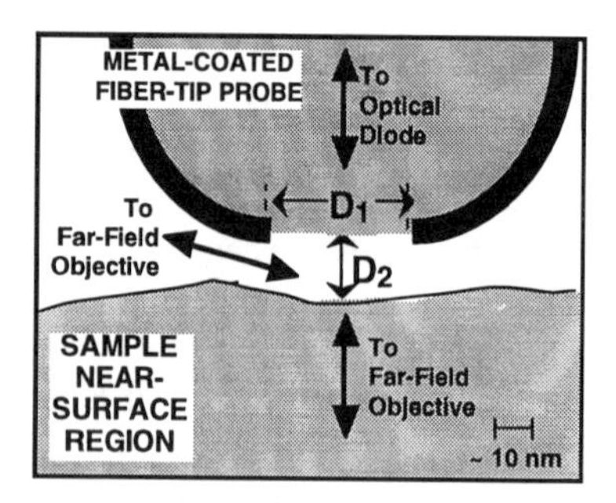

NFOT MODE	SOURCE	COLLECTION
1)	Probe	Far Field
2)	Far Field	Probe
3)	Probe	Probe
4)	Sample Emission	Probe

Figure 7-26 Schematic of high-resolution Near-Field Optical Thermometry (NFOT), which was developed for transient thermal mapping of interconnects by Goodson and Asheghi (1996). The use of an optical fiber to deliver or receive radiation promises resolution considerably better than conventional far-field techniques. The calibration and interpretation of NFOT signals requires further research, in particular to determine the influence of variations of the radiation orifice diameter D_1 and the tip-surface separation D_2 during a transient change in the surface temperature.

the surface. Goodson and Asheghi (1996) explored the feasibility of overcoming the diffraction-limited spatial resolution through the use of near-field irradiation of the sample surface through an optical fiber. The modes of near-field optical thermometry that are feasible using a scanning optical fiber are depicted in Fig. 7-26. These authors used mode 1 to observe large transient temperature variations near the edge of an interconnect-contact structure that are consistent with the far-field thermometry data shown in Fig. 7-4. This approach needs refinement, in particular to determine the dominant contribution to the optical signal as a function of the heating timescale and the sample thermal and thermomechanical properties.

Laser-transmittance thermometry. The temperature dependence of the index of refraction of a material can be monitored either through changes in the power or through deflection of a transmitted laser beam. Deboy et al. (1996) measured the average temperature along the optical path through semiconductor devices by monitoring the deflection of the beam. The average device temperature was deduced using independent data for the temperature dependence of the index of refraction and carrier-induced absorption. Problems with this approach include the arduous sample-preparation process, which allowed optical access to both sides of the device, and the difficulty of performing a precise calibration. This technique is poorly suited for active regions because of the impact of photogeneration on device operation.

Thermometry using the Raman effect. Raman spectroscopy examines inelastic photon-phonon scattering processes in a crystal through laser illumination. The scattering of the incident photons by phonons in a crystal leads to a shift in energies of many of the outscattered photons. The inelastically scattered photons have energies that are either increased or decreased by the energy of an optical phonon. The associated scattering processes require the creation and annihilation of an optical phonon, respectively, the probability of which is related to temperature by means of the solution to the harmonic oscillator problem in quantum mechanics (e.g., Kittel, 1986). If the outscattered radiation intensity is plotted as a function of the radiation wavelength, the maxima in the intensity corresponding to phonon creation and annihilation are called the *Stokes* and *anti-Stokes lines*. The ratio of the Stokes and anti-Stokes intensities is $\exp(-E_o/k_B T)$, where E_o is the energy of optical phonons in the crystal. If the energy of optical phonons is known with reasonable precision, as is the case for silicon and gallium arsenide, then the ratio provides an absolute measure of temperature. Micro-Raman spectroscopy uses highly focused laser light to perform diffraction-limited interrogation of photon-phonon scattering, which yields resolution better than 1 μm. Brugger et al. (1991) and Ostermeir et al. (1992) employed the micro-Raman spectroscopy to acquire temperature distributions in gallium arsenide and silicon semiconductor devices respectively. This technique offers a unique opportunity to measure the temperature of optical phonons, which interact most strongly with high-energy electrons-semiconductor devices as discussed in Section 7-4-3. The drawbacks associated with this technique include the unknown influence of photoelectric effects on the electrical behavior of the transistors.

Photothermal displacement measurements. Thermal expansion of solids can be measured with high precision using optical interferometry, which is coupled to the temperature field through thermal expansion. Martin and Wickramasinghe (1987) studied current distributions in microcircuits using this principle. Claeys et al. (1993) measured the deflection at the surface of an electrically biased compact FET using a similar approach. Since thermal expansion is a function of the temperature rise throughout the transistor and neighboring substrate, this technique yields relatively poor spatial resolution. The problem of deconvolution is quite difficult for most semiconductor devices, which consist of multilayers of vastly differing thermal expansion coefficients.

Liquid crystal thermometry. Liquid crystal thermometry has been available for more than two decades (e.g., Caroll, 1973; Meier et al., 1975) and has been used predominantly for the qualitative detection of hotspot temperatures in integrated circuits (e.g., Beck, 1986). The most common liquid crystals operate at temperatures up to a few hundred degrees Celsius and yield a low uncertainty only through exceedingly careful calibration procedures. Below a critical temperature, liquid crystals consist of elongated molecules that are directionally oriented in a manner that is denoted by their phase (e.g., Beck, 1986). Liquid crystals have been used for thermometry in the cholesteric phase, which induces a temperature-dependent peak in the wavelength dependence of the reflectivity. Fergason (1968) discussed the possibility of a spatial resolution of 25 μm and temperature resolution of 0.1 K using this approach. Stephens and Sinnadurai (1974) proposed a second approach, shown in Fig. 7-27, which uses the transition of crystals between the nematic and the isotropic phases. The nematic phase changes the polarization state of reflected light, and the isotropic phase does not. Polarization-sensitive optics allows regions coated with the nematic and isotropic crystals to be distinguished with spatial resolution of a few micrometers (Aszodi et al., 1981). The use of the nematic/isotropic phase transition suffers from the drawback that only one isotherm can be obtained.

Infrared (IR) thermography. The measurement of temperature by means of the radiation power emitted from a surface requires careful calibration of the emissivity, which depends strongly on the surface topography and the wavelengths that are interrogated (e.g., Bennett and Briles, 1989). The spatial resolution is limited by diffraction to the order of the wavelength contributing most strongly to the sensitivity. The wavelengths

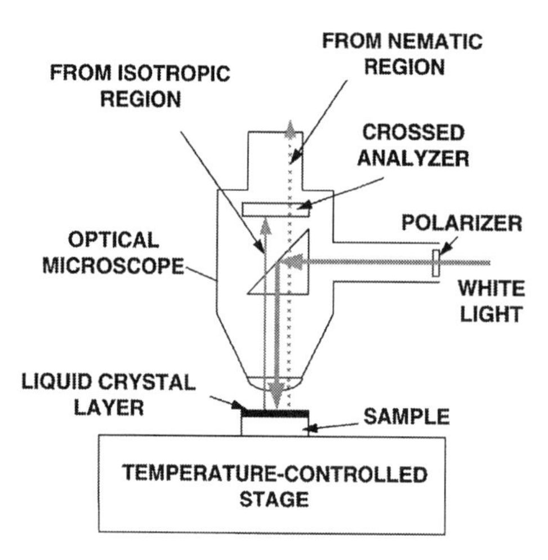

Figure 7-27 Apparatus for high-resolution liquid-crystal thermometry. A thin layer of a liquid crystal is deposited on the surface of a sample, which is subsequently illuminated with polarized white light. A crossed analyzer separates the light based on its polarization state. Regions with temperature above the phase transition temperature of the liquid crystal appear bright and regions with temperature below the phase transition temperature appear dark.

that dominate IR thermometry are near the maximum in the Planck distribution function for the radiation emissive power of a blackbody, which is described by means of Wien's displacement law. The dominant wavelength at room temperature is near 10 μm. This resolution renders IR thermography appropriate for qualitative thermal mapping over relatively large regions of an integrated circuit to identify the formation of hotspots. Kondo and Hinode (1995) employed IR thermometry to study the long timescale failure of interconnect structures.

Fluorescence thermal imaging. Fluorescence thermometry uses the temperature dependence of the radiation emitted from a surface that is irradiated from an independent source. The emitted radiation is stimulated by excitations near the surface that are induced by the incident radiation. The measurement requires the use of a fluorescing thin-film coating, such as europium thenoyltrifluoroacetonate (EuTTA) (Kolodner and Tyson, 1982). Incident radiation at a wavelength near 360 nm induces emission predominantly at 612 nm, the intensity of which varies strongly with the film temperature. The fluorescence quantum efficiency of EuTTA varies exponentially with temperature, which simplifies calibration of the emitted radiation power (Barton and Tangyunyong, 1996). Kolodner and Tyson (1983) demonstrated a spatial resolution of 0.7 μm on a flat surface. The temporal resolution has not been investigated. But the resolution limit must necessarily be long compared to the time for decay of the fluorescence, which is of the order of 200 μs for EuTTA excited near 360 nm. Disadvantages of this technique include the practical difficulties associated with sample preparation and the time degradation of the fluorescence properties of the EuTTA films during lengthy exposure to the ultraviolet radiation (Barton and Tangyunyong, 1996).

7-6 SUMMARY AND RECOMMENDATIONS

There is a wealth of tools available for the simulation and experimental investigation of micro- and nanoscale electrical transport in semiconductor devices and interconnects. Until recently, there has been comparatively little interest in studying thermal transport in devices and interconnects with a similar level of precision and rigor. This problem is being remedied by the theoretical and experimental research reviewed in this chapter.

There remain several problems for fundamental research. The scattering of phonons in semiconducting layers is poorly understood. The problem is most acute for scattering on electrons and holes, since the effects of these carriers are difficult to decouple from the influence of impurity scattering. In the active regions of a semiconductor device, electrons and holes exist with concentrations and energies that differ dramatically from those that prevail in equilibrium at the lattice temperature. The accuracy of the relatively complex simulation methods described in Section 7-4, such as the use of multiple Boltzmann equation moments, needs to be improved. The problem at present is due to the lack of precise knowledge of the phenomenological parameters, such as the energy and momentum relaxation times governed by electron-phonon scattering. The number of imprecisely known parameters increases with the number of governing equations. This problem can be remedied in part through measurements that provide greater leverage for determining the range of possible parameter values. The thermometry data described in Section 7-5 are useful for this purpose. Direct measurements of the *electron* temperature using an optical method would be especially helpful because of its extreme spatial gradients. Such a measurement must take much care to ensure that incident photons do not influence the behavior of active regions.

At a practical level, the greater complexity of thermal modeling and experimentation is most warranted for the device and interconnect technologies summarized in Section 7-1-2. For these technologies, microscale thermal phenomena can strongly influence performance and reliability figures of merit. Applied research is needed, for example, to determine the impact of short-timescale Joule heating on the electromigration and thermomechanical reliability of metallization in multilayer interconnect structures. The calculations of Hunter (1995) suggest that self-heating will limit the maximum current densities of future interconnect technologies. Another area for applied research is on the thermal failure of electrostatic discharge (ESD) protection circuitry. The measurement of transistor temperature fields during transient ESD failure is an important challenge that may be overcome through the use of rapid optical techniques described in Section 7-5-2. Thermal engineering research should not overlook the ultimate goal of this applied work, which is to identify and motivate design improvements in the materials or geometry of semiconductor devices and interconnects.

ACKNOWLEDGMENTS

This work was sponsored by the Semiconductor Research Corporation under contracts 96-PJ-357 and 97-SJ-461. Many of the concepts presented here developed during research-related discussions with the Device and Flow Physics Group of Texas Instruments, Inc., the Materials Research Group of Daimler-Benz AG, and Y. K. Leung and Professor Simon Wong of the Stanford University Department of Electrical Engineering. Katsuo Kurabayashi and Maxat Touzelbaev of the Stanford University Microscale Thermal and Mechanical Characterization Laboratory provided helpful comments. K. E. G. appreciates support from a Hewlett-Packard Terman Fellowship, the

Office of Naval Research Young Investigator Program, and the National Science Foundation through the CAREER Program and Grant CTS-9622178.

NOMENCLATURE

A_j	cross-sectional area in bipolar transistor, m^2
C	heat capacity per unit volume, $J\ m^{-3}\ K^{-1}$
C_{TR}	thermoreflectance coefficient, K^{-1}
$C_{carrier}$	effective heat capacity of chip carrier, $J\ K^{-1}$
C_{device}	effective heat capacity of device region, $J\ K^{-1}$
C_i	interconnect/insulator/substrate electrical capacitance, F
$C_{interconnect}$	effective heat capacity of interconnect, $J\ K^{-1}$
C_n	electron heat capacity per unit volume, $J\ m^{-3}\ K^{-1}$
C_{ox}	gate/insulator/channel electrical capacitance per unit area in a silicon FET, $F\ m^{-2}$
C_s	phonon heat capacity per unit volume, $J\ m^{-3}\ K^{-1}$
C_s'	phonon heat capacity per unit volume and nondimensional frequency, $J\ m^{-3}\ K^{-1}$
$C_{substrate}$	effective heat capacity of substrate, $J\ K^{-1}$
D_n	electron diffusivity, $m^2\ s^{-1}$
D_p	hole diffusivity, $m^2\ s^{-1}$
d_d	silicon device thickness in SOI transistor, m
d_{gb}	characteristic grain dimension, m
d_i	interconnect thickness, m
d_{ins}	thickness of insulator separating an interconnect from a substrate, m
d_o	thickness of buried silicon-dioxide for SOI transistor, m
d_s	thickness of semiconductor layer, m
$\mathbf{E}$	electric field, $V\ m^{-1}$
E	energy, J
E_c	electron energy at conduction band edge, J
E_F	Fermi energy, J
E_{Fn}	quasi Fermi level for electrons, J
E_{Fp}	quasi Fermi level for holes, J
E_g	bandgap, J
E_{imp}	electron energy loss due to impact ionization, J
E_o	optical phonon energy, J
F	mean free path reduction ratio due to boundary scattering, Eq. (7-33)
f_{EQ}	Fermi-Dirac equilibrium distribution function
f_n	electron distribution function
$f_{n,EQ}$	equilibrium electron distribution function
G_D	absolute transistor electrical conductance, $A\ V^{-1}$
g_D	transistor differential conductance, $A\ V^{-1}$

g_T	transistor differential transconductance, A V^{-1}
h_P	Planck's constant = 6.625×10^{-34} J s
I_D	electrical current from drain to source of field-effect transistor, A
$I_{D,sat}$	drain-source electrical current of saturated field-effect transistor, A
I_G	electrical current along gate in FET thermometry structure, A
I_b	electrical current into base of bipolar transistor, A
I_c	electrical current out of collector of bipolar transistor, A
I_e	electrical current into emitter of bipolar transistor, A
$\mathbf{j}_n$	electrical current density vector due to electrons, A m^{-2}
$\mathbf{j}_p$	electrical current density vector due to holes, A m^{-2}
j	electrical current density, A m^{-2}
j_s	saturation electrical current density for reverse-biased diode, A m^{-2}
$\mathbf{k}$	wavevector, m^{-1}
k	thermal conductivity, W m^{-1} K^{-1}
k_B	Boltzmann constant = 1.381×10^{-23} J K^{-1}
$k_{a,eff}$	effective thermal conductivity along a layer, W m^{-1} K^{-1}
k_{bulk}	thermal conductivity of material in bulk, dense form, W m^{-1} K^{-1}
k_{int}	thermal conductivity within layer, W m^{-1} K^{-1}
k_n	electron thermal conductivity, W m^{-1} K^{-1}
$k_{n,eff}$	effective thermal conductivity for conduction normal to a layer, W m^{-1} K^{-1}
k_s	phonon thermal conductivity, W m^{-1} K^{-1}
L_{Dn}	recombination length for minority electrons near p-n junction diode, m
L_{Dp}	recombination length for minority holes in near p-n junction diode, m
L_d	channel-interconnect separation in SOI FET, m
L_c	channel length of a field-effect transistor, m
L_i	interconnect length, m
L_o	Lorenz number = 2.45×10^{-8} W Ω K^{-2}
M	atomic mass of host atom, kg
m	mass of a free electron, kg
m_n^*	electron effective mass, kg
m_p^*	hole effective mass, kg
N	phonon distribution function
N_A	concentration of acceptor impurities, m^{-3}
N_D	concentration of donor impurities, m^{-3}
N_{EQ}	Bose-Einstein distribution function at the lattice temperature
N_c	effective density of electron states at the conduction band edge, m^{-3}
N_v	effective density of electron states at the valence band edge, m^{-3}
n	electron number density, m^{-3}
n_i	imperfection concentration, m^{-3}
P	power, W
$\mathbf{p}_n$	electron momentum density, kg m^{-2} s^{-1}

p	hole number density, m^{-3}
P_n	electron thermoelectric power, $V\ K^{-1}$
P_p	hole thermoelectric power, $V\ K^{-1}$
p_{sr}	specular reflection coefficient
q	electron charge magnitude $= 1.602 \times 10^{-19}$ C
q'''	volumetric heat-generation rate, $W\ m^{-3}$
R	electrical resistance, Ω
R_B	thermal boundary resistance , $K\ m^{-2}\ W^{-1}$
R_c	channel-substrate thermal resistance for FET, $K\ W^{-1}$
$R_{carrier\text{-}room}$	thermal resistance between chip carrier and the environment, $K\ W^{-1}$
$R_{device\text{-}substrate}$	thermal resistance between device and substrate, $K\ W^{-1}$
R_i	electrical resistance along interconnect, Ω
$R_{interconnect\text{-}substrate}$	thermal resistance between interconnect and substrate, $K\ W^{-1}$
R_{opt}	optical reflectance
R_s	sheet resistance, Ω
$R_{substrate\text{-}carrier}$	thermal resistance between a substrate and carrier, $K\ W^{-1}$
$\mathbf{r}$	position vector, m
r_g	rate of generation of electron-hole pairs, s^{-1}
r_r	rate of recombination of electron-hole pairs, s^{-1}
S	separation between emitter and collector of a bipolar transistor, m
S_i	scattering cross section due to localized imperfections, m^2
S_R	net rate of radiation power from semiconductor, $W\ m^{-3}$
T	temperature, K
T_c	average channel temperature in FET, K
$T_{carrier}$	temperature of chip carrier, K
T_{device}	temperature of device, K
$T_{interconnect}$	temperature of interconnect, K
T_j	transistor junction temperature, K
T_n	electron temperature, K
T_P	peak processing temperature of layer, K
T_p	hole temperature, K
T_{room}	room temperature, K
T_s	lattice temperature, K
$T_{substrate}$	temperature of substrate, K
T_0	temperature beneath silicon dioxide layer in SOI substrate, K
t	time, s
t_d	capacitive delay, s
V	volume of host atom, m^{-3}
V_D	drain-source voltage drop in a field-effect transistor, V
$V_{D,sat}$	saturation value of the drain-source voltage drop in a field-effect transistor, V
V_G	gate-substrate bias voltage in a field-effect transistor, V
V_T	threshold gate-substrate voltage of a MOSFET, V

V_{bi}	built-in voltage of a p-n junction diode, V
V_{be}	base-emitter bias in a bipolar transistor, V
V_{cb}	collector-base bias in a bipolar transistor, V
V_{pn}	external bias voltage between the p-type and n-types sides of a diode, V
$\mathbf{v}_n$	average electron velocity, m s^{-1}
v_F	electron Fermi velocity, m s^{-1}
v_s	phonon velocity, m s^{-1}
$v_{n,\mathbf{k}}$	velocity of electrons with a wavevector **k**, m s^{-1}
$v_{s,\mathbf{k}}$	velocity of phonons with a wavevector **k**, m s^{-1}
W	width of the active region, m
W_n	electron energy density, J m^{-3}
W_p	hole energy density, J m^{-3}
w_d	transistor width, m
w_e	extra width in FET thermometry structure, m
w_i	width of an interconnect, m
x, y, z	transistor coordinates, given in Fig. 7-4
x_ω	dimensionless phonon frequency $= h_P\, \omega/2\pi k_B T$
β	gain of a bipolar transistor
Δf	frequency bandwidth, s^{-1}
ΔM	mass difference between the impurity and the host atom, kg
Δp	difference in the minority hole concentrations, m^{-3}
ΔR_{opt}	change in surface reflectance
ΔT	temperature change, K
δ	dimensionless layer thickness $= d_s/\Lambda_{s,NB}$
ε	permittivity of semiconductor, F m^{-1}
ε_{ins}	permittivity of insulator, F m^{-1}
γ	Sommerfeld parameter, J m^{-3} K^{-2}
η	characteristic dimension of interface roughness, m
$\Lambda_{s,NB}$	phonon mean free path in the absence of boundary scattering, m
λ	phonon wavelength, m
μ_n	low-electric-field mobility of electrons, m^2 s^{-1} V^{-1}
μ_p	low-electric-field mobility of holes, m^2 s^{-1} V^{-1}
Θ	Debye temperature, K
ξ	argument for integration
ρ	electrical resistivity, Ω m
σ	electrical conductivity, Ω^{-1}
τ_m	recombination time for electrons, s
τ_n	relaxation time for electrons, s
τ_{n-a}	relaxation time for electron scattering with acoustic phonons, s
τ_{n-gb}	relaxation time for electron-grain boundaries collisions, s
τ_{n-i}	relaxation time for scattering with impurities, s
$\tau_{n,M}$	electron momentum relaxation time, s
τ_{n-o}	relaxation time for electron scattering with optical phonons, s

τ_{n-s}	relaxation time for phonon-electron collisions, s
$\tau_{n-s,E}$	electron-phonon energy relaxation time, s
τ_p	relaxation time for holes, s
$\tau_{p-s,E}$	hole-phonon energy relaxation time, s
τ_{rp}	recombination time for holes, s
τ_s	relaxation time for phonons, s
τ_{s-i}	relaxation time for phonon scattering on imperfections, s
$\tau_{s,NB}$	phonon relaxation time in the absence of boundary scattering, s
τ_{s-p}	relaxation time for phonon scattering on holes, s
τ_{s-s}	relaxation time for phonon scattering on phonons, s
ω	phonon angular frequency, rad s^{-1}

REFERENCES

Adams, A. C., "Dielectric and polysilicon film deposition," in *VLSI Technology*, S. M. Sze (ed.), McGraw-Hill, New York, pp. 233–271, 1988.

Agrawal, G. P., and N. K. Dutta, *Semiconductor Lasers*, Van Nostrand Reinhold, New York, 1993.

Amerasekera, A., M. C. Chang, J. A. Seitchik, A. Chatterjee, K. Mayaram, and J. H. Chern, "Self-heating effects in basic semiconductor structures," *IEEE Trans. Electron Dev.*, vol. 40, pp. 1836–1843, 1993.

Amerasekera, A., W. Van Abeelen, L. Van Roozendaal, M. Hannemann, and P. Schofield, "ESD failure modes: Characteristics mechanisms, and process influences," *IEEE Trans. Electron Dev.*, vol. 39, pp. 430–436, 1992.

Amerasekera, A., and J. Verwey, "ESD in integrated circuits," *Qual. Reliab. Eng. Int.*, vol. 8, pp. 259–272, 1992.

Apanovich, Y., P. Blakey, R. Cottle, E. Lyumkis, B. Polsky, and A. Shur, "Numerical simulations of ultra thin SOI transistor using non-isothermal energy-balance model," *Proceedings of the IEEE International SOI Conference*, Nantucket, MA, October 3–6, pp. 33–34, 1994.

Apanovich, Y., P. Blakey, R. Cottle, E. Lyumkis, B. Polsky, A. Shur, and A. Tcherniaev, "Numerical simulation of submicrometer devices including coupled nonlocal transport and nonisothermal effects," *IEEE Trans. Electron Dev.*, vol. 42, pp. 890–898, 1995.

Arnold, E., H. Pein, and S. P. Herko, "Comparison of self-heating effects in bulk-silicon and SOI high-voltage devices," *IEDM Tech. Dig.*, pp. 813–816, 1994.

Asheghi, M., M. N. Touzelbaev, K. E. Goodson, Y. K. Leung, and S. S. Wong, "Temperature-dependent thermal conductivity of single-crystal silicon layers in SOI substrates," presented at the *International Mechanical Engineering Congress and Exposition*, Atlanta, GA, November 17–22, DSC-Vol. 59, in *Micro-Electro-Mechanical Systems (MEMS)*, C. T. Avedisian et al. (eds.), pp. 83–91, 1996. *ASME J. Heat Transfer*, in press.

Aszodi, G., J. Szabon, I. Janossy, and V. Szekely, "High resolution thermal mapping of microcircuits using nematic liquid crystals," *Solid State Electron.*, vol. 24, pp. 1127–1133, 1981.

Aung, W. (ed.), *Cooling Techniques for Computers*, Hemisphere, New York, 1991.

Banerjee, K., A. Amerasekera, and C. Hu, "Characterization of VLSI circuit interconnect heating and failure under ESD conditions," *Proceedings of International Reliability Physics Symposium*, pp. 237–245, 1996.

Bar-Cohen, A., and A. D. Kraus (eds.), *Advances in Thermal Modeling of Electronic Components and Systems*, vol. 2, ASME Press, New York, 1990.

Barton, D. L., and P. Tangyunyong, "Fluorescent microthermal imaging—theory and methodology for achieving high thermal resolution images," *Microelectron. Eng.*, vol. 31, pp. 271–279, 1996.

Beck, F., "Use of liquid-crystal thermography for defect location on semiconductor devices," *Qual. Reliab. Eng. Int.*, vol. 2, pp. 143–151, 1986.

Beebe, S., F. Rotella, Z. Sahul, D. Yergeau, G. McKenna, L. So, Z. Yu, K. C. Wu, E. Kan, J. Mcvittie, and R. W. Dutton, "Next generation stanford TCAD—PISCES 2ET and SUPREM 007," *Proceedings of the IEEE International Electron Devices Meeting*, San Francisco, CA, Dec. 11–14, 1994.

Bennett, G. A., and S. D. Briles, "Calibration procedure developed for IR surface-temperature measurements," *IEEE Trans. Components, Hybrids Manufacturing Technol.*, vol. 12, pp. 690–695, 1989.

Berman, R., *Thermal Conduction in Solids*, Oxford University Press, Oxford, 1976.

Berman, R., E. L. Foster, and J. M. Ziman, "Thermal conduction in artificial sapphire crystals at low temperatures," *Proc. R. Soc. London A*, vol. 231, pp. 130–144, 1955.

Blackburn, D. L., "A review of thermal characterization of power transistors," *Proceedings of IEEE Semiconductor Thermal and Temperature Measurement Symposium*, pp. 1–7, 1988.

Bloetekjaer, K., "Transport equations for electrons in two-valley semiconductors," *IEEE Trans. Electron Dev.*, vol. ED-17, pp. 38–47, 1970.

Bose, B. K., "Power electronics—a technology review," *Proc. IEEE*, vol. 80, pp. 1303–1334, 1992.

Brugger, H., "Raman spectroscopy for characterization of lazered semiconductor materials and devices," in *Light Scattering in Semiconductor Structures and Superlattices*, K. J. Lockwood and J. F. Young (eds.), Plenum Press, New York, pp. 259–274, 1991.

Bunyan, R. J. T., M. J. Uren, J. C. Alderman, and W. Eccleston, "Use of noise thermometry to study the effects of self-heating in submicrometer SOI MOSFET's," *IEEE Electron Dev. Lett.*, vol. 13, pp. 279–281, 1992.

Cahill, D. G., "Heat transport in dielectric thin films and at solid–solid interfaces, " *Microscale Thermophys. Eng.*, vol. 1, pp. 85–109, 1997.

Callaway, J., "Model for lattice thermal conductivity at low temperatures," *Phys. Rev.*, vol. 113, pp. 1046–1051, 1959.

Cain, B. M., P. A. Goud, and C. G. Englefield, "Electrical measurement of the junction temperature of an RF power transistor," *IEEE Trans. Instrum. Measure.*, vol. 41, pp. 663–665, 1992.

Caroll, P., *Cholesteric Liquid Crystals: Their Technology and Applications*, Ovum, London, 1973.

Celler, G. K., and A. E. White, "Buried oxide and silicide formation by high-dose implantation in silicon," *MRS Bull.*, vol. 17, pp. 40–46, June, 1992.

Chen, G., "A comparative study on the thermal characteristics of vertical-cavity surface-emitting lasers," *J. Appl. Phys.*, vol. 77, pp. 4251–4258, 1995.

Chen, G., "Heat transfer in micro- and nanoscale photonic devices," *Annu. Rev. Heat Transfer*, in press, 1996a.

Chen, G., "Heat transport in the perpendicular direction of superlattices and periodic thin-film structures," *International Mechanical Engineering Congress and Exposition*, Atlanta, GA, November 17–22, in *Micro-Electro-Mechanical Systems (MEMS)*, C. T. Avedisian et al. (eds.), DSC-Vol. 59, pp. 13–24, 1996b.

Chen, G., and C. L. Tien, "Thermal conductivity of quantum-well structures," *J. Thermophys. Heat Transfer*, vol. 7, pp. 311–318, 1993.

Chen, G., C. L. Tien, X. Wu, and J. S. Smith, "Measurement of thermal diffusivity of GaAs/AlGaAs thin-film structures," *J. Heat Transfer*, vol. 116, pp. 325–331, 1994.

Chen, G., and X. Y. Yu, "Thermal diffusivity of GaAs/AlAs superlattices," *6th AIAA/ASME Joint Thermophysics and Heat Transfer Conference*, AIAA 94-1964, 1994.

Claeys, W., S. Dilhaire, V. Quintard, J. P. Dom, and Y. Danto, "Thermoreflectance optical test probe for the measurement of current-induced temperature changes in microelectronic components," *Reliab. Eng. Int.*, vol. 9, pp. 303–308, 1993.

Colinge, J. P., *Silicon-on-Insulator Technology: Materials to VLSI*, Kluwer Academic, Boston, 1991.

Deboy, G., G. Sölkner, E. Wolfgang, and W. Claeys, "Absolute measurement of transient carrier concentration and temperature gradients in power semiconductor devices by internal IR-laser deflection," *Microelectron. Eng.*, vol. 31, pp. 299–307, 1996.

Epperlein, P.-W., "Micro-temperature measurements on semiconductor laser mirrors by reflectance modulation: A newly developed technique for laser characterization," *Jpn. J. Appl. Phys.*, vol. 32, pp. 5514–5522, 1993.

Estreich, D. B., "A DC technique for determining GaAs MESFET thermal resistance," *Proceedings of the Fifth IEEE SEMI-THERM Symposium*, pp. 136–139, 1989.

Fergason, J. L., "Liquid crystals in nondestructive testing," *Appl. Opt.*, vol. 7, pp. 1729–1737, 1968.

Fletcher, L. S., "A review of thermal enhancement techniques for electronic systems," *IEEE Trans. Components, Hybrids, Manufacturing Technol.*, vol. 13, pp. 1012–1021, 1990.

Fushinobu, K., A. Majumdar, and K. Hijikata, "Heat generation and transport in submicron semiconductor devices," *ASME J. Heat Transfer*, vol. 117, pp. 25–31, 1995.

Goodson, K. E., "Impact of CVD diamond layers on the thermal engineering of electronic systems," *Annu. Rev. Heat Transfer*, vol. 6, pp. 323–353, 1995.

Goodson, K. E., "Thermal conduction in nonhomogeneous CVD diamond layers in electronic microstructures," *ASME J. Heat Transfer*, vol. 118, pp. 279–286, 1996.

Goodson, K. E., and M. Asheghi, "Near-field optical thermometry," *Microscale Thermophys. Eng.*, vol. 1, pp. 225–235, 1997.

Goodson, K. E., Y. S. Ju., M. Asheghi, O. W. Käding, M. N. Touzelbaev, Y. K. Leung, and S. S. Wong, "Microscale thermal characterization of high-power silicon-on-insulator transistors," *Proceedings of the 31st ASME National Heat Transfer Conference*, vol. 5, R. L. Mahajan et al. (ed.), Houston, TX, pp. 1–9, August 3–6, 1996.

Goodson, K. E., and P. T. Cooper, "The effect of high-energy electrons on lattice conduction in semiconductor devices," *Proceedings of the Symposium on Thermal Science and Engineering in Honor of Chancellor Chang-Lin Tien*, Richard O. Buckius (ed.), Berkeley, CA, pp. 153–159, Nov. 14, 1995.

Goodson, K. E., and M. I. Flik, "Effect of microscale thermal conduction on the packing limit of silicon-on-insulator electronic devices," *IEEE Trans. Components, Hybrids, Manufacturing Technol.*, vol. 15, pp. 715–722, 1992.

Goodson, K. E., and M. I. Flik, "Solid-layer thermal-conductivity measurement techniques," *Appl. Mechan. Rev.*, vol. 47, pp. 101–112, 1994.

Goodson, K. E., M. I. Flik, L. T. Su, and D. A. Antoniadis, "Annealing temperature dependence of thermal conductivity of CVD silicon-dioxide layers," *IEEE Electron Dev. Lett.*, vol. 14, pp. 490–492, 1993a.

Goodson, K. E., M. I. Flik, L. T. Su, and D. A. Antoniadis, "Annealing-temperature dependence of the thermal conductivity of CVD silicon-dioxide layers," *Proceedings of the 29th ASME National Heat Transfer Conference*, Atlanta, GA, Aug. 8–11, in *Heat Transfer on the Microscale*, F. M. Gerner and K. S. Udell (eds.), ASME HTD-vol. 253, pp. 21–28, 1993b.

Goodson, K. E., M. I. Flik, L. T. Su, and D. A. Antoniadis, "Prediction and measurement of the thermal conductivity of amorphous dielectric layers," *ASME J. Heat Transfer*, vol. 116, pp. 317–324, 1994.

Goodson, K. E., M. I. Flik, L. T. Su, and D. A. Antoniadis, "Prediction and measurement of temperature fields in silicon-on-insulator electronic circuits," *ASME J. Heat Transfer*, vol. 117, pp. 574–581, 1995a.

Goodson, K. E., O. W. Käding, M. Rösler, and R. Zachai, "Experimental investigation of thermal conduction normal to diamond-silicon interfaces," *J. Appl. Phys.*, vol. 77, pp. 1385–1392, 1995b.

Graebner, J. E., "Thermal conductivity of CVD diamond: Techniques and results," *Diamond Films Technol.*, vol. 3, pp. 77–130, 1993.

Greason, W. D., *Electrostatic Discharge in Electronics*, Wiley, New York, 1992.

Holland, M. G., "Analysis of lattice thermal conductivity," *Phys. Rev.*, vol. 132, pp. 2461–2471, 1963.

Hunter, W. R., "The implications of self-consistent current density design guidelines comprehending electromigration and joule heating for interconnect technology evolution," *Proceedings of the IEEE IEDM Conference*, pp. 19.2.1–19.2.4, 1995.

Ismail, M., and T. Fiez, *Analog VLSI: Signal and Information Processing*, McGraw-Hill, New York, 1994.

Ju, Y. S., O. W. Käding, K. E. Goodson, Y. K. Leung, and S. S. Wong, "Transient thermal mapping of SOI LDMOS transistors," submitted to *IEEE Electron Dev. Lett.*, 1996.

Ju, Y. S, and K. E. Goodson, "Short-timescale thermometry and reliability studies of metal interconnects in VLSI circuits," presented at the *International Mechanical Engineering Congress and Exposition*, Atlanta, GA, Nov. 17–22, 1996.

Käding, O. W., H. Skurk, and K. E. Goodson, "Thermal conduction normal to metallized silicon-dioxide layers on silicon," *Appl. Phys. Lett.*, vol. 65, pp. 1629–1631, 1994.

Kassakian, J. G., M. F. Schlecht, and G. C. Verghese, *Principles of Power Electronics*, Addison-Wesley, New York, 1994.

Kittel, C., *Introduction to Solid State Physics*, Wiley, New York, Chaps. 6 and 7, 1986.

Kolodner, P., and J. A. Tyson, "Microscopic fluorescent imaging of surface temperature profiles with 0.01 C resolution," *Appl. Phys. Lett.*, vol. 40, pp. 782–784, 1982.

Kolodner, P., and J. A. Tyson, "Remote thermal imaging with 0.7-μm spatial resolution using temperature-dependent fluorescent thin films," *Appl. Phys. Lett.*, vol. 42, pp. 117–119, 1983.

Kondo, S., and K. Hinode, "High-resolution temperature measurement of void dynamics induced by electromigration in aluminum metallization," *Appl. Phys. Lett.*, vol. 67, pp. 1606–1608, 1995.

Lai, J., M. Chandrachood, A. Majumdar, and J. P. Carrejo, "Thermal detection of device failure by atomic force microscopy," *IEEE Electron Dev. Lett.*, vol. 16, pp. 312–314, 1995.

Lai, J., and A. Majumdar, "Concurrent thermal and electrical modeling of sub-micrometer silicon devices," *J. Appl. Phys.*, vol. 79, pp. 7353–7361, 1996.

Lee, S. M., and D. G. Cahill, "Heat transport in thin dielectric films," *Microscale Thermophys. Eng.*, vol. 1, pp. 47–52, 1997.

Leung, Y.-K., Y. Suzuki, K. E. Goodson, and S. S. Wong, "Self-heating effect in lateral DMOS on SOI," *Proceedings of The 7th International Symposium of Power Semiconductor Devices and ICs*, Yokohama, Japan, pp. 136–140, 1995.

Liou, L. L., and B. Bayraktaroglu, "Thermal stability analysis of AlGaAs/GaAs heterojunction bipolar transistors with multiple emitter fingers," *IEEE Trans. Electron Dev.*, vol. 41, pp. 629–636, 1994.

Liu, W., and B. Bayraktaroglu, "Theoretical calculations of temperature and current profiles in multi–finger heterojunction bipolar transistors," *Solid State Electron.*, vol. 36, pp. 125–132, 1993.

Liu, W., and A. Yuksel, "Measurement of junction temperature of an AlGaAs/GaAs heterojunction biploar transistor operating at large power densities," *IEEE Trans. Electron Dev.*, vol. 42, pp. 358–360, 1995.

Luo, K., Z. Shi, J. Lai, and A. Majumdar, "Nanofabrication of sensors on cantilever probe tips for scanning multiprobe microscopy," *Appl. Phys. Lett.*, vol. 68, pp. 325–327, 1996a.

Luo, K., Z. Shi, J. Lai, and A. Majumdar, "Internal temperature distribution of a vertical-cavity surface-emitting laser measured by scanning thermal microscopy," presented at the *ASME International Mechanical Engineering Congress and Exposition*, Atlanta, GA, Nov. 17–22, 1996b.

Majumdar, A., J. P. Carrejo, and J. Lai, "Thermal imaging using the atomic force microscope," *Appl. Phys. Lett.*, vol. 62, pp. 2501–2503, 1993.

Majumdar, A., K. Fushinobu, and K. Hijikata, "Effect of gate voltage on hot-electron and hot-phonon interaction and transport in a submicrometer transistor," *J. Appl. Phys.*, vol. 77, pp. 6686–6694, 1995.

Majumdar, A., J. Lai, M. Chandrachood, O. Nakabeppu, Y. Wu, and Z. Shi, "Thermal imaging by atomic force microscopy using thermocouple cantilever probes," *Rev. Sci. Instrum.*, vol. 66, pp. 3584–3592, 1995.

Maloney, T. J., and N. Khurana, "Transmission line pulsing techniques for circuit modeling of ESD phenomena," *Proceedings of EOS/ESD Symposium*, pp. 49–54, 1985.

Manduteanu, G. V., "A new device — A power semiconductor diode with an integrated thermal sensor," *IEEE Trans. Electron Dev.*, vol. 35, pp. 700–703, 1988.

Martin, Y., and H. K. Wickramasinghe, "Study of dynamic current distribution in logic circuits by joule displacement microscopy," *Appl. Phys. Lett.*, vol. 50, pp. 167–168, 1987.

Mastrangelo, C. H., and R. S. Mäller, "Thermal diffusivity of heavily doped low-pressure chemical vapor deposited polycrystalline silicon films," *Sensors Mater.*, vol. 3, pp. 133–141, 1988.

Mautry, P. G., and J. Trager, "Self-heating and temperature measurement in sub-μm-MOSFETs," *Proceedings of the IEEE International Conference on Microelectronic Test Structures*, vol. 3, pp. 221–226, 1990.

Meier, G., E. Sackmann, and J. G. Grabmeier, *Applications of Liquid Crystals*, Springer Verlag, Berlin, 1975.

Miklos, A., and A. Lorincz, "Transient thermoreflectance of thin metal films in the picosecond regime," *J. Appl. Phys.*, vol. 63, pp. 2391–2395, 1988.

Murarka, S. P., *Metallization*, Butterworth-Heinemann, Boston, 1993.

Nagasima, N., "Structure analysis of silicon dioxide films formed by oxidation of silane," *J. Appl. Phys.*, vol. 43, pp. 3378–3386, 1972.

Nonnenmacher, M., and H. K. Wickramasinghe, "Scanning probe microscopy of thermal conductivity and subsurface properties," *Appl. Phys. Lett.*, vol. 61, pp. 168–170, 1992.

Ostermeir, R., K. Brunner, G. Abstreiter, and W. Weber, "Temperature distribution in Si-MOSFET's studied by Micro-Raman spectroscopy," *IEEE Trans. Electron Dev.*, vol. 39, pp. 858–863, 1992.

Paul, O. M., J. Korvink, and H. Baltes, "Thermal conductivity of CMOS materials for the optimization of microsensors," *J. Micromechan. Microeng.*, vol. 3, pp. 110–112, 1993.

Paul, O. M., J. Korvink, and H. Baltes, "Determination of the thermal conductivity of CMOS IC polysilicon," *Sensors Actuators A (Physical)*, vol. A41, no. 1–3, pp. 161–164, 1994.

Peters, L., "SOI takes over where silicon leaves off," *Semicond. Int.*, vol. 16, pp. 48–51, 1993.

Qiu, T. Q., C. P. Grigoropoulos, and C. L. Tien, "Novel technique for noncontact and microscale temperature measurements," *Exp. Heat Transfer*, vol. 6, pp. 231–241, 1993.

Redman-White, W., M. S. L. Lee, B. M. Tenbroek, M. J. Uren, and R. J. T. Bunyan, "Direct extraction of MOSFET dynamic thermal characteristics from standard transistor structures using small signal measurements," *Electron. Lett.*, vol. 29, pp. 1180–1181, 1993.

Slack, G. A., "Thermal conductivity of pure and impure silicon carbide and diamond," *J. Appl. Phys.*, vol. 35, pp. 3460–3466, 1964.

Shur, M., *Physics of Semiconductor Devices*, Prentice Hall, Englewood Cliffs, NJ, 1990.

Smolinsky, G., and T. P. H. F. Wendling, "Measurements of temperature dependent stress of silicon dioxide films prepared by a variety of CVD methods," *J. Electrochem. Soc.*, vol. 132, pp. 950–954, 1985.

Sondheimer, E. H., "The mean free path of electrons in metals," *Adv. Phys.*, vol. 1, pp. 1–42, 1952.

Stephens, C. E., and F. N. Sinnadurai, "A surface temperature limit detector using nematic liquid crystals with an application to microcircuits," *J. Phys. E*, vol. 7, pp. 641–643, 1974.

Su, L. T., D. A. Antoniadis, N. D. Arora, B. S. Doyle, and D. B. Krakauer, "SPICE model and parameters for fully-depleted SOI MOSFET'S including self-heating," *IEEE Electron Dev. Lett.*, vol. 15, pp. 374–376, 1994.

Sze, S. M., *Physics of Semiconductor Devices*, Wiley, New York, 1981.

Sze, S. M., "Introduction: Growth of the industry," in *VLSI Technology*, S. M. Sze (ed.), McGraw-Hill, New York, pp. 1–8, 1988.

Tai, Y. C., C. H. Mastrangelo, and R. S. Müller, "Thermal diffusivity of heavily doped low-pressure chemical vapor deposited polycrystalline silicon films," *J. Appl. Phys.*, vol. 63, pp. 1442–1447, 1988.

Touloukian, Y. S., R. W. Powell, C. Y. Ho, and P. G. Klemens, "Thermal conductivity of metallic elements and alloys," in *Thermophysical Properties of Matter*, vol. 1, New York: IFI/Plenum, p. 339, 1970.

Tsividis, Y. P., *Operation and Modeling of the MOS Transistor*, McGraw-Hill, New York, pp. 102–167, pp. 148–150, pp. 168–216, 1987.

Veendrick, H. J. M., *MOS ICs: From Basics to ASICs*, VCH, New York.

Von Arx, M., and H. Baltes, "A microstructure for measurement of thermal conductivity of polysilicon thin films," *J. Microelectromechan. Syst.*, vol. 1, pp. 193–196, 1992.

Von Arx, M., O. M. Paul, and H. Baltes, "Determination of the heat capacity of CMOS layers for optimal CMOS sensor design," *Sensors Actuators A (Physical)*, vol. A47, no. 1-3, pp. 428–31, 1995.

Wang, C. T., "A new set of semiconductor equations for computer simulation of submicron devices," *Solid State Electron.*, vol. 28, pp. 783–788, 1985.

Williams, C. C., and H. K. Wickramasinghe, "Scanning thermal profiler," *Appl. Phys. Lett.*, vol. 49, pp. 1587–1589, 1986.

Wilson, S. R., C. J. Tracy, and J. L. Freeman, *Handbook of Multilevel Metallization for Integrated Circuits*, Noyes, Park Ridge, NJ, 1993.

Yang, E. S., *Microelectronic Devices*, McGraw-Hill, New York, pp. 287–290, 1988.

Yu, X. Y., G. Chen, A. Verma, and J. S. Smith, "Temperature dependence of thermophysical properties of GaAs/AlAs periodic structure," *Appl. Phys. Lett.*, vol. 67, pp. 3554–3556, 1995.

Yu, Z., D. Chen, L. So, and S. W. Dutton, *PISCES-2ET—Two-Dimensional Device Simulation for Silicon and Heterostructures*, Stanford University, 1994.

Wachutka, G. K., "Rigorous thermodynamic treatment of heat generation and conduction in semiconductor device modeling," *IEEE Trans. Computer-Aided Design*, vol. 9, pp. 1141–1149, 1990.

Wolbert, P. B., G. K. Wachutka, B. H. Krabbenborg, and T. J. Mouthaan, "Nonisothermal device simulation using the 2-D numerical process/device simulator TRENDY and application to SOI-devices," *IEEE Trans. Computer-Aided Design of Integrated Circuits*, vol. 13, pp. 293–302, 1994.

Ziman, J. M., *Electrons and Phonons*, Oxford University Press, Oxford, 1960.

Zweidinger, D. T., R. M. Fox, J. S. Brodsky, T. Jung, and S. Lee, "Thermal impedance extraction for bipolar transistors," *IEEE Trans. Electron Dev.*, vol. 43, pp. 342–346, 1996.

CHAPTER

EIGHT

MICRO HEAT PIPES

G. P. Peterson

Department of Mechanical Engineering, Texas A&M University, College Station, TX 77843

L. W. Swanson

Department of Mechanical Engineering, Texas A&M University, College Station, TX 77843

Frank M. Gerner

Department of Mechanical, Industrial and Nuclear Engineering, University of Cincinnati, Cincinnati, OH 45221

8-1 FUNDAMENTAL OPERATING PRINCIPLES

Fundamentally, the operating principles of micro heat pipes are quite similar to those of larger, more conventional heat pipes. The heat applied to one end of the heat pipe vaporizes the working fluid in that region, forcing it to the cooler end where it condenses and gives up the latent heat of vaporization. This vaporization and condensation process causes the liquid–vapor interface in the sharp corner regions, which serve as liquid arteries, to change continually along the pipe (see Fig. 8-1) and results in a capillary pressure difference between the evaporator and condenser regions. This capillary pressure difference promotes the flow of the working fluid from the condenser back to the evaporator through the sharp-edged corner regions (Peterson, 1994). To date, investigations have been conducted on many different types of relatively small heat pipes (Dean, 1976; Basiulis et al., 1987; Murase et al., 1987; Babin and Peterson, 1990; Sotani et al., 1990; Badran et al., 1993; Longtin et al., 1994; Khrustalev and Faghri, 1994). Most of these devices, however, are in reality only miniaturized versions of larger more conventional heat pipes.

The concept of very small "micro" heat pipes incorporated into semiconductor devices to promote more uniform temperature distribution and improve thermal control was first introduced by Cotter (1984). To better understand what the term "micro heat pipe" implies, Babin et al. (1990) expressed Cotter's initial definition of a micro heat pipe mathematically as

$$K \propto \frac{1}{r_h} \tag{8-1}$$

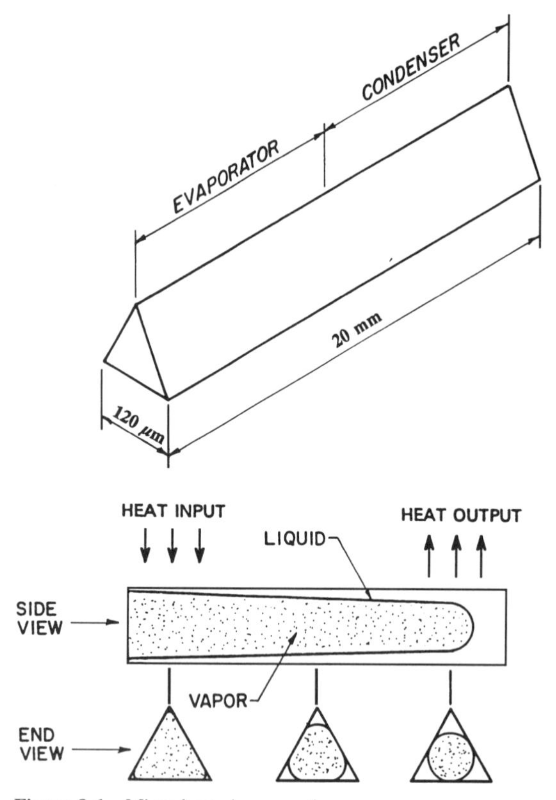

Figure 8-1 Micro heat pipe operation.

where K is the mean curvature of the liquid–vapor interface and r_h is the hydraulic radius of the flow channel. Then by assuming a constant of proportionality of one and multiplying both the mean curvature of the liquid–vapor interface and the hydraulic radius by the capillary radius, r_c, a dimensionless expression was developed (Peterson, 1988a, 1994). This expression,

$$\frac{r_c}{r_h} \geq 1 \tag{8-2}$$

better defines a micro heat pipe and helps to differentiate between small versions of conventional heat pipes and the more recently developed "micro heat pipes." Longtin et al. (1992, 1994) defined a micro heat pipe in a slightly different fashion, indicating that the capillary radius is of the same order of magnitude as the hydraulic radius. By examining the Bond number effect, gravitational body forces were found to be significant for heat pipes with a hydraulic diameter larger than approximately 1 mm.

While the fundamental operating principles of micro heat pipes are similar to those occurring in larger, conventional heat pipes, the effect of microscale phenomena on the operation of these devices is considerably more important. Also, while the initial application proposed by Cotter (1984) involved the thermal control of semiconductor devices, a wide variety of other uses have been investigated or proposed. These include the removal of heat from laser diodes (Mrácek, 1988) and other small localized heat-generating devices (Peterson, 1988b; Stulc and Horváth, 1988), the thermal control of photovoltaic cells (Peterson 1987a, 1987b), the removal or dissipation of heat from the leading edge of hypersonic aircraft (Camarda et al., 1995), and applications involving the nonsurgical treatment of cancerous tissue through either hyper- or hypothermia (Anon., 1989; Fletcher and Peterson, 1993). Other applications for micro heat pipes include space applications in which heat pipes are embedded in silicon radiator panels. Utilizing a liquid metal (mercury, sodium, potassium) as the working fluid, micro heat pipes can be used to dissipate large amounts of the waste heat generated (Badran et al., 1993).

The following review presents a summary of the current state of the art in micro heat pipes and consolidates information available for both individuals new to the field and those already working in this area. For clarity, this review is divided into three major sections: modeling micro heat pipe performance, experimental investigations, and construction techniques and issues (Peterson, 1994).

8-2 MODELING MICRO HEAT PIPE PERFORMANCE

The early attempts to model micro heat pipes began with simple steady-state pressure balance models and proceeded to the development of transient 3-D models of micro heat pipe arrays.

8-2-1 Steady-State Modeling of Individual Micro Heat Pipes

In the initial micro heat pipe concept, Cotter (1984) presented the first steady-state model, assuming a uniform cross-sectional area and no-slip boundary conditions at the wall. Shortly thereafter, Peterson (1988a) and Babin and Peterson (1990) developed a steady-state model for a trapezoidal micro heat pipe with a cross-sectional area similar to the one illustrated in Fig. 8-2. Using the conventional steady-state modeling techniques outlined by Chi (1976), the maximum heat transport capacity of this heat pipe was found to be primarily governed by the capillary pumping pressure.

Gerner et al. (1989, 1990) provided a critique of these two steady-state models and indicated that the major contributions of the model developed by Babin et al. (1990) were the inclusion of the effect of gravity and the recognition of the significance of the vapor pressure losses. However, the assumption that the pressure gradient in the liquid flow passages was similar to that occurring in Hagen-Poiseuille flow was questioned. As a result, Gerner et al. (1990) presented a force balance and scaling argument for the liquid pressure drop and estimated the average film thickness to be approximately one-fourth the hydraulic radius. This resulted in a capillary limitation of

$$\Delta P_c = \rho_\ell g L \sin\phi + \frac{8192}{3} \frac{V_v \dot{m}_v L}{\pi d^4}. \tag{8-3}$$

In addition to this modification to the capillary limit, it was hypothesized that the capillary limit may in fact never be reached due to Kelvin-Helmholtz-type instabilities occurring at the liquid–vapor interface. The question of which limit is most appropriate—the traditional capillary limit proposed by Babin et al. (1990), or the interfacial instability

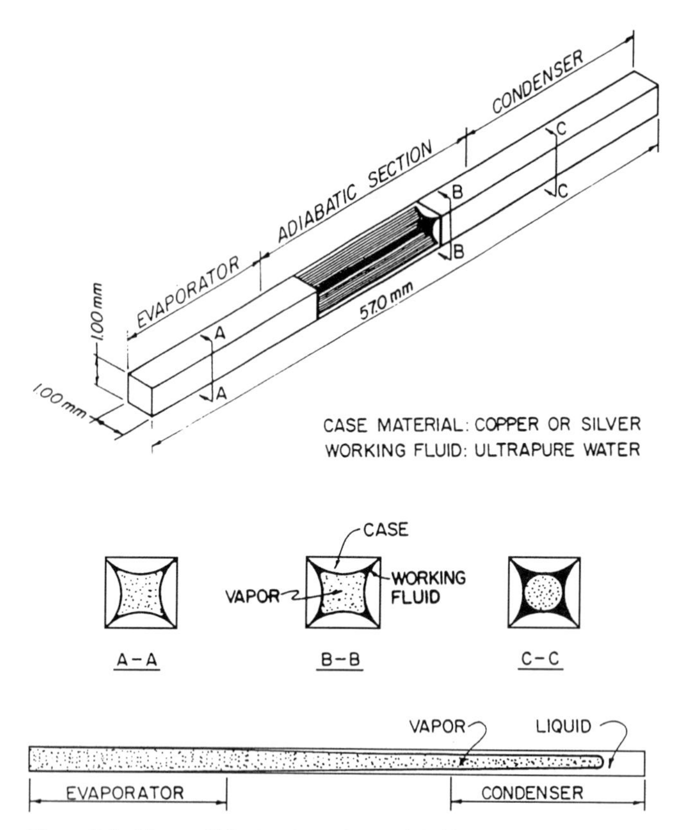

Figure 8-2 Trapezoidal micro heat pipe modeled by Babin and Peterson (1990).

limit proposed by Gerner (1990)—can be resolved by evaluating the shape and physical dimensions of the specific micro heat pipe being considered. Longtin et al. (1994) developed a mathematical, steady-state model to predict the fluid thermal behavior of micro heat pipes including the effect of the working fluid charge, the optimum operating condition, and the maximum heat load for a single micro heat pipe. This one-dimensional model indicated that the maximum heat transport capability varies with the inverse of the length and the cube of the hydraulic diameter. Badran et al. (1993) developed a conjugate model to account for the transport of heat within the heat pipes and the conduction within the silicon. The former is used to calculate the maximum heat load for a single micro heat pipe, and the latter is used to predict the energy transfer within the silicon substrate and the effective thermal conductivity. Mercury, sodium, or potassium was used as the working fluid in micro heat pipes embedded in a silicon substrate. Khrustalev and Faghri (1994) developed another model for a triangular micro heat pipe, which emphasized the interactions between the amount of the liquid fill, the wetting angle, and the maximum heat transport capacity.

More recently, an investigation by Ma et al. (1995) indicated that for a typical micro heat pipe with sharp V-shaped corners, the maximum heat transport capacity cannot be determined by simply assuming the minimum capillary radius. Instead, a true capillary limit, which considers the combined effects of the capillary pumping pressure, the liquid viscous pressure losses, and the vapor/liquid interaction, must be used. As a result, an analytical model was developed to predict the true capillary radius and the resulting maximum heat transport in micro heat pipes. In this model, a theoretical minimum meniscus radius was determined and coupled with the combined effects of the capillary pumping pressure, the liquid viscous pressure losses and the vapor–liquid interaction. A control volume technique was employed to determine the flow characteristics of micro heat pipes with no wicks, and to incorporate the effects of the frictional vapor–liquid interaction on the liquid flow. This model was then used to calculate the maximum heat transport limit based on the physical characteristics of the working fluid and the groove geometry. This work was complimented by a model developed by Ha and Peterson (1994) designed to predict the dryout point of a wetting liquid in triangular grooves.

The model of Ma et al. (1995) considered for the first time the true characteristics of micro heat pipes to determine the minimum meniscus radius and the maximum heat transport capacity. The calculated results indicated that for wickless micro heat pipes, the heat transport capacity is strongly dependent on the channel angle of the liquid arteries, the contact angle of the liquid flow, the length of the heat pipe, the vapor flow, and the tilt angle. In addition, the model provided a mechanism which, for a given specified set of conditions, allowed the geometry to be optimized and a micro heat pipe with the maximum heat transport for a given cross-sectional area, designed.

8-2-2 Transient Modeling of Individual Micro Heat Pipes

To further understand the steady-state models, Wu and Peterson (1991) developed a transient model, which utilized the relationship initially developed by Collier (1981) and later expanded by Chang and Colwell (1985), to determine the free molecular flow mass flux of evaporation. Based upon this work, the evaporation/condensation rate was assumed to be proportional to the liquid–vapor interface area in each section of the heat pipe, and the amount of heat released or absorbed in any section was determined from an energy balance. The single boundary condition utilized in the evaporator section was the time-dependent heat flux, and for a specific input heat flux the saturation pressure at a given location was obtained by using a combination of the mass flux expression and the energy conservation equation. In the adiabatic section, the governing equations for each element were solved by assuming no temperature drop at the liquid/container boundary (i.e., $T_\ell = T_b$) (Peterson, 1994).

The sensitivity of the evaporation/condensation rate to changes in the vapor pressure necessitated the use of a predictor-corrector treatment in order to obtain a stable solution. Initially, the evaporation rate was assumed to be equal to the condensation rate, and the vapor pressure was calculated. This vapor pressure was then used to compute the change in the mass of vapor and hence the difference between evaporation and condensation. The original assumption of equal evaporation and condensation rates was then checked and the values updated. The solution procedure consisted of initializing the variables and arrays and computing by iteration, the boundary, liquid, and vapor temperatures, along with the evaporation and condensation rates, the pressure drops, and the flowrates,

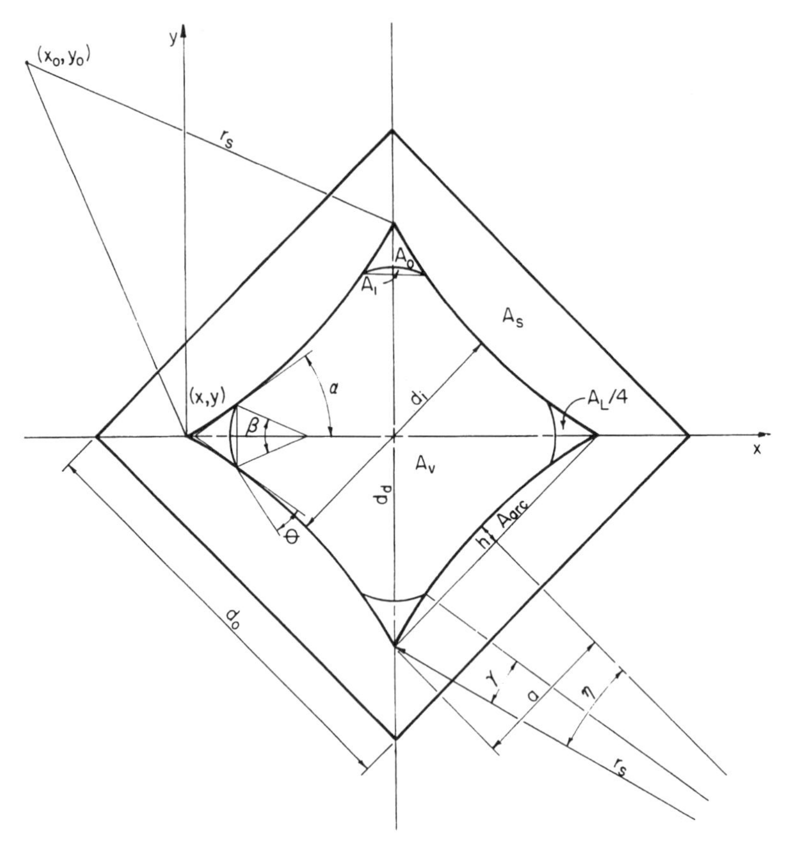

Figure 8-3 Cross-sectional geometry of the micro heat pipe investigated by Babin et al. (1990).

for each section. The resulting radii of curvature for the liquid–vapor interface was then computed, and the value for liquid–vapor mass flow rate was corrected using the resulting vapor pressure.

Utilizing the heat pipe originally modeled by Babin et al. (1990), shown in Fig. 8-2, the transient analytical model was verified. The heat pipe characteristics were defined in terms of the geometry, as shown in Fig. 8-3. The results indicated several interesting phenomena. First was that the vapor temperature variations were proportional to the heat input and that without dryout or flooding, the effective conductivity of the heat pipe was independent of time shortly after startup. In contrast, a fairly long period of time was required for the other thermal hydraulic parameters to reach their steady-state values. For example, the time required for the outer case to reach steady-state ranged from 10 to 30 seconds after full heat input. This time was found to be strongly dependent on the wetting angle, with smaller wetting angles and higher temperatures resulting in shorter time intervals required (Wu and Peterson, 1991).

Of perhaps greater interest from an operational perspective was the time-dependent behavior of the liquid mass flow rate. This behavior is illustrated in Figs. 8-4, and 8-5 and compares the changes in the liquid mass flow rate and liquid cross-sectional area,

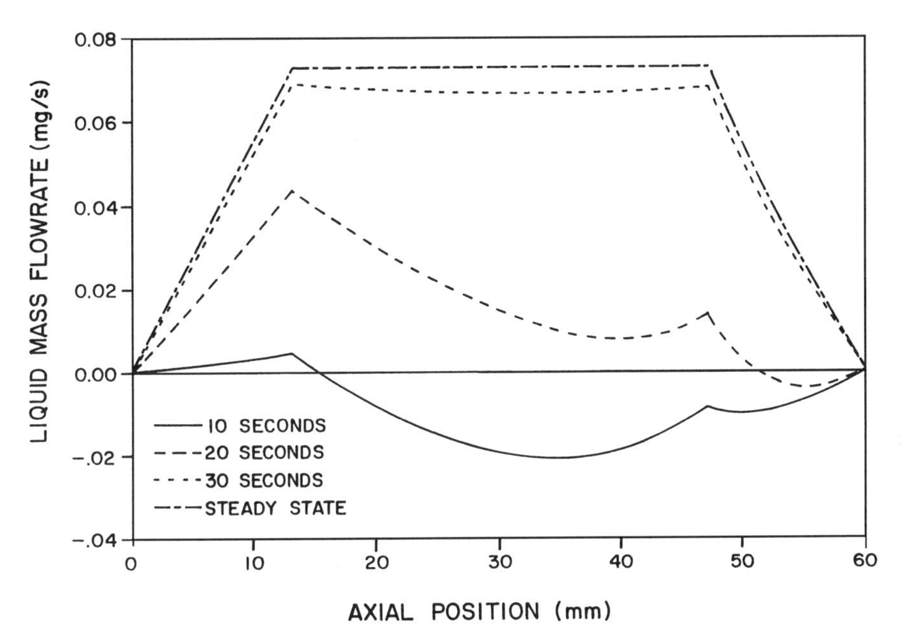

Figure 8-4 Liquid mass flow rate as a function of time and the axial position (Wu and Peterson, 1991).

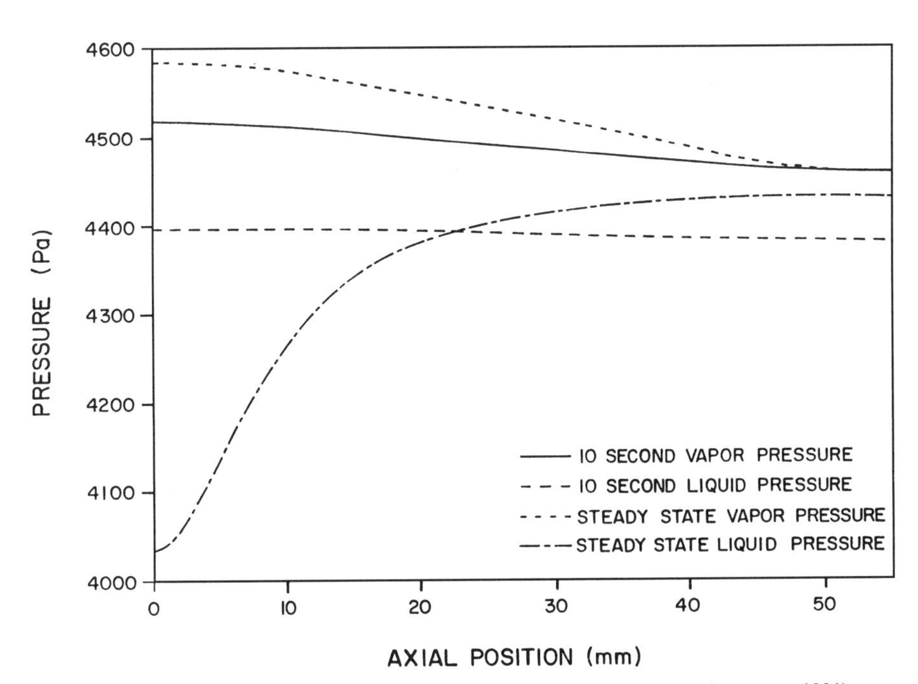

Figure 8-5 Liquid presssure as a function of time and the axial position (Wu and Peterson, 1991).

respectively, as a function of time. As illustrated, during the time immediately after the application of the heat to the evaporator, the liquid and vapor flow in the evaporator are both in the same direction. This reverse liquid flow is the result of an imbalance in the total pressure drop and occurs because the evaporation rate does not provide an adequate change in the liquid–vapor interfacial curvature to compensate for the pressure drop. As a result, the increased pressure in the evaporator causes the meniscus to recede into the corner regions, forcing liquid out of the evaporator and into the condenser. Once the heat input reaches full load, the reverse liquid flow disappears and the liquid mass flow rate into the evaporator gradually increases until a steady-state condition is reached. At this time the change in the liquid mass flow rate is equal to the change in the vapor mass flow rate for any given section (Wu and Peterson, 1991). This is more apparent in Fig. 8-5, which compares the pressure as a function of axial position for ten seconds after startup and for steady state. As illustrated, ten seconds after startup, the pressure of both the liquid and vapor are higher in the evaporator and gradually decrease with position, promoting flow away from the evaporator. At steady state, the liquid pressure is much lower in the evaporator and increases dramatically with position, promoting flow back to the evaporator.

8-2-3 Modeling of Heat Pipe Arrays

While the modeling of individual micro heat pipes is of interest, the original concept proposed by Cotter (1984) was to incorporate heat pipes as part of semiconductor devices. The previous analyses, while providing insight to the performance limitations and operational characteristics of individual micro heat pipes, do not address how the incorporation of an array of these devices into semiconductor devices might affect the overall temperature distribution and thermal performance. In an effort to determine the potential advantages of this concept, Mallik and Peterson (1991) developed a 3-D numerical model capable of predicting the thermal performance of an array of parallel micro heat pipes constructed as an integral part of semiconductor chips. For comparative purposes, several different thermal loading configurations were modeled both with and without heat pipes in order to determine the reduction in the maximum surface temperature, the mean chip temperature, and the maximum temperature gradient across the chip. Using the steady-state and transient analyses of Babin et al. (1990), Wu and Peterson (1991), and Wu et al. (1991), the effective thermal conductivity of the region occupied by the micro heat pipes was found to be approximately ten times that of the silicon substrate if methanol was used as the working fluid. The model of Badran et al. (1993), for micro heat pipes in space applications, indicates that the specific thermal conductivity (effective thermal conductivity divided by the density) for micro heat pipes embedded in space radiator panels can be as high as 200 times that of copper or 100 times of that of a Gr/Cu composite material.

As a follow on to this work, Mallik and Peterson (1995) developed a steady-state three-dimensional lumped capacitance model capable of predicting the temperature profile of the surface of a silicon wafer into which an array of micro heat pipes had been fabricated. In this model, each wafer was divided into forty equally spaced grids along the x-axis, with $\Delta x_1 = 0.5$ mm. Because of the symmetry of the problem, the wafer was divided into two halves 4×20 mm, midway along the x-axis. The control volume nodes were placed midway between the neighboring grid points, with grid points of zero control volume at the boundaries. At the interface between surfaces, a harmonic mean value was used to obtain the effective thermal conductivity. Solution of the finite difference model was accomplished using an Alternating Direction Implicit (ADI) method with no relaxation (Patankar, 1980). For the steady-state analysis the temperature convergence between two successive iterations was less than 0.0025%, with an energy balance of 98.0%. The numerical code was vectorized (87%) and run on a Cray YM-P supercomputer.

The space along the z-axis was divided into two regions. The region of the cover glass slip (125 μm) was divided into 5 equally spaced grids, with $\Delta z_1 = 0.025$ mm. For the plain wafer, the wafer thickness (355 μm) was divided into five equally spaced grids, with $\Delta z_2 = 0.071$ mm. For the wafer containing the array, the wafer was divided into four equally spaced grids, with thickness $\Delta z_2 = 0.0825$ mm and a single grid containing the heat pipes with width $\Delta z_3 = 0.025$ mm. The thickness of the vapor deposited copper layer was assumed to be 0.030 mm thick. The region between each heat pipe consisted of a single grid with $\Delta y_1 = 0.204$ mm or $\Delta y_1 = 0.095$ mm, and a single grid at the edges with $\Delta y_2 = 0.209$ mm and 0.0875 mm, respectively. The micro heat pipes were modeled using a single grid with $\Delta y_3 = 0.025$ mm (Mallik et al., 1992).

The numerical model utilized the thermal conductivity obtained from the steady-state experiments. The mean surface temperature of the silicon wafer, coated surface, and glass obtained from earlier iterations were used to evaluate the thermal properties for silicon, copper, and glass, respectively. For the wafer with the micro heat pipe array, the individual micro heat pipes were modeled as homogeneous, solid regions embedded in the wafer with an equivalent thermal conductivity different from that of the wafer. The value of the effective thermal conductivity of the micro heat pipes was determined from the steady-state analytical model of Babin et al. (1990). This value was incorporated into the numerical model using a lumped capacitance approximation. Results from this analysis are compared with experimental data in Section 8-3.

The steady-state numerical model of Mallik and Peterson (1995) was extended to include transient response characteristics by Peterson and Mallik (1995). This transient numerical model was used to predict the temperature profile of the wafer surface and was compared with the experimental data. Each wafer was divided into forty equally spaced grids along the x-axis, with $\Delta x_1 = 0.5$ mm. Because of the symmetry of the problem, the wafer was divided into two halves 4×20 mm, midway along the x-axis. The control volume nodes were placed midway between the neighboring grid points, with grid points of zero control volume at the boundaries. At the interface between surfaces, a harmonic mean value was used to obtain the effective thermal conductivity. Again, the solution of the transient finite difference model was accomplished using an ADI method with no relaxation, and the space along the z-axis was divided into two regions. The results of this model will be compared with the experimental results in Section 8-3.

8-2-4 Microscopic Models

One shortcoming of the model developed by Mallik et al. (1991) was that it utilized the macroscopic models of Babin et al. (1990), Wu and Peterson (1991), and Wu et al.

(1991), which are concerned primarily with bulk properties and behavior, to estimate an effective thermal conductivity of each individual heat pipe in the array. And even though these models had been shown to be reasonably accurate for heat pipes on the order of 1 mm in diameter, there was no assurance that continued reductions in size would yield compatible results. For this reason, development of a microscopic model, focusing on the behavior of thin films and particularly the evaporation and condensation occurring in the thin film region, was necessary.

The precise location of this transition is not yet clear; however, for micro heat pipes with diameters greater than 100 μm, the percentage of the liquid–vapor interface in the thin film region is still relatively small, and the liquid flow retains the characteristics of the bulk fluid. For micro heat pipes with diameters less than 100 μm, a majority of the liquid–vapor interface occurs over a region where the liquid would be considered in the thin film regime, and the fluid behavior may begin to deviate from the bulk liquid behavior. This indicates that the transition occurs somewhere near the point where the mean passage diameter is approximately 100 μm. As a result, accurate prediction of the thermal behavior of micro heat pipes smaller than 100 μm in diameter will require the development of models that utilize knowledge of the thin film behavior and include the effects of the wetting angle, the disjoining pressure, and van der Waals forces.

Previous investigations by Wayner (1976), Mirzamoghadam and Catton (1988), and Xu and Carey (1990) have studied evaporation and condensation in small triangular channels. In a more recent study, Swanson and Peterson (1992) investigated the behavior of the intrinsic meniscus in triangular channels as directly applied to micro heat pipes. In this investigation a mathematical model of the evaporating intrinsic meniscus in a V-shaped channel was developed to determine the effect of wedge half-angle and vapor mass transfer on the meniscus morphology, fluid flow, and heat transfer behavior. The Navier-Stokes and energy equations, as well as the interfacial conditions, were scaled using the lubrication approximation. The one-sided formulation applied to gas phase was constant. This reduced the mathematical complexity of the problem and allowed the use of a simple mass transfer coefficient at the vapor–liquid interface. The scaled normal stress interfacial condition was examined in detail to compare the magnitudes of the surface tension forces and van der Waals forces both near the interline and at the edge of the intrinsic meniscus. The scaled equation set showed that the hydrostatic pressure acting in both the radial and azimuthal directions can be neglected. Second, flow in the axial direction was found to dominate in the intrinsic meniscus.

The analysis of the normal stress interfacial condition, which was based on a copper/methanol/air system and the nonretarded form of the disjoining pressure, indicated that the disjoining pressure term could be neglected for thin films greater than approximately 500 Å. This film thickness roughly defined the edge of the intrinsic meniscus. Furthermore, near the interline ($< \sim$250 Å), the magnitude of the surface tension term was found to be much less than that for the disjoining pressure. The scaled equation set was solved numerically to determine the Nusselt number, interface morphology, axial pressure gradient profile, and locus of axial dryout points as functions of the wedge half-angle and the surface mass transfer number. The endpoints defining the range of the nonretarded disjoining pressure for a copper/methanol/air system were used as parameters. The results showed that the Nusselt number was independent of the surface mass transfer number and passed through minimum points near $\alpha = 18°$ and $21°$ for the limiting values of the nondimensional disjoining pressure. The magnitude of the Nusselt number at any given wedge half-angle was always largest for the larger disjoining pressure. This behavior was attributed to a stronger van der Waals attraction near the interline, which forced the overall thickness of the intrinsic meniscus to decrease. The decrease in layer thickness caused a mild increase in the interface temperature and produced a larger evaporation surface area; both of these effects caused an enhancement in heat transfer. The axial pressure gradient decreased linearly downstream, with a slope that increased with increasing surface mass flux. The locus of liquid dryout locations were estimated based on points where the piezometric pressure was equal to zero: this represented a point of zero flow in the channel. As expected, the points where dryout occurred moved toward the inlet (or evaporator) as the surface mass flux (or heat input) was increased.

In a later study, Swanson and Peterson (1995) developed a thermodynamic model of the vapor–liquid interface for triangular micro heat pipes. The primary goal of this study was to define the interfacial boundary conditions necessary for coupling transport phenomena between the vapor and liquid phases. The general model included axial pressure and temperature differences, changes in local interfacial curvature, Marangoni effects, and dispersion forces. The model did not account for the effects of interfacial shear induced by bulk flow in the vapor and liquid phases. Here it was found that when the dispersion forces were neglected and the intrinsic meniscus was the sole driving force for liquid flow, the heat transfer rate was strongly dependent on the minimum evaporator capillary radius. Allowing the evaporator capillary radius to go to zero produced a thermocapillary heat pipe limitation, whose magnitude was extremely large and not likely attainable for most practical applications of micro heat pipes.

Including the dispersion forces in the thermodynamic formulation resulted in a relationship that could be applied to nonevaporating superheated films. These films form a bridge between the menisci in adjacent grooves and cover the entire dry portion of the extended meniscus when dryout occurs in the evaporator. Under dryout conditions, the mean curvature in the evaporator was asymptotic to a maximum value corresponding to a minimum capillary radius. The minimum capillary radius decreases as the wall superheat increased and can be used to determine the capillary heat transfer limitation for a given wall superheat.

Khrustalev and Faghri [1994] developed a comprehensive model describing the heat and mass transfer processes for a copper-water micro heat pipe with triangular-shaped channels. The model accounts for the change in curvature of the vapor–liquid interface and interfacial shear stresses due to vapor–liquid frictional interaction. They applied a hemispherical vapor–liquid interfacial condition at the end of the condenser, with liquid completely filling the pipe beyond the interface in a so-called liquid blocking zone. Their numerical results agreed very well with Wu and Peterson's experimental data for a copper-water micro heat pipe. They discussed how the distribution of liquid, liquid charge, wetting contact angle, interfacial shear stresses, and applied heat load can affect the thermal characteristics of micro heat pipes. They concluded that the interfacial frictional interaction significantly influences the maximum heat transfer capacity of the heat pipe and causes an increase in the length of the liquid blocking zone in the condenser region. Near the maximum heat load, most of the liquid pressure drop occurs in the evaporator section and near the evaporator end of the adiabatic section. Under these conditions, the liquid cross section in the evaporator can be several times smaller than that in the condenser. The dominant thermal resistances are those of the bulk vapor and liquid in the evaporator and condenser. They also found that the liquid charge and minimum wetting contact angle has a strong effect on the performance of the micro heat pipe.

8-3 EXPERIMENTAL INVESTIGATIONS

Unlike the analytical or numerical investigations discussed previously, experimental investigations have progressed fairly systematically from relatively large micro heat pipes, approximately 3 mm in diameter, to micro heat pipes in the 30 μm diameter range. These investigations have included steady-state and transient investigations along with investigations of arrays of micro heat pipes included as an integral part of silicon wafers.

8-3-1 Steady-State Experimental Investigations

Murakami et al. (1987) conducted one of the earliest experimental investigations of micro heat pipes in conjunction with the study of a flat plate heat pipe several millimeters thick. Two types of heat pipe augmentation were evaluated: one was a conventional flat plate heat pipe that used a series of rectangular grooves machined into the upper and lower plates as the wicking structure; the other was composed of a series of triangular grooves machined into the plate to form an array of micro heat pipes, 3.2 mm on a side. In addition to the experimental tests, an analytical model similar to those described previously was developed. For comparison, the maximum heat transfer rate as predicted by this model is also shown in Fig. 8-6. As illustrated, the predicted values compare quite favorably with the experimental results.

Shortly thereafter, Babin et al. (1990) investigated the thermal characteristics and transport capacity of micro heat pipes each approximately 1 mm in diameter. The primary purpose of this investigation was to determine the accuracy of the previously described steady-state modeling techniques, to verify the micro heat pipe concept, and to determine the maximum heat transport capacity. The test articles used in this investigation were constructed by Itoh Research and Development Company, Osaka, Japan (Itoh, 1988). The general size and shape of these test articles are illustrated in Fig. 8-2. A total of four test articles were evaluated, two constructed from silver and two from copper. The investigation focused on steady-state tests at tilt angles that both assisted and hindered the return of liquid to the evaporator, and all four test pipes were evaluated at a total of six adiabatic wall temperatures. As illustrated in Fig. 8-7, the steady-state model overpredicted the experimentally determined heat transport capacity at operating temperatures below 40°C, and underpredicted it at operating temperatures above 60°C.

8-3-2 Transient Experimental Investigations

The previous analytical model developed by Babin et al. (1990) was shown to predict the steady-state performance limitations and operational characteristics of the trapezoidal heat pipe reasonably well for operating temperatures between 40 and 60°C; however, no

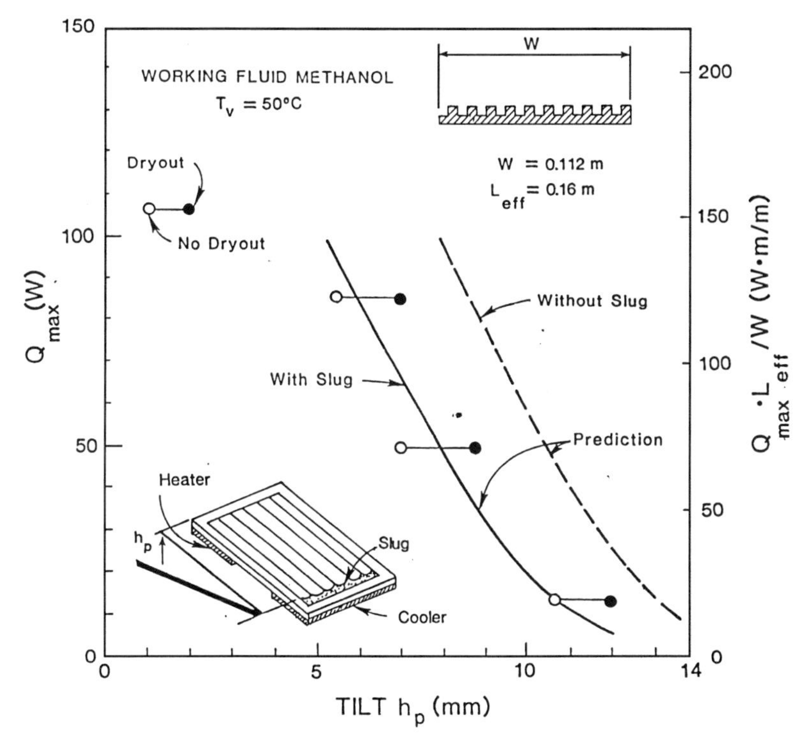

Figure 8-6 Comparison of experimental and analytical results of the flat plate heat pipe of Murakami et al. (1987) with triangular micro heat pipes.

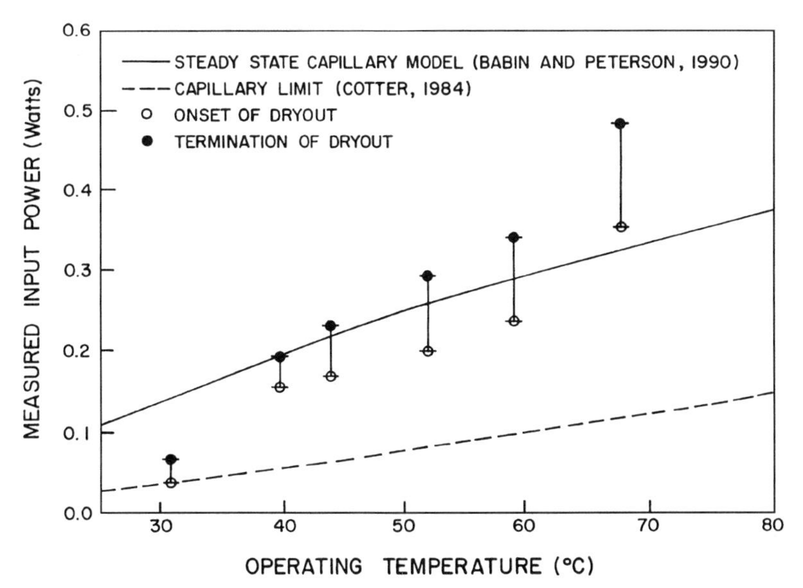

Figure 8-7 Comparison of the maximum heat transport capacity of a trapezoidal micro heat pipe as a function of the operating temperature (copper, 0.0032 gm H_2O) (Babin et al., 1990).

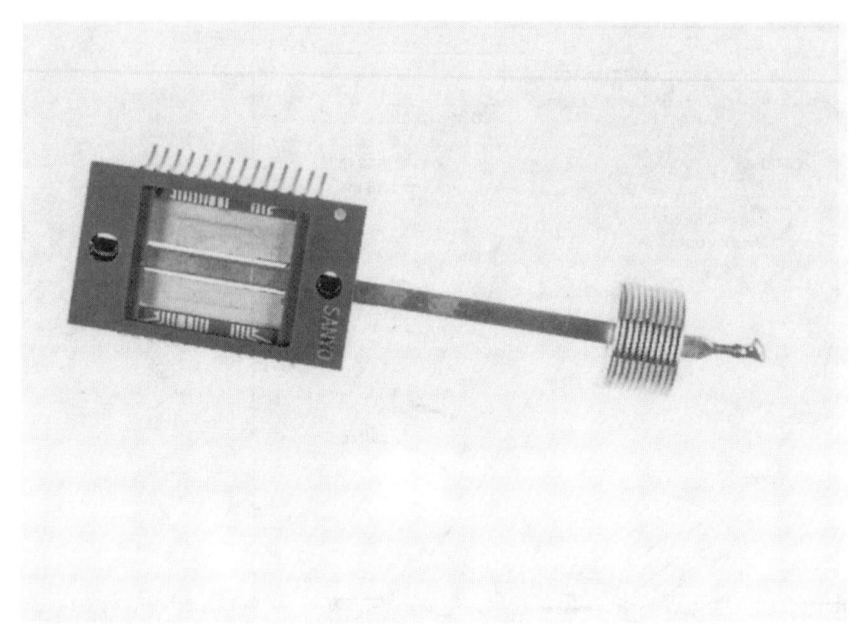

Figure 8-8 Tapered micro heat pipe (Wu et al., 1991).

transient information was obtained. As a result, investigations of the transient characteristics were conducted by Wu and Peterson (1991). These tests again utilized a micro heat pipe test article developed by Itoh (1988). However, this test pipe was specifically designed for use in the thermal control of ceramic chip carriers (Wu et al., 1991) and was slightly different in shape. As shown in Fig. 8-8, the heat pipe was designed to fit securely under the chip and was attached to the chip carrier. Fins were located at the condenser end of the heat pipe to assist the condenser heat rejection through free or forced convection.

This investigation utilized the same test facility as the one used for the trapezoidal heat pipe; however, because the transient response was of interest, a thin nichrome strip heater located on the top portion of the heat pipe was used for heat input. Further, in addition to a series of nine thermocouples attached to the underside of each test pipe, an infrared thermal measurement system was used to monitor the temperature distribution in the heat pipe during both startup and transient operation. To determine the accuracy of the numerical model at steady-state conditions, the experimentally measured steady-state temperature distribution was compared with the distribution predicted by the numerical model for a power level of 0.12 W. As illustrated in Fig. 8-9, the temperature data closely follow the predicted trend and although some heat rejection occurs in the adiabatic section, the condenser temperature remains relatively constant. Throughout the entire length the predicted and measured temperature distributions are extremely close to the predicted value, with the largest deviation being less than 0.3°C. Startup and transient tests were then conducted.

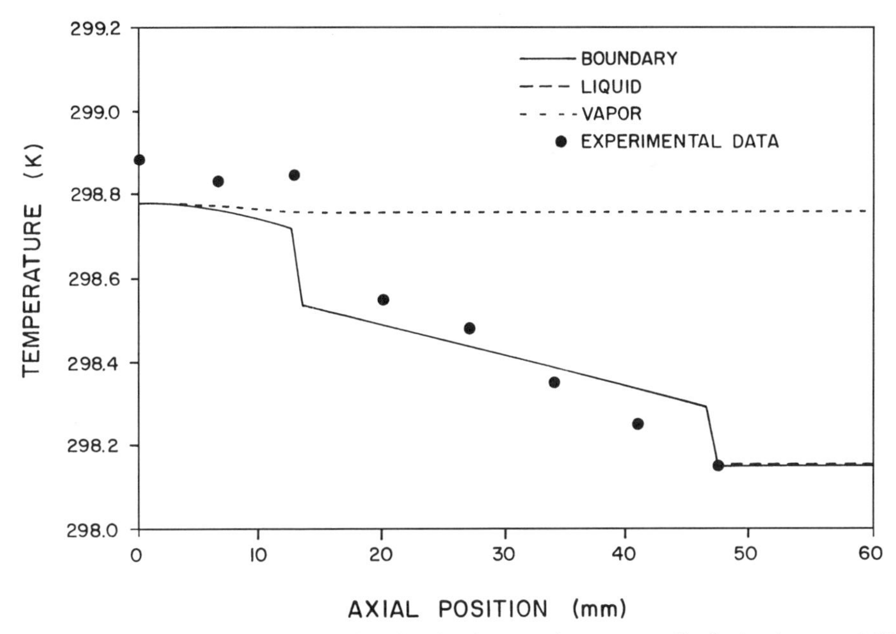

Figure 8-9 Steady-state comparison of predicted and measured temperature distribution (power = 0.12 W, T_c = 24.5°C, no tilt) (Peterson, 1988a).

In the transient tests, the time-dependent temperature distributions were measured to determine the transient response characteristics of the heat pipe as a function of incremental power increases, tilt angle, and mean operating temperature. Figure 8-10 compares the predicted and measured temperature difference for the tapered micro heat pipe shown in Fig. 8-8 as a function of time. As illustrated, the transient model predicted a much more rapid response than was actually measured. However, as time increased, the measured temperature drop between the evaporator and the condenser approached and exceeded the predicted value by approximately 15%. As discussed by Wu and Peterson (1991), the initial deviation of the actual response from that predicted may be due in part to the time constant associated with the nichrome heater. It is interesting to note that the reverse liquid flow apparent in Figs. 8-4 and 8-5 is also apparent in Fig. 8-10 (the slight inflection at 12 seconds) but is not apparent in the experimental results.

Itoh and Polásek (1990) conducted an extensive experimental investigation of micro heat pipes ranging in size from 1 to 3 mm in diameter and from 30 to 150 mm in length. Although most of the heat pipes evaluated used cross-sectional configurations similar to those presented previously, several additional pipes with conventional internal wicking structures were evaluated (Polásek, 1990; Fejfar et al., 1990). Some of these pipes were as small as 1.2 mm in diameter with lengths up to 30 mm. All of the test pipes investigated utilized ultra-pure water as the working fluid and either silver, copper, or a copper-beryllium alloy as the case material. The tests were conducted to determine

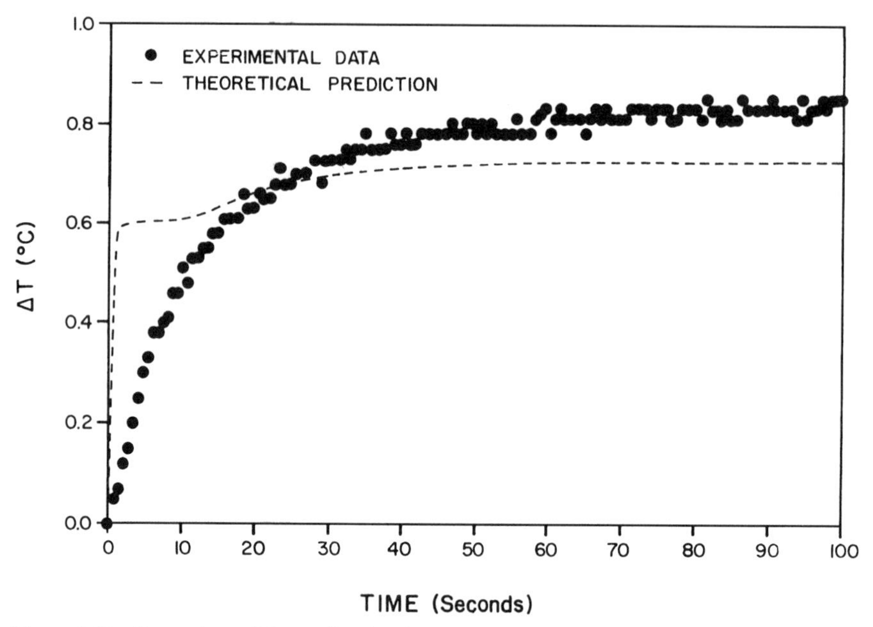

Figure 8-10 Comparison of the predicted and measured transient response for the tapered micro heat pipe (Peterson, 1988a).

the maximum transport capacity of the various types of pipes under adverse tilt conditions.

Neutron radiography tests were also conducted to determine the distribution of the working fluid within the heat pipes (Itoh and Polásek, 1990a, 1990b; Ikeda, 1990). Two types of neutron radiography testing were conducted; one involved a static film method, and the other utilized real-time neutron television (Ikeda, 1990). Using these techniques, the amount and distribution of the working fluid within the heat pipes could be evaluated during real-time operation. In addition to the location of the working fluid, the location and behavior of noncondensible gases could be observed, along with boiling and/or reflux flow. The results of these tests indicated several important results: i) as is the case for conventional heat pipes, the maximum heat transport capacity is principally dependent upon the mean adiabatic vapor temperature and the inclination angle of the pipe, ii) micro heat pipes with smooth inner surfaces were found to be more sensitive to overheating than those with grooved capillary systems, iii) the wall thickness of the individual micro heat pipes had greater effect on the thermal performance than did the casing material, and iv) the maximum transport capacity of heat pipes utilizing axial channels for return of the liquid to the evaporator were found to be superior to those utilizing a formal wicking structure (Peterson, 1994).

8-3-3 Testing of Micro Heat Pipe Arrays

In the first reported investigation of micro heat pipe arrays fabricated as an integral part of a semiconductor wafer, Peterson et al. (1991) conducted an experimental investigation to verify the micro heat pipe concept. Several silicon wafers were fabricated with distributed heat sources on one side and an array of micro heat pipes on the other. These micro heat pipe arrays were evacuated and charged with a predetermined amount of methanol. In this first series of tests, a series of 39 parallel rectangular channels, 30 μm wide, 80 μm deep, and 19.75 mm long, were machined into a silicon wafer 2 cm square and 0.378 mm thick, with an interchannel spacing of 500 μm.

In an extension of this work, Peterson et al. (1993) fabricated a series of micro heat pipes in 20 mm $\times$ 20 mm silicon wafers. In this second fabrication procedure, an array of parallel triangular channels 120 μm wide and 80 μm deep were fabricated into silicon wafers 0.5 mm thick using an anisotropic etchant process. Both of these two fabrication procedures utilized a clear glass cover plate bonded to the surface in order to close off the micro heat pipe array, and the heat pipe arrays were charged using the procedure developed by Peterson et al. (1991) and discussed later in Section 8-4. Gerner et al. (1994) fabricated a 125 micro heat pipe array on a 2-inch silicon wafer with a thickness of 280 μm. The pipes are about 1 inch long, 100 μm wide, and 70.72 μm deep, and are spaced 100 μm apart. They showed that the effective thermal conductivity (assuming a one-dimensional temperature distribution) of the water-silicon wafer sample is four times that of the pure silicon.

These fabrication procedures both resulted in a cross-sectional area porosity of approximately 1.87%, which was close to the optimum values predicted by Mallik et al. (1991). The objectives of these two early investigations were

- to develop the necessary fabrication techniques;
- to provide definitive experimental data to verify the micro heat pipe concept; and
- to determine the extent to which an array of micro heat pipes could reduce the maximum wafer temperature, decrease the temperature gradient across the wafer, increase the effective thermal conductivity, and decrease the number and intensity of localized hot spots.

A comparative technique was used to evaluate the steady-state and transient behavior of the micro heat pipe arrays. In this technique, several sets of silicon wafers, each set consisting of one wafer with the micro heat pipe array, and one without the micro heat pipe array were attached to a common copper heat sink as illustrated in Fig. 8-11. Two heaters, one on each chip, were connected in parallel to insure that a constant power was supplied to each wafer and the voltage and current were measured using digital multimeters. Heat was removed from the wafers by a common copper heat sink. The temperature distribution across the surface of the two wafers was obtained using a Hughes Probeye TVS Model 3000 Infrared Thermal Imaging System. The resolution of this system as specified by the manufacturer was $\pm 0.05°C$, allowing precise determination of the temperature gradients across the chip surface.

Throughout all of the tests, the IR camera was situated so that both the wafer with the micro heat pipe array and the plain ungrooved wafer were in the field of view. In this manner, the temperature gradient and maximum wafer temperature for both wafers could be observed simultaneously for different power levels, compared, and the effectiveness of the micro heat pipe array determined. Using this technique, variations in the maximum

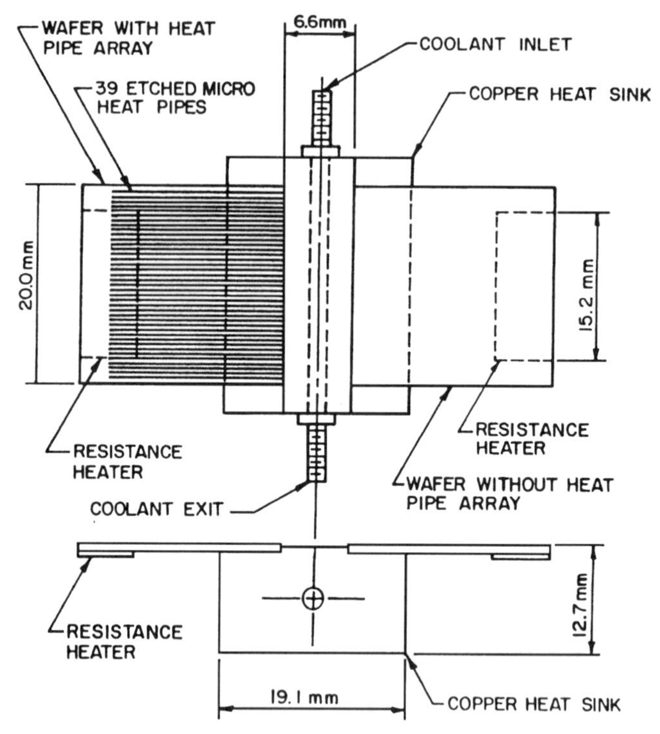

Figure 8-11 Experimental test facility (Peterson et al., 1991).

chip temperature, the effective thermal conductivity, and the transient response were readily apparent.

Figure 8-12 illustrates typical steady-state temperature profiles for the wafers with arrays of rectangular and triangular micro heat pipes evaluated at an input power of 4 watts and a bath temperature of 15°C. Also shown for comparison is the steady-state temperature profile for a wafer without a micro heat pipe array. As expected, the maximum temperature for all of the wafers tested occurred in the region directly over the heaters, with the maximum wafer temperature and temperature gradients for the wafers with the micro heat pipe arrays considerably smaller than those obtained for the plain ungrooved wafers. At an input power of 4 watts, the maximum steady-state temperatures for the wafers with the micro heat pipe arrays were 68.0°C and 59.2°C for the rectangular and triangular arrays, respectively, while for the plain silicon wafer the maximum wafer temperature was 82.1°C. This resulted in decreases in the temperature gradients of approximately 11.4°C and 16.2°C, respectively. Finally, as shown, incorporating an array of 39 micro heat pipes into a 2 cm by 2 cm wafer reduced the temperature gradient or slope of the temperature profile by as much as 18% for the array of rectangular micro heat pipes and 30% for the array of triangular micro heat pipes.

The maximum wafer temperature is shown in Fig. 8-13 and as illustrated, there is a significant reduction in the maximum wafer temperature for the plain wafer, the wafer with the array of rectangular micro heat pipes and the wafer with the array of

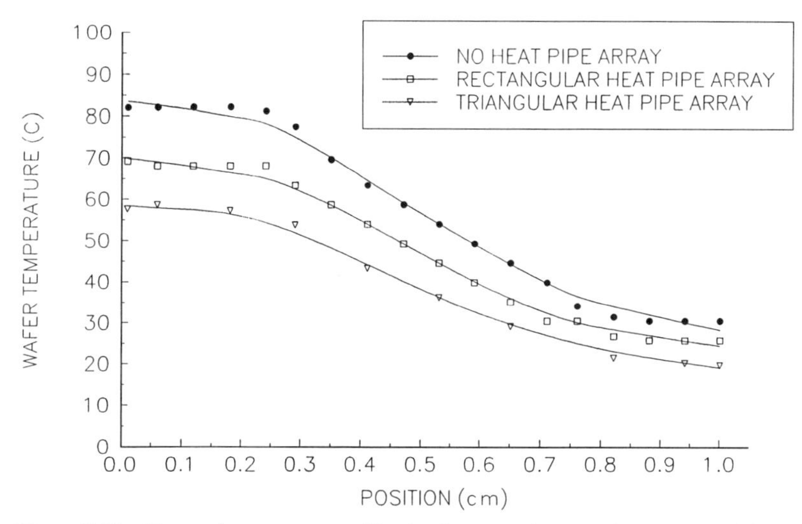

Figure 8-12 Measured temperature profiles for the test wafers at an input power of 4.0 watts and a sink temperature of 15°C (Peterson et al., 1993).

triangular micro heat pipes. In an attempt to better quantify the experimental results, additional data were taken from which an effective wafer conductivity could be computed using Fourier's law. The axial heat flux (the heat transported through the wafer in the direction of the heat pipes) was computed by dividing the input power by the cross-sectional area of the wafer. This value was then divided by the average temperature gradient, which was calculated from Fig. 8-13. As illustrated in Fig. 8-14, the effective thermal conductivity of the plain ungrooved silicon wafer decreased from approximately

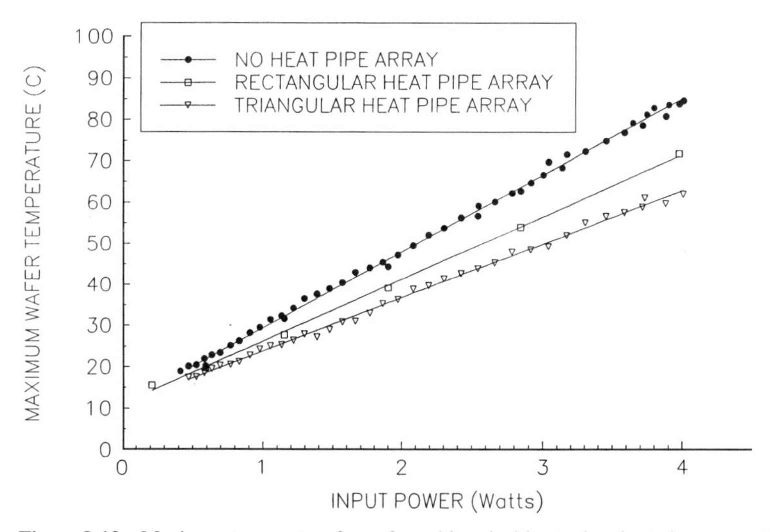

Figure 8-13 Maximum temperature for wafers with and without micro heat pipe arrays at an input power of 4.0 watts and a sink temperature of 15°C (Peterson et al., 1993).

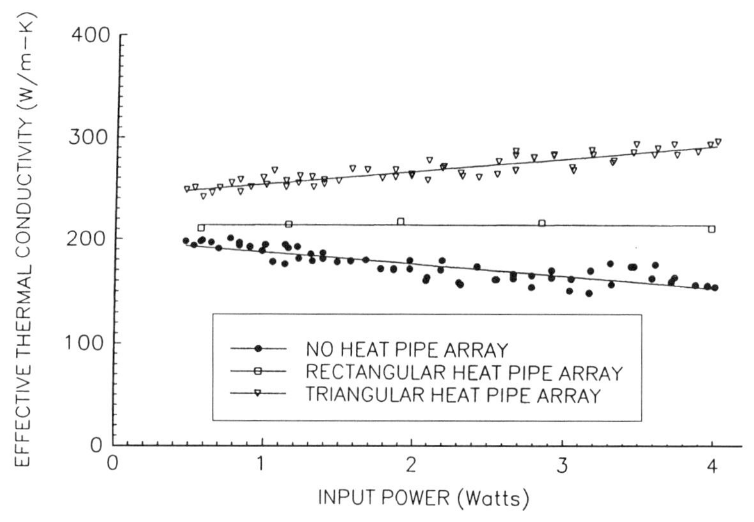

Figure 8-14 Variation of the effective thermal conductivity as a function of the input power (Peterson et al., 1993).

200 W/m-°C at an input power of 0.5 watts to a value of approximately 160 W/m-°C at an input power of 4.0 watts. These values correspond quite well with the thermal conductivity data for silicon available in the open literature. Although the effective thermal conductivity for the plain ungrooved wafers shown in Fig. 8-14 is presented as a function of input power, it is clearly apparent that increases in wafer temperature result in decreases in the effective thermal conductivity. This trend is also consistent with the published thermophysical property data for silicon. In contrast, measurement of the effective thermal conductivity of the wafer with the array of rectangular micro heat pipes resulted in a nearly constant value for the effective thermal conductivity of 210 W/m-°C, while the wafer with the array of triangular micro heat pipes resulted in a linearly increasing value of approximately 250 W/m-°C at input powers of 0.5 watts to 290 W/m-°C at input powers of 4.0 watts. The increasing trend observed in the array of triangular micro heat pipes results from the increased driving potential (i.e., temperature gradient) occurring at higher input powers, which makes the heat pipes perform more effectively. The increased effective thermal conductivity of the wafers with micro heat pipes amounts to an increase of nearly 31% for the array of rectangular micro heat pipes and 81% for the array of triangular micro heat pipes when compared to the plain silicon wafer. The large differences observed between the machined and etched micro heat pipe arrays may be the result of a combination of the rounded corner regions, which reduces the capillary pumping pressure and increases the liquid pressure drop in the channels and the rough scaly deposits occurring in the rectangular machined micro heat pipes.

Somewhat later, an experimental investigation was undertaken by Mallik and Peterson (1995) to determine the steady-state performance enhancement resulting from the incorporation of a series of vapor deposited micro heat pipes. Wafers were fabricated

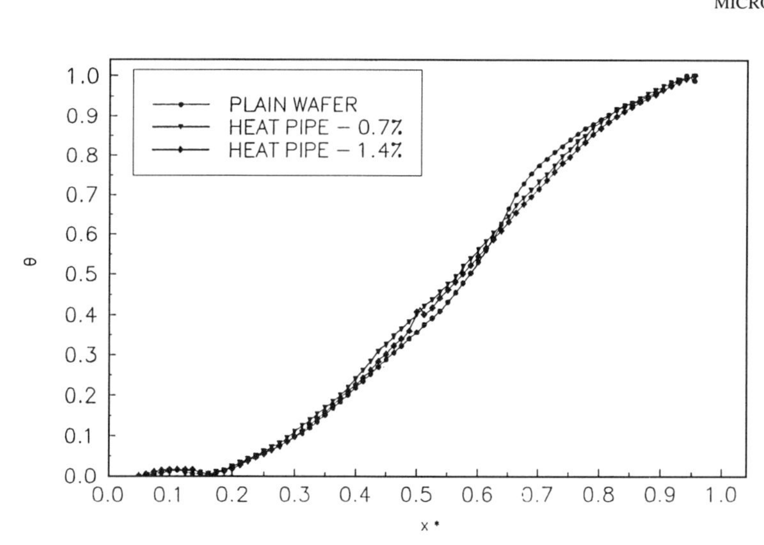

Figure 8-15 Centerline dimensionless temperature profile (Mallik and Peterson, 1995).

with an array of 34 or 66 vapor-deposited micro heat pipes, each with a width of 25 μm and an altitude of 55 μm and using methanol as the working fluid. These two arrays resulted in a percentage of the cross-sectional area occupied by the heat pipes of 0.75% and 1.45%, respectively. The investigation involved the steady-state measurement of wafers both with and without the micro heat pipe arrays to determine the surface temperature while varying the power input, and the array density. The effectiveness of the arrays (i.e., the resulting increase in thermal conductivity) was determined by comparing the results from the wafers with and without the heat pipe arrays to determine the reduction in the maximum wafer surface temperature, the reduction in the maximum temperature gradient on the wafer surface, and the increase in the effective thermal conductivity of the wafer/heat pipe combination.

The effective thermal conductivity of the wafers was computed using Fourier's law and the axial heat flux in the direction of the heat pipe array, which was computed using the heat input and the total cross-sectional area of the wafer. The "maximum" temperature (T_h) was calculated as the arithmetic mean of nine temperature readings located along a line 4 mm from the edge containing the heater. Similarly, the "minimum" temperature (T_c) was calculated as the mean of nine temperature readings located along a line 3 mm from the edge containing the heat sink. The average temperature gradient was evaluated as the ratio of $(T_h - T_c)$ and the linear distance $(x_h - x_c)$. This resulted in an expression for the effective thermal conductivity of

$$k_{eff} = 3.385 * 10^3 \left(\frac{q}{T_h - T_c} \right) \tag{8-4}$$

where k_{eff} is in W/m-K and q is the heat input in watts. Figure 8-15 illustrated the variation of the dimensionless temperature gradient as a function of the dimensionless position. From this data, the effective thermal conductivity can be represented as a function of the power input. These results are consistent with those of Peterson et al. (1993) obtained for an array of etched micro heat pipes.

The large reductions in the maximum surface temperature and increases in the effective thermal conductivity for the wafers containing the micro heat pipe arrays are an indication of the efficiency with which the micro heat pipe array functions as a heat spreader. To better understand the effect of these devices, the experimental data were nondimensionalized using a dimensionless temperature θ, defined as $(T - T_c)/(T_h - T_c)$, and dimensionless position x^*, defined as $(x - x_c)/(x_h - x_c)$. Use of these parameters resulted in the coalescence of the experimental data into a single curve shown in Fig. 8-15. As illustrated, for all cases, the variation of the dimensionless temperature θ as a function of dimensionless distance x^* remained flat in the condenser region and increased linearly in the adiabatic region.

Figures 8-16a through 8-16f compare the results of the experimental investigation with those predicted by the numerical model. As shown in Fig. 8-16a, the predicted maximum and mean surface temperatures of the plain wafer, for a heat flux of 2.34 W/cm^2 were 56.16°C, 34.02°C, respectively. For a heat flux of 3.91 W/cm^2 these values were 89.30°C, 51.18°C (Fig. 8-16b). For both of these cases the predicted temperature profiles were within ±9.0% of the experimentally measured values.

The predicted temperature and mean surface temperature of the wafer with the array of 34 charged micro heat pipes were 45.21°C and 28.91°C, for a heat flux of 2.34 W/cm^2 (Fig. 8-16c). The reduction in the maximum and mean wafer surface temperatures were 19.5% and 15.0%, respectively. The predicted temperature profile was within 7.2% of the experimental temperature profile. For a heat flux of 3.91 W/cm^2, the predicted maximum and mean surface temperatures were 68.36°C and 40.63°C, which represented

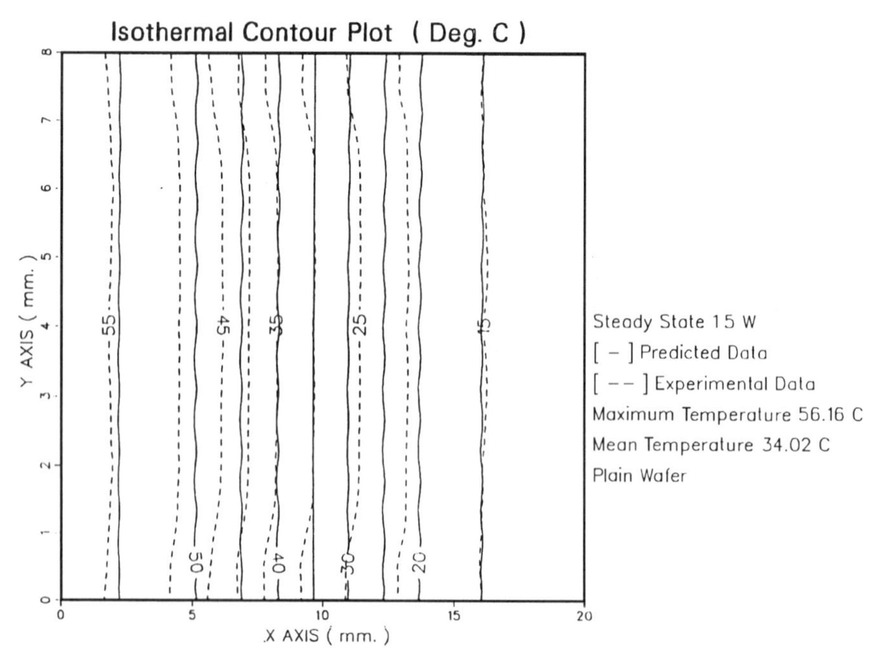

Figure 8-16a Comparison of the predicted steady-state temperature distribution on the plain silicon wafer with the experimental data (heat flux = 2.34 W/cm^2) (Mallik and Peterson, 1995).

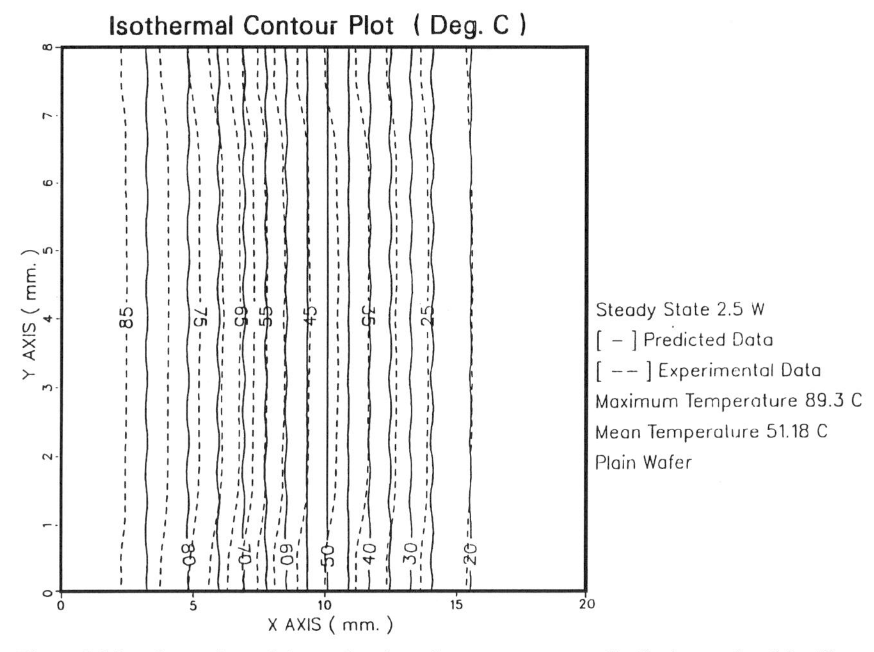

Figure 8-16b Comparison of the predicted steady-state temperature distribution on the plain silicon wafer with the experimental data (heat flux = 3.91 W/cm^2) (Mallik and Peterson, 1995).

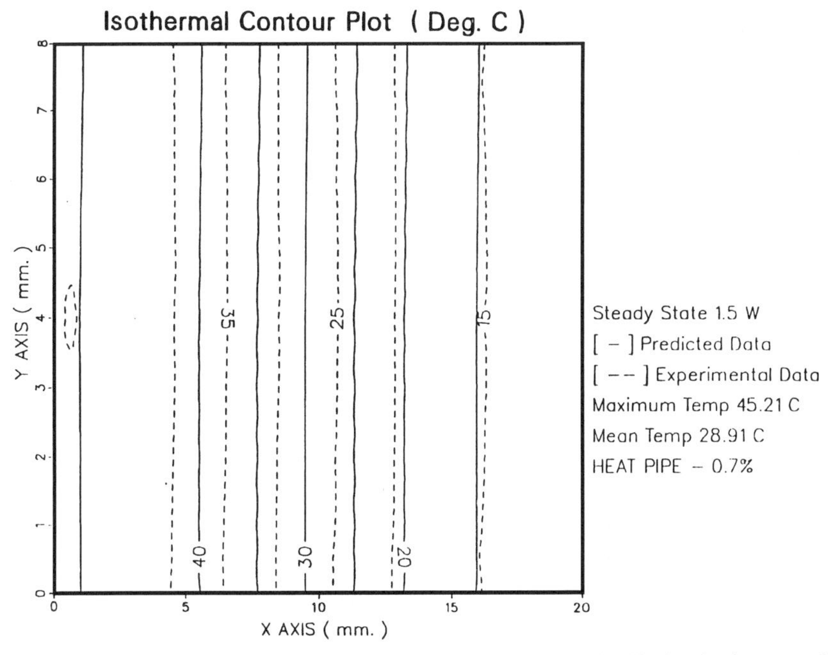

Figure 8-16c Comparison of the predicted steady-state temperature distribution for the array of 34 vapor-deposited micro heat pipes (heat flux = 2.34 W/cm^2) (Mallik and Peterson, 1995).

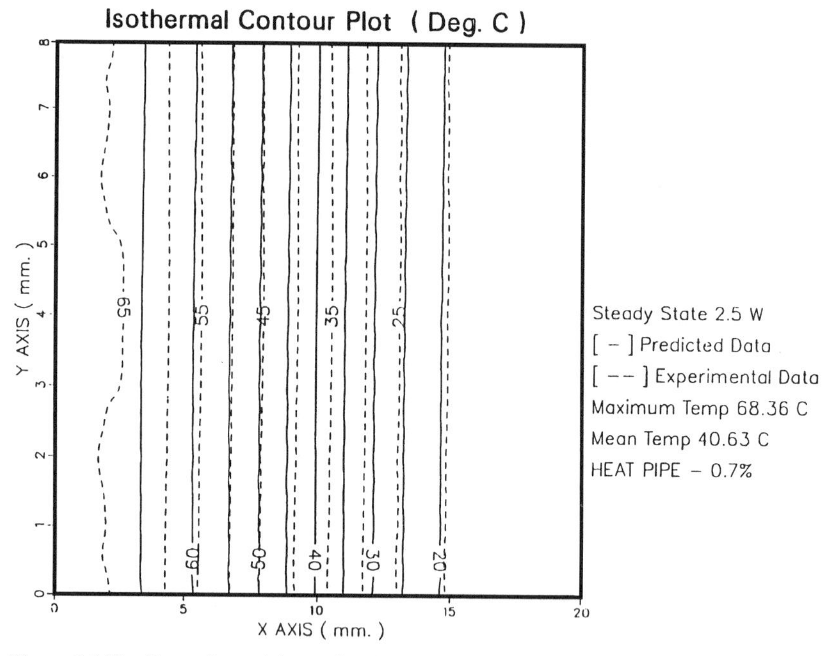

Figure 8-16d Comparison of the predicted steady-state temperature distribution for the array of 34 vapor-deposited micro heat pipes (heat flux = 3.91 W/cm^2) (Mallik and Peterson, 1995).

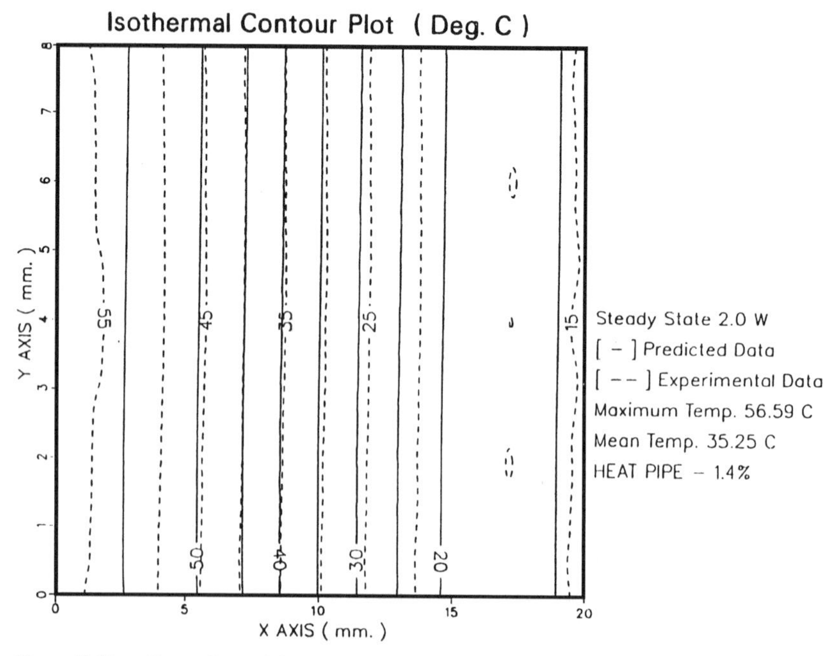

Figure 8-16e Comparison of the predicted steady-state temperature distribution for the array of 66 vapor-deposited micro heat pipes (heat flux = 2.34 W/cm^2) (Mallik and Peterson, 1995).

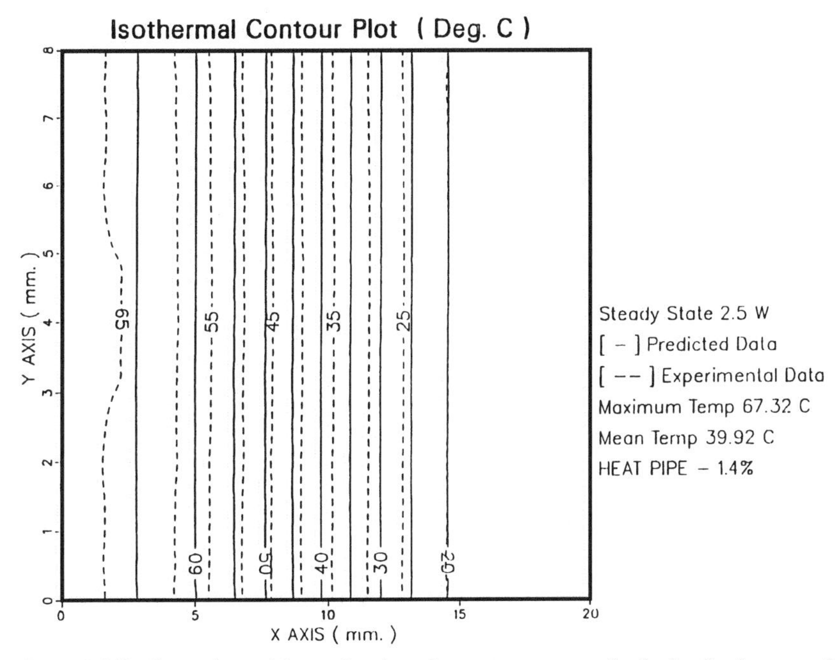

Figure 8-16f Comparison of the predicted steady-state temperature distribution for the array of 34 vapor-deposited micro heat pipes (heat flux = 3.91 W/cm^2) (Mallik and Peterson, 1995).

a reduction of 23.5% and 20.6%, respectively, as shown in Fig. 8-16d. The predicted temperature profile was within ±7.3% of the experimental temperature profile.

The predicted temperature and mean surface temperature for the wafer with the array of 66 charged micro heat pipes were 56.59°C and 35.25°C, respectively, for a heat flux of 3.13 W/cm^2 (Fig. 8-16e), and the reductions in the maximum and mean wafer surface temperatures were 22.4% and 16.9%, respectively. For this case, the predicted temperature profile was within 8.1% of the experimental temperature profile. For a heat flux of 3.91 W/cm^2, the predicted maximum and mean surface temperature was 67.32°C and 39.92°C, which represented reductions of 24.6% and 22.0%, as shown in Fig. 8-16f. Here, the predicted temperature profile was within ±6.5% of the experimental profile.

Using the technique described previously by Mallik and Peterson (1995), Peterson and Mallik (1995) conducted a transient analysis to provide insight into the transient response characteristics of wafers with vapor deposited micro heat pipe arrays. The wafers with and without arrays were analyzed experimentally by subjecting the wafers to a step function power input between 1.5 and 3.25 W, in increments of 0.25 W and observing the thermal response at discrete locations on the wafer surface. The accuracy of the time measurement was 0.05 seconds. As described earlier, the sensitivity of the IR camera was varied for the same power level, to improve the accuracy of the transient response data.

To compare the accuracy of the previously developed transient numerical model, Figs. 8-17a, 8-17b, and 8-17c illustrate the two-dimensional temperature distribution of the plain wafer at an input power of 2 Watts, the wafer with the array of 34 vapor

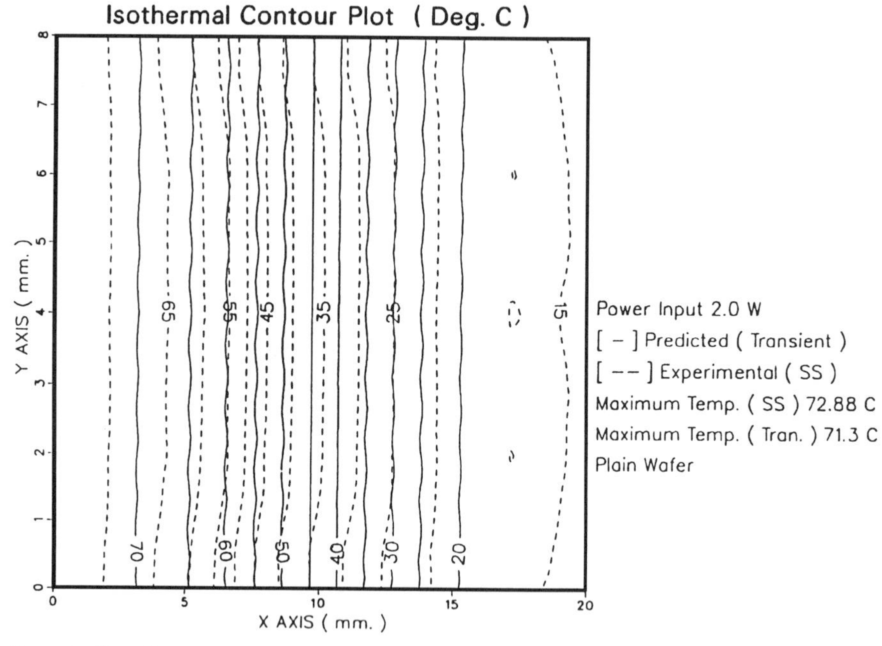

Figure 8-17a Comparison of the predicted transient thermal response and the experimental data for a plain wafer, an input power of 2 Watts, at an elapsed time of 7.5 seconds (Peterson and Mallik, 1995).

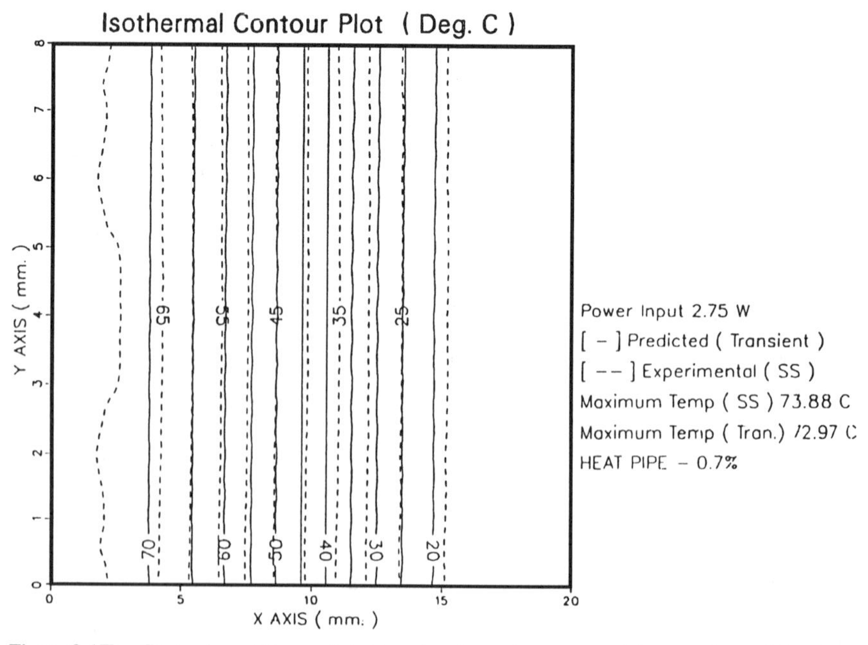

Figure 8-17b Comparison of the predicted transient thermal response and the experimental data for a wafer with an array of 34 micro heat pipes, an input power of 2.75 Watts, at an elapsed time of 10 seconds (Peterson and Mallik, 1995).

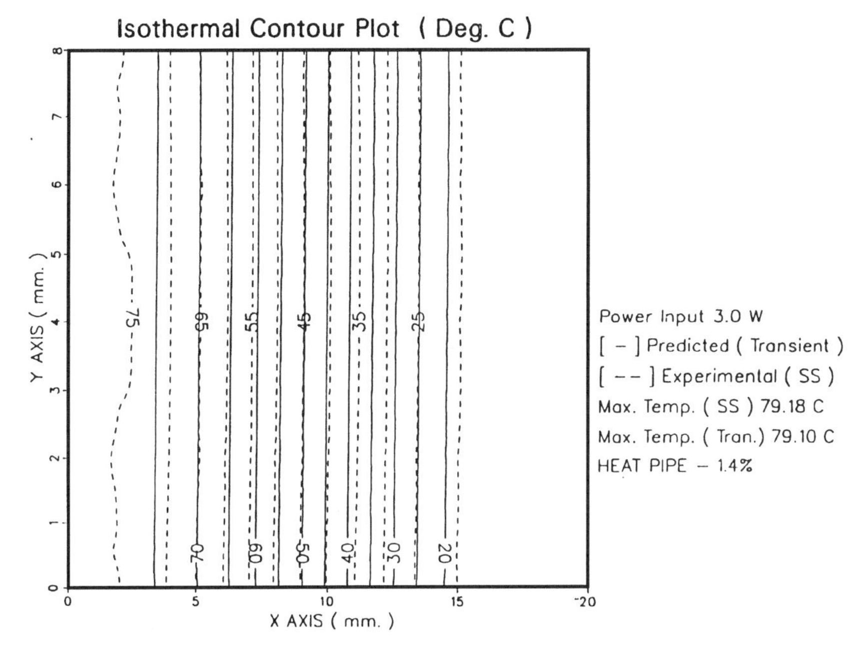

Figure 8-17c Comparison of the predicted transient thermal response and the experimental data for a wafer with an array of 66 micro heat pipes, an input power of 3.0 Watts, at an elapsed time of 12.5 seconds (Peterson and Mallik, 1995).

deposited micro heat pipes at an input power of 2.75 Watts, and the array of 66 vapor deposited micro heat pipes at an input power of 3.0 Watts, respectively, at various times. For the plain wafer at an input power of 2 Watts and an elapsed time of 7.5 seconds, the measured and predicted temperatures are quite close with the small variations as a result of the difference between the actual temperature dependent thermal conductivity and the constant value used in the numerical model.

For the wafer with the array of 34 vapor deposited micro heat pipes at an input power of 2.75 Watts and an elapsed time of 7.5 seconds, the predicted time constant was 4.24 seconds, while the experimental time constant was equal to 3.25 seconds. The predicted maximum surface temperature after 7.5 seconds for the wafer was 72.97°C, compared to a measured value of 73.88°C. And for the array of 66 vapor deposited micro heat pipes at an input power of 3.0 Watts and an elapsed time of 12.5 seconds, the predicted time constant was 4.30 seconds, while the experimental time constant was equal to 3.38 seconds. The predicted maximum surface temperature after 12.5 seconds was 79.10°C, compared to the measured value of 79.18°C.

To more clearly illustrate the accuracy of the numerical model, Figs. 8-18a, 8-18b, and 8-18c show the temperature distribution for the three previous cases for a single point located on the surface of the wafer. As shown, the measured values shown by the solid symbols and the predicted values shown by the solid line are quite close over the entire time domain with the largest variation occurring in the intermediate time range (i.e., the range where the time constant is calculated, hence the large variation). The final values for all three cases were extremely close with no noticeable variation.

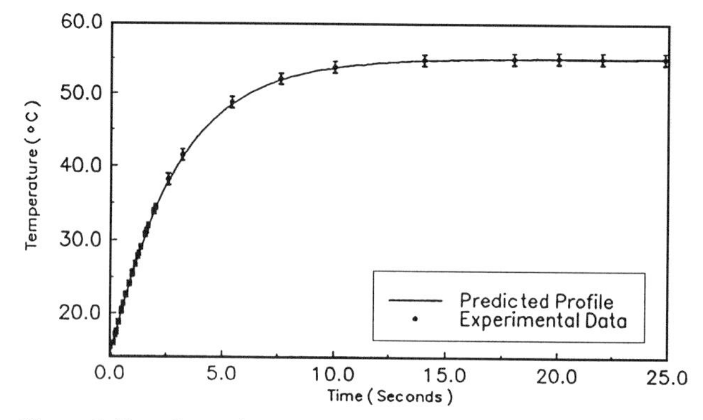

Figure 8-18a Comparison of the predicted transient thermal response with the experimental data for the plain wafer and a heat flux of 2.34 W/cm^2 (Peterson and Mallik, 1995).

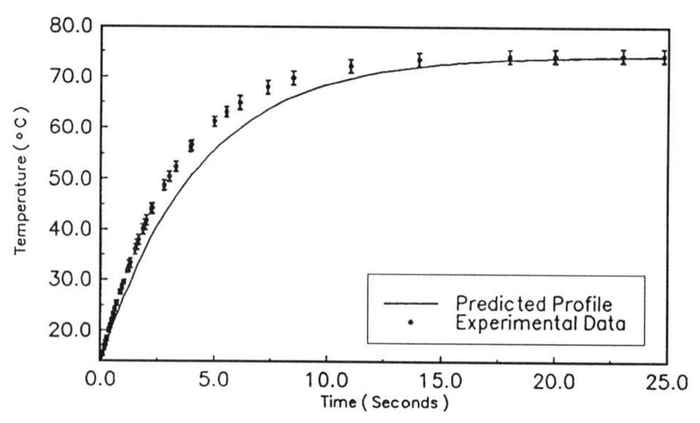

Figure 8-18b Comparison of the predicted transient thermal response with the experimental data for a wafer with an array of 34 micro heat pipes, and a heat flux of 4.30 W/cm^2 (Peterson and Mallik, 1995).

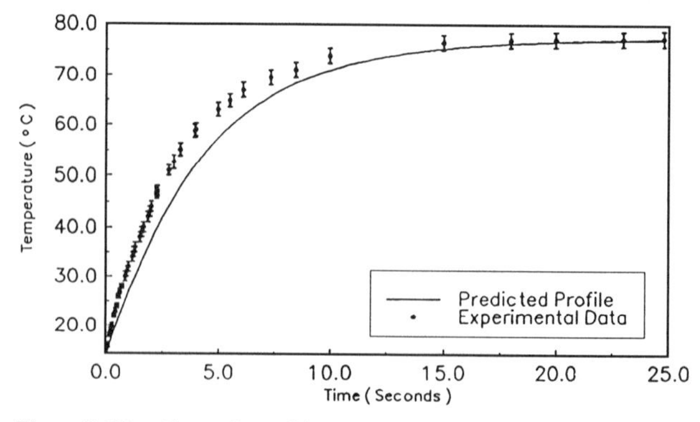

Figure 8-18c Comparison of the predicted transient thermal response with the experimental data for a wafer with an array of 66 micro heat pipes, and a heat flux of 4.69 W/cm^2 (Peterson and Mallik, 1995).

8-4 CONSTRUCTION TECHNIQUES AND ISSUES

Previously developed numerical models developed by Mallik et al. (1991) have demonstrated the potential advantages of constructing an array of micro heat pipes as an integral part of semiconductor chips, with the principal function being to act as a highly efficient heat spreader. The heat generated by a single active device on the chip surface can be dissipated over the entire chip area or transported to a condenser region where it can be rejected to an external heat sink. In this type of application, micro heat pipes serve to eliminate localized hotspots, reduce temperature gradients, and improve the reliability of the chip. The results of the modeling effort indicated that significant reductions in the maximum chip temperature, thermal gradients, and localized heat fluxes could be obtained through the incorporation of these heat pipe arrays.

8-4-1 Fabrication Methods

A number of different techniques have been proposed for fabricating these arrays of micro heat pipes and incorporating them as an integral part of silicon or gallium arsenide wafers. A majority of these have focused on the fabrication, development, and testing of these devices (Weichold et al., 1993; Peterson et al., 1991; Gerner, 1990). Several fabrication techniques have been proposed including the formation of a series of parallel triangular grooves using directionally dependent etching (Peterson, 1988a; Gerner, 1990) or a fabrication technique that utilizes a multisource vapor deposition process (Mallik et al., 1991).

Anisotropic or orientation-dependent etching techniques have long been used to produce micro channels in silicon wafers (Bean, 1978; Kendall, 1979). In the investigations of Peterson (1988b), Gerner (1989, 1990), and Peterson et al. (1991), a series of V-shaped grooves were etched using a series of oxidation, photolithography, anisotropic wet chemical etching, diffusion, and various bonding processes (Peterson, 1991; Mallik and Peterson, 1991; Ramadas et al., 1993; and Ramadas, 1993). These processes are similar to those used to fabricate ordinary microcircuits, except that a third dimension is added by the etching process. The first step in the fabrication of micro heat pipes is a sequence of dry-wet-dry oxidation to form silicon dioxide on the silicon wafer. Photolithography is then used to precisely align the heat pipe pattern edges with the [110] direction for anisotropic etching, such that the etched angle is 54.74° to the (100) surface, with walls formed by {111}-type planes which serve as etching barriers due to the atomic bond density in that direction. Anisotropic etchants such as KOH (potassium hydroxide), EDP (ethylene diamine pyrocatechol), and hydrazine are used to remove the exposed silicon without attacking the silicon dioxide. All of these etchants were successfully used, but KOH is preferred here for both quality and safety in processing. These etchants are selective to the different atomic planes, attacking the (100) plane quite vigorously while treating the defining {111}-type planes as etching barriers. The pipes may be kept bare or preferably be coated with silicon dioxide to enhance the surface wetting characteristics, since silicon is hydrophobic while silicon dioxide is hydrophilic. The final process is the bonding of the cover plate, which can be accomplished through ionic bonding (for silicon to silicon), ultraviolet bonding (for silicon to glass), or electrostatic bonding (also for glass to silicon).

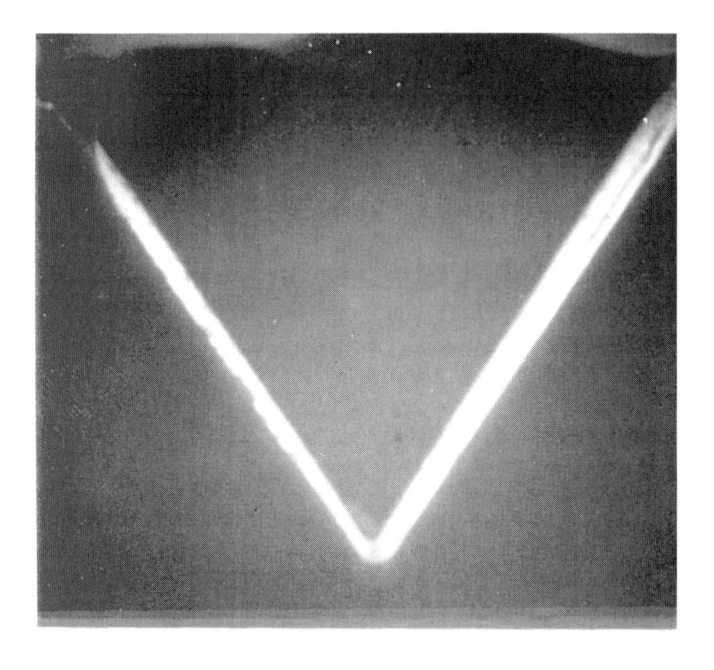

Figure 8-19a SEM of etched triangular groove (Peterson et al., 1991).

In the first successful demonstration of the operation of an array of micro heat pipes, Peterson et al. (1991) fabricated a series of nineteen V-shaped channels with a depth of 80 μm and a width of approximately 84 μm, a SEM of which is shown in Fig. 8-19a. As discussed in Section 8-3, a vapor-deposited resistive heater was used to provide thermal power to the wafer, and temperature distributions were obtained using an IR camera. Gerner et al. (1994) utilized a similar technique to construct an array of 120 micro heat pipes shown in Fig. 8-19b, which was designed for use as an efficient thermal spreader. These pipes were V-grooved, 100 μm wide (or 250 μm), 1$''$ long, and 70.72 μm (or 176.8 μm) deep and were spaced at 100 μm intervals. Integrated resistive heaters of appropriate size were fabricated in the back of the silicon wafer near the evaporator end by thermal diffusion of p-type (boron) ions into the n-type substrate. A series of p-n junctions (diodes) were also fabricated into the silicon to measure the axial temperature distribution profile. While information on the fabrication of these arrays has been presented, no actual experimental test results for these arrays have ever been presented in the literature.

Because of the problems associated with having the working fluid directly in contact with the silicon, Weichold et al. (1993) proposed an alternative technique which utilized a dual-source vapor-deposition process. The process begins with the fabrication of a series

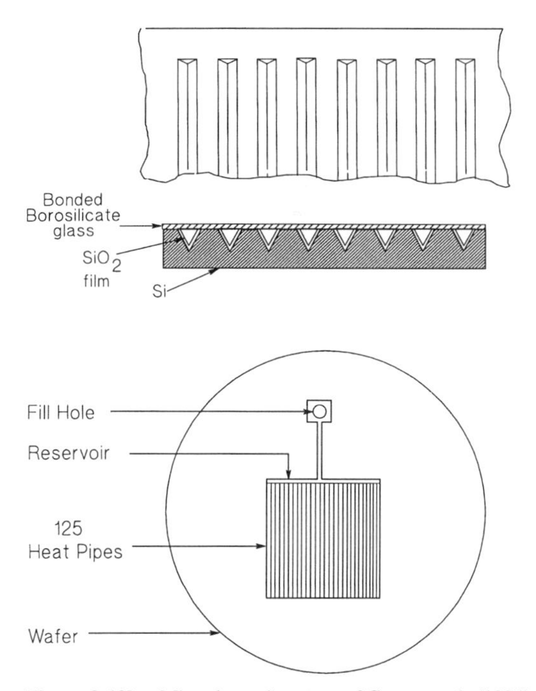

Figure 8-19b Micro heat pipe array of Gerner et al. (1994).

of square or rectangular grooves either machined or etched in a silicon wafer, as shown in Fig. 8-20a and 8-20b, respectively. Then using a dual E-Beam vapor-deposition process the grooves are closed creating an array of long narrow channels of triangular cross section and open on both ends. Figure 8-21 shows a photo micrograph of the end view of a vapor-deposited micro heat which has not quite been completely closed at the top. As shown, the micro heat pipes are lined with a thin layer of copper, significantly reducing problems associated with the migration of the working fluid (methanol) throughout the semiconductor material. Using copper as the material for the vapor deposited heat pipes on silicon substrate, ensured less than 2.0% differential expansion during the deposition process. A computer code "Simulation and Modelling of Evaporated Deposition Profiles" (SAMPLE) was used to simulate the metal step coverage and predict the final heat pipe cross-sectional profile (Sung, 1979). The shape of the vapor-deposited micro heat pipe illustrated in Fig. 8-21a is shown in Fig. 8-21b as predicted by this model. As shown the correlation is quite good. Using this model, the vapor deposition process was optimized and construction of an array of vapor deposited micro heat pipes was successfully completed in grooves with an aspect ratio of 1.0 (Mallik et al., 1991).

8-4-2 Charging Procedures

Regardless of the fabrication technique used, it was necessary to charge the heat pipes with the correct amount of working fluid. Several charging methods have been proposed (Peterson et al., 1993). The most reliable of these is to place the micro heat pipe array

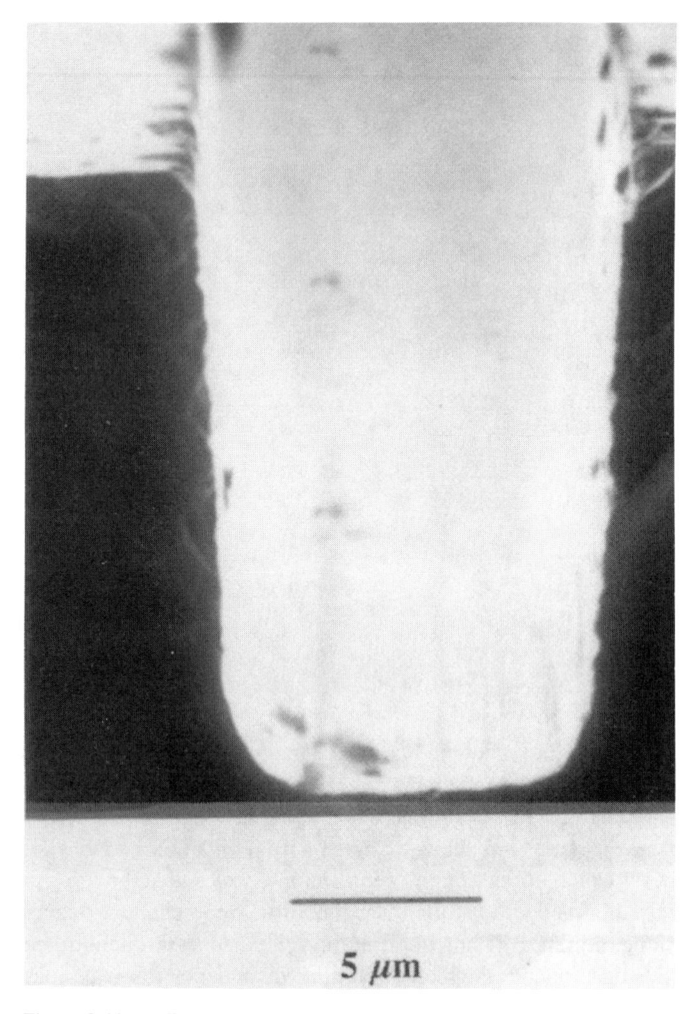

Figure 8-20a SEM of machined rectangular grooves in silicon (Peterson, 1994).

in a high-pressure chamber. After evacuating the chamber, a predetermined amount of working fluid is added and the chamber is heated to a point above the critical temperature of the working fluid. In this state all of the working fluid exists as a vapor and is uniformly distributed throughout the chamber and micro heat pipe array. The heat pipes are then sealed using an ionic or ultraviolet bonding process. When the chamber is then opened and the temperature reduced, a portion of the working fluid condenses. By controlling the volume of the pressure vessel and the quantity of the working fluid added after initial evacuation, the final amount of fluid in each heat pipe can be precisely controlled.

An alternative method, which is simpler but somewhat less accurate, has been developed by Peterson (1992) and consists of sealing one end of the micro heat pipe array and positioning the wafer in a vacuum chamber vertically with the open end of the

Figure 8-20b SEM of etched rectangular grooves in (110) silicon (Peterson, 1994).

micro heat pipes approximately 1 mm above a small charging trough. After evacuating the vacuum chamber, a small amount of working fluid is slowly injected into the charging trough using a micro syringe. As the chamber pressure increases and approaches the saturation pressure corresponding to the temperature of the chamber, the working fluid forms a pool in the charging trough and wicks up into the heat pipe channels, effectively sealing off the channel and trapping vapor in the upper portion of the heat pipe. The micro heat pipe array can then be removed from the vacuum chamber and the open ends sealed. The final amount of liquid and vapor present in the microgrooves after sealing can be controlled by varying the temperature of the wafer, the height of the wafer above the bottom of the charging trough, and the rate at which the working fluid is injected into the vacuum chamber.

A third charging method has been proposed by Gerner (1990). This method uses one hole for evacuating and filling the micro heat pipe array through a common reservoir connecting their condenser sections. Three valves are used in the system. The first one is to isolate the vacuum system, the second one is used for backfilling the working fluid into the pipe array, and third one, located above the array, is to isolate the micro heat pipe

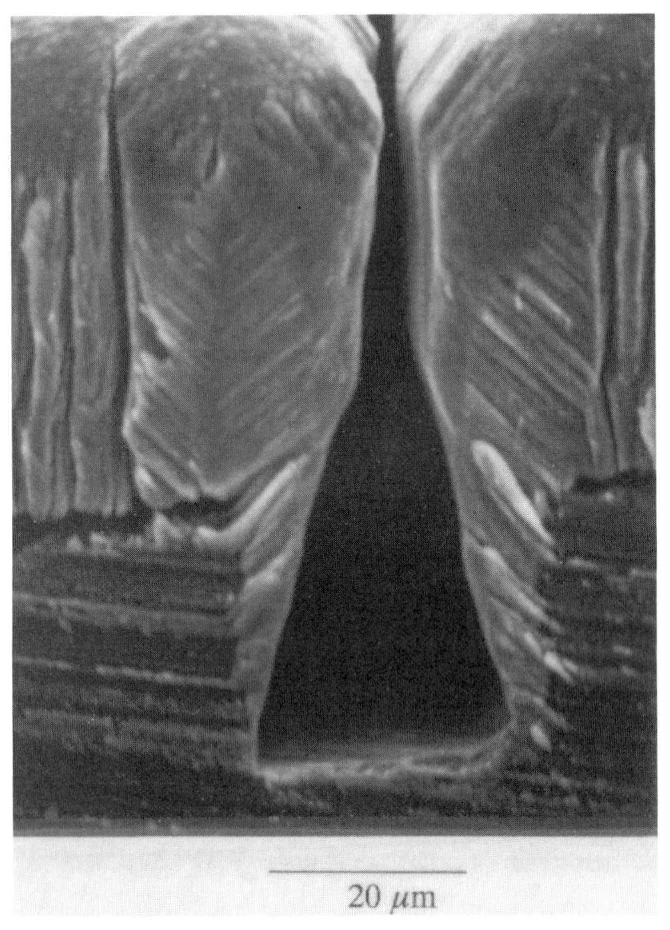

Figure 8-21a SEM of vapor-deposited micro heat pipe fabricated using a dual vapor-deposition process and an etched rectangular groove (Mallik et al., 1995).

array from the system. The latter can be used for two purposes, one to control the initial amount of liquid fill, and a second to isolate the micro heat pipe array from the system. Once the proper vacuum pressure is achieved in the pipes, a controlled amount of the working fluid is introduced to the array using a micro flow meter positioned between the array and the filling inlet valve (second valve). The charge of the liquid inside the pipes can be changed by opening the shutoff valve gradually (third valve) and increasing the heat load. Increasing the heat load increases the vapor pressure inside the array. Consequently, the working fluid is pushed outside the array.

8-4-3 Charge Optimization

In a majority of the modeling discussed thus far, the liquid–vapor interface was assumed to exist with a constant radius of curvature in both the evaporator and condenser regions.

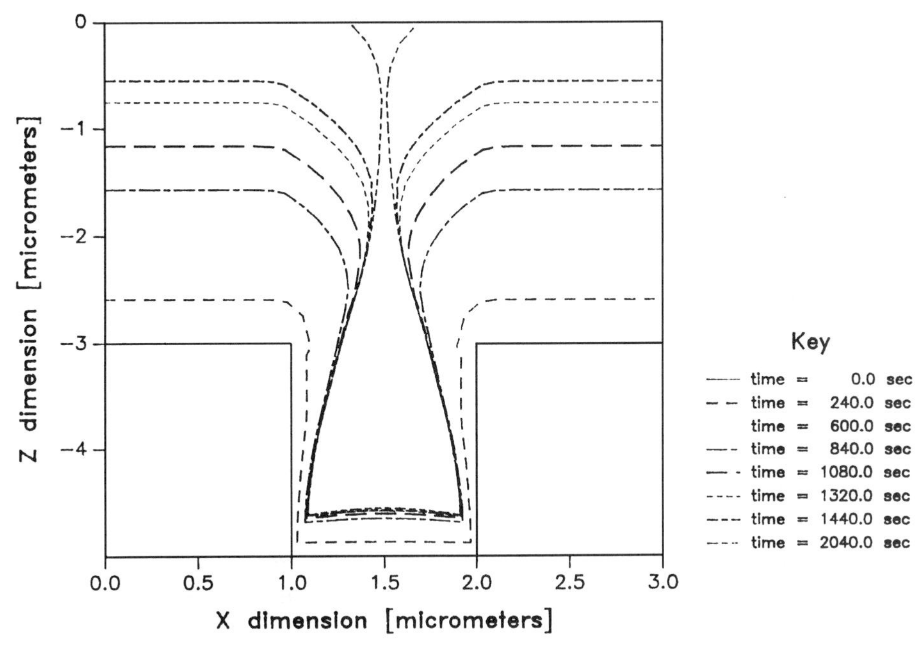

Figure 8-21b Predicted shape of vapor deposited micro heat pipe using a dual vapor-deposition process and an etched rectangular groove (Mallik et al., 1995).

Duncan and Peterson (1995) developed an analytical model to determine the operating conditions and performance limitations of triangular micro heat pipes similar to those that might be incorporated as an integral part of silicon wafers. This analytical model was the first to predict not only the capillary limit of an array of triangular micro heats pipes fabricated in silicon wafers, but also the radius of the liquid–vapor interface in the evaporator as a function of input power. This latter capability allows this model to be used to calculate the optimal liquid charge of micro heat pipes as a function of the design conditions (i.e., the operating temperature and input power).

The analysis of the values of the pressure drop associated with the flow of the liquid and vapor phases required certain assumptions in order to define the configuration of the liquid phase throughout the micro heat pipe. As mentioned previously, the liquid–vapor interface was assumed to exist with constant radii of curvature in both the evaporator and condenser regions. In the evaporator region, the radius of curvature was allowed to vary between a value approaching zero and the value of the maximum possible radius of curvature that was assumed to exist in the condenser. The radius of curvature of the liquid–vapor interface in the adiabatic region of the micro heat pipe was assumed to increase linearly from that assumed in the evaporator region to the maximum allowable radius of curvature. In this manner, the flow areas of both the liquid and vapor phases were defined. In order to accurately predict the pressure drop terms associated with the flow of the liquid and vapor phases, the adiabatic region was divided into a number of finite regions, and the sum of the pressure drops associated with each of these regions was used to approximate the total pressure drop in the adiabatic section of the micro heat

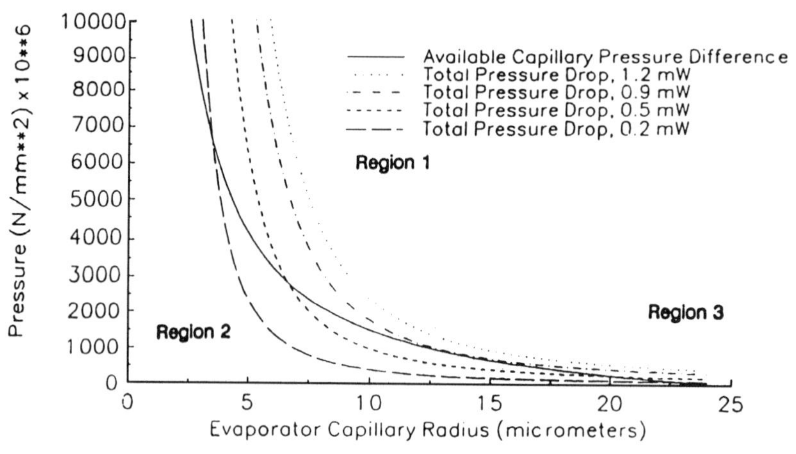

Figure 8-22 Comparison of the available capillary pressure and the total pressure drop as a function of capillary radius (Duncan and Peterson, 1995).

pipe. Each of the regions utilized liquid and vapor phase flow areas that are representative of the actual flow area based on the assumed linear distribution of the radius of curvature of the liquid–vapor interface.

Figure 8-22 illustrates the operating characteristics of the triangular etched micro heat pipe as predicted by the analytical model of the capillary limit. The solid line in Fig. 8-22 represents the available capillary pressure difference, predicted by varying the radius of curvature of the liquid–vapor interface in the evaporator region from zero to the maximum possible value in the condenser. It should be noted that as this radius of curvature approaches zero, the available capillary pressure difference becomes unbounded.

Region 1 of Fig. 8-22 represents the "non-operational" liquid–vapor configurations for the micro heat pipe. In Region 1, the total pressure drop (i.e., the sum of the hydrostatic pressure drop and the pressure drops associated with the flow of the liquid and vapor phases, corresponding to the three greatest input power levels [1.2 mW, 0.9 mW, and 0.5 mW]) exceed the available capillary pressure difference. However, as the evaporator capillary radius increases, leading into Region 3, only the greatest value of input power (1.2 mW) continues to produce a pressure drop in excess of the available capillary pressure difference of the micro heat pipe. This indicates that the input power of 1.2 mW exceeds the capillary limit of the micro heat pipe. The curve representing the total pressure drop associated with an input power of 0.9 mW intersects the curve representing the available capillary pressure difference at a point that corresponds to a radius of curvature in the evaporator region of approximately 13.7 μm. This, in effect, defines the entire configuration of the liquid in the micro heat pipe for that input power. For an input power of 0.5 mW, the evaporator radius of curvature required for operation is approximately 20.5 μm and for 0.2 mW is approximately 22.5 μm.

Region 2 of Fig. 8-22 represents operating parameters for which the available capillary pressure difference exceeds the total pressure drop associated with the two lower input power values (0.2 mW and 0.5 mW). Note, however, that these two curves do again intersect the curve representing the available capillary pressure difference in this region. In other words, there are two possible mathematical solutions to Eq. (8-1) for input power values that do not exceed the capillary limit of the micro heat pipe. The slope of the curves representing total pressure drop in Region 2 indicates, however, that the solutions in this region are unstable by nature and implies that the system will seek the more stable configuration, which corresponds to the greater radius of curvature in the evaporator region. The computer model used to investigate the capillary limit of the etched micro heat pipes determined the physical properties of the working fluid based on a single operating temperature. This operating temperature is typically representative of the temperature in the adiabatic region of the micro heat pipe. Initially, an investigation was conducted to determine the optimal operating temperature using methanol as the working fluid. To do this both the radius of curvature of the liquid–vapor interface in the evaporator region and the operating temperature were varied.

The optimum charge was also found to be a function of the operating temperature, and in this investigation resulted in maximum heat transport capacity of approximately 0.9 mW at 385 K. This operating temperature was utilized in the subsequent portions of the analytical investigation to determine the maximum performance characteristics of the micro heat pipes. Using the information provided by the capillary limit analysis, the radius of curvature in the evaporator region of the micro heat pipe can be predicted for all input power values less than the capillary limit. With the known radius of curvature in the evaporator, adiabatic, and condenser regions, the optimal liquid charge can be predicted as a function of input power. Figure 8-23 illustrates the resulting optimal liquid charge as a function of input power. As shown, this liquid charge varies from approximately 25% of the total micro heat pipe volume at input power values approaching zero to approximately 16% at the capillary limit of the micro heat pipe.

To validate the analysis conducted by Duncan and Peterson (1995), an experimental investigation, similar to one previously conducted by Peterson et al. (1993), was undertaken. As discussed earlier, high heat flux values may induce a condition where such a small amount of liquid exists in the evaporator that the required flow velocities are very high and, hence, the viscous forces become dominant, preventing the liquid from

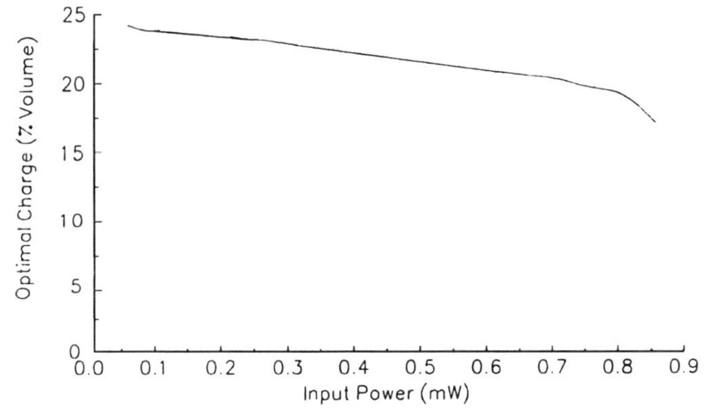

Figure 8-23 Determination of the optimal liquid charge (Duncan and Peterson, 1995).

reaching the evaporator. This condition is referred to as the capillary limit or dryout and, as indicated by Babin et al. (1990), is the primary performance limit in micro heat pipes. Dryout typically can be observed experimentally by either a sudden increase in evaporator temperature or a decrease in effective thermal conductivity as power input to the evaporator is increased.

In an earlier investigation, arrays of triangular micro heat pipes were tested to verify the micro heat pipe concept and determine the performance limits using an infrared thermal imaging system (Peterson et al., 1993). At the conclusion of the initial experimental tests and concept verification, it was apparent that a different heat source would be required to produce dryout conditions in the micro heat pipes. As a result, in the current investigation, a new higher-power distributed heat source was used, which could provide significantly higher heat flux values so evaporator dryout could be achieved.

The current investigation utilized an anisotropic etching process to produce a series of 59 parallel, triangular channels each 120 μm wide and 80 μm deep, in silicon wafers. Once fabricated, a clear pyrex cover plate was bonded to the top surface of each wafer using an ultraviolet bonding technique to form the micro heat pipe arrays. These micro heat pipe arrays were then evacuated and charged with varying amounts of methanol and evaluated using an infrared thermal imaging unit to measure the temperature gradients and maximum localized temperatures. From these data, an effective thermal conductivity could be computed. For comparison purposes, plain silicon wafers were also evaluated in a similar manner.

The experimental fabrication and test procedures as well as the experimental apparatus used in the current investigation were similar to those of the earlier investigation of Peterson et al. (1993). Observations were made in order to determine the effective thermal conductivity of a 20 mm $\times$ 20 mm silicon wafer containing an array of 59 micro heat pipes as a function of input power (operating temperature) and the initial liquid charge.

Figure 8-24 presents the experimental results for the four different fluid charges evaluated, 50, 30, 20, and 10% charge by volume, respectively. For comparison, the

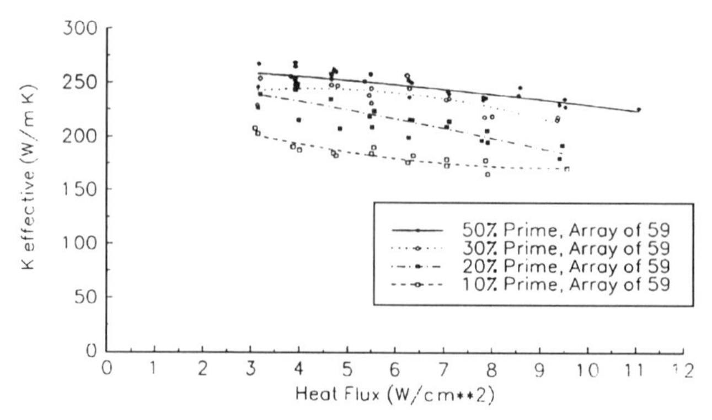

Figure 8-24 Effective thermal conductivity of the micro heat pipe array with various liquid charges (Duncan and Peterson, 1995).

Table 8-1 Measured heat transport at onset of dryout: input power = 2.00 W, predicted heat transport at onset of dryout = 0.9 mW

% Liquid charge	Measured heat transport
10%	1.02 mW
20%	3.38 mW
30%	3.76 mW
50%	5.53 mW

measured thermal conductivity of the plain wafers are also shown as a function of heat flux. As illustrated, the experimental results consistently indicated the gradual onset of dryout (indicated by the slight decrease in effective thermal conductivity) over the entire range of input power levels utilized in this investigation.

Considering the silicon and micro heat pipes as parallel thermal resistances subjected to the input power from the distributed heat source, the heat transported by each micro heat pipe can be determined from the experimental results. A heat flux value of 3.15 W/cm^2, which corresponds to a total heat input of 2.0 W, was the smallest input power value utilized in the current investigation. If this input power is considered to be that which corresponds to the onset of dryout, then dryout actually occurred at input power levels which were four to six times those predicted by the analytical model as illustrated in Table 8-1.

The difference between the predicted and measured capillary limit in this investigation can be attributed to three primary factors. First, the thermal and spatial resolution characteristics of the thermal imaging system were not high enough to observe and indicate the precise location and power level at which dryout occurred. Although the system was capable of detecting the temperature gradients, which were used to calculate the effective thermal conductivity of the wafers, the system was not capable of producing spatial resolution on the scale of 1 μm or less, which would be required in order to observe the progressive dryout of an individual micro heat pipe. Second, the analysis used to develop the model of the capillary limit relied solely on the assumptions associated with the mass and momentum transport characteristics of bulk fluids. Previous investigations have shown that thin liquid films such as those that exist in the evaporator section of micro heat pipes have tremendously enhanced evaporative characteristics when compared to the evaporative characteristics of bulk fluids. Finally, the analytical model does not account for the actual process of gradual dryout, wherein the wetted liquid front progresses toward the condenser, allowing heat pipe operation to continue though decreasing the effective heat pipe length.

One other factor may provide a partial explanation for this difference and this has to do with the initial charge. Excess liquid in the system would tend to move the actual point of dryout closer to the evaporator region of the heat pipe. This would, in effect, allow partial operation of the heat pipe to occur, increasing the effective thermal conductivity of the wafer/heat pipe array composite. For this reason, identification of the precise dry out heat flux and location will require a significantly different experimental approach (Ha and Peterson, 1993).

8-5 SUMMARY

The analytical and experimental results summarized in the preceding discussion clearly indicate that micro heat pipes are a feasible method of heat removal and distribution from both an experimental and a theoretical perspective. Both steady-state and transient models are available that are capable of predicting the operational limits and performance characteristics of micro heat pipes with diameters less than 100 μm with a high degree of reliability. These models are currently being expanded for use in both individual heat pipes and also with arrays of heat pipes constructed as an integral part of semiconductor devices.

From an experimental perspective, individual micro heat pipes on the order of 1 mm in diameter have been tested and shown to be effective in dissipating and transporting heat from localized heat sources. The results of tests conducted on these heat pipes have provided the background for the adoption of these heat pipes in commercialized applications. In addition, arrays of micro heat pipes on the order of 35 μm have been successfully fabricated, charged, and tested. When incorporated as an integral part of semiconductor devices, these heat pipe arrays have been shown to be an effective method for dissipating localized heat fluxes, eliminating localized hotspots, reducing the maximum wafer temperatures and thereby improving the wafer reliability.

Several obstacles still remain that are currently preventing the wide spread use of micro heat pipes. These include performance degradation over time, the effect of the presence of the working fluid on the chip operation, large-scale fabrication issues, and the determination of the effect of the precise size and shape of the micro heat pipe and the array density. However, it is clear from these results that incorporating relatively large micro heat pipes (1 mm diameter) or arrays of micro heat pipes as an integral part of silicon chips presents a feasible alternative cooling scheme that merits serious consideration for a number of heat transfer applications.

NOMENCLATURE

d	Vapor space hydraulic diameter
g	Acceleration due to gravity
K	Mean curvature of the vapor–liquid interface in a groove
L	Total heat pipe length
k_{eff}	Effective thermal conductivity
$\dot{m}$	Mass flow rate
r_c	Capillary radius of the vapor–liquid interface in a groove
r_h	Hydraulic radius of the flow channel
T_b	Boundary temperature
T_c	Minimum temperature based on the arithmetic mean along a line 3 mm from the edge containing the heat sink

T_h Maximum temperature based on the arithmetic mean along a line 4 mm from the edge containing the heater
T_ℓ Liquid temperature
V_v Vapor velocity
x Axial coordinate
x^* Dimensionless axial coordinate
x_c Axial position in the middle of the condenser
x_h Axial position in the middle of the evaporator

α Wedge channel half angle
ΔP_c Capillary pressure
Δx_i x-position increment
Δy_i y-position increment
Δz_i z-position increment
θ Dimensionless temperature
ρ_ℓ Liquid density
ϕ Tilt angle

REFERENCES

Anon, "Application of micro heat pipes in hyperthermia," Annual Report of the Itoh R and D Laboratory, Osaka, Japan, 1989.

Babin, B. R., and G. P. Peterson, "Experimental investigation of a flexible bellows heat pipe for cooling discrete heat sources," *ASME J. Heat Transfer*, vol. 112, pp. 602–607, 1990.

Babin, B. R., G. P. Peterson, and D. Wu, "Steady-state modeling and testing of a micro heat pipe," *ASME J. Heat Transfer*, vol. 112, pp. 595–601, 1990.

Badran, B., F. M. Gerner, P. Ramadas, H. T. Henderson, and K. W. Baker, "Liquid metal micro heat pipes," *29th National Heat Transfer Conference, HTD*, vol. 236, pp. 71–85, Atlanta, GA, 1993.

Basiulis, A., H. Tanzer, and S. McCabe, "Heat pipes for cooling of high density printed wiring boards," *Proc. 6th International Heat Pipe Conference*, Grenoble, France, pp. 531–536, 1987.

Bean, K. E., "Anisotropic etching of silicon," *IEEE Trans. Electron Dev.*, vol. ED-25, no. 10, pp. 1185–1193, 1978.

Camarda, C. J., D. R. Rummler, and G. P. Peterson, "Multi heat pipe panels," *NASA Tech. Briefs*, LAR-14150, in press.

Chang, W. S., and G. T. Colwell, "Mathematical modeling of the transient operating characteristics of a low-temperature heat pipe," *Numer. Heat Transfer*, vol. 8, pp. 169–186, 1985.

Chi, S. W., *Heat Pipe Theory and Practice*, McGraw-Hill, 1976.

Collier, J. G., *Convective Boiling and Condensation*, McGraw-Hill, New York, 1981.

Cotter, T. P., "Principles and prospects of micro heat pipes," *Proc. 5th International Heat Pipe Conference*, Tsukuba, Japan, pp. 328–335, 1984.

Dean, D. S., "An integral heat pipe package for microelectronic circuits," *Proc. 2nd International Heat Pipe Conference*, ESTEC, Noordwijk, The Netherlands, pp. 481–502, 1976.

Duncan, A. B., and G. P. Peterson, "Charge optimization of triangular shaped micro heat pipes," *AIAA J. Thermophys. Heat Transfer*, vol. 9, pp. 365–367, 1995.

Fejfar, K., F. Polásek, and P. Stulc, "Tests of micro heat pipes," Annual Report of the SVÚSS, Prague 1990, 1990.

Fletcher, L. S., and G. P. Peterson, *A Micro Heat Pipe Catheter for Local Tumor Hyperthermia*, U.S. Patent No. 5,190,539, issued March 2, 1993.

Gerner, F. M., "Flow limitations in micro heat pipes," AFSOR Final Report No. F49620-88-6-0053, Wright-Patterson AFB, Dayton, OH, 1989.

Gerner, F. M., "Micro heat pipes," AFSOR Final Report No. S-210-10MG-066, Wright-Patterson AFB, Dayton, OH, 1990.

Gerner, F. M., B. Badran, H. T. Henderson, and P. Ramadas, "Silicon-water micro heat pipes," *Thermal Sci. Eng.*, vol. 2, pp. 90–97, 1994.

Gerner, F. M., J. P. Longtin, P. Ramadas, T. H. Henderson, and W. S. Chang, "Flow and heat transfer limitations in micro heat pipes," 28th National Heat Transfer Conference, San Diego, CA, Aug. 9–12, 1992.

Ha, J. M., and G. P. Peterson, "Analytical prediction of the axial dryout of an evaporating liquid film in triangular micro channels," *ASME J. Heat Transfer*, vol. 116, pp. 498–503, 1994.

Ikeda, Y., "Neutron radiography tests of Itoh's micro heat pipes," private communications of the Nagoya University to F. Polásek, 1990.

Itoh, A., "Micro heat pipes," *Prospectus of the Itoh R and D Laboratory*, Osaka, Japan, 1988.

Itoh, A., and F. Polásek, "Micro heat pipes and their application in industry," *Proc. of the Czechoslovak-Japanese Symposium on Heat Pipes*, Rícany, Czechoslovakia, 1990a.

Itoh, A., and F. Polásek, "Development and application of micro heat pipes," *Proc. 7th International Heat Pipe Conference*, Paper No. 123, Minsk, USSR, May 21–25, 1990b.

Kendall, D. L., "Vertical etching of silicon at very high aspect ratios," *Annu. Rev. Mater. Sci.*, pp. 373–403, 1979.

Khrustalev, D., and A. Faghri, "Thermal analysis of a micro heat pipe," *ASME J. Heat Transfer*, vol. 116, pp. 189–198, 1994.

Longtin, J. P., B. Badran, and F. M. Gerner, "A one-dimensional model of a micro heat pipe during steady-state operation," *ASME J. Heat Transfer*, vol. 116, pp. 709–715, 1994.

Longtin, J. P., B. Badran, and F. M. Gerner, "A one-dimensional model of a micro heat pipe during steady-state operation," *Proc. 28th National Heat Transfer Conference*, San Diego, CA, Aug. 9–12, 1992.

Longtin, J. P., B. Badran, and F. M. Gerner, "A one-dimensional model of a micro heat pipe during steady-state operation," *J. Heat Transfer*, vol. 116, pp. 709–715, 1994.

Ma, H. B., G. P. Peterson, and X. J. Lu, "The influence of the vapor–liquid interactions on the liquid pressure drop in triangular microgrooves," *Int. J. Heat Mass Transfer*, vol. 37, pp. 2211–2219, 1995.

Mallik, A. K., G. P. Peterson, and M. H. Weichold, "Fabrication of vapor deposited micro heat pipes arrays as an integral part of semiconductor devices," *ASME J. Micromechan. Syst.*, vol. 4, pp. 119–131, 1995.

Mallik, A. K., and G. P. Peterson, "Steady-state investigation of vapor deposited micro heat pipe arrays," *ASME J. Electron. Packaging*, vol. 117, 1995, pp. 75–81, 1995.

Mallik, A. K., G. P. Peterson, and M. H. Weichold, "On the use of micro heat pipes as an integral part of semiconductor devices," *ASME J. Electron. Packaging*, vol. 114, pp. 436–442, 1992.

Mallik, A. K., and G. P. Peterson, "On the use of micro heat pipes as an integral part of semiconductors," *3rd ASME-JSME Thermal Engineering Joint Conference Proc.*, vol. 2, Reno, NV, pp. 394–401, 1991.

Mallik, A. K., G. P. Peterson, and W. Weichold, "Construction processes for vapor deposited micro heat pipes," *10th Symposium on Electronic Materials Processing and Characteristics*, Richardson, TX, 1991.

Mirzamoghadam, A., and I. Catton, "A physical model of the evaporating meniscus," *ASME J. Heat Transfer*, vol. 110, pp. 201–208, 1988.

Mrácek, P., "Application of micro heat pipes to laser diode cooling," Annual Report of the VÚMS, Prague, Czechoslovakia, 1988.

Murakami, M., T. Ogushi, Y. Sakurai, H. Masumoto, M. Furukawa, and R. Imai, "Heat pipe heat sink," *Proc. 6th International Heat Pipe Conference*, vol. 2, Grenoble, France, pp. 537–542, 1987.

Murase, T., S. Tanaka, and S. Ishida, "Natural convection type long heat pipe heat sink 'POWERKICKER-N' for the cooling of GTO thyristor," *Proc. 6th International Heat Pipe Conference*, vol. 2, Grenoble, France, pp. 537–542, 1987.

Patankar, S. V., *Numerical Heat Transfer and Fluid Flow*, Hemisphere, Washington, DC, 1980.

Peterson, G. P., and A. K. Mallik, "Transient response characteristics of vapor deposited micro heat pipe arrays," *ASME J. Electron. Packaging*, vol. 117, pp. 82–87, 1985.

Peterson, G. P., A. B. Duncan, and M. H. Weichold, "Experimental investigation of micro heat pipes fabricated in silicon wafers," *ASME J. Heat Transfer*, vol. 115, pp. 751–756, 1993.

Peterson, G. P., A. B. Duncan, A. K. Ahmed, A. K. Mallik, and M. H. Weichold, "Experimental investigation of micro heat pipes in silicon devices," 1991 ASME Winter Annual Meeting, ASME, vol. DSC-32, Atlanta, GA, pp. 341–348, 1991.

Peterson, G. P., "Analysis of a heat pipe thermal switch," *Proc. 6th International Heat Pipe Conference*, vol. 1, Grenoble, France, pp. 177–183, 1987a.

Peterson, G. P., "Heat removal key to shrinking avionics," *Aerospace Am.*, no. 8, October, pp. 20–22, 1987b.

Peterson, G. P., "Heat pipes in the thermal control of electronic components," Invited Paper, *Proc. 3rd International Heat Pipe Symposium.*, Tsukuba, Japan, pp. 2–12, 1988b.

Peterson, G. P., "Investigation of miniature heat pipes," Final Report, Wright Patterson AFB, Contract No. F33615-86-C-2733, Task 9, 1988a.

Peterson, G. P., "Analytical and experimental investigation of micro heat pipes," *Proc. 7th International Heat Pipe Conference*, Paper No. A-4, Minsk, USSR, 1990.

Peterson, G. P., "An overview of micro heat pipe research," *Appl. Mechan. Rev.*, vol. 45, pp. 175–189, 1992.

Peterson, G. P., *An Introduction to Heat Pipes: Modeling, Testing and Applications*, John Wiley & Sons, New York, 1994.

Polásek, F., "Testing and application of Itoh's micro heat pipes," Annual Report of the SVÚSS, Prague, Czechoslovakia, 1990.

Ramadas, P., "Silicon micromachined heat pipe fabrication technology for electronic and space applications," M.S. Thesis, University of Cincinnati, Cincinnati, OH, 1993.

Ramadas, P., B. Badran, F. M. Gerner, T. H. Henderson, and K. W. Baker, "Liquid metal micro heat pipes incorporated in waste-heat radiator panels," presented in the Tenth Symposium on Space Power and Propulsion, Albuquerque, NM, 1993.

Sotani, J., Y. Susa, S. Tanaka, K. Sato, and Y. Kimura, "Micro heat pipe," *Proc. 7th Interanational Heat Pipe Conference*, Paper No. A-41, Minsk, USSR, 1990.

Stulc, P., and L. Horváth, "Heat pipes for cooling of electronic elements," Annual Report of the SVÚSS Prague, 1988.

Sung, C., "Simulation and modelling of evaporated deposition profiles," Ph.D. Thesis, Department of Electrical Engineering and Computer Sciences, University of California, Berkeley, CA, 1979.

Swanson, L. W., and G. P. Peterson, "The evaporating extended meniscus in a V-shaped channel," *AIAA J. Thermophys. Heat Transfer*, vol. 8, pp. 172–181, 1994.

Swanson, L. W., and G. P. Peterson, "The interfacial thermodynamics of micro heat pipes," *ASME J. Heat Transfer*, vol. 117, pp. 195–201, 1995.

Wayner, P. C., Jr., Y. K. Kao, and L. V. LaCroix, "The interline heat-transfer coefficient of an evaporating wetting film," *Int. J. Heat Mass Transfer*, vol. 19, pp. 487–492, 1976.

Weichold, M. H., G. P. Peterson, and A. Mallik, *Vapor Deposited Micro Heat Pipes*, U.S. Patent No. 5,179,043, issued January 12, 1993.

Wu, D., and G. P. Peterson, "Investigation of the transient characteristics of a micro heat pipe," *AIAA J. Thermophysics and Heat Transfer*, vol. 5, pp. 129–134, 1991.

Wu, D., G. P. Peterson, and W. S. Chang, "Transient experimental investigation of micro heat pipes," *AIAA J. Thermophys. Heat Transfer*, vol. 5, pp. 539–545, 1991.

Xu, X., and V. P. Carey, "Film evaporation from a micro-grooved surface—an approximate heat transfer model and its comparison with experimental data," *AIAA J. Thermophys. Heat Transfer*, vol. 4, pp. 512–520, 1990.

CHAPTER

NINE

MICROSCALE HEAT TRANSFER IN BIOLOGICAL SYSTEMS AT LOW TEMPERATURES

Boris Rubinsky

Department of Mechanical Engineering,
University of California at Berkeley,
Berkeley, CA 94720

9-1 INTRODUCTION

Life has evolved in relation to the temperature of our planet. The relation between life processes (metabolism), temperature, and heat transfer is the single most fundamental property of life. Normal mammalian life is possible only in a narrow range of temperatures of several degrees centigrade, and survival is possible in a range of only about ten degrees. However, on the surface of our planet the temperature of the atmosphere and oceans varies from low subfreezing temperatures to above boiling temperatures. Evolution has made it possible for life to thrive from the low temperatures of the Antarctic to the high temperatures of the ocean ridge volcanic vents. The study of heat transfer and life in biological systems can be roughly divided into three areas of temperature: a) normal mammalian physiological temperatures, b) above normal mammalian physiological temperatures, and c) below normal physiological temperatures. Each of these areas involves exciting and relatively unexplored topics of microscale heat transfer, such as the thermodynamics of normal life processes; the response to elevated temperatures, including production of heat shock proteins and high-temperature enzymes; and the response of life to low temperatures. This chapter will deal with the area of below normal physiological temperatures.

Because all life processes involve water, this temperature range can be further divided into two ranges: a) temperatures above the freezing temperature of water, and b) temperatures below the freezing temperature of water. This chapter will focus on two main topics, the first dealing with fundamental aspects of life processes at low temperatures, and the second with applications of low temperatures in medicine. Life processes occur in three length scales: nanoscale, relating to the behavior of organic molecules;

microscale, relating to the behavior of single cells in the organism; and macroscale, relating to the behavior of the whole organism. This chapter will focus primarily on studies related to nanoscale heat transfer and microscale heat transfer in biological systems.

9-2 LIFE AT LOW TEMPERATURES ABOVE THE FREEZING TEMPERATURE OF WATER

Low temperatures above the freezing temperature of water affect biological systems at both the nanoscale (molecular) and microscale (cellular) levels. A *cell* is the smallest unit that still retains at least some of the organized function of the organism. Cells are made of a phospholipid bilayer membrane that surrounds the intracellular solution (cytoplasm), which can contain also several organelles. Cells range in dimension from a few micrometers (spermatozoa and red blood cells) to several hundred micrometers (oocytes), and are mostly a few tens of micrometers in size. The cell phospholipid bilayer membrane, which is primarily impermeable, incorporates ion channels and other proteins that facilitate mass transfer between the intracellular and the extracellular solutions. The membrane is several angstroms thick; its functions are to separate the controlled environment inside the cell from the uncontrollable exterior and to regulate the mass transfer between the interior of the cell and its exterior.

One of the most important sites affected by temperature is the cell membrane. The phospholipid bilayers that form the cell membrane exhibit thermotropism (Morris and Watson, 1984), that is, they undergo an abrupt change from a disordered fluid or crystalline state to a highly ordered hexagonal lattice of fatty acyl chains (the gel state) over a specific temperature range. This process is akin to the solid-liquid phase transformation process, commonly encountered in heat transfer analysis of inorganic compounds. The temperature at which this phase transition occurs is known as either the *lipid phase transition temperature* or the *order–disorder transition*. This phase transition temperature is strongly dependent on the composition of the phospholipids, their chain length, and the degree of saturation (Hazel, 1984). Obviously the phase transition of the phospholipid bilayer in the cell membrane will affect its ability to effectively separate the interior of the cell from the extracellular solution and control the mass transfer. The kinetics of lipid phase transition as a function of temperature is studied in cells or liposomes with a variety of techniques, including low-temperature scanning electron microscopy (Holt and Northon, 1984), X-ray diffraction (Nakajama et al., 1983), light microscopy, and mathematical modeling (McGrath, 1993). Understanding and controlling this phase transformation process is probably one of the most important areas of microscale research in bio-heat transfer. How animals survive in nature the change in mass transfer properties of membranes undergoing phase transition intrigues many researchers. For instance, how do hibernating mammals survive at reduced temperatures? How do cold-blooded animals like reptiles, fish, insects, and plants survive seasonal changes in temperature (Hochahchka, 1986)? The mechanism is not understood yet. Among other theories, it was proposed that the composition of the membrane changes seasonally in such a way that the phase transition is depressed (Hochachka, 1986). Crowe and his group (Crowe et al., 1987) have also found that certain compounds such as trehalose or proline stabilize membrane bilayers. Understanding how membranes are naturally protected from this mode of damage has important applications in developing techniques for preservation of mammalian cells and organs in a refrigerated state.

Temperature reduction has additional effects. It also affects the weak bond interactions that determine protein structure, cytoskeleton structure, or enzyme-ligant interactions (Hochachka and Somero, 1984). Probably one of the most important energy-related effects of reducing temperatures is the change in the rate with which ATP-driven ion pumps transport ions across cell membranes. The ATP ion pumps are membrane protein, which traverse the membrane with the function of keeping up with the ionic movement across ion channels in the membrane and regulating the internal composition of the cell. However, as the temperature is reduced, the diffusion across ion channels is affected much less than the energetic efficiency of the ionic pumps. Consequently, the intracellular composition changes. Major changes can occur, such as uncontrolled influx of Ca^{2+} and Na^{+}, consequent denaturation of proteins, and collapse of membrane potential difference. However, it is obvious that nature can overcome these problems, since many mammals survive hibernation and many cold-tolerant animals and plants survive exposure to reduced temperatures. Understanding how this is accomplished in nature is another area in which much work remains to be done in low-temperature biology (Storey, 1990).

Probably one of the most interesting and important areas of research in low-temperature biology deals with a family of compounds known as *antifreeze proteins* (AFP), also sometimes referred to as *thermal hysteresis proteins*. Evidence of the presence of an unusual compound in cold-tolerant animals was found first by Scholander and his collaborators (Scholander et al., 1957) in fish in the Labrador. They noticed that these fish survive, apparently in a thermodynamically supercooled state, in ocean waters that are colder than the colligative freezing temperature of their blood serum. DeVries and Wohlschlag (1969) discovered that a certain protein in the body fluids of Antarctic fish is responsible for the unusual thermal behavior of the fish. They found that these proteins, which they named *antifreeze glycoproteins* (AFGP), have the ability to depress the freezing temperature of a solution hundreds of times more than the colligative effect would have predicted. The freezing point depression is concentration dependent and in fish can reach $-1.6°C$. However, the proteins depressed only the freezing temperature without affecting the melting temperature, and this is why they are sometimes referred to as thermal hysteresis proteins. These proteins allow the fish to survive in what appears to be a thermodynamically supercooled state, many times in the presence of nucleating ice crystals which they ingest in oceans that are colder than the colligative freezing temperatures of their blood. The structure of the AFGP was determined to be a repeating tripeptide unit, Ala-Ala-Thr, with threonine glycosidically linked to the disaccharide galactosyl-N-acetlagalactosamine (Feeney and Yeh, 1993). The body fluids of these fish contain various fractions of these proteins with eight different molecular weights ranging from 2.7 kDa to 33 kDa. Subsequently, other groups found that North Atlantic fish also have AFGP (Hew et al., 1981). The group of Hew and Fletcher also began finding antifreeze proteins in a large variety of other fish living in cold waters. These proteins are different from the AFGP, contain no carbohydrates, and are therefore referred to as AFPs (e.g., Fletcher et al., 1981; Slaughter et al., 1981). The proteins are of three general types. Their composition and properties are described in several reviews, such as that by Ananthanarayanan (1989). Perhaps of particular interest to the heat transfer readers is the protein's unusual ability to affect ice crystals and phase transformation temperatures. These proteins completely modify the structure of ice crystals, causing them to grow in special crystallographic orientations. Figures 9-1a and 9-1b illustrate the effect of these proteins. They show the freezing interface morphology of pure water (9-1a) in comparison with the morphology of a solution of pure water with a negligible amount of antifreeze proteins, 10^{-3} M (9-1b). Several theories have been proposed for the mechanism by which these proteins depress the freezing temperature. It is thought that they bind to various facets of the ice crystals, inhibiting the growth of these facets and thereby depressing the freezing temperature and producing the structures in Fig. 9-1b. However, the mechanism is not known yet, and much work remains to be done in this area (Ananthanarayanan, 1989; Hew and Yang, 1992; Feeney and Yeh, 1993).

Proteins with "antifreeze" or "thermal hysteresis" properties are found not only in fish. Ramsay and his group have found already in 1968 a thermal hysteresis protein in the gut of an overwintering insect, the larvae of the beetle *Tenebrio molitor* (Grimstone et al., 1968). Since then, in the last decade, tens of overwintering insects have been shown to produce antifreeze proteins in the winter (Duman and Horwath, 1983; Duman, 1993). It is interesting that these antifreeze proteins have a relatively wide range of thermal hysteresis, from fractions of degrees centigrade up to as high as five degrees centigrade (some anecdotal reports suggest ten degrees centigrade). These proteins have been found in both cold-tolerant species as well as freeze-tolerant species (a topic that will be discussed next). In addition to insects, a large number of plants that are cold and freeze tolerant also produce proteins with unusual thermal hysteresis properties (Griffith, 1992, 1995). The proteins in insects, plants, and animals share both the unique ability to depress the freezing-point temperature noncolligatively without affecting the melting temperature and the ability to interact with ice crystals and modify their structure. It has been also found that these proteins inhibit the recrystallization of ice crystals (Knight et al., 1984). These proteins are produced when the biological organisms are exposed to low temperatures and then protect the organisms from these thermal conditions. It is taught that the mechanism of protection relates to depression of freezing temperature in cold-tolerant animals, inhibition of recrystallization in freeze-tolerant animals, or membrane protection from cold injury (which will be discussed next). However, the mechanism of protection is still under debate, and so is the mechanism by which the proteins interact with ice crystals and modify the freezing temperature. While the mechanism is not yet understood, numerous engineering applications of these proteins have already begun, including their use in inhibiting recrystallization in frozen foods such as ice cream, the development of cooling systems that use water or water/ice slurry in a liquid form at low subzero temperatures, and the de-icing of airplanes (Feeney and Yeh, 1993).

Rubinsky and his group have found new properties and functions of these antifreeze proteins. Some of these findings have generated significant controversy that is not yet resolved. They have shown that in addition to affecting the structure of ice, the antifreeze proteins have the ability to confer cold resistance to mammalian cells, presumably by protecting the cell membrane from the various modes of cold injury discussed earlier, including the detrimental effect of the lipid phase transition and the ATP ionic leakage.

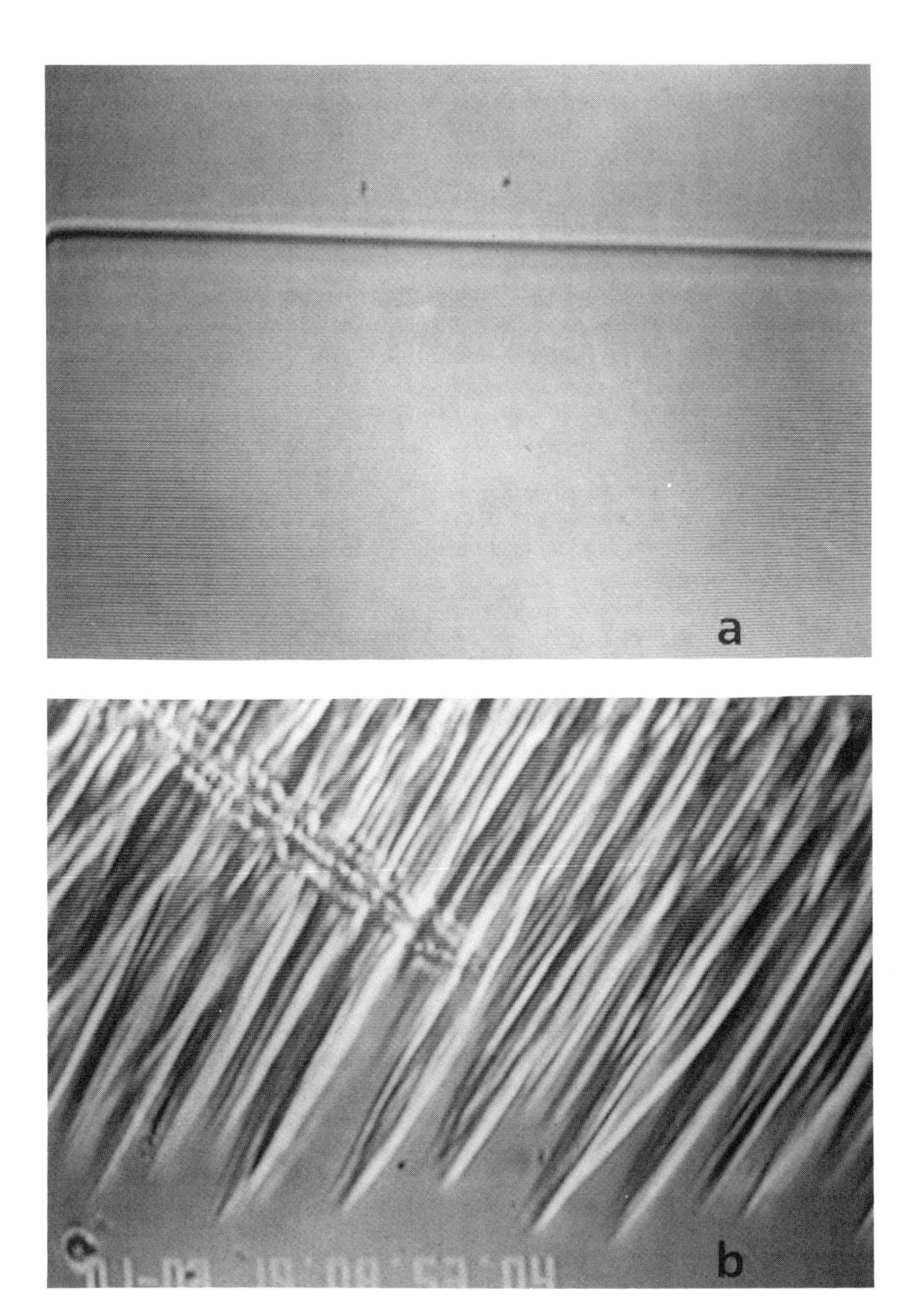

Figure 9-1 Freezing interface morphology during freezing of (a) pure water and (b) pure water with 20 mg/ml AFGP. Note the planar horizontal morphology of the ice water interface (dark horizontal line in (a)) in comparison to the spicular ice crystal morphology that is produced by the antifreeze proteins in (b). The typical dimensions of the spicules are about 10 μm.

Using pig oocytes, they have shown that the oocytes can survive cold in the presence of antifreeze proteins (Rubinsky et al., 1991). Measuring voltage across cell membranes (Rubinsky et al., 1990), performing patch clamp tests (Rubinsky et al., 1992a) and tests with fluorescent dyes (Negulescu et al., 1993), they have shown that the antifreeze proteins inhibit ionic leakage across cell membranes. In a related study with liposomes, Crowe and his group (Hays et al., 1993) have shown that antifreeze proteins inhibit leakage across cooled liposomes without actually affecting the lipid phase transition process. In another related study, Ewart et al. (1992) have shown that one type of antifreeze protein is homologous to C-type lectins, which are proteins known to interact with cell membranes, and that lectins in turn can modify the structure of ice crystals in a similar form to antifreeze proteins (Rubinsky et al., 1992b). Prior to the work of Rubinsky and his group, it was thought that the only function of the antifreeze proteins is in relation to their effect on ice crystals, namely inhibiting ice formation by depressing freezing temperatures or inhibiting recrystallization once ice has formed. No other function was attributed to these proteins. If future research confirms that antifreeze proteins have functions related to protecting membranes from cold injuries, this may open a new area of research and may have practical applications in developing new species of animals and plants that are cold resistant.

9-3 LIFE AT LOW TEMPERATURES BELOW THE FREEZING TEMPERATURE OF WATER

Freezing is more detrimental to biological systems than cooling alone. Biological tissues are made primarily of between 65% and 95% water. The water is in a solution with ions and organic molecules. When biological systems freeze, the water in the organism is removed as ice, modifying both the composition and the physical state of the system. Fundamental studies on the process of freezing in biological systems will be described in the first part of this section, followed by a summary of studies on freeze tolerant animals and plants.

9-3-1 Freezing of Cells and Tissues

A panel gathered in 1966 to review and summarize the research on cryopreservation of cells. It concluded that the "... the temperature measurement most universally relevant to freezing injury in animal tissue (is) the rate of temperature fall (cooling rate, °C/min.) following the initiation of freezing" [1966]. The experiments that led to this conclusion were performed with cells in suspensions, which were frozen to different temperatures with different cooling rates, kept at those temperatures for variable periods of time, and then thawed with different warming rates to physiological temperatures. After thawing the cell, survival was evaluated using different tests, such as morphological integrity and the ability of cells to grow in vitro (Meryman, 1966).

Studies on the effects of cooling rate on frozen cell survival show that the correlation between cell survival and cooling rates takes the shape of an "inverse U" (Meryman, 1966; Mazur et al., 1970; Mazur, 1970; McGrath et al., 1975). Cell survival is greatest for optimal cooling rates, underneath the peak of the "inverse U shape." The survival decreases with either an increase or a decrease in cooling rates. Experiments show that while most of the cells have the inverse U shape–dependence on cooling rates, different types of cells have different optimal cooling rates. For example, in red blood cells the optimal cooling rate is on the order of 1000 C/min. (Meryman, 1966); in yeast the optimal cooling rate is 10 C/min. (Meryman, 1966); and in HeLa cells the optimal cooling rate is 40 C/min. (McGrath, 1975). The temperature and the time of frozen storage also affects the survival of frozen cells (Meryman, 1966; Mazur et al., 1970; Mazur, 1984). The damage increases with an increase in the subzero temperature of storage and with the time of storage. Most of the storage damage occurs at temperatures higher than −60°C. The survival of cells frozen in suspension also depends on the heating rate during thawing. In general, a slower warming rate is more detrimental than rapid thawing (Meryman, 1966; Mazur et al., 1970)

The results of the experimental and analytical studies on freezing of cells were summarized in the generally accepted "two factor" theory proposed by Mazur. Mazur proposed that since the probability for an ice crystal to form at any temperature is a function of volume (Turnbull, 1969), during freezing of cells in a cellular suspension, ice will form first in the much larger extracellular space, before each individual cell freezes. Since ice does not incorporate solutes, the ice that forms in the extracellular space will reject the solutes into the remaining unfrozen solution. The concentration of solutes in the extracellular solution will consequently increase. The volume of the intracellular solution is small, and there is a low probability for ice nucleation to occur inside the cell. Therefore, if the freezing is with sufficiently low cooling rates, the intracellular solution can remain supercooled and unfrozen when the extracellular solution already begins to freeze and becomes hypertonic. To equilibrate the difference in chemical potential between the intracellular and the extracellular solution, the water will leave the cell membrane, which is permeable to water but impermeable to ions and other organic solutes. Therefore, as the temperature of the solution is lowered and additional ice forms in the extracellular solution, water will leave the cell to equilibrate the intracellular and the extracellular concentration, and the cell will dehydrate and shrink. The intracellular solution will remain unfrozen and become hypertonic.

It has been shown originally by Lovelock (1953) that cells exposed to hypertonic solutions experience a similar mode of damage as do cells frozen with low cooling rate. This damage is associated with the high concentration of intracellular solutes and the consequent denaturation of intracellular proteins. This mode of damage is known as *solute damage*. Mazur has expanded on the work of Lovelock and has suggested that during freezing with low, suboptimal cooling rates, the damage to cells is chemical, due to the hypertonicity of the intracellular solution (Mazur, 1963, 1970, 1984). Since chemical damage is a function of time and temperature, the damage will increase with lower cooling rates. Because water transport is a rate-dependent process, faster freezing with higher cooling rates decreases the amount of time a cell is exposed to the chemically damaging conditions, and thus increases survival. This explains the increase in cell viability with an increase in cooling rates toward an optimum. However, increasing cooling rates also results in a more rapid decrease in temperature. The unfrozen water in the cells will therefore experience a greater thermodynamic supercooling. The supercooled intracellular solution is thermodynamically unstable, and after reaching a certain value it will nucleate and freeze. It is thought that the intracellular ice formation damages the cells (Mazur, 1963; Diller and Cravalho, 1970; McGrath et al., 1975; Toner and Cravalho, 1990). The probability for intracellular ice formation increases with an increase in cooling rates, and consequently the survival of the frozen cells will decrease with the increase in cooling rates. These two modes of damage—chemical at low cooling rates and due to intracellular ice formation at high cooling rates—form the basis of the "two factor" theory of cellular damage proposed by Mazur. Viability of cells is optimal during freezing with thermal conditions in which these two conflicting modes of damage are minimized. The effects of the storage temperature and warming rates can be also explained through the mechanisms of chemical damage and ice crystal thermodynamics. Obviously, keeping partially dehydrated cells in hypertonic solutions longer will increase the chemical damage. Therefore, storing cells for longer periods of time at high subfreezing temperatures or thawing with lower heating rates should increase the chemical damage in unfrozen cells surrounded by hypothermic solutions. Ice crystals also have a tendency to recrystallize, that is, to coalesce (to reduce their free energy). The process of recrystallization is a direct function of temperature and results in large ice crystals. These large ice crystals could cause mechanical damage to the cells entrapped between the crystals.

The cryomicroscope is one of the most important tools in studying the effect of freezing on biological materials on a microscale level and has provided much of the current understanding of the mechanisms of damage during freezing. Earlier cryomicroscopes have used stages with a spatially uniform temperature that was cooled or warmed with constant, controlled rates (Diller, 1982). However, when we began studying the process of freezing in tissue, it became evident that the process of freezing in large volumes cannot be accurately represented by a stage with a spatially uniform temperature (Rubinsky, 1985; Rubinsky and Ikeda, 1985). In any freezing process, including cellular suspensions, ice crystals form at a particular nucleation site and then propagate in the direction of the temperature gradients. Freezing does not occur in domains at uniform temperatures and is a directional process with ice crystals propagating from a high to low temperatures. To study this process of freezing, we have developed a "directional solidification" cryomicroscope that will be described briefly, next.

The microscope stage of the directional solidification cryomicroscope is shown in a schematic form in Fig. 9-2. The microscope stage consists of two constant temperature bases, separated by a gap. Each base can be maintained at a constant temperature, one base at a temperature above the freezing temperature and the other at a temperature below the freezing temperature of the sample. A microslide rests on these bases. The system is designed in such a way that a linear temperature distribution is generated in the microslide between the high- and the low-temperature base. In typical experiments the sample is placed on the microslide, which is moved, using a motor, with a constant velocity across the bases. Moving the sample from the high-temperature base to the low-temperature base, it can be frozen with constant velocities between predetermined temperatures. The stage is set in the focal path of the microscope and the process of freezing can be observed and recorded. After freezing the frozen samples can be studied by various techniques such as freeze substitution or low-temperature scanning electron microscopy, and the findings can be correlated to the known thermal history experienced

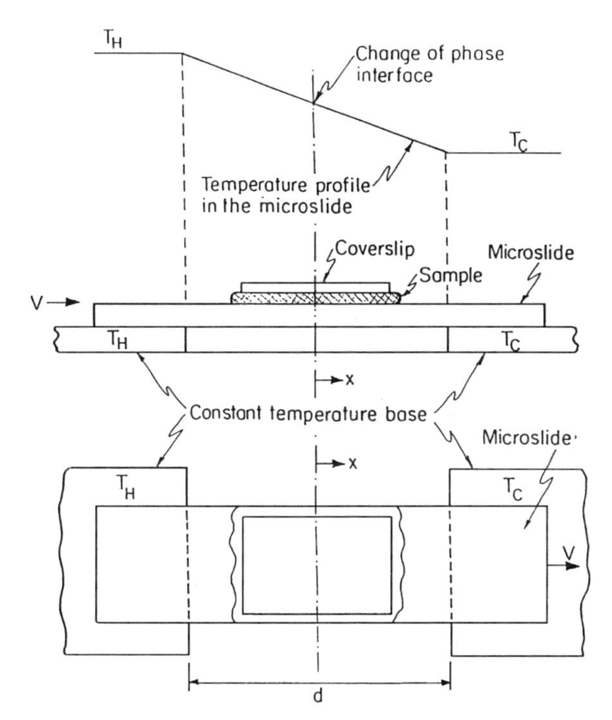

Figure 9-2 Schematic of a directional solidification stage.

by the sample. By moving the microslide in reverse the samples can be also studied after thawing and the effects of the complete freezing/thawing protocol can be determined.

Using the directional solidification stage to freeze cells, it was observed that during freezing finger-like ice crystals form and interact with the cells (Rubinsky and Ikeda, 1985; Ishiguro and Rubinsky, 1994). The dimensions of the ice crystals are determined by such thermal parameters as the velocity of propagation of the freezing interface and the temperature gradients near the freezing interface (Rubinsky and Ikeda, 1985; Ishiguro and Rubinsky, 1994). These parameters are related to the cooling rate: the product of the velocity of the freezing interface and the temperature gradients is equal to the cooling rate. Recent experiments with the directional solidification stage show that in addition to the parameters discussed earlier the viability of cells frozen in suspension is also affected by the structure of the ice crystals that form during freezing and by their interaction with the cells (Ishiguro and Rubinsky, 1994). The effect of the ice crystal structure is of such a nature that different survival values can be obtained for the same cooling rates if the ice crystals are different. It is thought that the ice crystals have a damaging mechanical effect related to the shear of the cell membrane that adds to the other mechanisms of damage due to chemical effects and due to intracellular freezing.

To illustrate the microscale phenomena which occur during freezing of cells, Fig. 9-3 shows a sequence of events during freezing of red blood cells on the directional solidification stage (Ishiguro and Rubinsky, 1994). This sequence is typical of the various microscale heat and mass transfer processes which occur during freezing of cells. The

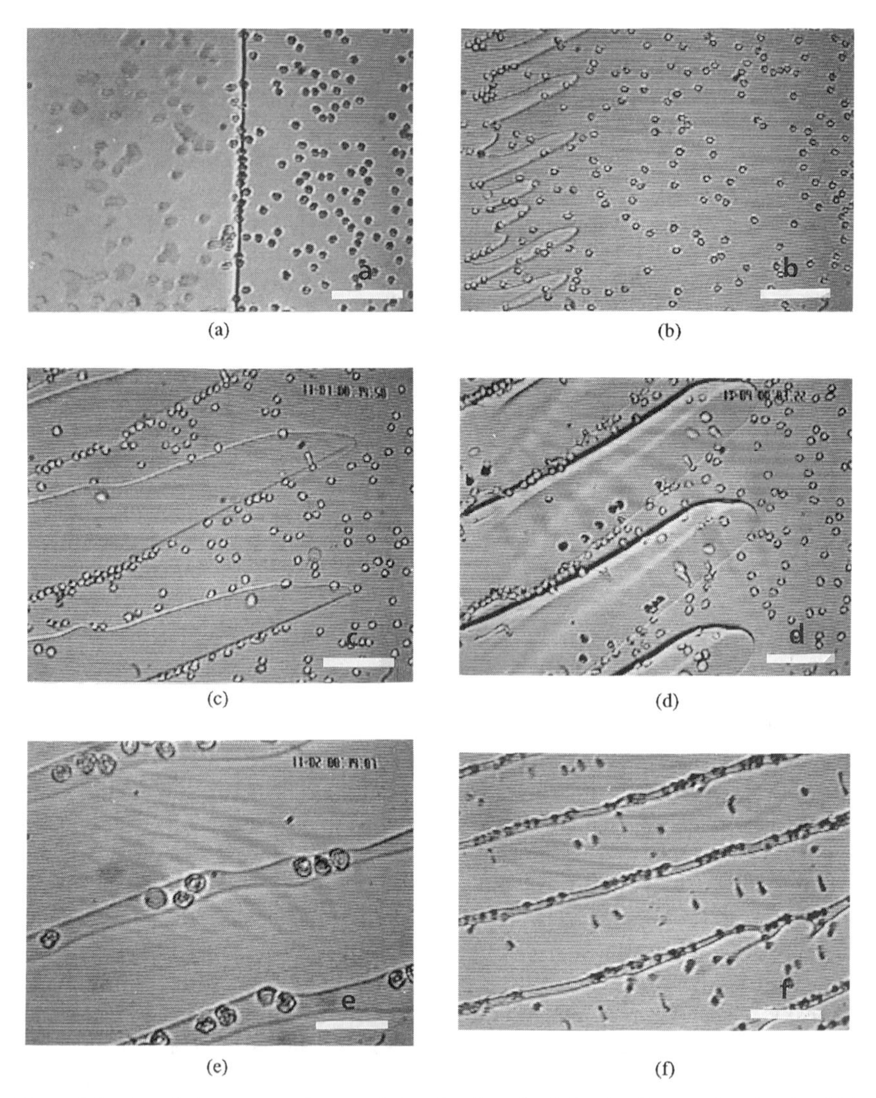

Figure 9-3 A typical sequence of events during freezing of red blood cells on a directional solidification stage. Freezing is from left to right. Scale bar 50 μm, in Figures (a) to (d), and 20 μm in figures (e) and (f).

vertical line in the middle of Fig. 9-3a is the freezing interface, which separates the frozen region on the left from the unfrozen region of physiological saline containing round red blood cells on the right. The process of freezing is strongly affected by the presence of solutes in the physiological solution. Because ice cannot incorporate solutes, they are rejected in front of the freezing interface and accumulate on the interface. The accumulated solutes depress the freezing temperature on the interface, which ultimately becomes thermodynamically unstable and forms the dendritic finger-like ice structures seen in Figs. 9-3b and 9-3c (Ishiguro and Rubinsky, 1994). Apparently the cells are also rejected by the freezing interface and they become entrapped in the channels that form between the finger-like ice crystals, Fig. 9-3d. These channels accumulate the solutes that have been rejected during freezing and the solutions in these channels are hypertonic. The water in the red blood cells is unfrozen and therefore is thermodynamically supercooled relative to the surrounding solution which is in equilibrium with the ice. Therefore the water in the cells surrounded by the hypertonic solution leaves through the cell membrane, which is permeable for water but impermeable to ions and organic molecules. The driving force is the difference in Gibbs free energy between the interior of the cell that is unfrozen and the extracellular ice-solution mixture that is in thermodynamic equilibrium. As the cells lose water to the extracellular space, they dehydrate and shrink (see Fig. 9-3f). At this stage the cells of the intracellular proteins become denaturated and damaged.

The study of the microscale heat and mass transfer processes during freezing of cells was made possible by the use of the cryomicroscope. However, tissue, which is usually opaque, cannot be studied with light microscopy. The limited number of studies on tissue freezing have dealt with tissues that were frozen with controlled cooling rates and then examined frozen with freeze substitution or electron microscopy. In general, it was observed that tissue cells also freeze intracellularly or dehydrate and freeze extracellularly as a function of cooling rates (Bank, 1973; Hunt et al., 1983).

The need for a better understanding of the microscale processes that occur during freezing of tissues have led us to develop a new technique, using the directional solidification stage described earlier (Rubinsky et al., 1987). The experimental procedure uses tissue that is cut in thin slices, placed on a normal microscope slide, and exposed on a directional solidification stage to various freezing protocols in which the cooling rates and the final freezing temperatures are varied. After freezing, the samples are thrown into liquid nitrogen, and the frozen structure is studied with either low-temperature scanning electron microscopy (Rubinsky et al., 1987) or freeze substitution (Bischof et al., 1993). This procedure allows the freezing of the whole sample with the same cooling rates between temperatures predetermined by the temperature of the stage and subsequently the correlation of the various changes in tissue morphology during freezing with the thermal history. Our studies show that different tissues have different microscopic freezing patterns which depend on the microstructure of the tissue. For instance, in the liver the freezing is affected by the morphology of the small blood vessels (Rubinsky et al., 1987); in cancerous tumors it is affected by the organization of the blood vessels around the tumor (Bischof et al., 1993); in the kidney it is affected by the structure of the glomerulae and the nephron (Bischof et al., 1990); and in the breast it is affected by the presence of collagen fibers and fat cells (Hong and Rubinsky, 1994).

The microscopic process of freezing in biological systems is tissue specific, and each tissue requires a separate study of the microscale heat and mass transfer events that occur during the freezing of that tissue. The microscale heat transfer analysis of the great majority of tissues has not been done yet. Typical results, obtained by freezing of a normal liver, are shown in Fig. 9-4 (Rubinsky and Pegg, 1988). Figure 9-4a shows the morphology of liver tissue frozen with low cooling rates of about 4 C/min. to high subzero temperatures ($-10°$C). The image of the frozen tissue was taken with a low-temperature scanning electron microscope in which ice is distinguished by a smooth fracture and other tissue components by a rough fracture. The most interesting feature of the results is the

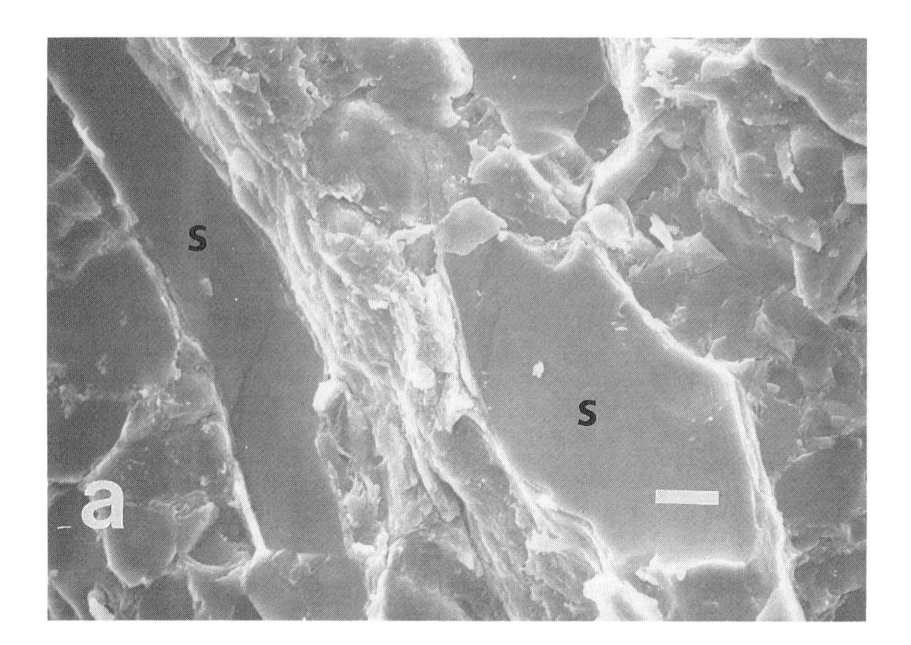

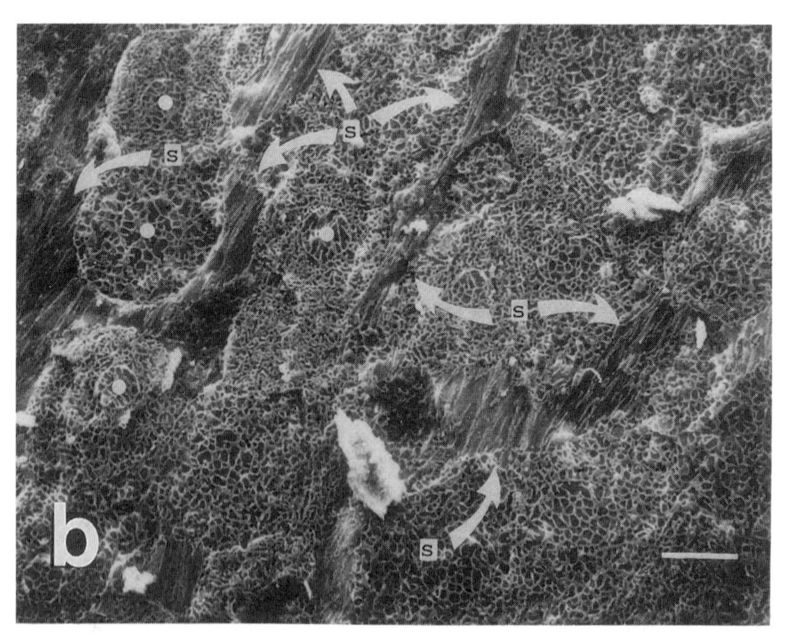

Figure 9-4 Low-temperature scanning electron micrograph of liver tissue frozen with (a) low cooling rates (about 4 °C/min.) to high subzero temperatures (about −10 °C), and (b) high cooling rates by quenching in liquid nitrogen. Ice can be identified from the smooth features of the fracture, while cellular components have rough features. In (a), ice is found primarily in the expanded sinusoids and the cells are dehydrated; and in (b), small ice crystals are seen inside the hepatocytes. (s)—sinusoids, (h)—hepatocytes, scale bar 100 μm.

observation that ice forms within the blood vessels (sinusoids) and appears to propagate along the blood vessels. During freezing with low cooling rates, the blood vessels are expanded and the hepatocytes surrounding the sinusoids are dehydrated. Figure 9-4b shows tissue after rapid freezing, this time by quenching in liquid nitrogen. Here ice is found within the blood vessels, but the cells also contain numerous small intracellular ice crystals. These figures suggest that in the liver ice forms first in the extracellular space, which is comprised primarily of blood vessels, because this space has a larger volume and a greater probability for ice nucleation than each individual cell (Turnbull, 1969). Once ice has formed in the blood vessels, nothing inhibits its propagation along temperature gradients, and therefore it propagates along the blood vessel in the general direction of the temperature gradients. The cells surrounding the blood vessels remain unfrozen, as during freezing of cells in cellular suspensions, and dehydrate to accommodate the difference in Gibbs free energy between the interior of the cell and its exterior. In the liver this process causes the blood vessels to expand, as seen in Fig. 9-4a. Increasing the cooling rate will increase the probability of intracellular ice formation in the cells prior to complete dehydration, as seen in Fig. 9-4b. These experiments show that while the freezing of cells in tissue resembles that described earlier for cells in suspensions, this process is modulated by the microscopic structure of the tissue. Furthermore, the freezing events will affect not only the cells but also the microscopic structural integrity of the tissue. Obviously, in the liver the expansion of the blood vessels during freezing will impair the structural integrity of the organ and promote further damage after freezing. It is important to emphasize that different types of tissue will experience entirely different modes of freezing and consequently different modes of structural damage. For instance, during freezing of breast tissue, additional damage can result from water loss from the collagen fibers of the breast (Hong and Rubinsky, 1994), while in the prostate additional freezing damage results from the disconnection between the epithelial lining and the stroma (Onik et al., 1994).

Several mathematical models have been developed over the years to study the microscale heat transfer in cells and tissues (Mazur, 1963; Ling and Tien, 1970; Rubinsky and Pegg, 1989; Rubinsky, 1989; Toner and Cravalho, 1990; Bischof and Rubinsky, 1993; Hayashi et al., 1994). The models combine equations dealing with the formation of ice inside the cells and outside, the mass transfer between the interior of the cells and the exterior, and thermodynamics of nucleation and phase transformation. A simple set of general equations derived from the studies referenced above are given below. The governing equations are the conservation of mass in the system. These include the conservation of water given by

$$\frac{dV(t)}{dt} = P \cdot A \cdot (c_c(t) - c_e(t)) \tag{9-1}$$

where V is the volume of the cells, P is the permeability of the cells to water, A is the surface area of the cells exposed to the extracellular space, t is time, c is molar concentration, and the subscripts c and e stand for inside and outside the cell, respectively. (Although the use of permeability in analyzing mass transfer is acceptable, it is more precise to use the Kedem-Katchalsky irreversible thermodynamics relations for describing mass transfer across cell membranes.)

Solutes also are conserved in the system. In most of the models, the cell membrane is assumed to be impermeable to solutes, and therefore the amounts of solutes inside the cells are considered to remain constant during the process of freezing. The conservation of mass for the solutes in the cells is given at time t relative to the initial physiological composition, by

$$c_t(t) = c_t(0)\left[\frac{V(t)}{V(0)}\right]. \tag{9-2}$$

The third equation (9-3) correlates the temperature history experienced by the cells or the tissues with the extracellular concentration, through the phase diagram relation:

$$T(t) = 273.15 - 1.86 \cdot \phi \cdot c_e(t) \tag{9-3}$$

where the temperature T is given as a function of time in Kelvin.

The probability of intracellular nucleation is related to the composition of the cell, its volume, and temperature. In homogeneous nucleation the probability P for homogeneous nucleation in a volume V at time t is given as a function of time, volume, and temperature T by (Turnbull, 1969)

$$P(V, t) = \int_t \int_V I(T(x, y, z, t))dVdt. \tag{9-4}$$

I is the nucleation frequency given by

$$I(T) = \frac{k}{\eta(T)} \exp\left(\frac{-b \cdot \alpha^3 \cdot \beta \cdot T_{ph}^3}{T(T_{ph} - T)^2}\right) \tag{9-5}$$

where the subscript ph stands for the phase transition temperature, and $\eta(T)$ is the viscosity of the solution. The rest of the terms are more complicated expressions related to the thermodynamics of the solution and can be found in the book by Franks (1972). (The homogeneous nucleation equation was used here for convenience; however, the heterogeneous nucleation equations developed by Toner and Cravalho (1990) are probably more relevant. Nevertheless, there is little doubt that studying nucleation in biological materials is an important topic with many unknowns.)

Combining Eqs. (9-1) to (9-5) produces expressions for the change in cell volume, intracellular composition, supercooling, and intracellular nucleation as functions of temperature during freezing (Rubinsky and Pegg, 1989). Furthermore, combining these equations with a macroscopic heat transfer model results in new energy equations that can be used for the macroscopic study of heat transfer with phase transformation in biological systems (Rubinsky, 1989; Rubinsky and Eto, 1990; Bischof and Rubinsky, 1993).

Figures 9-5, 9-6, and 9-7 were obtained from Eqs. (9-1) to (9-5) and illustrate a typical behavior of cells during freezing (Rubinsky and Pegg, 1989). They compare the physiological phenomena during freezing of cells (hepatocytes) in suspension and in the liver. Figure 9-5 shows the intracellular concentration as a function of temperature during freezing with different cooling rates. The intracellular concentration is indicative of chemical damage. The figure shows that at any temperature cells will have higher intracellular concentrations when the cooling rates are lower. For example, at $-10°C$

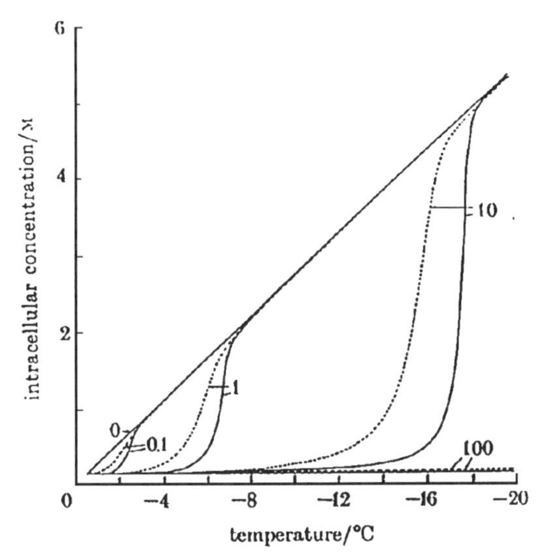

Figure 9-5 Analytically derived curves showing the intracellular concentration as a function of hepatocyte temperature in hepatocytes cooled with the different cooling rates listed in units of °C/min. adjacent to each curve. The dashed lines represent the results for hepatocytes frozen in cellular suspensions, and the straight lines represent the results of calculations performed for hepatocytes frozen in the liver. (Rubinsky and Pegg, 1989).

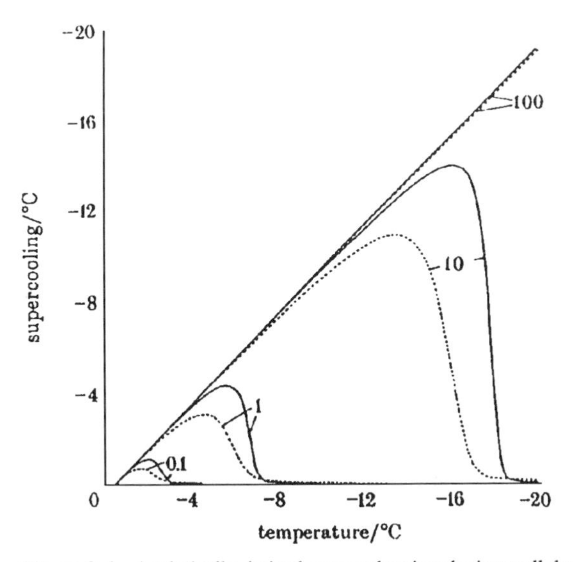

Figure 9-6 Analytically derived curves showing the intracellular supercooling as a function of hepatocyte temperature in hepatocytes cooled with the different cooling rates listed in units of °C/min. adjacent to each curve. The dashed lines represent results for hepatocytes frozen in suspensions, and the straight lines are for hepatocytes frozen in the liver. (Rubinsky and Pegg, 1989).

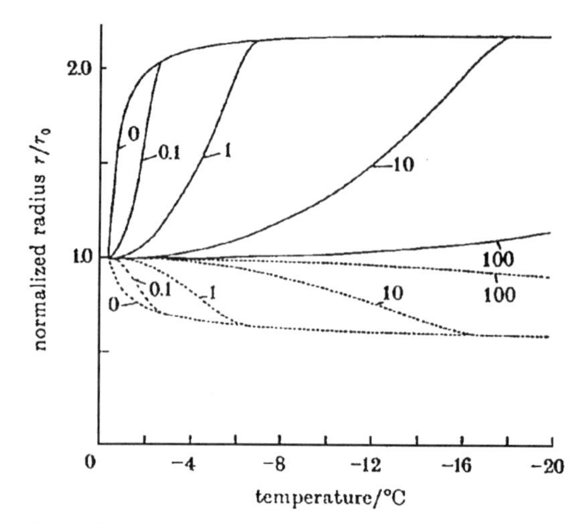

Figure 9-7 Normalized radius (r/r_0) of the sinusoids in the liver (solid line) and of an isolated hepatocyte (interrupted line) as a function of temperature for freezing with different cooling rates. (Rubinsky and Pegg, 1989).

the intracellular concentration of cells frozen with 10 °C/min. will be isotonic, while that of cells frozen with a cooling rate of 1 °C/min. will be 2.1 M. It is obvious from Fig. 9-5 that freezing with lower cooling rates will increase the intracellular concentration and the consequent chemical damage. The figure also shows that cells frozen in tissue will experience less chemical damage than cells frozen in cellular suspension, but the differences are relatively small.

Figure 9-6 shows the intracellular supercooling as a function of temperature during freezing with different cooling rates. The intracellular supercooling is indicative of the probability for intracellular ice formation. It is obvious from Fig. 9-6 that higher cooling rates increase the intracellular supercooling and the probability for ice forming inside the cell. The figure also compares the intracellular supercooling experienced by cells during freezing in suspension to that which the same cells experience during freezing in tissue. It is evident that in tissue the probability for intracellular ice formation is somewhat greater, but the difference is not large. Therefore, it may be anticipated that the survival of cells frozen in tissue or in cellular suspensions will depend on the same mechanisms and with similar parameters.

It should be noted that the transport of water across the cell membrane is a function of cell membrane permeability to water. The cell membrane permeability varies between different cell types, which explains why the optimal cooling rate for survival varies between different cells. However, the general response of the different cells will be qualitatively similar to that depicted in Figs. 9-5 and 9-6 for hepatocytes.

Another important result of the study on freezing of liver tissue is given in Fig. 9-7. The figure compares the expansion of the blood vessels (sinusoids) caused by water leaving the cells to the dehydration of the cells. While the decrease in the radius of the cells during freezing is only 30%, the expansion of the blood vessels is 100% for the same freezing protocol. This suggests that in freezing of tissues, the structural integrity of the blood vessels may be also impaired, leading to the eventual destruction of the cells after thawing from ischemia.

In summary, during freezing the tissue experiences a wide variety of modes of damage, some of them related to the cells and others to the structural integrity of the tissue. The complexity and variety of phenomena that occur during freezing of tissue clearly demonstrate that much more work remains to be done to generate a better understanding of these modes of damage.

9-3-2 Mechanisms of Freeze Survival in Nature

The previous section on the mechanisms of freezing and damage in biological cells and tissue would suggest that biological organisms cannot survive freezing. Therefore, at first glance, it is most surprising to study life processes in nature and to find that a wide variety of animals and plants employ freeze tolerance as a natural survival strategy for sustaining life when the temperatures of the environment fall below the freezing temperature. On the other hand, a second thought would predict that animals and plants that are freeze resistant should exist. A great part of our planet experiences temperatures below freezing for extended periods of time. Warm-blooded animals have developed primarily to gain independence from these reduced exterior temperatures. However, warm-blooded animals are a latter stage in the evolution of life. The more primitive animals and all the plants must have developed means to survive the reduced temperatures on our planet. For species that survive in cold climates, there are three options. The first option is to elude low temperatures by behavioral tactics such as digging deep in the ground or living at the bottom of ponds that do not freeze. Indeed, salamanders, snakes, and some frogs choose this strategy (Gregory, 1982). The second option is to endure subzero temperatures without freezing by depressing their freezing temperatures. The Antarctic and Arctic fish discussed in the previous section produce antifreeze proteins that depress the freezing temperature to the temperature of the surrounding water. Certain insects such as the caterpillars of the gall moth produce glycerol up to 40 percent of their body fluids and thereby depress their freezing temperature to −30°C (Storey and Storey, 1990). The third option is to develop a strategy for surviving freezing. To survive freezing, freeze-tolerant animals and plants must overcome the various modes of damage related to freezing that were discussed in the previous section. They must ensure a) that freezing is only extracellular and that their cells do not freeze, b) that their cells do not dehydrate during freeze below a critical minimal value, and c) that the cellular constituents such as cell membranes and proteins withstand the reduced temperatures. Indeed, freeze-tolerant animals that have developed mechanisms to avoid these modes of damage have been found.

Although freeze-tolerant insects have been studied for many years (Salt, 1961), the first scientific paper on the survival of vertebrates in a frozen state in nature was published by Schmid only in 1982. He found that two hibernating frogs, the wood frog (*Rana sylvatica*) and the gray frog (*Hyla versicolor*), survive long periods of time in a frozen state. Subsequently, Storey and Storey (1984) confirmed his findings and also found two other frogs that survive freezing. Berman et al. (1984) found that Siberian salamanders (*Hynobius keyserlingi*) were also freeze tolerant, surviving temperatures as low as −35°C. The first freeze-tolerant reptile, the painted turtle (*Chrysemys picta marginata*), was reported in 1988 by Storey et al. At the same time, it was reported that the garter snake can also survive limited freezing (Constanzo et al., 1988). Extensive studies over the last few years have shown that these animals have indeed developed mechanisms to overcome the modes of freezing damage discussed earlier. These mechanisms have important aspects that are biochemical in nature (Storey, 1990). However, here I will focus only on the aspects of the survival mechanism relating to heat transfer.

Water can be easily supercooled by tens of degrees centigrade; then freezing and nucleation becomes a statistically random event (Franks, 1985). However, the freeze-tolerant animals and plants cannot afford to have ice forming at random in their cells, since cells frozen intracellularly are destroyed. Therefore, they have developed unique proteins that have ice-nucleating capability and are therefore referred to as *ice-nucleating proteins*. These proteins appear in the extracellular compartments of the organism and have been found in insects (Duman, 1982), plants (Wolber, 1993), and in the blood serum of freeze-resistant frogs (Wolanczyk et al., 1990) and the painted turtles (Storey et al., 1991). These proteins ensure that the freeze-tolerant animals will not supercool before freezing and that freezing will be in the extracellular space and not inside the cells. Many freeze-tolerant animals, insects, and plants produce in addition to ice nucleating proteins also antifreeze (or thermal hysteresis) proteins similar in structure to those found in fish (Knight and Duman, 1986; Duman et al., 1991; Griffith, 1992, 1995). The role of these proteins in freeze-tolerant species is still not clear. Since these proteins obviously do not function here to inhibit freezing, their function may be to inhibit recrystallization of the ice crystals at the high subzero temperatures at which the animals stay or to protect the cell membranes from cold damage.

Studies with freeze-tolerant animals have demonstrated that they can survive freezing only to a certain lower limit in temperature, and that below this limit they die. For instance, the wood frog cannot survive freezing below about −7°C. In addition, all freeze-tolerant animals produce chemicals known as *cryoprotectants*, such as glycerol or glucose. It is plausible that both these aspects of freeze tolerance are related. The cryoprotectants are chemicals that can penetrate inside cells. They depress the freezing temperature colligatively and thereby can increase the intracellular volume as the animal cells dehydrate during freezing. The lower limit in survival is related to the maximal dehydration that the animal frozen cells can withstand. The function of the cryoprotectants is to make this limit occur at lower temperatures. To study these aspects of the freezing process in freeze-tolerant animals, we have used the directional solidification technique described earlier. Tissues from freeze-tolerant frogs (Storey et al., 1992) and turtles (Rubinsky et al., 1994) were frozen on the directional solidification stage to different subzero temperatures. Low-temperature scanning electron micrographs of the frozen tissues have shown that the cells in tissues frozen to temperatures at which the animals survive retained a certain amount of water. Tissues frozen to temperatures at which the animals did not survive freezing were completely dehydrated. This demonstrates the importance of retaining a certain volume of intracellular water in freeze tolerance.

Unlike inanimate matter, it is difficult to study the process of freezing inside live animals. To overcome this problem we have recently developed a new technique for studying freezing in live biological tissues and applied the technique to study

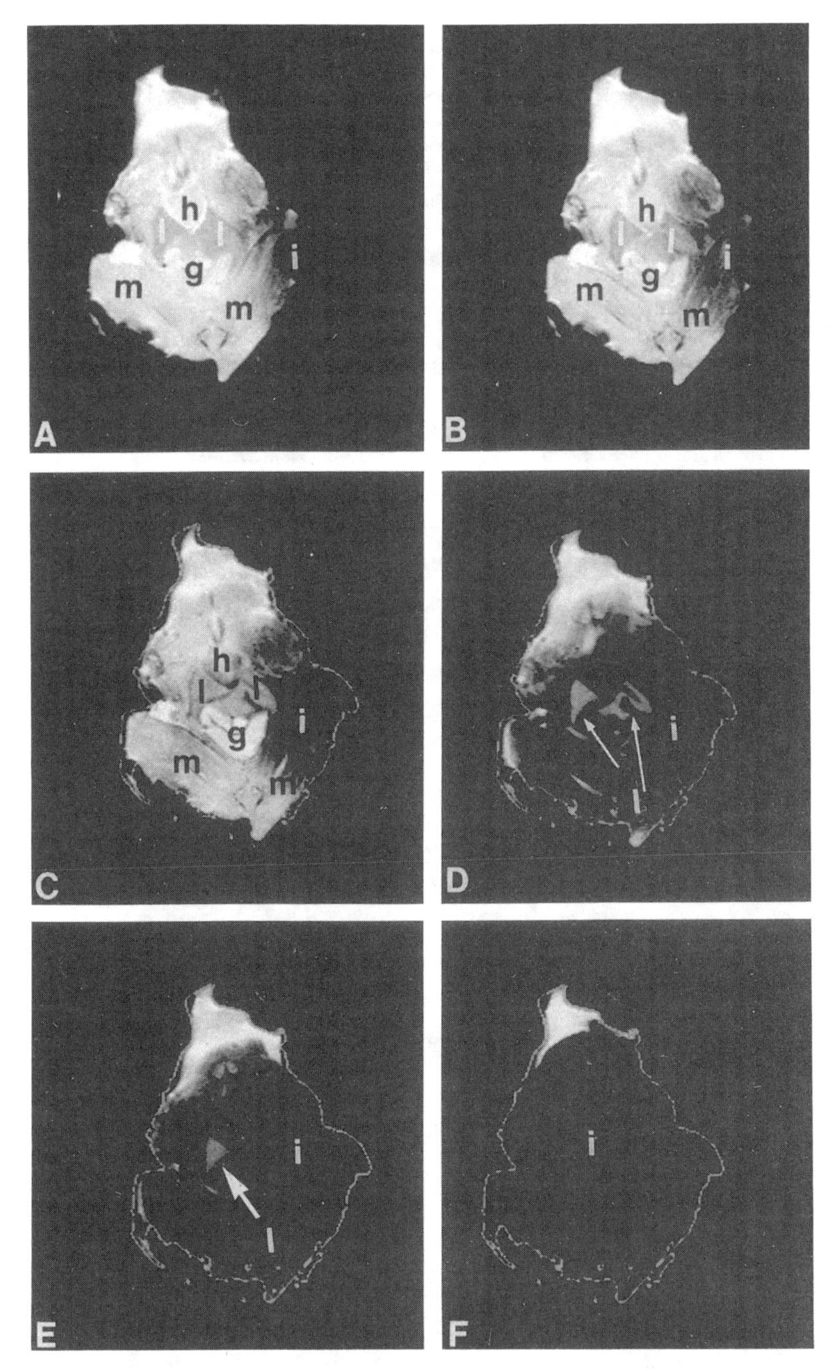

Figure 9-8 Dorsal cross section through a wood frog at 6 times during course of freezing at $-7°C$, showing leg skeletal muscle (m), liver (l), gut (g), and heart (h). Ice (i) is not visible under proton MRI, appears darker than the unfrozen areas. Photos are full size. Timed from initiation of freezing, images were taken at 1 h 16 min (A), 1 h 28 min (B), 1 h 48 min (C), 2 h 10 min (D), 2 h 23 min (E), and 4 h 24 min (F). (Rubinsky et al., 1994).

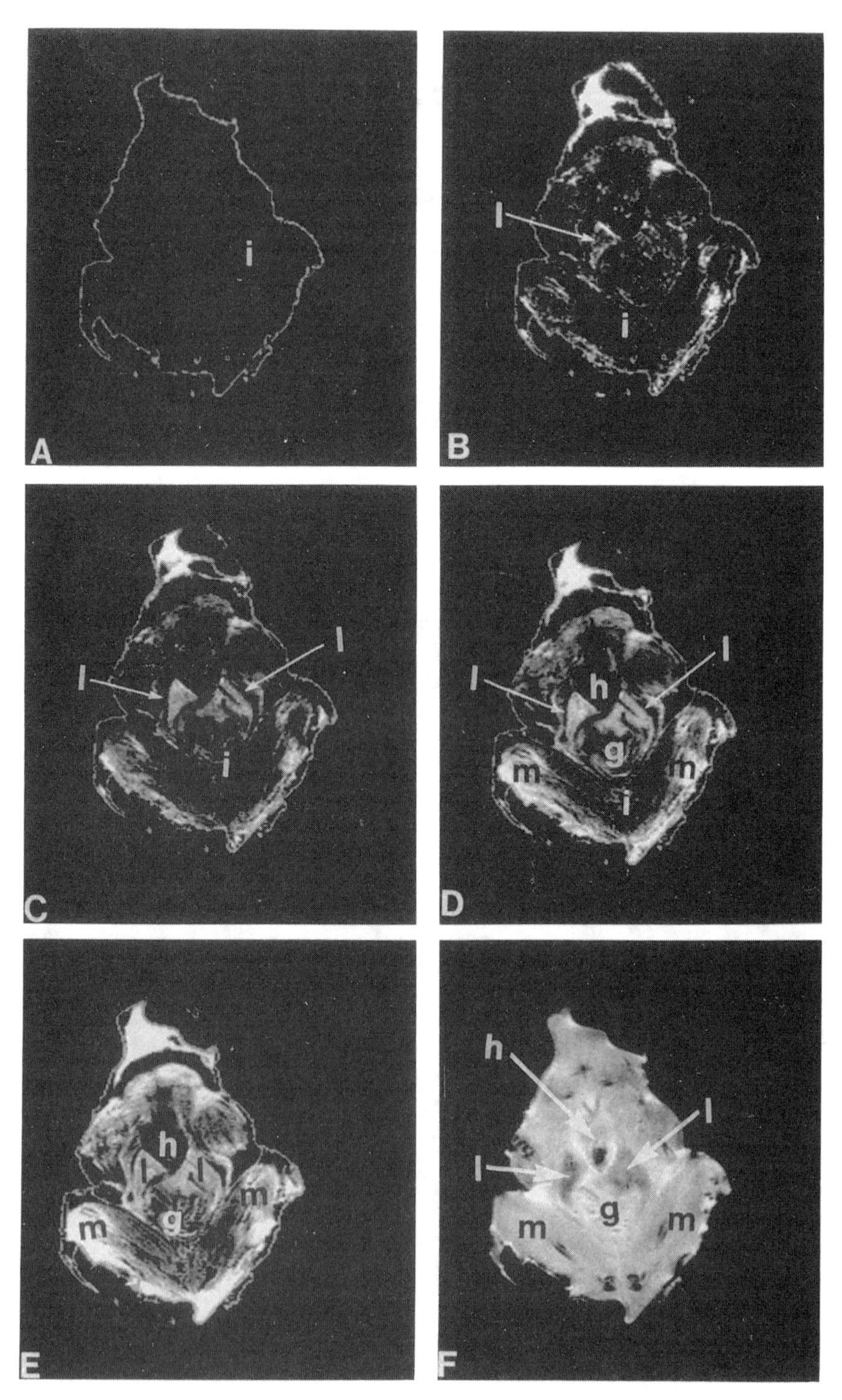

Figure 9-9 Progress of thawing in wood frog showing the same cross section as in Fig. 9-8. Timed from when temperature was raised to 4°C, images were taken at 0 min (A), 34 min (B), 52 min (C), 1 h 8 min (D), 1 h 27 min (E), and 3 h 8 min (F) (Rubinsky et al., 1994).

freeze-tolerant animals: frogs (Rubinsky et al., 1994a) and turtles (Rubinsky et al., 1994b). The technique uses magnetic resonance imaging (MRI). MRI works by causing protons in water to emit radio-frequency signals that contain information about the spatial distribution of the nuclear magnetization and proton distribution within the object. The signal is characterized by two relaxation times, T1 and T2. In solid ice the relaxation time T2 is close to zero, and therefore any MRI sequence that can image biological tissue will show that the frozen region emits no signal. Because conventional proton MRI provides three-dimensional images of tissue, this imaging technique can also provide three-dimensional images of the frozen region in live biological tissue. We have developed the technique originally for monitoring the process of freezing during cryosurgery (Rubinsky et al., 1993). Our current system can image the process of freezing with a resolution of 200 μm (Pease et al., 1995), which is suitable to study microscale heat transfer with phase transformation in biological systems.

Studies of the process of freezing in freeze-tolerant animals has revealed intriguing aspects of the heat transfer processes that occur inside the animal. Figures 9-8 and 9-9 are brought here to illustrate some of the observations (Rubinsky et al., 1994). Figure 9-8 shows sequential images of the process of freezing in a cross section through the central region of a frog. Ice (dark in this MR image) is seen propagating from the outer limb of the frog along the muscles (marked "m"). It is interesting to notice that as the whole body of the animal freezes, the liver ("l") remains unfrozen and surrounded by ice. Figure 9-9 shows the process of thawing. Here, the liver begins to thaw (takes a light appearance) before the rest of the tissue. A similar behavior was observed with other important organs of the animal, in particular the brain. The brain also freezes last and thaws first. This obviously is an important survival strategy, because it would be desirable for the animal to become functional as soon as its limbs thaw and not to be in a situation in which the limbs are unfrozen while the circulatory system is yet not functional. The explanation for the observed phenomena is related to the production of large quantities of glucose in the liver of the freezing frog and the distribution of the glucose to the important organs during freezing. This yields a greater concentration of glucose in the important organs. Recently we have shown that this uneven distribution of cryoprotectant induces a large variation in the change of phase temperature in the tissues of the frogs and is responsible for the observed patterns of freezing and thawing. The special mode of thawing observed in the frogs can be correlated to a typical Biot number for the frog (Hong and Rubinsky, 1995).

This brief review should illustrate the wealth of heat transfer problems at the microscale in freeze-tolerant animals. Much work remains to be done in this relatively young field.

9-4 APPLICATIONS

The studies reported in the previous sections show that low temperatures and freezing have two opposite effects on biological system. On one hand, they damage cells and tissue. On the other hand, when properly modified, organisms and plants can be preserved for long periods of time at reduced temperatures or in a frozen state. A method for protecting cells from freezing damage has been discovered, by accident, much earlier than the discovery of the existence of freeze-tolerant organisms. In 1949, Polge, Smith, and Parkes discovered that the addition of glycerol to spermatozoa facilitates the survival of the frozen cells, which cannot survive freezing in the absence of glycerol. From the perspective of our current knowledge, it is now evident that the mechanisms by which glycerol protects the cells is by penetrating inside the cell. The glycerol remains in the cells during freezing and thereby reduces the chemical damage due to the dehydration of the cells. Today a large variety of single cells are preserved in a frozen state by utilizing cryoprotective agents such as glycerol, ethhylene glycol, and others that protect against freezing damage. Red blood cells, reproductive cells (which are the basis for in vitro fertilization), and in fact all the thousands of cells lines produced by genetic engineering or studied in laboratories are currently all preserved in a frozen state. Organ preservation by hypothermia is an important ingredient in successful transplantation procedures. Cryopreservation and hypothermic preservation are important in medicine, and significant research is being done now to develop new methods for preserving cells and organs by hypothermia and freezing (e.g., Jamieson et al., 1988). However, perhaps in great part as a legacy of the serendipitous discovery of Polge, Smith, and Pares, the research in this field is primarily empirical. Studies with cold- and freeze-tolerant animals, which are just at the beginning, offer the potential for finding new methods for long-term preservation of mammalian cells and organs by hypothermia or in a frozen state. Recently we have tried to apply the understanding gained from studies on the mechanisms by which frogs and cold-tolerant animals survive cold and freezing to preservation of mammalian organs. In experiments with rat livers exposed to 4°C for 24 hours, we have shown that the addition of antifreeze proteins significantly improves the survival of the liver as measured by the production of bile after preservation (Lee et al., 1992). Subsequently, we were successful in preserving a rat liver for 6 hours in a frozen state at −3°C using antifreeze proteins, glycerol, and a freezing method that mimics the process of freezing in the freeze-tolerant frogs (Rubinsky et al., 1994). While these results are preliminary, it is quite evident that a fundamental understanding of the processes that occur during freezing and cooling, including the microscale heat transfer processes, are important in developing these new applications.

The ability of freezing to destroy cells and tissues also has important applications in the field of cryosurgery. *Cryosurgery* is a surgical technique in which undesirable tissues are frozen in the hope that freezing alone will destroy the tissue. It is performed using internally cooled cryosurgical probes. The probes are inserted in the tissue, and freezing begins from the outer surface of the cryosurgical probe outward. In a typical cryosurgical procedure, freezing is continued until the frozen region has encompassed the undesirable tissue. The frozen tissue is left after thawing in situ to be removed by the body's immune system. Cryosurgery has numerous advantages that have promoted the small scale, but there has been a steady use of cryosurgery for the last 150 years, since the first description of the method by J. Arnott in 1845. One of these advantages is the ease with which this procedure can be applied and the minimal trauma to the patient. For example, in modern prostate cryosurgery, five liquid nitrogen–cooled probes are inserted in the prostate percutaneously (Onik et al., 1993). This is the extent to which surgical trauma is induced to the patient. In comparison, resection surgery of the prostate is a major surgery, with significant bleeding, morbidity, mortality and lengthy recovery. In another advantage, the cryosurgical probes can be applied focally, to treat only the undesirable region, thereby sparing much of the healthy tissue. This aspect of the procedure has found important applications in liver cryosurgery (Onik et al., 1991). In resection surgery the extent of the tissue removed is strongly related to other considerations related to the resection strategy such as integrity of the blood supply and the functionality of the tissue remaining after surgery. Often, this requires removal of significant amounts of healthy tissue or even whole organs. In contrast, during cryosurgery only the undesirable tissue is frozen, even if it has irregular margins and shape, and the healthy tissue remains. Cryosurgery is a tissue-sparing procedure. Furthermore, after resection surgery it is often very difficult to retreat the tissue if the disease recurs. In cryosurgery the tissue can be and is routinely retreated. This aspect of the procedure makes cryosurgery advantageous also over other modern localized treatment modalities such as laser, hyperthermia, or radiation therapy.

Traditionally, cryosurgery was inferior to resection surgery in two aspects: a) the ability to identify the extent of the tissue that is affected by the surgical procedure, and b) the ability to control the extent to which the undesirable tissue is being removed from the body by the procedure. With respect to the first aspect, in resection surgery, the surgeon identifies the extent of the tissue he affects using his senses, primarily tactile and visual. However, in cryosurgery, the frozen region propagates from the probe surface into the interior of the body, which is opaque and inaccessible. Therefore the surgeon cannot evaluate the extent of the region that he affects with this procedure. The second disadvantage of cryosurgery is as important. When a surgeon removes the undesirable tissue from the body, that tissue is not there anymore at the end of the procedure. In contrast, after cryosurgery, the undesirable tissue is left in the body and the user of cryosurgery does not know if the undesirable tissue has been extirpated. Therefore, in cryosurgery the surgeon has no means for confirming the success of his procedure immediately after the procedure is terminated, which leads to further lack of confidence.

For cryosurgery to become widely acceptable, the two main disadvantages of this method vís-a-vís resection surgery must be eliminated. The user of cryosurgery must have precise knowledge of: a) the extent of the region that is frozen during cryosurgery, and b) how much of the frozen region is actually destroyed by freezing.

During the last decade, important advances were made in the field of cryosurgery related to the development of imaging guided cryosurgery (Onik et al., 1984; Rubinsky, 1986; Rubinsky and Onik, 1986). This development resolves the problem of monitoring the extent of the frozen region during cryosurgery and has contributed most to the recent vigorous revival of this field. In the early 1980s, Onik and Rubinsky recognized the potential value of novel whole body imaging techniques for imaging the extent of the frozen region during cryosurgery and introduced the use of ultrasound to monitor the position of the freezing interface during cryosurgery (Onik et al., 1984; Rubinsky, 1986; Rubinsky and Onik, 1986).

Ultrasound operates using piezoelectric elements that send sound waves into the tissue. These waves travel through the tissue with a known velocity and are reflected from boundaries at which changes in acoustic impedance occurs. Between water and ice there is a very large mismatch in acoustic impedance that causes most of the ultrasound wave to be returned from the freezing interface. Measuring the time the ultrasound wave travels to the returning interface and back, coupled with knowledge of the wave velocity in tissue, allows the calculation of the position of the freezing interface relative to the ultrasound transducer. This facilitates real-time monitoring of the process of freezing during cryosurgery. Following preliminary experiments with in vitro animal tissue, gels, and in vivo animal tissue, clinical applications of ultrasound-monitored cryosurgery began in the mid-1980s (Onik et al., 1991). Over the past ten years, the clinical use of ultrasound-monitored cryosurgery has become increasingly accepted, first in the liver (Onik et al., 1991) and then in the prostate (Onik et al., 1993), as witnessed by the fact that prostate cryosurgery is now almost universally carried out under transrectal ultrasound guidance. During the past two years the number of centers performing prostate cryosurgery in the United States has grown to over one hundred, with over two thousand patients treated.

While ultrasound is an ideal modality for monitoring freezing in the liver, this imaging modality has certain drawbacks when used with other tissues. Specifically, because there is such a significant difference in acoustic impedance between the frozen and unfrozen regions, and because the interface returns the sound waves so effectively, ultrasound can be used to monitor only the freezing interface facing the ultrasound transducer. The back of the frozen zone is in shade, and the position of the freezing interface in the back cannot be determined. This does not present a problem in surgical procedures in which the ultrasound transducer can be moved around the cryosurgical probe to visualize the frozen region from all sides. However, in many applications, such as prostate cryosurgery or in particular brain cryosurgery, there is no easy access for the ultrasound. To overcome this problem we have recently developed the technique for monitoring freezing with magnetic resonance imaging (MRI). MRI-guided cryosurgery in the prostate and in the brain was described first in Rubinsky et al. (1993).

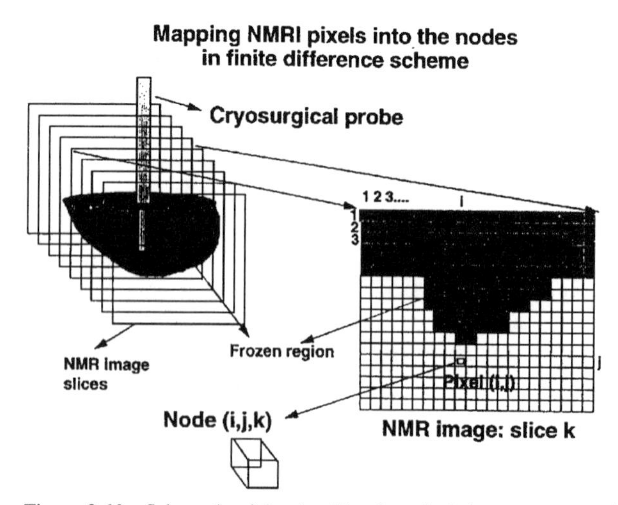

Figure 9-10 Schematic of the algorithm for calculating temperatures in the frozen region with MRI (Hong et al., 1994).

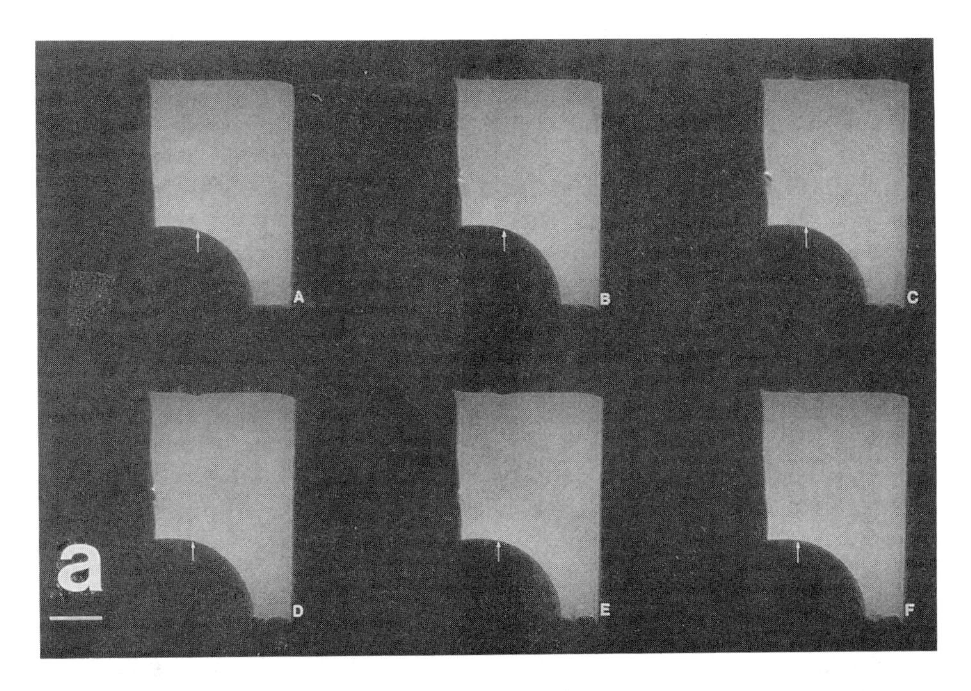

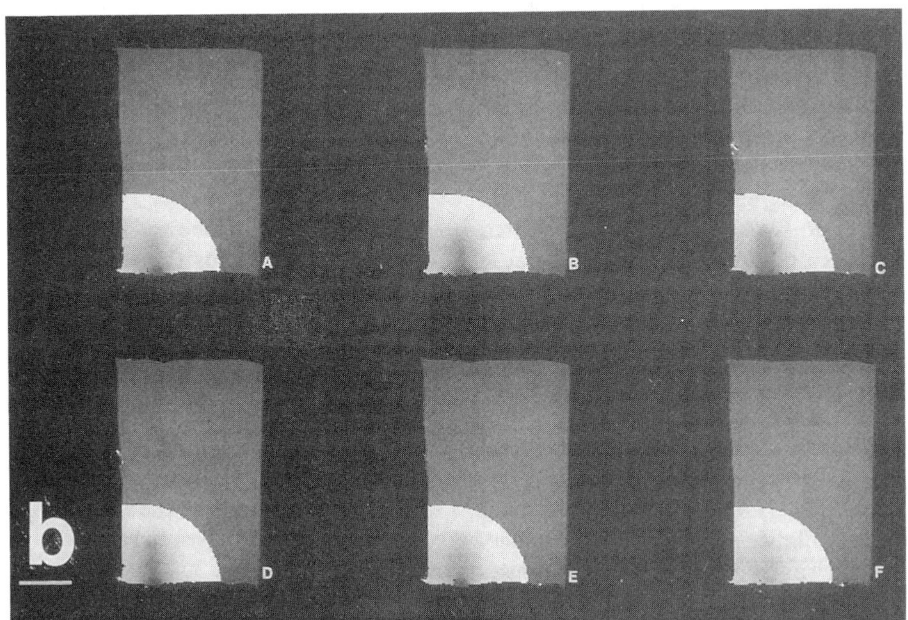

Figure 9-11 (a) An MRI image of a frozen region (dark area) surrounded by unfrozen tissue (light area), and (b) the same image after the temperature in the frozen region was calculated and displayed as shades of gray (darker gray represents lower temperatures) (Hong et al., 1994).

Magnetic resonance imaging can be also used to resolve the second problem related to cryosurgery, that is, determining which part of the tissue is actually destroyed by freezing. This can be done by using MRI information to calculate the temperature distribution in the frozen region. The method used to determine the temperature distribution in the frozen region from an MRI image has been described in detail in a recent paper by Hong et al. (1994), and is illustrated in Fig. 9-10. Briefly, the typical MRI voxels, which create an image, are used in this method as control volumes to solve the energy equation using a finite difference formulation. Because only the dark voxels correspond to frozen tissue, it is sufficient to solve this equation only for the region that was identified with MRI as frozen (dark under MRI), subject to measured temperatures on the cryosurgical probe and to the specified phase transition temperature on the outer edge of the frozen region and the proper energy equation in the frozen region. Figure 9-10 illustrates results obtained with this method.

The temperature history obtained from MRI can be combined with fundamental information on the relation between tissue thermal history and freezing damage and can provide surgeons with details on the extent of tissue destruction. While this technique appears to be quite promising, correlations between microscale heat transfer phenomena and tissue damage of the type described earlier for the normal liver are not readily available. A limited number of studies were done, primarily by the author of this chapter and his group (Bischof et al., 1993; Hong and Rubinsky, 1994; Onik et al., 1994). However, fundamental information on the thermal parameters that produce tissue damage in a large variety of tissues and tumors is missing. The mechanisms of tissue damage are primarily at the cellular and molecular scale. Determining the thermal parameters that destroy malignant tissues is urgently needed and is probably one of the greatest areas of ongoing research microscale heat transfer in bioengineering.

REFERENCES

Summary of the panel discussion "Symposium on cooling rate," *Cryobiology*, vol. 2, p. 210, 1966.

Ananthanarayanan, V. S., "Antifreeze proteins: Structural diversity and mechanism of action," *Life Chem. Rep.*, vol. 7, pp. 1–32, 1989.

Arnott, J., "On the present state of therapeutically inquiry," *J. Churchill Publ.*, London, 1845.

Bank, H., "Visualization of freezing damage, structural alteration during warming," *Cryobiology*, vol. 10, pp. 157–170, 1973.

Berman, D. I., A. N. Leirikh, and E. I. Mikhailova, "Winter hibernation of the Siberian salamander Hynobius keyserlingi," *J. Evol. Biochem. Physiol.*, vol. 3, pp. 323–327, 1984 (In Russian with English summary).

Bischof, J., C. J. Hunt, B. Rubinsky, A. Burgess, and D. E. Pegg, "The effect of cooling rate and glycerol concentration on the structure of the frozen kidney: Assessment by cryo-scanning electron microscopy," *Cryobiology*, vol. 27, pp. 301–310, 1990.

Bischof, J. C., and B. Rubinsky, "Microscale heat and mass transfer of vascular and intracellular freezing," ASME Trans., *J. Heat Transfer*, Aug., 1993.

Bischof, J., K. Christov, and B. Rubinsky, "A morphological study of cooling rate response in normal and neoplastic human liver tissue: Cryosurgical implications," *Cryobiology*, vol. 30, pp. 482–492, 1993.

Constanzo, J. P., D. I. Claussen, and R. E. Lee, "Natural freeze tolerance in a reptile," *Cryo-Letters*, vol. 9, pp. 380–385, 1987.

Crowe, J. H., L. M. Crowe, J. F. Carpenter, and C. A. Winstrom, "Stabilization of dry phospholipid bilayers and proteins by sugars," *Biochem. J.*, vol. 242, pp. 1–10, 1987.

DeVries, A. L., and D. E. Wohlschlag, "Freezing resistance in some Antarctic fishes," *Science*, vol. 163, pp. 1073–1075, 1969.

Diller, K. R., and E. G. Cravalho, "An experimental study of freezing and thawing in biological cells," *Cryobiology*, vol. 7, pp. 191–199, 1971.

Diller, K. R., "Quantitative low temperature optical microscopy of biological systems," *J. Microsc.*, vol. 126, pp. 9–28, 1982.

Duman, J. G., "Insect antifreezes and ice-nucleating agents," *Cryobiology*, vol. 19, pp. 613–627, 1982.

Duman, J. G., and K. Horwarth, "The role of hemolymph proteins in the cold tolerance of insects," *Annu. Rev. Physiol. 1983*, vol. 45, pp. 261–270, 1983.

Duman, J. G., L. Xu, G. Neven, D. Tursman, and D. W. Wu, "Hemolymph proteins involved in insect subzero-temperature tolerance: ice nucleators and antifreeze proteins," in *Insects at Low Temperature,* R. E. Lee and D. L. Denlinger (eds.), Chapman & Hall, New York, pp. 94–127, 1991.

Duman, J. G., "Adaptation of poikilotherms for maintaining the liquid state of water at subzero temperatures," *Cryobiology*, vol. 30, pp. 205–206, 1993.

Ewart, K. V., B. Rubinsky, and G. L. Fletcher, "Structure and functional similarity between fish antifreeze proteins and calcium-dependent lectins," *Biochem, Biophys. Res. Commun.*, vol. 185, pp. 335–340, 1992.

Feeney, R. E., and Y. Yeh, "Antifreeze proteins: Properties, mechanism of action and possible mechanism," *Food Technol.*, pp. 82–89, Jan., 1993.

Fletcher, G. I., "Effects of temperature and photoperiod on the plasma freezing point depression, Cl^- concentration and protein 'antifreeze' in winter flounder," *Can. J. Zool.*, vol. 59, pp. 193–201, 1981.

Franks, F., *Water: A Comprehensive Treatise*, Plenum Press, New York, 1972.

Franks, F., *Biophysics and Biochemistry at Low Temperature*, Cambridge University Press, Cambridge, 1985.

Griffith, M., P. Ala, D. S. Yang, W.-C. Hon, and B. A. Moffatt, "Antifreeze protein produced endogenously in winter rye leaves," *Plant Physiol.*, vol. 100, pp. 593–596, 1992.

Griffith, M., Personal communication, 1995.

Grimstone, A. V., A. M. Mullinger, and J. A. Ramsay, "Further studies on the rectal complex of the meal worm Tenebrio molitor," *Phil. Trans. R. Soc. Ser. B*, vol. 253, pp. 342–382, 1968.

Gregory, P. T., "Reptilian hibernation," in *Biology of the Reptilia*, vol. 13, C. Gans and F. H. Pough (eds.), Academic Press, New York, pp. 53–154, 1982.

Hayashi, Y., N. Momose, and Y. Tada, "Micro-freezing of biological material," *Thermal Sci. Eng.*, vol. 2, pp. 85–89, 1994.

Hays, L., R. Feeney, L. M. Crowe, and J. H. Crowe, "Interaction of antifreeze glycoproteins with lipososmes," *Biophys. J.*, vol. 64, p. 8296, 1993.

Hazel, J. R., "Effects of temperature on the structure and metabolism of cell membranes in fish," *Am. J. Physiol.*, vol. 246, pp. R460–R470, 1984.

Hew, C. I., D. Slaughter, G. I. Fletcher, and S. B. Joshi, "Antifreeze proteins in the plasma of Newfoundland Atlantic cod (*gadus morhua*)," *Can. J. Zool.*, vol. 59, pp. 2186–2192, 1981.

Hew, C. L., and D. S. C. Yang, "Protein interaction with ice," *Eur. J. Biochem*, vol. 203, pp. 33–42, 1992.

Hochachka, P. W., "Defense strategies against hypoxia and hypothermia," *Science*, vol. 231, pp. 234–241, 1986.

Hochachka, P. W., and G. N. Somero, *Biochemical Adaptation*, Princeton University Press, Princeton, 1984.

Holt, W. V., and R. D. North, "Thermotropic phase transitions in the plasma membrane of ram spermatozoa," *J. Repr. Fert.*, vol. 78, pp. 447–457, 1987.

Hong, J. S., and B. Rubinsky, "Freezing of normal and malignant breast tissue," *Cryobiology*, vol. 31, pp. 109–120, 1994.

Hong, J. S., S. Wong, G. Pease, and B. Rubinsky, "MR imaging assisted temperature calculations during cryosurgery," *Magn. Reson. Imaging,* vol. 12, pp. 1021–1031, 1994.

Hong, J. S., and B. Rubinsky, "Phase transformation in materials with non-uniform phase transition temperatures," *ASME J. Heat Transfer*, vol. 117, pp. 803–805, 1995.

Hunt, C. J., M. J. Taylor, and D. E. Pegg, "Freeze substitution and isothermal freeze fixation studies to ellucidate the pattern of ice formation in smooth muscle at 252 K," *J. Microsc.*, vol. 125, pp. 177–186, 1982.

Ishiguro, H., and B. Rubinsky, "Mechanical interactions between ice crystals and red blood cells during directional solidification," *Cryobiology*, vol. 31, pp. 483–500, 1994.

Knight, C. A., A. L. DeVries, and L. D. Oolman, "Fish antifreeze proteins and the freezing and recrystalization of ice," *Nature*, vol. 308, pp. 295–296, 1984.

Knight, C. A., and J. G. Duman, "Inhibition of recrystalization of ice by insect thermal hysteresis proteins: A possible cryoprotective role," *Cryobiology*, vol. 23, pp. 256–262, 1986.

Ling, G. R., and C. L. Tien, "An analysis of cell freezing and dehydration," *ASME J. Heat Transfer,* vol. 92, pp. 393–398, 1970.

Lee, C. Y., B. Rubinsky, and G. L. Fletcher, "Hypothermic preservation of whole mammalian organs with antifreeze proteins," *Cryo-Letters*, vol. 13, pp. 59–66, 1992.

Lovelock, J. E., "The mechanism of the protective action of glycerol against haemolysis by freezing and thawing," *Biochim. Biophys. Acta*, vol. 17, pp. 28–36, 1953.

Mazur, P., "Kinetics of water loss from cells at subzero temperatures and the likelihood of intracellular freezing," *J. Gen. Physiol.*, vol. 47, pp. 347–369, 1963.

Mazur, P., S. P. Leibo, J. Farrant, H. H. Chu, M. G. Hanna, and L. M. Smith, "Interaction of cooling rate, warming rate and protective additives on the survival of frozen mammalian cells," *The Frozen Cell. J. and A*, Churchill, London, 1970.

Mazur, P., "Cryobiology: The freezing of biological systems," *Science*, vol. 168, pp. 939–949, 1970.

Mazur, P., "Freezing of living cells: Mechanisms and implications," *Am. J. Physiol.*, vol. 247, pp. C125–C142, 1984.

McGrath, J. J., E. G. Cravalho, and C. E. Huggins, "An experimental comparison of intracellular ice formation and freeze thaw survival of Hela S-3 cells," *Cryobiology*, vol. 12, pp. 540–550, 1975.

McGrath, J. J., "Low temperature injury processes," *HTD*, vol. 268. "Advances in bioheat and mass transfer: microscale analysis of thermal injury processes," ASME, pp. 125–131, 1993.

Meryman, M. T. (ed.), *Cryobiology*, Academic Press, London, 1966.

Morris, G. J., and P. F. Watson, "Cold shock injury—a coprehensive bibliography," *Cryo-Letters*, vol. 5, pp. 352–357, 1984.

Nakajam, H., M. Goto, K. Ohki, T. Hitsui, and Y. Nozaawa, "An x-ray diffraction study of phase transition temperatures of various membranes isolated from Tetrahymena pyriformis cells grown at different temperatures," *Biochim. Biophys. Acta*, vol. 730, pp. 17–24, 1983.

Onik, G., C. Cooper, H. I. Goldenberg, A. A. Moss, B. Rubinsky, and M. Christianson, "Ultrasonic characteristics of frozen liver," *Cryobiology*, vol. 21, pp. 321–328, 1984.

Onik, G. M., B. Rubinsky, R. Zemel, L. Weaver, D. Diamond, C. Cobb, B. Porterfield, "Ultrasound guided hepatic cryosurgery in the treatment of metastatic colon carcinoma," *Cancer*, vol. 67, pp. 901–907, 1991.

Onik, G. M., J. K. Cohen, G. D. Reyes, B. Rubinsky, Z. H. Chang, and J. Baust, "Transrectal ultrasound-guided percutaneous radical cryosurgical ablation of the prostate," *Cancer*, vol. 72, pp. 1291–1299, 1993.

Onik, G., B. Rubinsky, G. Watson, and R. J. Ablin, *Percutaneous Prostate Cryoablation*, Quality Medical Publishing, St. Louis MO, 1994.

Polge, S., A. U. Smith, and A. S. Parkes, "Revival of permatozoa after vitrification and dehydration at low temperatures," *Nature*, vol. 164, p. 666, 1949.

Negulescu, P., B. Rubinsky, G. L. Fletcher, and T. E. Machen, "Fish antifreeze proteins block Ca entry into mammalian cells," *Am. J. Physiol.*, vol. 263 (*Cell Physiol.*, vol. 32), pp. C1310–C1313, 1992.

Pease, G. R., B. Rubinsky, S. T. S. Wong, M. S. Roos, J. C. Gilbert, and A. Arav, "An integrated probe for magnetic resonance imaging monitored skin cryosurgery," *J. Biomechan. Eng. ASME Trans.*, vol. 117, pp. 59–63, 1995.

Rubinsky, B., *Controlled Freezing of Biological Materials Using Directional Solidification,* U.S. Patent No. 4,531,373, issued July 1985.

Rubinsky, B., and M. Ikeda, "A cryomicroscope using directional solidification for the controlled freezing of biological material," *Cryobiology*, vol. 22, pp. 55–68, 1985.

Rubinsky, B., "Cryosurgery imaging with ultrasound," *Mechan. Eng.*, vol. 108, pp. 48–51, 1986.

Rubinsky, B., C. Y. C. Lee, J. Bastacky, and T. L. Hayes, "The mechanism of freezing in biological tissue: The liver," *Cryo-Letters*, vol. 8, pp. 370–381, 1987.

Rubinsky, B., and D. E. Pegg, "A mathematical model for the freezing process in biological tissue," *Proc. R. Soc.*, vol. 234, pp. 343–358, 1988.

Rubinsky, B., "The energy equation for freezing of biological tissue," *ASME J. Heat Transfer*, vol. 111, pp. 988–996, 1989.

Rubinsky, B., A. Arav, M. Mattioli, and A. L. DeVries, "The effect of antifreeze glycoproteins on membrane potential changes at hypothermic temperatures," *Biochem. Biophys. Res. Commun.*, vol. 173, pp. 1369–1374, 1990.

Rubinsky, B., and K. T. Eto, "Heat transfer during freezing of biological materials," *Anu. Rev. Heat Transfer*, vol. III, pp. 1–38, 1990.

Rubinsky, B., and G. Onik, "Cryosurgery: Recent advances on the application of cold medicine," *Int. J. Refrig.*, vol. 14, pp. 1–10, 1991.

Rubinsky, B., A. Arav, and G. L. Fletcher, "Hypothermic protection—A fundamental property of antifreeze proteins," *Biochem. Biophys. Res. Commun.*, vol. 180, pp. 566–571, 1991.

Rubinsky, B., M. Mattioli, A. Arav, B. Barboni, and G. L. Fletcher, "Inhibition of Ca^{2+} and K^{+} currents by antifreeze proteins," *Am. J. Physiol.*, vol. 262 (*Reg. Int. Comp. Physiol.*), pp. R542–R565, 1992.

Rubinsky, B., R. Coger, K. V. Ewart, and G. L. Fletcher, "Ice crystals and lectins," *Nature*, vol. 360, pp. 114–155, 1992.

Rubinsky, B., J. C. Gilbert, G. M. Onik, M. S. Roos, S. T. S. Wong, and K. M. Brennan, "Monitoring cryosurgery in the brain and in the prostate with proton NMR," *Cryobiology*, vol. 30, pp. 191–199, 1993.

Rubinsky, B., A. Arav, J. S. Hong, and C. Y. Lee, "Freezing of mammalian livers with glycerol and antifreeze proteins," *Biochem. Biophys. Res. Commun.*, vol. 200, pp. 732–741, 1994.

Rubinsky, B., S. T. S. Wong, J.-S. Hong, J. Gilbert, M. Roos, and K. B. Storey, "^{1}H magnetic resonance imaging of freezing and thawing in freeze tolerant frogs," *Am. J. Physiol.*, vol. 266, pp. R1771–R1777, 1994.

Rubinsky, B., J.-S. Hong, and K. B. Storey, "Freeze tolerance in turtles: Visual analysis by microsopy and magnetic resonance imaging," *Am. J. Physiol.*, vol. 267, pp. R1078–R1088, 1994.

Salt, R. W., "Principles of insects cold hardiness," *Annu. Rev. Entomol.*, vol. 6, pp. 55–74, 1961.

Scholander, P. F., L. Van Don, J. W. Kanwisher, H. T. Hammel, and M. S. Gordon, "Supercooling and osmoregulation in Arctic fish," *J. Cell. Comp. Physiol.*, vol. 49, pp. 5–24, 1957.

Schmid, W. D., "Survival of frogs at low temperatures," *Science*, vol. 215, pp. 697–698, 1982.

Storey, K. B., and J. M. Storey, "Biochemical adaptation for freezing tolerance in the wood frog, *Rana sylvatica*," *J. Comp. Physiol.*, vol. 155B, pp. 29–36, 1984.

Storey, K. B., "Biochemistry of natural freeze tolerance in animals: Molecular adaptations and applications to cryopreservation," *Biochem. Cell Biol.*, vol. 68, pp. 687–698, 1990.

Storey, K. B., and J. M. Storey, "Frozen and alive," *Sci. Am.*, vol. 262, pp. 92–97, 1990.

Storey, K. B., "Life in a frozen state: Adaptive strategies for natural freeze tolerance in amphibians and reptiles," *Am. J. Physiol.*, vol. 258, pp. R559–R568, 1990.

Storey, K. B., D. G. McDonald, J. G. Duman, and J. M. Storey, "Blood chemistry and ice nucleating activity in hatchling painted turtles," *Cryo-Letters*, vol. 12, pp. 351–358, 1991.

Storey, K. B., J. Bischof, and B. Rubinsky, "Cryomicroscopic analysis of freezing in liver of freeze-tolerant wood frog," *Am. J. Physiol.*, vol. 263 (*Reg. Int. Comp. Physiol.* vol. 32), pp. R185–R194, 1992.

Slaughter, D., G. L. Fletcher, V. S. Ananthanarayanan, and C. L. Hew, "Antifreeze proteins from the sea raven," *Hemitripterus Americanus*," *J. Biol. Chem.*, vol. 256, pp. 2022–2026, 1981.

Toner, M., and E. G. Cravalho, "Thermodynamics and kinetics of intracellular ice formation during freezing of biological cells," *J. Appl. Phys.*, vol. 67, pp. 1582–1593, 1990.

Turnbull, D., "Under what conditions can a glass be formed?" *Contemp. Phys.*, vol. 10, pp. 473–488, 1969.

Wolanczyk, J. P., K. B. Storey, and J. G. Baust, "Nucleating activity in the blood of the freeze tolerant frog, *Rana sylvatica*," *Cryobiology*, vol. 27, pp. 328–335, 1990.

Wolber, P. K., "Bacterial ice nucleation," *Adv. Microb. Physiol.*, vol. 34, pp. 201–237, 1993.

CHAPTER

TEN

SILICON MICROMACHINED THERMAL SENSORS AND ACTUATORS

Norman C. Tien

School of Electrical Engineering, Cornell University, Ithaca, NY 14853

10-1 INTRODUCTION

In recent years, there has been rapidly growing interest in microelectromechanical systems (MEMS), particularly since the late 1980s when a spinning micromotor made from polycrystalline silicon, or polysilicon, was fabricated on a silicon chip by researchers at the University of California at Berkeley (Fan et al., 1989). Though MEMS products made from silicon such as Motorola's automobile manifold pressure sensor already existed, Berkeley's motor on a chip excited the imagination and made the idea of a "micromachine" on a chip a real possibility.

But what is considered a microelectromechanical system? Reviewing the literature in the field, one finds that MEMS includes a wide variety of devices, structures, and systems which can generally be categorized as microsensors, microactuators, or microtransducers. These components can be individual units or parts of a mechanical system with integrated electronics. They can be used to sense certain physical quantities such as heat, pressure, or electrical capacitance or use these properties as a means for actuation. These MEMS are fabricated using various micromachining techniques from those based on traditional machining and machine tools to those based on semiconductor integrated-circuit fabrication technology. The result is a wide range of MEMS products such as the micro-car—a Toyota the size of a grain of rice, made by Denso Corp. of Japan (Pollack, 1996)—to the ADXL-50, airbag accelerometer on a silicon chip, made by Analog Devices Inc. of Massachusetts. The miniature car, which has 24 parts and moves at speeds up to 5 cm/sec is an example of micromachining based on conventional manufacturing techniques of machining, grinding, and electroplating. Though complex, three-dimensional structures can be made, these micromachines are very difficult to assemble and are expensive. The accelerometer, on the other hand, is a mass-spring system made of polysilicon on a silicon chip with integrated electronics around it. Although these structures are made from thin-film technology and are generally confined to two dimensions, this device takes advantage of the very advanced technology that has been developed for the fabrication of integrated circuits. As with transistors, thousands or even millions of micro-mechanical devices can be batch-fabricated on a single thumbnail-sized chip, resulting in very low costs. Further, this approach allows mechanical devices to be integrated, on the same chip, with the electronic circuitry needed to control them or to process the information that is generated.

In the United States, the focus of research in MEMS has been on this semiconductor based approach. From one perspective, MEMS is just the next chapter in the silicon revolution by allowing silicon integrated circuits to break the confines of the electrical world and give it additional dimensions for interacting with its environment through sensors and actuators. MEMS is certainly emerging as an exciting new silicon technology that will potentially impact the economy and society every bit as much as have integrated circuits since their invention in the early 1960s. The key element for the success of MEMS will be the integration of electronics with mechanical components to create high-functionality, high-performance, low-cost, integrated microsystems. Rapid progress in this direction is occurring as exemplified by the Analog Devices airbag accelerometer and upcoming Texas Instruments Digital Mirror Device for projection video displays. These new integrated microsystems will have diverse applications. And as history of microelectronics demonstrates, we can expect the increased performance and complexity of microelectromechanical systems to bring about new, unforeseen markets. The combination of sensors, actuators, and electronics all on a silicon chip will usher in a new era of micromachines.

Among the devices and structures being made and developed are microsensors that either thermal quantities such as temperature or use thermal quantities to measure another physical quantity, such as flow or pressure. In addition, if complex integrated microsystems or micromachines are to be built, microactuators must also be developed, and a wide range of actuation techniques will be required to meet diverse applications, including those that use thermal energy. The realization of these microelectromechanical systems is the result of the highly developed state of silicon integrated circuit fabrication technology. The focus of this chapter will be to show how silicon micromachining can be applied to create microscale thermal sensors and actuators and to highlight some of these devices fabricated on silicon. Though many types of thermal microsensors and microactuators exist, their commercial success will likely depend on the ability to batch fabricate them at low cost and the best platform for this is on silicon.

10-2 MEMS TECHNOLOGY

The manufacture of silicon MEMS can leverage off of the highly developed technology used to fabricate integrated circuits. Not only does this help speed the realization of MEMS, it enables MEMS products to be produced at low cost by using batch fabrication. The use of common IC fabrication processes in addition to some new specialized techniques to create MEMS is often described as "micromachining." The diversity of

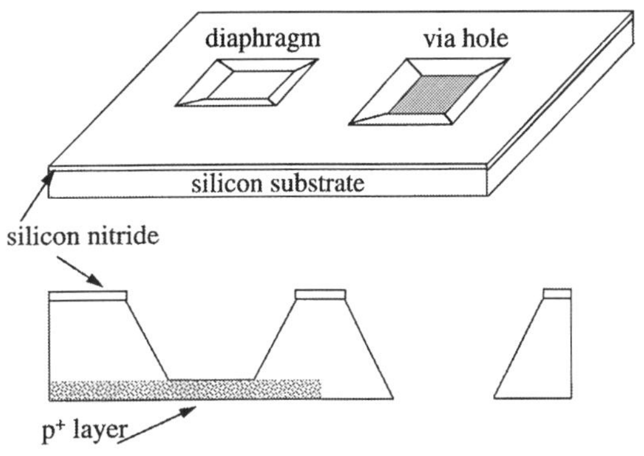

Figure 10-1 Bulk micromachined structures (diaphragm and via hole) in a silicon substrate. Deposited silicon nitride is the mask for the wet etch, and the p^+ silicon layer serves as an etch stop for the membrane formation.

sensors and actuators means that no single technology will be used to fabricate MEMS and in fact many exist. There are, however, two major technologies that have been developed, and they are bulk micromachining and surface micromachining.

10-2-1 Bulk Micromachining

Bulk micromachining was the first technology developed and has been used to fabricate sensors for about the past 20 years. Silicon bulk micromachining typically relies on the orientation-dependent, anisotropic etching of a single-crystal silicon substrate to form mechanical structures. In conjunction with masking films and etch-stopping techniques, a variety of structures such as a diaphragm can be formed. Two common etchants are EDP (ethylene diamine and pyrocatechol) and an aqueous solution of KOH (potassium hydroxide). Both of these etch silicon along the crystallographic orientation at different rates (slow in the $\langle 111 \rangle$ direction, faster in the $\langle 100 \rangle$ and $\langle 110 \rangle$ directions). Figure 10-1 is an illustration of the processing steps for micromachining a hole and a diaphragm in a wafer. Silicon nitride which is not etched by EDP or KOH can be used as an etch mask. To form the diaphragm, a p^+ etch stop, which is a region that is heavily doped with boron, is used. Here the etch rate of the etchants reduces significantly.

Because of the orientation dependence of the etches, care must be observed when designing the mask patterns, as well as when aligning the patterns to the crystallographic planes of the wafer. For example, structures with convex corners such as mesas are hard to create. Fast etching planes in silicon are easily exposed at convex corners, which can result in rounded or faceted etching at those areas or if etched long enough, the entire mesa could disappear. On the other hand, etching stops nicely at concave corners formed by the intersection of $\{111\}$ planes. Techniques have been developed (Bao et al., 1993; Puers and Sansen, 1990) to compensate for the convex corner, for instance, by altering the original shape of the design with the addition of different shape patches at the corners. In fact, simulation programs (Hubbard, 1994) have been developed that will determine the initial design pattern to produce a desired shape after etching an example of which is shown in Fig. 10-2.

Bulk micromachining is the most widely used technology at present, being the basis of products such as the pressure sensors in automobile manifolds and cantilever-

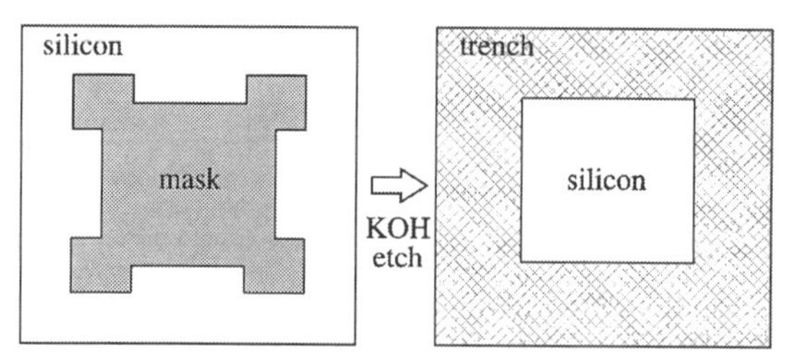

Figure 10-2 Etch compensation for convex corners in silicon bulk micromachining.

beam piezoresistive accelerometers. The fabrication process is not complex and does not require much elaborate equipment. In addition, the high purity and single crystalline nature of the silicon substrate leads to well-controlled and reproducible mechanical characteristics. However, it does have some disadvantages. One disadvantage is the use of alkaline etchants, which are incompatible with integrated-circuit manufacturing because they are a source of contaminants to integrated-circuit devices and the fabrication equipment. Also, bulk micromachining tends to use large amounts of wafer surface area, sometimes wasting expensive silicon real estate. The angular relationship between the {111} planes and the {100} wafer surface, which is 54.7 degrees, causes the etched holes to have openings $\sqrt{2}$ times the depth, with wafers often being 500 μm thick. However, new bulk micromachining techniques, such as SCREAM (Shaw et al., 1994; Zhang and MacDonald, 1992), which use a form of plasma etching, called reactive ion etching (RIE) of the silicon substrate, can produce submicron size features with high-aspect ratio and yet retain the benefits of using single-crystal silicon. A brief outline of the process flow for SCREAM, which is an acronym for single-crystal reactive etching and metallization, is shown in Fig. 10-3.

10-2-2 Surface Micromachining

Surface micromachining uses thin films deposited on the surface of the wafer to create mechanical elements. Surface micromachining gained attention with the fabrication of polysilicon micromotors at Berkeley in the late 1980s (Fan et al., 1989). Layers of polycrystalline silicon or polysilicon were used as the mechanical material with sacrificial material (silicon dioxide, SiO_2) sandwiched between the layers and between the polysilicon and the substrate. Commonly, the polysilicon and SiO_2 have been deposited using low-pressure chemical vapor deposition (LPCVD), though other processes such as plasma-enhanced chemical vapor deposition (PECVD) of SiO_2 and silicon nitride also have been used. The LPCVD process occurs at higher temperatures (roughly 600°C for polysilicon, 400°C for low-temperature oxides, and 800°C for silicon nitride) than for PECVD (<400°C). As a result, LPCVD film tends to be structurally stronger with fewer pinholes and point defects. To free the polysilicon structures, the sacrificial layers are removed with a wet etch. Hydrofluoric acid can etch SiO_2 at rates as high as several μm/min while leaving the polysilicon virtually untouched. Figure 10-4 depicts a typical process flow for a slider with a central guide rail.

Two layers of structural polysilicon are needed to fabricate this slider. The process begins with the deposition of the first sacrificial layer of LPCVD phosphosilicate glass

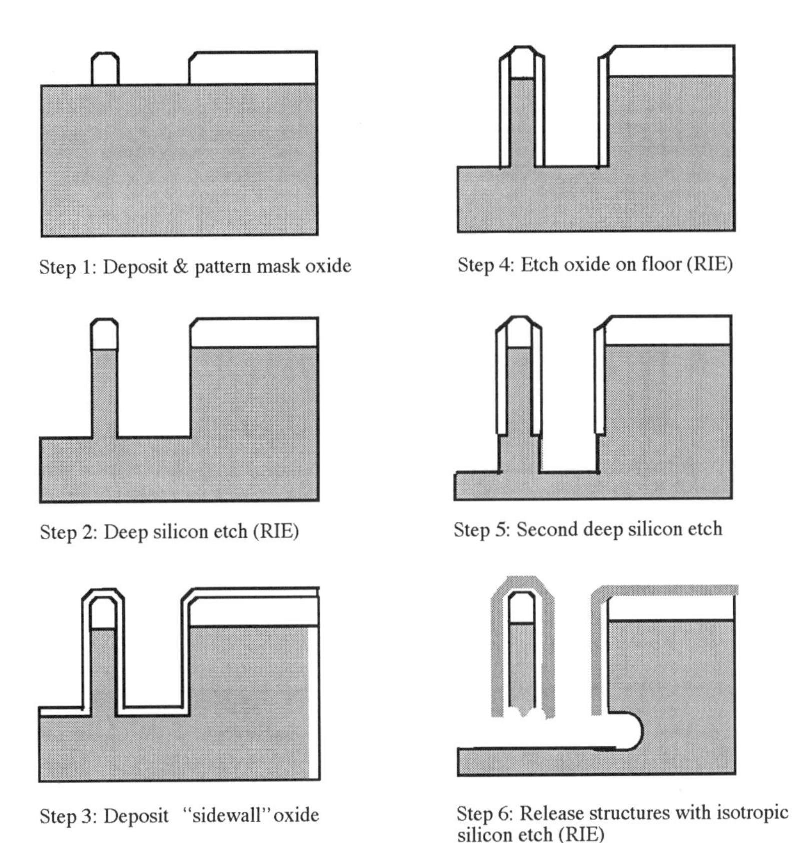

Figure 10-3 Typical process flow for SCREAM process.

(PSG), which hydrofluoric acid (HF) etches faster than undoped silicon dioxide. Depressions are wet-etched into the PSG using 5:1 HF to create dimples under the slider. Because the slider will drop to the surface of the substrate after release, dimples are used to lessen stiction between the structures and the substrate by reducing surface-to-surface contact. Next, the first structural polysilicon layer, which forms the slider, is deposited by LPCVD and patterned using a chlorine-based (Cl_2) plasma etch. A second sacrificial layer is then deposited, and openings are etched (usually in a fluorine-based (CF_4) plasma) through both sacrificial layers for anchors between the next structural polysilicon and the substrate. Deposition of the second structural polysilicon will fill these anchor openings, and this layer is patterned and etched to form the central guide rail. The polysilicon structural layers are 2 μm thick, and the sacrificial oxide layers range from 1 to 2 μm in thickness. The polysilicon is annealed at around 1000°C to relieve the stress in the film. If conductive structures are to be made, such as for an electrostatic actuator, the fabrication process would begin with sequential passivation layers of thermal silicon oxide and silicon nitride to isolate the surface micromachined components from the usually conductive silicon substrate.

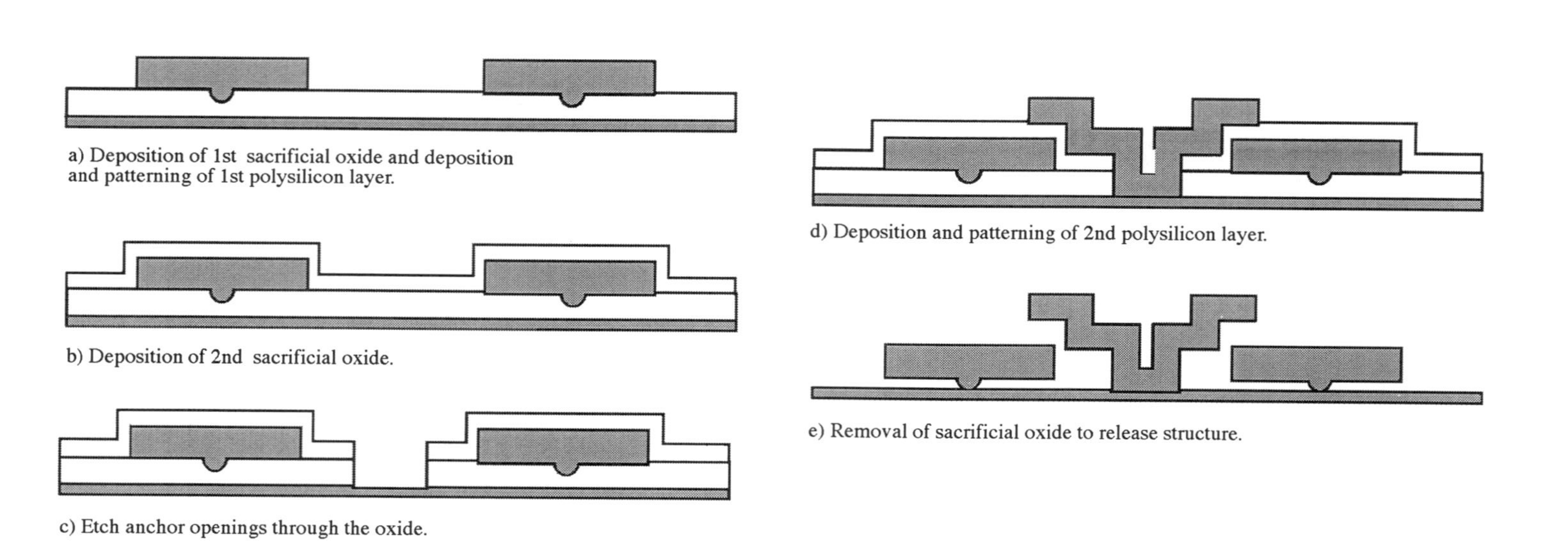

Figure 10-4 Polysilicon surface micromachining process flow for the fabrication of a slider with a central rail.

After the completion of the devices, the sacrificial layers are removed through wet etching in hydrofluoric acid to release the movable structures. During drying, the surface tension of the water will pull structures to the wafer surface. Some structures that are not designed to come into contact with the wafer, such as a cantilever or suspended plate, may fail because the restoring force is not large enough to overcome the adhesion forces, or *stiction*, between the structure and the substrate. To eliminate this problem, many different techniques have been developed. Techniques to reduce such device failure fall into three general categories: (1) preventing contact during release by circumventing drying in a rinse liquid and instead using such methods as critical point drying (Mulhern et al., 1993), sublimation (Guckel et al., 1989), or dry etching (Kobayashi et al., 1993); (2) altering the surfaces to lower adhesion such as by reducing contact area, or by creating stable hydrophobic surfaces to prevent the presence of trapped capillary liquids (Houston, 1995); (3) freeing structures stuck to the substrate using procedures such as rapid thermal annealing (Abe et al., 1995), electromagnetic pulses (Gogoi, 1995), or ultra-short laser pulses (Tien et al., 1996). An excellent critical review of stiction has been published by Maboudian and Howe (1997).

Other thin-film materials that have been surface micromachined include silicon nitride and aluminum (e.g., Texas Instruments' Digital Mirror Device). What surface micromachining offers is a high degree of compatibility with IC processing. Instead of potentially contaminating chemicals and unconventional processes, surface micromachining uses IC thin-film deposition and etching processes. This makes such structures more readily integratable with on-chip electronics, which is important if integration of electronics with mechanical systems will be key to the future success of MEMS. There is also no constraint to how small the devices can be other than that due to lithography technology limitations. Furthermore, structurally complex mechanical systems, such as free-standing and movable parts, can be created with surface micromachining by stacking multiple layers of material rather than with bulk micromachining which shapes the substrate. On the other hand, the material characteristics of the deposited polycrystalline or amorphous thin films are less reproducible than single-crystal silicon because of their dependence on deposition conditions. Like integrated-circuit technology, surface micromachining is a thin-film technology and creates essentially two-dimensional structures with typical layer thicknesses of 2 μm. However, with creative design of hinges (Pister et al., 1992) and flexures, structures can be fabricated that are three-dimensional, as is shown in the fold-up mirror (Tien et al., 1996) in Fig. 10-5. Lastly, because of their light weight and small size, the structures are also very susceptible to environmental factors such as humidity, and particulates.

10-2-3 Micromolding Techniques

Another micromachining technique that has attracted some attention is the creation of miniaturized or microscale parts through micromolding. The most common process is known as LIGA (Ehrfeld et al., 1987), which is a German acronym for "Lithographie, Galvanoformung, Abformung" (lithography, electroplating, and molding). This technique is more similar to conventional machining in concept, though it draws heavily on advanced integrated circuit lithography. In the basic LIGA process, a metal mold is

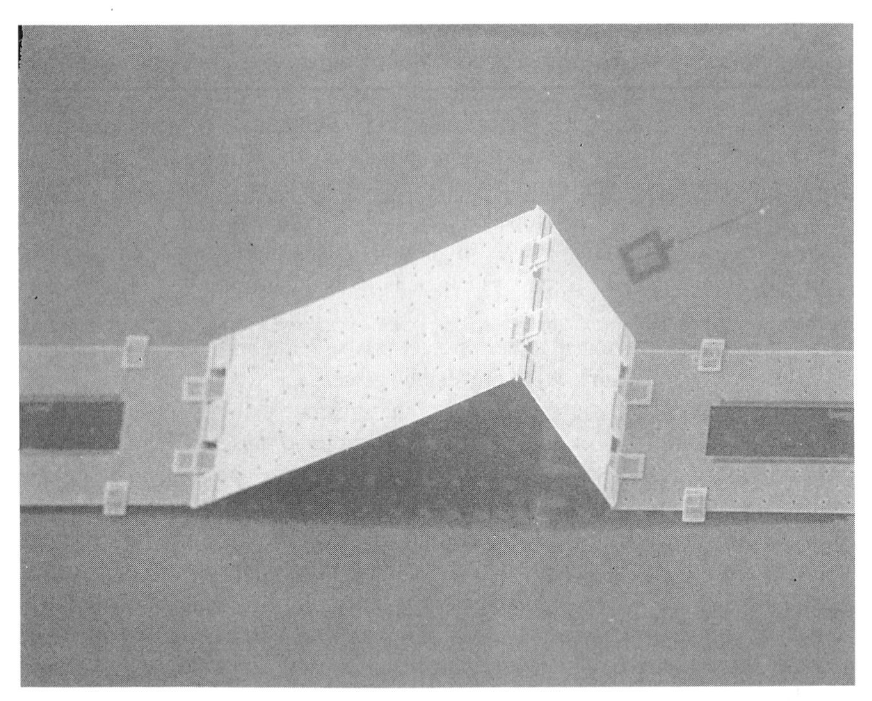

Figure 10-5 Photograph of a polysilicon micromirror. The structure consisits of two polysilicon plates between sliders connected by hinges that allow the mirror to fold up into position.

formed using lithographic techniques, which allows fine feature resolution. To form this mold, a very thick layer of photoresist (300–500 μm) on top of a conductive substrate is exposed and developed. Because of the photoresist thickness, x-ray lithography must be used for this process. After patterning, metal is electroplated from the substrate through the openings in the photoresist. The photoresist is then removed, and the resulting metal mold can be used as the mold insert in injection molding of plastic parts, for instance. A basic diagram of this process is shown in Fig. 10-6. The LIGA process can create three-dimensional structures with thicknesses of several hundred microns and with features several microns in size. However, assembly of these small parts and the integration of electronics is a challenge.

Alternative micromolding techniques have been developed including variations to LIGA such as SLIGA (LIGA with a sacrificial layer). Another approach which also includes a sacrificial layer is HEXSIL (Keller and Howe, 1995), which creates polysilicon components. In this process, the mold is formed by deep trench etching of the silicon substrate. A sacrificial layer of oxide is deposited in the silicon mold, which is then filled with deposited polysilicon. One can also sequentially fill the mold with different layers to obtain multiple layer structures. One advantage of this process is the ability to make thick (100 μm or more) polysilicon structures on which electronics can be integrated. Polysilicon thin-film transistors is an active area of research for flat-panel displays.

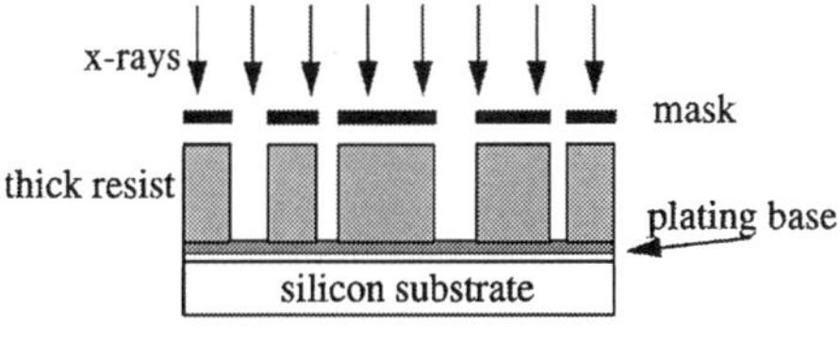

a) exposure and development of thick x-ray resist

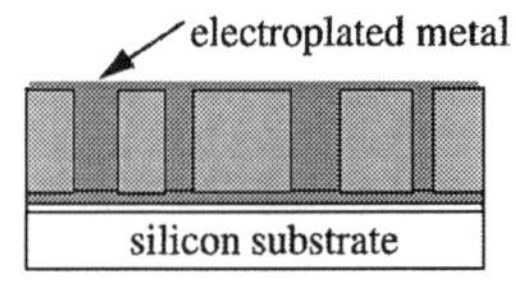

b) electroplating of metal (i.e. nickel) through the resist

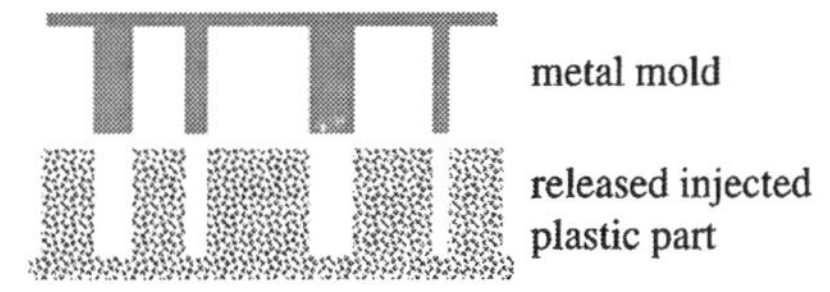

c) metal mold is released from the substrate and used to fabricate plastic parts

Figure 10-6 Basic LIGA process.

10-3 MICROSCALE THERMAL SENSORS

Thermal sensors measure a physical quantity by transducing their signal into thermal quantities that can then be measured by transduction into another quantity, such as an electrical signal, where the information can be easily processed. Silicon micromachining offers the possibility of microscale thermal sensors that are inexpensive and that can be integrated with electronics for improved performance. On silicon, three types of thermal structures are usually found: the closed membrane, the cantilever beam (and bridge, which is two beams connected together), and the floating membrane. In this section, examples of three types of sensors—flow sensor, infrared radiation sensor, and thermal conductivity sensor, which use these types of thermal structures—will be discussed.

10-3-1 Micromachined Flow Sensors

Measurement of the flow of a gas or liquid is important in many scientific and industrial applications. Several types of thermal flow sensors exist to perform this function, but one of them, the *anemometer*, has achieved significant benefit from silicon micromachining. In the anemometer, the measurement of the flow comes from the cooling of a hot object by that flow. The basic component of this sensor is a heated or hot element that preferably is thermally isolated. Furthermore, in order to increase the time response of such a sensor, the thermal mass should be small. It is not uncommon for large-scale bulk anemometers to have a time response on the order of several seconds. With silicon micromachining, small thermally isolated elements can be readily fabricated on a silicon substrate. Now, time responses on the order of milliseconds can be obtained.

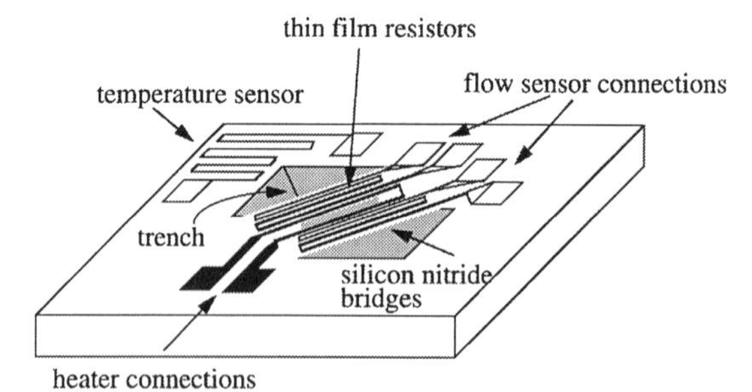

Figure 10-7 Schematic of double-bridge anemometer made by Honeywell.

As an example, Honeywell (Johnson and Higashi, 1987) has manufactured a double-bridge anemometer made out of 500-μm-long silicon nitride bridges over an etched pit in the silicon substrate. A schematic of this device is shown in Fig. 10-7. Each bridge has thin-film resistors which act as a separate heater and temperature sensor. When the bridges are equally heated, the temperature difference between the two bridges is a measure of the flow rate and the direction of the flow. The time response of this sensor is better than 5 msec.

A flat sensor, however, will have difficulty measuring the flow in the middle of a channel, and so three-dimensional structures are desirable for some anemometer applications. Using silicon surface micromachining, it is possible to build a hot wire anemometer which can measure the flow some distance away from the silicon surface. Figure 10-8 depicts a design (Pister, 1992) in which a doped polysilicon beam, which makes up the hot wire, is connected to two hinged plates that can be rotated up out of the plane of the wafer surface. The hot wire is now further thermally isolated from the substrate, thus improving the sensitivity of the device.

10-3-2 Infrared Radiation Detectors

There are two primary means for the detection of infrared radiation. One is the creation of electron-hole pairs in some sensor material; the other is the conversion of the thermal energy of the radiation into heat. Detectors of the first type are often cooled to cryogenic temperatures to reduce their sensitivity to thermal noise. The thermally based sensors can be uncooled but tend to have slower response times and are less sensitive. They are simple to operate and have a very flat response over the infrared spectrum. Two popular IR detectors are the thermopile and the bolometer. Both sensors convert the incident radiation into heat and then electrical signals. The thermopile, which consists of a closed membrane, measures the temperature difference between the middle of the membrane

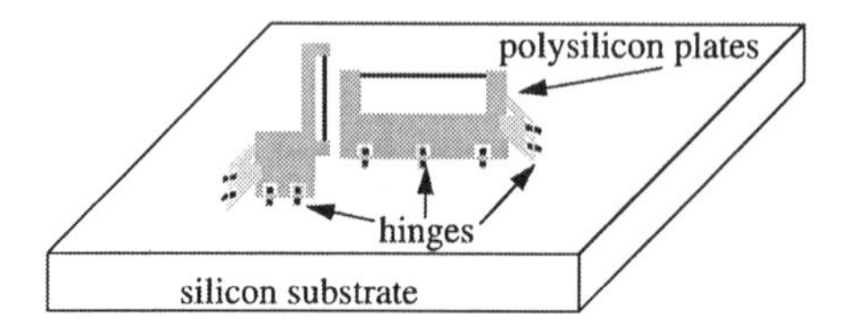

Figure 10-8 A schematic design of an out-of-plane two axis polysilicon hot-wire anemometer.

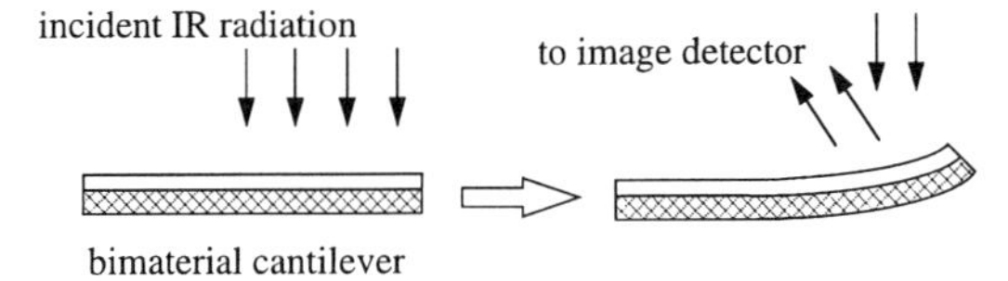

Figure 10-9 Incident infrared radiation causes a deflection of a bimorph cantilever which can subsequently be imaged.

and the rim, so it is self-generating without offset or biasing. The bolometer, on the other hand, has both a temperature-sensitive resistor that needs biasing and an offset, but it can be more sensitive than the thermopile.

Silicon micromachining makes the fabrication of arrays of these detectors very feasible and has generated much interest in infrared imagers, for a variety of applications such as night vision and automated process control. Inside these imagers is an array of pixels, where each pixel is an IR detector. Recently, a 1024-element thermopile-based bulk-micromachined uncooled infrared imager has been reported (Oliver, Baer, and Wise, 1995). This 32×32 element array has a pixel area of 12×12 mm, with a pixel size of 375×375 μm. The overall chip size including the CMOS signal conditioning and control electronics is 16×16 mm.

Research is also continuing on development of higher-resolution uncooled infrared detectors. Recently, a new type of detector, which is based on a bi-material (bimorph) cantilever beam has sparked interest. It is widely known that a temperature change in a bimorph will result in a deflection of the structure because of the mismatch in the thermal expansion coefficients of the two materials. Incident IR radiation on the cantilever causes a temperature rise, which results in a deflection of the beam, as is depicted in Fig. 10-9. The deflection can be measured with high precision using optical techniques or, if an electrical signal is desired, with piezoresistive or piezoelectric techniques (Oden et al., 1996). Optimization of these microfabricated beams have led to a resolution in the 10 pW range (Varesi et al., 1997).

10-3-3 Thermal Conductivity Sensors

Thermal conductivity sensors are popular for the measurement of vacuum. These devices measure pressure by sensing the change in the thermal resistance of the heated element to the ambient caused by a change in the (pressure-dependent) thermal conductivity of the gas around the sensor. As with other thermal sensors, miniaturization can improve the performance of the devices. Lower parasitic thermal losses and reduced thermal time constants will lead to lower power consumption and faster response. Structures that use a floating membrane (an example of which is shown in Fig. 10-10) have a high degree of thermal isolation between the interaction area and ambient.

The other benefit silicon micromachining brings is the ability to integrate on-chip circuitry with the sensor, or vacuum sensor in this case (Mastrangelo and Muller, 1991). As an illustration of the potential to mass-produce sensors at low cost, a thermal conductivity vacuum sensor with on-chip control electronics was developed, fabricated in a completely unmodified CMOS process at a foundry (Klaasen and Kovacs, 1997). The

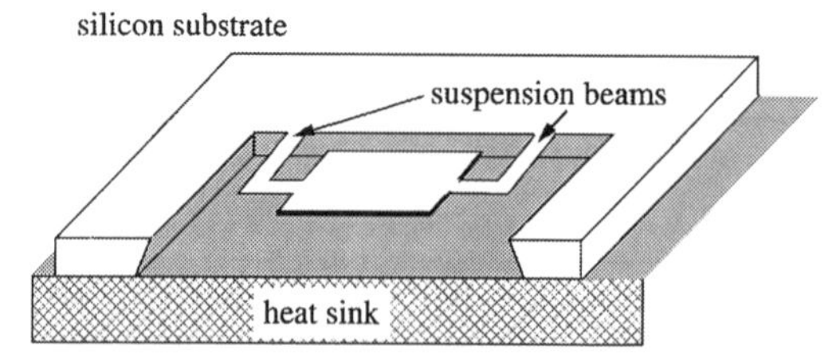

Figure 10-10 Floating membrane structure fabricated on a silicon wafer which is then bonded to heat sink.

vacuum sensor, whose sensing element is shown in Fig. 10-11, has a pressure range of 5 decades, is 0.3 mm^2, and can be fabricated inexpensively in a standard CMOS foundry.

10-4 MICROSCALE THERMAL ACTUATORS

Many types of actuation—magnetic, thermal, piezoelectric, pneumatic, electrostatic—have been used to move micromechanical systems. Each has its particular attributes and drawbacks. Electrostatic actuators are usually made from materials common to silicon integrated circuit fabrication, and because they often rely on capacitance, they can operate at low power. In addition, their time response is fast, and so they can be run at high frequencies. However, a major concern with these actuators is the weakness of the electrostatic force, particularly as distances increase to more than a couple of microns. To compensate, some electrostatic actuators are operated at high voltages (>100 V), which can lead to problems such as electrical shorts or the electrostatic attraction of foreign particles. Or they might be designed with increased sizes, which will result in a large surface area, particularly if one uses surface-micromachining. Magnetic actuators, on the other hand, can operate in environments unsuitable for electrostatic actuators and can deliver increased forces, but they generally require a cumbersome external magnetic source. A further concern is that the materials used are not compatible with IC-fabrication processes; in fact, this is often the case for many MEMS actuators.

Thermal actuators are capable of generating the large forces required to perform useful work in MEMS without the size penalty of electrostatic actuators. Though some thermal actuators use non-IC compatible materials, others have been fabricated from integrated circuit materials, such as polysilicon. They also do not require external sources and can operate in a current/voltage regime compatible with integrated circuits. There is a drawback, and that is the response time of thermal actuators, which is slow in comparison to other actuation techniques because of the time required for heating and cooling of the device. Most thermal actuators being developed are based on thermal expansion of materials, and examples of some of these are presented here. But to illustrate the realm

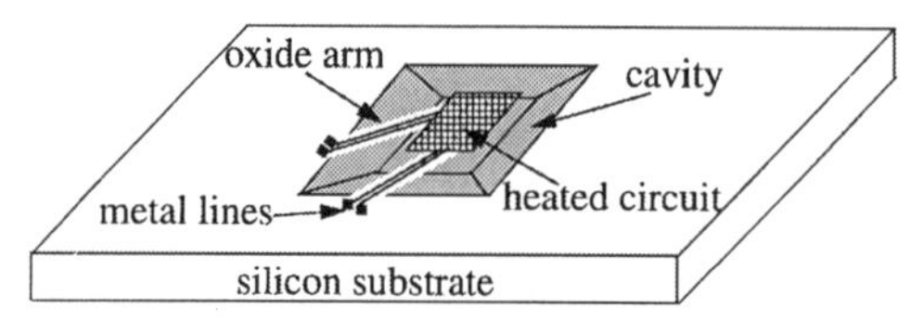

Figure 10-11 Schematic of a silicon vacuum sensor fabricated in a CMOS process.

of possible devices that can be micromachined on a chip, we conclude this section with an example of a microsteam engine on a silicon chip.

10-4-1 Single-Material Thermal Expansion Actuators

A thermal actuator made from a single material would be easy to fabricate, but the displacement from the thermal expansion of a simple beam, for example, is quite small. Use of mechanical leverage can result in larger displacements as was demonstrated in a thermally actuated micro-tweezer made from nickel (Keller and Howe, 1995) and later in polysilicon (Keller and Howe, 1997). In this device, a schematic of which is shown in Fig. 10-12, a large beam is resistively heated by the application of current, and the subsequent lengthening causes other beams in the linkages to rotate and open the tweezer tips. When cooled, the contraction of the thermal element closes the tweezer.

In another approach, single-material thermal actuators with an asymmetric structure (one with a "hot" side and a "cold" side) have been able to generate large deflections. An example of this type of deflection multiplying structure is seen in Fig. 10-13 and was first demonstrated by Guckel et al. in their thermomagnetic actuator fabricated in the LIGA process out of nickel (Guckel et al., 1992). This type of actuator has been used to move optical fibers for optical switching applications (Field et al., 1996). More recently, the lateral thermal actuator shown in Fig. 10-13 has been fabricated in polysilicon through surface micromachining (Comtois, Bright, and Phipps, 1995). Using the more resistive

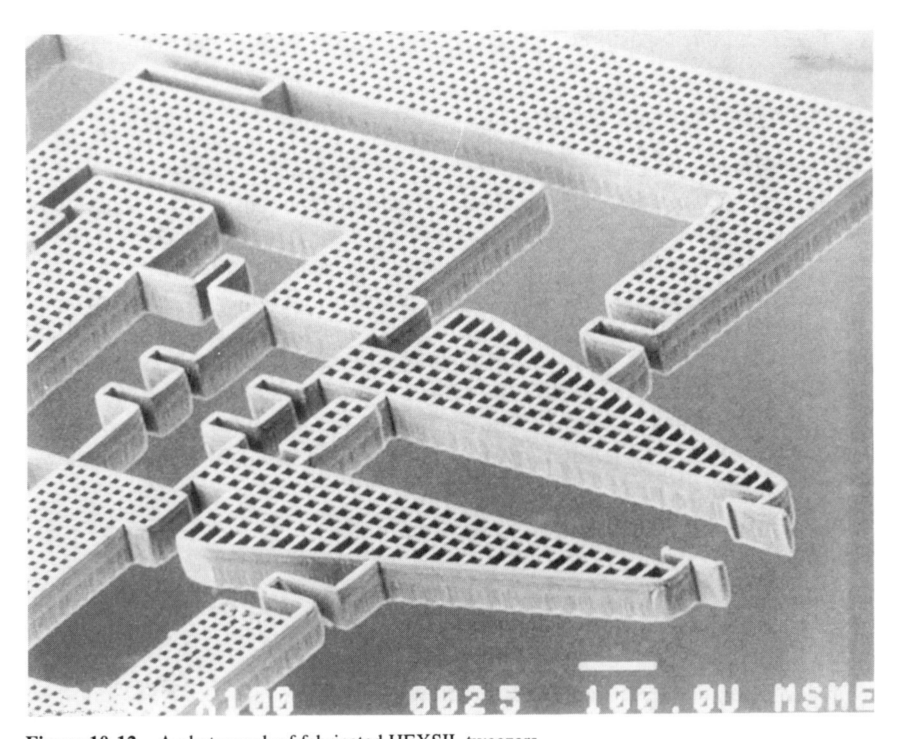

Figure 10-12 A photograph of fabricated HEXSIL tweezers.

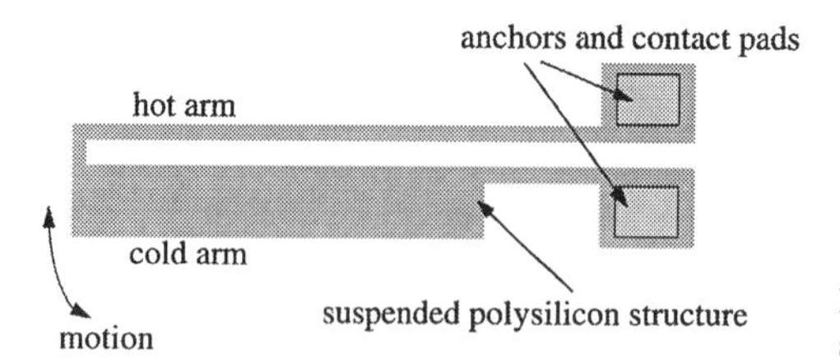

Figure 10-13 Schematic of a single-material as-symetric thermal actuator.

polysilicon instead of metals allows a smaller voltage and current (<10 V and <25 mA) to be used, which are more within the realm of standard IC operating conditions.

The actuator consists of two suspended conductive arms of differing widths that are connected at one end and anchored at the other. The arms are resistively heated by passing an electrical current through them. Because the cross-sectional area of the "hot" arm is smaller than the other "cold" arm, the hot arm heats up faster and has a greater thermal expansion, causing a deflection in the direction of the cold arm. For the actuator shown in Fig. 10-13, which is fabricated from 2-μm-thick polysilicon, the hot arm generated a force of 4.4 μN at an input power of 10.8 mW (2.94 V and 3.68 mA) and a deflected 8 μm (Comtois and Bright, 1996). Increasing the temperature difference between the two arms, for example through better thermal isolation (Field et al., 1996), will lead to greater deflections for a given current. These actuators are compact in size (a typical one having a dimension of 240 μm $\times$ 10 μm) and can be combined together in an array to increase the force output. Arrays of these thermal actuators have been used in stepper motors, micro-optical component positioners, and micro-grippers (Comtois and Bright, 1996).

10-4-2 Multiple-Material Thermal Expansion Actuators

Thermal expansion based actuators made from multiple materials are more common than those made from one material. The advantage of multiple materials is the availability of different thermal expansion coefficients for the design of the actuator. A common type is the bimorph thermal actuator, which consists of two materials sandwiched together. An example of which is shown in Fig. 10-14. The theory behind bimorph beam actuation is well developed (Soderkvist, 1993), and many such actuators have been demonstrated. These actuators are limited to a vertical curling motion, but they are useful in applications such as teleorobotics (Suh et al., 1997) and micropositioning (Ataka et al., 1993).

One of the difficulties of thermal bimorph actuators is the control of their position. Few actuators have been reported that have bistable states, but an example of one is shown

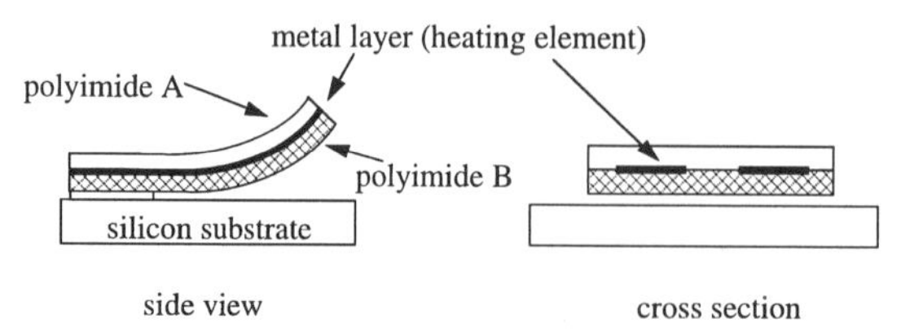

Figure 10-14 Thermal bimorph actuator made of two polyimide layers with different thermal expansion coefficients. A metal heating element is sandwiched between the layers.

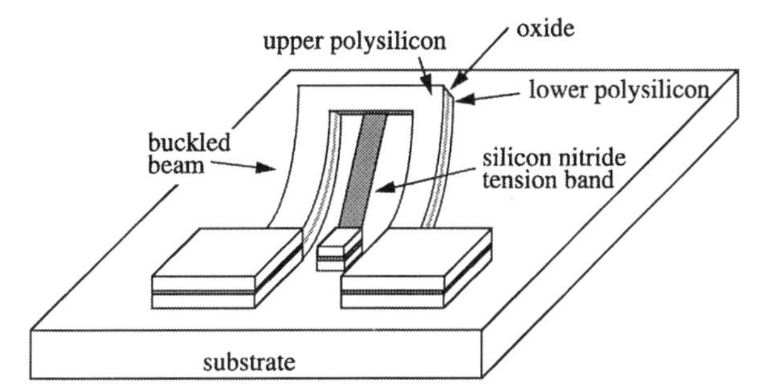

Figure 10-15 A bistable thermal actuator with a U-shaped buckling cantilever with a tension band in the open center area.

in Fig. 10-15, which was demonstrated by Matoba et al. (1994). The device consists of a U-shaped buckling cantilever with a tension band in the open center area. The cantilever is a three-layer sandwich of polysilicon, oxide, and polysilicon. The cantilever is caused to buckle by the tension band made out of silicon nitride. If the cantilever is in the up position, heating of the top layer of polysilicon resistively will cause a bending force in the downward direction. At some critical point, the cantilever will snap to the other stable direction, pointing downward. Heating the lower layer of polysilicon will reverse the process. This type of actuator can be useful for microrelays, optical mirror-based switches, and possibly memory.

10-4-3 Microsteam Engine

Though many of the thermal actuators rely on thermal expansion of materials, other means of thermal actuation are possible. One of these, the microsteam engine or microbubble powered actuator, helps illustrate the wide realm of possible systems that can be made through silicon micromachining. The microbubble-powered actuator consists of two primary components: a bubble generator, and a movable mechanical element to which the displacement of the growing bubble can be transferred. A schematic diagram of one such actuator, which is immersed in a nonconducting fluid, is shown in Fig. 10-16 (Lin, Pisano, and Lee, 1991). The polysilicon resistor heats the fluid and creates a bubble which pushes up against a plate that is at the end of a cantilever beam. A 400-μm-long cantilever has been displaced up to 140 μm in the vertical direction by the bubble.

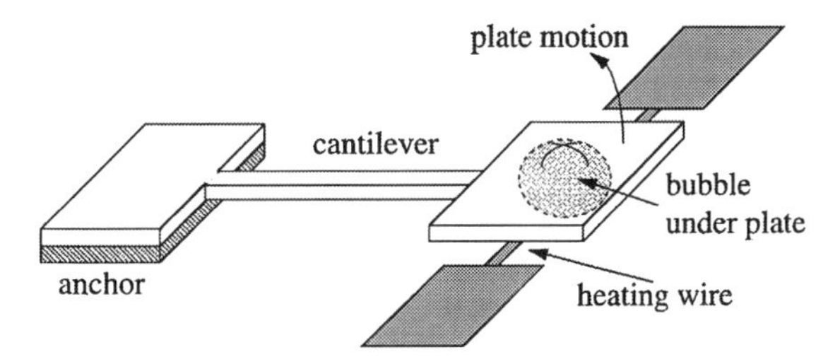

Figure 10-16 A bubble-powered thermal actuator. The plate at the end of the cantilever moves upward with the formation of a small bubble.

Motion parallel to the substrate has been demonstrated by Sandia National Laboratories (Sniegowski, 1993). The Sandia actuator, or micro-steam engine, consists of a polysilicon shell that is open at one end, where a polysilicon piston is found. Water inside the shell (or compression cylinder) is vaporized through resistive heating of the polysilicon and forces the piston out, while capillary forces pull the piston back in after the electrical current is turned off. Bubble-powered actuators can be used in valves and pumps in micro-fluidic systems (Evans et al., 1997).

10-5 CONCLUSION

The applications for MEMS built by silicon micromachining are numerous and increasing rapidly. In this chapter, only a small segment of the field has been introduced. In one view, the advanced technology developed for silicon integrated circuits allows many large, bulky, and expensive mechanical systems to be miniaturized, reduced in cost, and enhanced in performance, leading to previously unforeseen products, applications, and markets. In another view, this is just the next chapter in the integrated-circuit revolution, where MEMS will allow "the silicon chip" to break from the confines of the electrical world to make its mark on everything else.

REFERENCES

Abe, T., W. C. Messner, and M. L. Reed, "Effects of elevated temperature treatments in microstructure release procedures," *J. Microelectromechan. Syst.*, vol. 4, pp. 66–75, June, 1995.

Ataka, M., A. Omodaka, N. Takeshima, and H. Fujita, "Fabrication and operation of polyimide bimorph actuators for a ciliary motion system," *J. Microelectromechan. Syst.*, vol. 2, pp. 146–150, 1993.

Bao, M., C. Burrer, J. Esteve, J. Bausells, et al., "Etching front control of (110) strips for corner compensation," *Sensors Actuators A (Physical)*, vol. A37–A38, pp. 727–32, 1993.

Comtois, J. H., V. M. Bright, and M. W. Phipps, "Thermal microactuators for surface-micromachining processes," *Proc. SPIE Conference on Micromachined Devices and Components*, Austin, TX, vol. 2642, pp. 10–21, 1995.

Comtois, J. H., and V. M. Bright, "Surface micromachined polysilicon thermal actuator arrays and applications," *Solid-State Sensor and Actuator Workshop Technical Digest*, Hilton Head, SC, pp. 109–112, 1996.

Ehrfeld, W., P. Bley, F. Gotz, P. Hagmann, A. Maner, J. Mohr, H. O. Moser, D. Munchmeyer, W. Schleb, D. Schmidt, and E. W. Becker, "Fabrication of microstructures using the LIGA process," *Proc. IEEE Micro Robots and Teleoperators Workshop*, Hyannis, MA, pp. 1–11, 1987.

Evans, J., D. Liepmann, and A. P. Pisano, "Planar Laminar Mixer," *Proc. IEEE Workshop on Micro Electro Mechanical Systems*, Nagoya, Japan, pp. 96–101, 1997.

Fan, F.-S., Yu-Chong Tai, and R. S. Muller, "IC-processed electrostatic micromotors," *Sensors Actuators*, vol. 20, pp. 41–47, 1989.

Field, L. A., D. L. Burriesci, P. R. Robrish, and R. C. Ruby, "Micromachined 1*2 optical-fiber switch," *Sensors Actuators A*, vol. A53, pp. 311–316, 1996.

Gogoi, B. P., and C. H. Mastrangelo, "Adhesion release and yield enhancement of microstructures using pulsed Lorentz forces," *J. Microelectromechan. Syst.*, vol. 4, pp. 185–192, Dec., 1995.

Guckel, H., J. J. Sniegowski, T. R. Christenson, S. Mohney, and T. F. Kelly, "Fabrication of micromechanical devices from polysilicon films with smooth surfaces," *Sensors Actuators*, vol. 20, pp. 117–122, 1989.

Guckel, H., J. Klein, T. Christenson, and K. Skrobis, "Thermo-magnetic metal flexure actuators," *Solid-State Sensor and Actuator Workshop Technical Digest*, Hilton Head, SC, pp. 73–75, 1992.

Houston, M. R., and R. Maboudian, "Stability of ammonium fluoride-treated Si(100)," *J. Appl. Phys.*, vol. 78, pp. 3801–3808, Sept. 15, 1995.

Hubbard, T. J., and E. K. Antonsson, "Emergent faces in crystal etching," *J. Microelectromechan. Syst.*, vol. 3, pp. 19–28, 1994.

Johnson, R. G., and R. E. Higashi, "A highly sensitive silicon chip microtransducer for air flow and differential pressure sensing applications," *Sensors Actuators*, vol. 11, pp. 63–72, 1987.

Keller, C. G., and R. T. Howe, "Nickel-filled hexsil thermally actuated tweezers," *8th International Conference on Solid-State Sensors and Actuators and Eurosensors IX*, Digest of Technical Papers, Stockholm, Sweden, vol. 2, pp. 376–379, 1995.

Keller, C. G., and R. T. Howe, "Hexsil tweezers for teleoperated micro-assembly," *Proc. IEEE Workshop on Micro Electro Mechanical Systems*, Nagoya, Japan, pp. 72–77, 1997.

Klaassen, E. H., and G. T. A. Kovacs, "Integrated thermal-conductivity vacuum sensor," *Sensors Actuators A*, vol. A58, pp. 37–42, 1997.

Kobayashi, D., C.-J. Kim, and H. Fujita, "Photoresist-assisted release of movable microstructures," *Jpn. J. Appl. Phys.*, Part 2 (Letters), vol. 32, pp. L1642–1644, 1993.

Lin, L., A. P. Pisano, and A. P. Lee, "Microbubble powered actuator," *6th International Conference on Solid-State Sensors and Actuators*, Digest of Technical Papers, San Francisco, pp. 1041–1044, 1991.

Maboudian, R., and R. T. Howe, "Critical review: Adhesion in surface-micromechanical structures," *J. Vacuum Sci. Technol. B*, vol. 15, no. 1, 1997.

Mastrangelo, C. H., and R. S. Muller, "Fabrication and performance of a fully integrated mu-pirani pressure gauge with digital readout," *6th International Conference on Solid-State Sensors and Actuators*, Digest of Technical Papers, San Francisco, pp. 245–248, 1991.

Mulhern, G. T., D. S. Soane, and R. T. Howe, "Supercritical carbon dioxide drying of microstructures," *7th International Conference on Solid-State Sensors and Actuators Digest of Technical*, Yokohama, Japan, pp. 296–299, 1993.

Oden, P. I., P. G. Datskos, T. Thundat, and R. J. Warmack, "Uncooled thermal imaging using a piezoresistive microcantilever," *Appl. Phys. Lett.*, vol. 69, pp. 3277–3279, 1996.

Oliver, A. D., W. G. Baer, and K. D. Wise, "A bulk-micromachined 1024-element uncooled infrared imager," *8th International Conference on Solid-State Sensors and Actuators and Eurosensors IX*, Digest of Technical Papers, Stockholm, Sweden, vol. 2, pp. 636–639, 1995.

Pister, K. S. J., M. W. Judy, S. R. Burgett, and R. S. Fearing, "Microfabricated hinges," *Sensors Actuators A*, vol. 33, pp. 249–256, 1992.

Pister, K. S. J., "Hinged polysilicon structures with integrated CMOS thin film transistors," Ph.D. Thesis, Berkeley, CA, 1992.

Pollack, A., "Tiny Toyota utilizes new advances in micro-machine technology," *New York Times*, Aug. 20, 1996.

Puers, B., and W. Sansen, "Compensation structures for convex corner micromachining in silicon," *Sensors Actuators A (Physical)*, vol. A23, pp. 1036–1041, 1990.

Shaw, K. A., Z. L. Zhang, and N. C. MacDonald, "SCREAM I: a single mask, single-crystal silicon, reactive ion etching process for microelectromechanical structures," *Sensors Actuators A (Physical)*, vol. A40, pp. 63–70, 1994.

Sniegowski, J. J., "A microactuation mechanism based on liquid-vapor surface tension," *7th International Conference on Solid-State Sensors and Actuators*, Digest of Technical Papers Late News, Yokohama, Japan, pp. 12–13, 1993.

Soderkvist, J., "Similarities between piezoelectric, thermal and other internal means of exciting vibrations," *J. Micromechan. Microeng.*, vol. 3, pp. 24–31, 1993.

Suh, J. W., S. F. Glander, R. B. Darling, C. W. Storment, and G. T. A. Kovacs, "Organic thermal and electrostatic ciliary microactuator array for object manipulation," *Sensors Actuators A*, vol. A58, pp. 51–60, 1997.

Tien, N. C., S. Jeong, L. M. Phinney, K. Fushinobu, and J. Bokor, "Surface adhesion reduction in silicon microstructures using femtosecond laser pulses," *Appl. Phys. Lett.*, vol. 68, pp. 197–199, 1996.

Tien, N. C., O. Solgaard, M.-H. Kiang, M. J. Daneman, K. Y. Lau, and R. S. Muller, "Surface-micromachined mirrors for laser-beam positioning," *Sensors Actuators A*, vol. A52, pp. 76–80, 1996.

Varesi, J., J. Lai, T. Perazzo, Z. Shi, and A. Majumdar, "Photothermal measurements at picowatt resolution using uncooled micro-optomechanical sensors," *Appl. Phys. Lett.*, vol. 71, pp. 306–308, 1997.

Zhang, Z. L., and N. C. MacDonald, "A RIE process for submicron, silicon electromechanical structures," *J. Micromechan. Microeng.*, vol. 2, pp. 31–38, 1992.

INDEX

Ablation, 140
Absorption, 60–65, 122, 134, 136–137
Accelerometer, 370
Actuators, 369–384
ADI method, 302
Aerogels, 142
AFGP. *See* Antifreeze glycoproteins
AFM. *See* Atomic force microscope
AFP. *See* Antifreeze proteins
Alternating direction method, 302
Amorphous solids, 12–13, 70–74
Anemometer, 377
Anisotropies, 168
Antibonding orbital, 8
Antifreeze glycoproteins (AFGP), 341–344
Antifreeze proteins (AFP), 341–344
Anti-Stokes lines, 280–281
Apolar systems, 194–198
Atomic-force microscope (AFM), 273
ATP pumps, 341
Auger-Shockley-Read-Hall process, 240
Auger term, 92
Autocorrelation function, 156
Avalanche breakdown, 245, 248

Babin-Peterson model, 297, 300, 303
Badran model, 298, 302
Band gap, 44
Band index, 40
Bangham equation, 193
Batch fabrication, 370–371
BCC structure, 5
Beer's law
Bipolar transistors, 246–249
Bistability, 160
Blackbody distribution, 121
Bloch function, 39
Body diagonal, 17
Boltzmann theory, 75–83, 283
  accuracy of, 283
  Boltzman constant, 8, 10
  Boltzman distribution, 10
  commercial software for, 268
  conduction and, 103
  electrons and, 265–269
  Joule heating and, 87–88
  lasers and, 91
  phonons and, 265–269
  point defects and, 10
  semiconductors and, 229, 237, 258–260, 264
  transport theory, 74–83
Bonding, 326
  classification of, 7–8
  crystal structures, 5–13
Bose-Einstein distribution, 21, 28, 50, 65, 266
Bosons, 21
Bragg reflectors, 98
Brillouin zones, 16–17, 40
Built-in voltage, 247
Bulk films, 106
Bulk micromachining, 371–372

Cahill-Pohl model, 73–74
Callaway approach, 259

Capacitive delay, 253
Capillaries
  evaporation of, 218–219
  limitation of, 189, 297
  pressure and, 189
Carey equation, 190
Carriers
  band gap and, 44, 46
  density of, 44–47, 91
  energy conservation and, 91
  heat conduction, 70, 90–91
  intrinsic, 44, 46
  momentum equation of, 92
  nonequilibrium distributions, 66
  scattering and, 47–52, 68
  semiconductors and, 44–47
  transport theories, 67–69
  tunneling of, 68, 248
  See also specific types, effects
Casimir limit, 72, 83
Catalysis process, 167
Cattaneo equation, 78
Cells
  definition of, 340
  freezing of, 344–355
  membranes and, 340, 354
Cesium chloride structure, 5–6
Chapman-Enskog approximation, 80
Characteristic length, 65
Charge optimization, 328–334
Charge, transport and, 236–253
Chemical bonding, 5–13, 326
Closure problem, 91
Clusters, 170
  condensation and, 183–185
  equilibria and, 151–158
  formation rate of, 175
  free energy and, 182
  kinetics and, 173–176
  Lennard-Jones clusters, 152, 159–161
  magnetic properties and, 180
  melting and, 159–161
  molecular, 167–185
  nucleation and, 183–185
  phase-change and, 149, 159–161, 182
  reactivity and, 182–183
  specific heat and, 182
  thermodynamic properties of, 180–183
  thermophysical properties of, 177–183
Coherence length, 123
Coherent-based effects, 142
Collision, 68
Collodial suspensions, 141
Compact diodes, 246–249
Compact field-effect transistors (FET), 241–245, 250, 252, 271, 281
Condensation, 295
  clusters and, 183–185
  mechanism of, 176
  thin film and, 219
Conduction band, 41–42, 44, 232, 268
Contact line motion, 220
Contamination problem, 203
Continuum limit, 16
Continuum model, 16, 99, 188
Covalent bonds, 8
Cross-sectional area porosity, 311
Cryomicroscope, 346, 349
Cryopreservation, 356, 360
Cryosurgery, 359–361, 364
Crystals, 5–14
  diamonds, 6–7, 258
  imperfections in, 25–26
  phonons and, 14–28
  polycrystals, 9–10
  scattering by, 25–26
  thermal conductivity, 70–74
  vibrations in, 14–28
  wave propagation and, 15–19
Cubic structures, 5–7
CW technique, 134
Cylindrical interface equation, 201

Damping term, 59, 63
Debye model, 22, 104, 110
  assumptions in, 74
  crystal waves in, 73
  cutoff frequency, 23
  density, 114
  induction interaction, 188
  temperature of, 23, 263
  vibrational states and, 108
  wave coherence in, 73
Deformation, 50–51, 72–73, 155
Densification, 257
Derjaguin theory, 189
Diamonds, 6–7, 71, 258
Dielectrics
  dielectric constant, 54–56, 58
  Lyddane-Sachs-Teller relation, 62
  polarization and, 55
  saturable absorption and, 138
  thin films and, 95–116
Differential conductance, 243
Differential transductance, 243
Diffraction, 68

Diffuse mismatch model, 110–112, 133
Diffusion, 10, 68, 80, 99, 156, 238, 252
Dipole moments, 62
Directionally dependent etching, 323
Disjoining pressure, 209
Dislocations, 10–11
Disordered materials, 103–109
Dispersion relations, 16
DLP theory. *See* Dzyaloskinskii-Lifshitz-Pitaevskii theory
Dopant atoms, 240
Doping, 46, 59. *See also* Impurities
Dots, 14
Drift, 238
  diffusion and, 80, 265, 269–271
  region of, 250
  terms for, 75
  velocity of, 87
Drude model, 58–59
Dryout, 332
Dual-source vapor-deposition process, 324
Dulong-Petit law, 22, 73
Duncan-Peterson model, 329–331
Dzyaloskinskii-Lifshitz-Pitaevskii (DLP) theory, 194, 202, 204

Edge dislocations, 11
Effective pressure, 206
Ehrenfest pattern, 150
Einstein, A.
  Einstein-Bose theory, 21, 50, 65, 266
  Einstein model, 22
  energy transport and, 104
  heat conduction model of, 73–74
  oscillators and, 104
Electrically-insulating layers, 254–258
Electrical overstress (EOS), 233
Electrical resistance, 252
Electric field vector, 239
Electromagnetic waves, 54
Electromigration, 252
Electrons
  absorption and, 63–65
  band structure, 37–44
  Boltzmann equations and, 265–269
  concentration per unit volume, 239
  diffusion constant, 249
  Drude model, 58–59
  effective mass, 86
  electronic structures, 28–52
  emission spectroscopy, 179
  free, 31–37, 58–59
  gas of, 9
  heat capacity, 36, 71–74
  holes and, 65–67, 241
  impact method, 179
  mass of, 239
  mobility, 86
  motion of, 37–38
  phonons and, 53–57, 90
  photons and, 53–67
  recombination rates, 66
  relaxation time, 263
  scattering rate, 266
  shielding, 37
  solid structures and, 28
  speed of, 49
  structures, 28–52
  temperature, 268
  thermalized, 59
  See also Bonding; Ionization; specific structures
Electrostatic discharge, 229, 233, 283
Emissions, 122
Energy bands, 29–31
Energy carriers. *See* Carriers; specific types
Energy conservation, 63–64
Energy density, 81, 91
Energy gap, 41
Energy quantization, 19
Energy transfer process, 90
Energy transport equation, 88
Engineering surfaces, 203
Entropy, 63, 150
EOS. *See* Electrical overstress
Epitaxial films, 13
Equilibrium film pressure, 193
Ergodic system, 150, 157
Etching
  anisotropic, 332
  directionally dependent, 323
  micromachining and, 371–372
  techniques, 323
Evaporation
  capillaries and, 218–219
  condensation rate, 299
  local rate, 210
  grooves and, 218–219
  process of, 189
  thin film and, 219
Extinction coefficient, 124

Fcc structure, 5–7
Fermi-Dirac distribution, 34–37, 44, 237, 266
Fermi level, 33, 37, 237, 239–240, 245, 247, 263
FET. *See* Compact field-effect transistors
Fick's law, 10, 75, 77

Film temperature sensors, 99
Fine particles, 168
Flash diffusivity method, 99
Flat film thickness equation, 207
Floaters, 160
Fluidized beds, 140–141
Fluorescence thermal imaging, 282
Fourier's law, 4, 70, 75, 77, 88, 256, 313–315
Free electrons, 31–37, 58–59
Free energy, 182
Freezing
  animals and, 355–359
  melting and, 149–163
  microscale phenomena, 347
  substitution and, 349
Frenkel defect, 10
Frumkin-Deragin theory, 198

Gamma function, 66
Gamma point, 17
Gauss law, 88, 127–128
Genner method, 327–328
Gibbs-Duhem equation, 205
Gibbs energy, 150, 191, 204–205, 349, 351
Grain boundaries, 9–10
Grooves, 218–219
Gruinessen constant, 28
Gupta potential, 152

Hagen-Poiseuille flow, 297
Hamaker constant, 188, 193–198, 206, 208
  determination of, 200–203
  DLP theory and, 204
Harmonic oscillators, 16, 19, 280
Heat-diffusion equation, 241
Heat transport
  axial, 313–315
  charge transport and, 236–253
  dielectric thin films and, 95–116
  diffusion and, 241
  disordered materials and, 103–109
  heat sinks and, 215
  hyperbolic, 75, 78–79
  mismatch models and, 110–112
  solid-solid interfaces and, 95–116
  spectral distribution of, 100
  total energy and, 19–21
Heaviside function, 52, 64
Hepatocytes, 351
High-field electronic devices, 84–90
Highmers, 168
Holes, 41–42, 237, 239–240, 268
Holland model, 259
Hughes thermal imaging system, 311
Hybrid model, 59–60, 73
Hydrodynamic equations, 78–81, 267
Hydrogen bond, 8
Hydrophilicity, 199
Hydrophobic surfaces equation, 199
Hyperbolic heat equation, 77–78
Hypothermia, 360

IAI. *See* Image-analyzing interferometry
Ice formation, 346–347
Ice-nucleating proteins, 356
Image-analyzing interferometry (IAI), 202
Impurities
  concentrations of, 247
  doping, 46
  ionized, 49
  radiative recombinations and, 66
  scattering and, 48–49, 63
  semiconductors and, 46
Infrared thermometry, 281–282, 378–379
Initial charge, 333
Insects, freeze tolerant, 355
Insulators, 44
Integrated circuits, 229–232, 236
  insulators in, 257
  metal layers and, 263
  temperature and, 235
  VLSI and, 230, 236, 242, 252
Interaction energy equation, 194
Interband scattering, 48
Intercluster potential, 175
Interconnects, 229–284
Interfacial effects
  free energy, 189–194
  heat transfer coefficient, 211
  mass flux, 209–210
  pressure difference and, 213–215
  reflectivity, 56–57
  solid-liquid-vapor systems and, 190
  thin films and, 187–222
Interference, 68
Internal states, 178–180
Intracellular supercooling, 354
Inverse problem, 134
Inversion, 243
Ionization
  clusters and, 177
  defect scattering and, 49
  impact, 240, 267
  ionic bonds, 8
  lasers and, 139–140
  optical vibrations and, 19

potential of, 179–180
See also Carriers; Phonons; specific effects
Isothermal nonevaporating case, 201

Joule heating, 84–90

KC model. *See* Kelvin-Clapeyron model
Kedem-Katchalsky irreversibility, 351
Keesom orientation, 188
Kelvin-Clapeyron model, 188, 204–208, 213
Kelvin-Helmholtz instability, 297
Khrustalev-Faghri model, 298, 305
Kinetic theory, 69–74, 173–176
Kubo equation, 107

Lai relation, 89
Laplace transforms, 99
Lasers
ablation, 13
epitaxy and, 13
fluorescence of, 179
light and, 123
photons and, 90–91
pulse propagation of, 127–128
radiation and, 89–92, 121
reflectance thermometry and, 279–280
semiconductors and, 279
short-pulse, 127, 134
ultrafast interactions, 134
Lattices, reciprocal, 17
Lattice vacancy, 10
Lattice vibrations. *See* Phonons
Length scales, 68–69
Lennard-Jones clusters, 152, 159–161
Lewis interactions, 188
Lifshitz-van der Waals interactions, 188, 192, 194, 196, 219
Light dispersion, 53
Light-emitting devices, 67
Lindemann criterion, 151
Liouville equation, 150
Liquids
autophobic, 208
crystal thermometry and, 281
mass flow rate of, 300
phase transition, 340
solid interfacial model of, 192
vapor interface and, 205
London dispersion, 188, 215
Longtin model, 298
Lorenz bound-oscillator model, 55–60
Lorenz number, 77, 263
Low-electric field mobility, 243
Low-pressure chemical vapor deposition (LPCVD), 255, 372
Low-temperature scanning electron microscopy, 349
LPCVD. *See* Low-pressure chemical vapor deposition
Lubrication theory, 210
LW interactions. *See* Lifshitz-van der Waals interactions
Lydanne-Sachs-Teller relation, 62

Macroscopic convection, 232
Macroscopic heat transfer model, 352
Macroscopic interfacial force balance, 212
Madjumar equation, 26, 89
Magic numbers, 170, 182
Magnetic actuators, 380
Magnetic resonance imaging (MRI), 362–364
Mallik-Peterson experiment, 302–303, 314
Ma model, 299
Mantell-Wightman studies, 204
Marangoni effect, 219, 305
Mathiessen rule, 113
Matrix studies, 178–179
Maxwell equation, 53, 157–158, 267–268
Mazur theory, 345
MC method. *See* Monte-Carlo method
MD method. *See* Molecular dynamics method
Medium theory, 106
Melting, 149–163
MEMS, 370–371, 376–377
Meniscus properties, 204–219
MESFETs. *See* Metal-semiconductor field-effect transistors
Metal-oxide-semiconductor field-effect transistor, 84–86
Metal-semiconductor field-effect transistors (MESFETs), 84–86, 271, 273
Metals
free electron theory and, 31–36
layers in, 263
metallic bonds, 9
plasma wavelengths of, 59
thermal conductivity and, 263–264
See also specific effects, structures
Metastability, 158
Microelectronic devices, 3
Micro heat pipes, 295–334
Babin model, 300
capillary limit of, 329–330
charging procedures, 325–328
cross-sectional configurations, 309
defined, 296
fabrication methods, 323–325

Micro heat pipes (*cont.*)
liquid phase, 329–330
modeling of, 295–306
steady-state models, 306
testing of, 310–323
transient tests, 306–310
Micromachining
etching and, 371–372
flow sensors, 377–378
surfaces, 372–375
Micromolding techniques, 375–376
Micro-Raman spectroscopy, 281
Microscale regimes, 125–126, 134
Microscale thermal actuators, 380–384
Microscopic theories, 55–60
Microsteam engine, 383–384
Microstructures, 5–14
angstrom-scale, 256
dots, 14
fabricated, 13–14
narrow lines, 14
natural, 5–13
radiation and, 140–143
thin films, 13–15
Miniature car, 369
Miniaturized engineering systems, 3–5. *See also* specific systems, types
Mismatch models, 110–112
MO-LCAO method, 175
Molecular dynamics method, 173, 178, 181
Momentum equation, 64, 79, 89, 267
Monte-Carlo (MC) method, 88, 150, 178, 181, 264
MOSFETs, 84
MRI. *See* Magnetic resonance imaging

Narrow lines, 14
Navier-Stokes equation, 80, 88, 304
Near-equilibrium transport, 236–241
Negative mass, 42
Newton's laws, 44, 150
Noise-thermometry, 277
Noncontact methods, 99
Nonequilibrium energy transfer, 84–92
Nonevaporating superheated films, 305
Nonrigidity, 155
Nose method, 158
Nozzle beam method. *See* Supersonic free jet method
Nucleation
bubble type, 142
clusters and, 183–185
frequency, 352
homogeneous equation, 352
mechanism of, 176
microlayer boiling and, 216–217
probability of, 352
stages of, 167
Nusselt number, 304–305

Ohm's law, 75, 77
One-dimensional conduction solution, 210
Optical branch, 66
Optics
circuits and, 53, 230
lasers and, 134
molecules and, 180
thermometry and, 277–282
Optoelectronic devices, 66
Order-disorder transition, 340

Partially wet systems, 198
Particle transport theories, 67–83
Particle volume fraction, 141
Particle-wave duality, 20
Pauli exclusion principle, 29, 32
PECVD. *See* Plasma-enhanced chemical vapor deposition
Penetration depth, 65, 124
Pentamers, 168
Periodic potentials, 37–44
Perturbation theory, 263
Peterson experiment, 310
Phase transitions, 150
clusters and, 159–161, 182
liquid, 340
metastability and, 158
phase rule and, 157–158
small systems and, 156–157
thin liquid films and, 187–222
time scales and, 156–157
Phonons, 4, 19–28, 53–67
absorption and, 60–63, 65
Boltzmann equations and, 265–269
Bose-Einstein theory, 28
bottlenecks, 89
Casimir limit, 83
conduction and, 268
confinement of, 258
conservation equations, 64
crystal vibrations and, 14–28
defined, 4, 19
dispersion relation, 19
dominant, 523
electrons and, 53–67
emission of, 65
energy quantization and, 19

generation of, 268
heat capacity, 19
LO type, 61, 86–87
optical scattering, 51–52
optical types, 61–87
photons and, 53–61
polar scattering, 52
production of, 67
quantum theory and, 21
radiation and, 24, 112–114
relaxation time and, 28
scattering, 25–28, 113, 260–262
temperature and, 64
total energy, 19
transport, 258
TO type, 61
Wien law, 24
See also specific effects, types
Photoacoustic methods, 99
Photoelectronic spectroscopy, 179
Photo-ionization, 179
Photolithography, 101
Photons, 61
devices and, 233
electrons and, 53–67
emission rate of, 65
holes and, 66
lasers and, 90–91
phonons and, 62–63
quantum mechanics and, 60–66
wavevector of, 63
Photothermal displacement, 281
Piezoelectric effect, 51
Pinch-off condition, 244
Planck distribution, 24, 65, 119, 121, 281
Plasma-enhanced chemical vapor deposition (PECVD), 256, 372
Plasma frequency, 56, 58
Plasmons, 59
Point defects, 10
Poisson equation, 88, 240
Polarization, 55
Polar optical scattering, 52
Polar systems, 19, 188, 198–200
Porosity, 96, 311
Prandtl number, 81
Propagation delay, 253

Quantum mechanics
origins of, 119
photons and, 60–66
statistics of, 21
transition probabilities, 63
Quasi-equilibrium, 77
Quasi-thermodynamics, 209–210

Radiation
excitations, 121
heating and, 89–92
impurities and, 66
lasers and, 89–92
length scales, 122–124
microscale, 119–143
microstructures and, 140–143
propagation time, 128–129
relaxations, 121
thermal engineering, 120–122
transfer equation, 83
Raman effects, 156, 179
thermometry and, 280–281
scattering, 61, 82
Rate equations, 136
Rayleigh law, 26, 108, 262
Reciprocal lattice, 17
Recombination rate, 66
Reflectivity, 54
Refractive index, 56
Relaxation time, 28, 68, 75, 83, 129
Resection surgery, 361
Resistive heating, 241
Restricted liquid, 152
Reynolds number, 88
Roughness, 188

SAMPLE model, 325
Saturation
density and, 248
intensity and, 138–139
regime of, 244
Scanning thermal microscopy (SThM), 273–274
Scattering, 68, 75, 81
boundary, 260–263
BTE and, 75
categories of, 48
carrier-carrier, 48
charge carriers and, 47–52
defect, 48–49, 72–73
dependent, 141
drift and, 75
elastic, 48–50
independent, 141
inelastic, 26, 48, 51
interband, 48
intervalley, 48
intravalley, 48
mean collision time, 48

Scattering (*cont.*)
normal process of, 27
phonons and, 25–28, 49–52, 89, 113, 260–262
Rayleigh rate and, 108
relaxation time and, 28, 68, 75, 83, 89, 129
temperature and, 87
Umklapp process and, 27
Schottky defect, 10
Schrage equation, 209
Schrodinger equation, 31, 39, 126, 175, 264
SCREAM technique, 372
Screw dislocations, 11
Self-heating effect, 245
Semiconductors, 14
band structure of, 41–44, 46
charge carrier densities and, 44–47
degenerate, 47
direct gap, 41
doping and, 47
impurities and, 46, 49, 66, 247
indirect gap, 41
interconnects and, 229–284
intrinsic, 44, 463
laser reflectance thermometry and, 279
layers, 254, 263
MESFETs, 84
n-doped, 46
near-equilibrium transport and, 236–241
quantum devices, 13
thermal conduction and, 258
thermal phenomena and, 229–284
thin films and, 13
See also specific topics, types
Separation by Implantation with Oxygen (SIMOX), 255
Shockley equation, 247
Short-channel effect, 245
Silicon micromachined thermal sensors, 369–384
Silicon-on-insulator circuits (SOI), 232, 250
SIMOX. *See* Separation by Implantation with Oxygen
Simulation and Modeling of Evaporated Deposition Profiles (SAMPLE), 325
Sinusoidal power, 102
Size effect, 124
Size parameter, 141
Slip direction, 11
Small systems, 156–157
SOI. *See* Silicon-on-insulator circuits
Solid-liquid vapor systems, 190
Solid-solid interfaces, 95–116
Solutes, 345, 348, 352
Sommerfeld parameter, 263
Specific heat, 182
Spectroscopy, 156–157, 179
Specular reflection coefficient, 260
Spreading, 191–194, 203
Sputtering, 13
Steady-state experiments, 306
Stefan-Botlzmann law, 24
Stephan-Busse model, 217
SThM. *See* Scanning thermal microscopy
Stiction, 375
STM. *See* Scanning tunneling microscope
Stokes lines, 280–281
Supercooling, 12, 354
Superheat equation, 221
Super lattices, 114
Supersonic free jet method, 178
Surface excess conventions, 191
Swanson-Peterson model, 305
Symmetry argument, 162

Temporal delay, 253
Temporally extrapolated absorbance method, 134
Temporal microscale, 127, 130–139
Tetramers, 168
Thermal conductivity, 70–74
acoustic phonons and, 238
crystals and, 70–74
equation for, 315
Fourier law and, 4
limits in, 107
measuring of, 254
metals and, 263–264
minimum, 103
semiconductors and, 258
sensors, 379–380
stabilized zirconia and, 106
temperature and, 4
thermoreflectance and, 101–102
thin films and, 100–103, 108–109
variations in, 256
wafers and, 314
Thermal diffusivity, 96–100, 128
Thermal energy conservation equation, 81
Thermal hysteresis proteins. *See* Antifreeze proteins
Thermal insulators, 142
Thermalization, 122
Thermalized electrons, 59
Thermal resistance, 232, 256
Thermal sensors microscale, 377–379
Thermal transport properties, 253–264
Thermionic emission, 35
Thermocapillary limitation, 219, 305
Thermodynamics

clusters and, 180
internal pressure, 187
thin films and, 189–208
Thermoelectric currents, 238
Thermometry, 271–282
electrical parameters of, 275–277
electrical-resistance to, 271
liquid crystal method, 281
noise-based, 277
optical methods, 277–282
Raman effect and, 280–281
Thermoreflectance approach, 101–102
Thermotropism, 340
Thin films
bulk films and, 106
condensation and, 219
defined, 13
evaporation and, 219
interfacial forces and, 187–222
measurement techniques, 96–100
meniscus properties and, 204–208
microstructures and, 13
phase changes and, 187–222
thermal conductivity and, 100–103, 108–109
thermal diffusivity and, 96–100
thermal properties of, 96
thermodynamics and, 189–210
Thomson heating, 241
Three-dimensional disjoining pressure model, 198
Three-dimensional intermolecular force field, 188
Threshold photo-electron spectroscopy, 179
Time constant, 63
Time scales, 68–69, 87, 156–157
Total energy, 19–21
energy quantization and, 19
heat capacity and, 19–21
Transconductance, 249
Transient experiments, 306–310
Transistors, 246–251, 269
Transport laws, 67–69
Transverse optical phonons, 61
Trimers, 168
Truong-Wagner equation, 195
TSEP thermometry, 277
Tunneling, 68

Ultraviolet bonding, 326
Umklapp process, 27–28, 268
Unconventional electrical insulators, 257
U-process. *See* Umklapp process

Vacuum ultraviolet absorption spectroscopy, 179
Valence band, 41
Van der Waals theory, 8, 191
contact distance, 194
forces, 180, 304–305
molecules, 170
Vapor deposition, 13, 323
Vaporization, 295
Vapor pressure, 204–208
Velocity saturation, 245
Vertical cavity surface emmitting lasers, 95
Very-large-scale-integrated circuits (VLSI), 230, 236, 242, 252
VLSI. *See* Very-large-scale-integrated circuits

Wachutka equation, 241
Wafers, 314–315
Wave propagation, 15–19
Wayner model, 220
Wetting, 194
Wiedemann-Franz law, 77, 263, 267
Wien's law, 24, 281
Work function, 34
Work hardening, 11
Wu-Peterson model, 303

Young-Dupre equation, 193
Young-Laplace equation, 200–21, 204–205, 213, 219

Zero-bias condition, 247
Zinc-sulfide structure, 7
Zweidinger approach, 275–277